Performance-Based Urban Design: Integrated Urban Analytics, Simulation and Climate-Responsive Design

Performance-Based Urban Design: Integrated Urban Analytics, Simulation and Climate-Responsive Design

Collection Editors

Peng Du

Rahman Azari

MDPI

Basel • Beijing • Wuhan • Barcelona • Belgrade • Novi Sad • Cluj • Manchester

Collection Editors
Peng Du
College of Architecture & The Built Environment
Thomas Jefferson University
Philadelphia
USA

Rahman Azari
College of Arts and Architecture
Pennsylvania State University
University Park
USA

Editorial Office
MDPI AG
Grosspeteranlage 5
4052 Basel, Switzerland

This is a reprint of the Topical Collection, published open access by the journal *Buildings* (ISSN 2075-5309), freely accessible at: https://www.mdpi.com/journal/buildings/topical_collections/perfor_bas_urba_des.

For citation purposes, cite each article independently as indicated on the article page online and as indicated below:

Lastname, A.A.; Lastname, B.B. Article Title. *Journal Name* **Year**, *Volume Number*, Page Range.

ISBN 978-3-7258-5211-6 (Hbk)
ISBN 978-3-7258-5212-3 (PDF)
https://doi.org/10.3390/books978-3-7258-5212-3

Contents

About the Editors

Peng Du

Dr. Peng Du is an Associate Professor and Director of the Master of Urban Design – Future Cities (MUD) at the College of Architecture and the Built Environment at Thomas Jefferson University. He is also the Assistant Provost for Research and Scholarship at the university's East Falls Campus, managing applied research activities, supporting interdisciplinary collaboration, and enhancing research infrastructure across the university. Dr. Du's research focuses on analyzing the performance of tall buildings and high-density cities, and also developing data-driven, computational design, and analytical methods at both urban and building scales. Prior to joining Thomas Jefferson University, Dr. Du taught at the Illinois Institute of Technology and Texas Tech University. He was a visiting professor at Università Iuav di Venezia (IUAV) in 2024, supported by the European Union's Erasmus Grant. Dr. Du has also served as the Asia Regional Director at the Council on Tall Buildings and Urban Habitat (CTBUH). Dr. Du is a LEED-Accredited Professional and WELL-Accredited Professional.

Rahman Azari

Rahman Azari, Ph.D., is an Associate Professor of Architecture and Director of Professional Graduate Programs at the Pennsylvania State University Department of Architecture, where he also leads the Resource and Energy Efficiency Lab (RE2 Lab). Azari is also affiliated with the Penn State Department of Architectural Engineering and is a co-funded faculty member at the Penn State Institute of Energy and the Environment (IEE). Azari's research on carbon-neutral buildings and cities has been supported by research grants from the US Department of Energy (DOE), the American Institute of Architects (AIA), the Pennsylvania Department of Environmental Protection, the Council on Tall Buildings and Urban Habitat (CTBUH), and Penn State. Azari's publications include edited books, book contributions, conference papers, and journal articles published in *Nature Communications*, *Energy Policy*, *Energy and Buildings*, *Building and Environment*, the *Journal of Building Engineering*, and the *Journal of Management in Engineering*. Azari has served as a jury member for the AIA+ACSA COTE Top Ten Competition and has coached award-winning student teams in various architectural competitions, such as the US DOE Solar Decathlon Design Challenge and Build Challenge, and AIA+ACSA COTE Top Ten competitions. Before working at Penn State, Azari served as an Assistant Professor at the Illinois Institute of Technology and the University of Texas San Antonio.

buildings

MDPI

Article

Balancing Construction Costs and Environmental and Social Performances in High-Rise Urban Development: A Generative Urban Design Approach

Peng Du [1,*], Geoffrey Little [2] and Erick Romero [1]

[1] College of Architecture and the Built Environment, Thomas Jefferson University, Philadelphia, PA 19144, USA; erick.romero@students.jefferson.edu
[2] Genesis AEC, Blue Bell, PA 19422, USA; littlegeoff2@yahoo.com
* Correspondence: peng.du@jefferson.edu

Abstract: The urban design process is complex and interdisciplinary, especially in the context of high-density cities with high-rise buildings. The design of high-rise buildings requires input from a variety of stakeholders in the city, who often represent conflicting requirements and interests. However, conventional design approaches struggle to address this complexity. This study introduced a generative urban design approach and applied it to a case study for high-rise urban development in Guangzhou, China. Specifically, 7500 urban forms were generated with variables such as street networks, block offset, building typology, and green space, and then were evaluated and optimized via multi-objective evolutionary algorithms. A total of 30,000 performance values were generated through various simulations. This study also presented a two-round, quantitative evaluation process against eight performance objectives from environmental, social, and economic aspects, including urban density, green space area, Daylight Potential, construction cost, Heat Stress, Green Space Accessibility, View Interest, and Shadow Impact.

Keywords: generative urban design; performance-based design; evolutionary algorithm; simulation

1. Introduction

Cities are facing significant challenges due to rapid urbanization, climate change, and resource depletion. Therefore, designing future cities is becoming extremely complex and requires interdisciplinary collaborations. One solution that cities have been investigating is increasing density. Developing high-rise buildings is not the only way to increase urban density, but has become a viable solution, especially in fast-growing cities in Asia. The concentration of people in high-rise developments has made significant impacts on the environment, in both positive and negative ways, e.g., greater overall operational energy efficiency and lower carbon emissions vs. greater air pollution and decreased daylight access [1,2]. In addition, tall buildings present significant social challenges, such as reduced community interaction and access to green spaces. Many studies have examined both the environmental and social performances of high-rise developments qualitatively and quantitively, but very few were able to bring economic indicators, such as construction cost, into the equation when assessing the performance of high-rise developments. Particularly, designers often do not prioritize construction costs enough in early design stages, which might lead to potential cost overruns. Estimating the construction costs for tall buildings is even more complicated due to the complexity of the projects and limited data availability.

Urban design is a very complex process, especially in the context of high-density cities with high-rise buildings, so it requires participation from a variety of stakeholders who might represent different requirements and even conflicting interests [3,4]. However, conventional design approaches struggle to address this complexity. Specifically, only a very limited number of design proposals are often developed and refined through manual iteration and experience without rigorously testing the full range of possible design schemes by quantifying different performance objectives, such as land use, zoning, density, transportation, building morphology, energy consumption, and carbon emissions [5,6]. Therefore, new design approaches and tools are very much needed to better equip urban designers to address various challenges from environmental, social, and economic aspects.

In recent years, the generative design approach, which is a framework that uses computational design methods integrating simulation tools and evolutionary algorithms, was developed to generate, evaluate, and evolve a very large number of design options at various scales [7,8]. The study introduced in this paper presented a generative urban design approach to generate urban forms and evaluate and balance their environmental, social, and economic performance. Specifically, a case study for high-rise urban development in Guangzhou, the capital and largest city of Guangdong Province in southern China, was conducted using this approach.

2. Literature Review

2.1. Construction Cost of High-Rise Buildings

It is generally accepted that the construction costs increase as the building height rises on a floor area basis. However, very few empirical studies on the relationship between construction cost and the height of tall buildings have been conducted. Recent studies have found that the relationship between the construction cost and building height was represented by a U-shaped curve, instead of a simplistic linear regression model [9–11]. The U-shaped curve suggested that the construction cost per square meter decreased initially as the number of stories increased, but it started to rise after passing the bottom point.

The construction cost of high-rise buildings is greater than low-rise buildings mainly because of the greater costs for vertical transportation, mechanical systems, tower cranes, on-site labor, etc. [12,13]. The number of studies on high-rise construction costs, especially for buildings over 100 m tall, is extremely limited, largely due to data availability. Two recent studies on the relationship between high-rise construction costs and building height confirmed a U-shaped curve [11,14]. D. H. Picken and Ilozor examined 24 residential buildings in Hong Kong ranging from 3 to 39 stories (height ranging from 9 m to 112 m) completed in the early 1990s, and found that the construction cost declined until the height reached 100 m, and then raised again [11]. Blackman and Picken analyzed 36 residential buildings in Shanghai ranging from 2 to 37 stories (height ranging from 6 m to 115 m), completed between 2000 and 2007, and found that the construction cost declined until the height reached 24 m, and then raised again [14]. The raw data available in the literature were used to predict the construction cost in this study.

2.2. Multi-Objective Evolutionary Algorithms

Evolutionary algorithms are population-based metaheuristic optimization models that generate design solutions paired with competing performance metrics. There are two main types of design optimization in evolutionary algorithms in the generative design process: single-objective optimization and multi-objective optimization [7]. However, multi-objective evolutionary algorithms have been proven to be more effective in more complex urban contexts. A successful optimization run using objective evolutionary algorithms relies on the relationships between three key metrics: the parameters informing the morphology (genes), the morphological characteristics being generated (the phenotype), and the performance objectives (fitness functions) [15].

An increasing amount of generative urban design research has applied multi-objective evolutionary algorithms to generate and optimize urban forms paired with a set of quantifiable performance objectives, and to conduct and visualize comparative analysis on different design solutions [16–18]. However, most of the studies conducted using multi-objective evolutionary algorithms ended up with a one-round selection process from the algorithm's output. However, this one-round selection analysis limits the understanding of the design proposals within pre-determined performance objectives, which can only be a few given the capacity of the tools that are currently available. Some very recent research included two-round selection stages, but the performance metrics in the second stage were evaluated subjectively. Therefore, this study examined more performance objectives via multi-objective evolutionary algorithms, by integrating a two-round selection process with two different sets of quantifiable objectives to find optimal solutions. Also, most of the studies focused on social and environmental performances, such as energy, carbon, daylight, mobility, and comfort, but very few were able to integrate economic indicators into the overall design performance. Therefore, the research methods introduced in the Research Methodology section below will fill these two significant research gaps by integrating construction costs and a wide range of environmental and social performance indicators into a generative urban design workflow.

3. Research Methodology

3.1. Urban Form and Performance Objectives

This study began with the creation of a parametric urban model using Urbano, an urban mobility modeling plugin for Rhino/Grasshopper, with data loaded from GIS and OpenStreetMap (OSM) [19,20]. Then, an algorithm was created to generate the urban form, with variables such as street networks, block offsets, buildings, and green spaces, and meanwhile examining the performance of these urban forms through various Grasshopper plugins. This study conducted a two-round, quantitative evaluation process. Specifically, the study integrated 4 performance objectives, including urban density, green space area, Daylight Potential, and construction costs, into the evolutionary algorithms in the first-round selection process, followed by another round of evaluation by using a different set of 4 performance objectives, including Heat Stress, Green Space Accessibility, View Interest, and Shadow Impact (Table 1).

Table 1. The performance objectives used in the two-round, quantitative evaluation process.

No.	First-Round Evaluation		Second-Round Evaluation	
	Performance Objective	**Measurement**	**Performance Objective**	**Measurement**
1	Urban Density	Floor Area Ratio (FAR)	Heat Stress	Universal Thermal Climate Index (UTCI) on the street level
2	Green Space Area	Green Area Ratio (GAR)	Green Space Accessibility (GSA)	The walking distances connecting the green spaces within the site from the surrounding neighborhoods
3	Daylight Potential	The ratio of the building facades with direct sun exposure	View Interest	ratio of the building facades with direct view access to the river

Table 1. *Cont.*

No.	First-Round Evaluation		Second-Round Evaluation	
	Performance Objective	**Measurement**	**Performance Objective**	**Measurement**
4	Construction Cost	The total cost of constructing all the buildings	Shadow Impact	Direct sun exposure to the buildings outside the site boundary

3.2. Construction Cost Estimates

The research used data from the study on Hong Kong high-rise construction costs [11] introduced in the Literature Review section due to the geographical proximity to the study area in Guangzhou, and then conducted a quadratic regression analysis, which is a statistical method to model a relationship between two sets of variables. Specifically, a quadratic regression model ($y = 0.0809x^2 - 16.13x + 2062.6$, $R^2 = 0.32$) was created to predict the estimated construction costs of the buildings, with a height up to 250 m. The development of buildings taller than 250 m has been strictly restricted since 2022 by the Chinese government. Therefore, the research used 250 m as the height limit of buildings (Figure 1). As the Hong Kong study was conducted in 2001, the construction cost growth over time was factored in based on the Building Works Tender Price Index (BWTPI) [21], and then the prices were converted to United States Dollars (USD) from Hong Kong Dollars (HKD). The Hong Kong study was conducted in April 2001, so this study used 2001 Quarter 1 and the latest available 2023 Quarter 3 data from the Building Works Tender Price Index (BWTPI) to project the estimated construction costs for today (Hong Kong Architectural Services Department, n.d.). The study used 0.13 as the conversion rate from Hong Kong Dollars (HKD) to United States Dollars (USD).

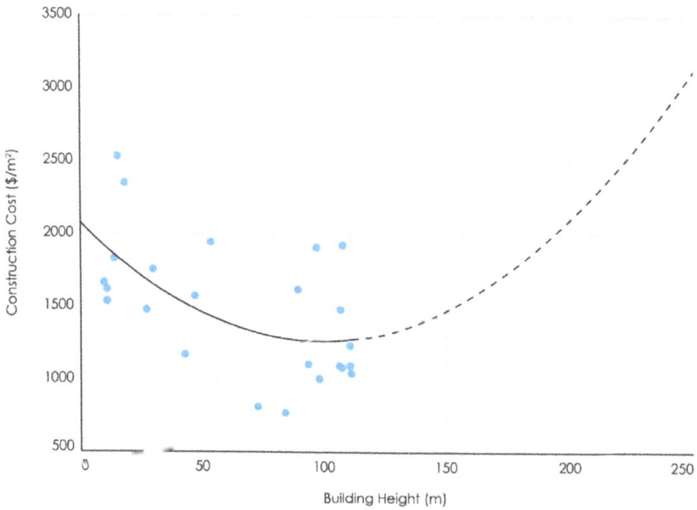

Figure 1. The quadratic regression model of the construction cost (USD/m^2) vs. building height (m), with the plots representing data from the Hong Kong study [11].

3.3. Multi-Objective Optimization

A multi-objective evolutionary algorithm was developed using Wallacei, an evolutionary and analytic plugin for Rhino/Grasshopper. Through this algorithm, the performance of each design objective of every single iteration was evaluated and optimized. Then, the pareto front solutions were used, which are a set of solutions that are non-dominated, representing the fact that the optimal trade-offs between different objectives when improving one objective cannot be obtained without worsening another. Selecting pareto front solutions helped the selection process focus on the groups of efficient options and make tradeoffs within this set, rather than considering the full-size population (the total number of iterations). Further, a K-means clustering algorithm was applied to the pool of the pareto front solutions, so these solutions were organized into clusters based on their similarity. In this process, the solution closest to the center of each cluster was selected. Lastly, a few additional solutions were selected, including the most optimal solution for each performance objective, the "Average Solution" (with the closet equal weighting between all the performance objectives) and the "Ideal Solution" (with the closest optimal result for all the performance objectives). The "Average Solution" is considered a non-existent optimal solution in the objective space, with the closest equal weighting between all the design objectives. Similarly, the "Ideal Solution" is also considered a non-existent optimal solution in the objective space, and thus the selected solution is the one nearest to this point based on a Euclidean measure between the two [15].

4. Case Study: High-Rise Urban Development in Guangzhou, China

4.1. Site Analysis

The site is in the International Business District, a fast-growing, high-density area in Guangzhou, China. Situated alongside the Pearl River in the south, the site is also adjacent to a canal in the East (Figure 2). The area of the site is approx. 73,000 m^2 and is expected to be developed into a high-rise urban complex, with a building height limit of 250 m. Therefore, the key focus of the design was to increase the urban density but also balance environmental and social implications. One of the major challenges for a high-rise, mixed-use development is the shadow caused by the buildings, so increasing daylight access was one of the key design strategies. Both the river and canal are important ecological and environmental assets to the site, so the design strategies also included an increase in the visual connections with the water as well as the accessibility to open spaces. Due to accelerated climate change, heat has significantly impacted our society, including a rise in heat-related deaths, for many cities nationally and internationally [22], and this is particularly true for cities with hot and humid climatic conditions, such as Guangzhou. Thus, this study also implemented strategies to mitigate the heat impact. There are very limited green spaces in the surrounding urban context, so the design also focused on creating more green spaces, not only on the ground level but also through the rooftops.

4.2. Form Generation and Performance Evaluation

The main elements of the urban form included streets, parcels, buildings, and open spaces. A 20 m-wide green corridor was created alongside the canal to be a recreational and ecological space on the waterfront. A total of 3 building types were created based on the height and program: Building Type 1 for office and mix-used with a height range of 175–250 m, Building Type 2 for residential used with a height range of 75–150 m, and Building Type 3 for retail and entertainment with a height range of 25–50 m. In order to reduce the pressure of high-rise buildings on the street level, and to create rooftop spaces for activities, all three building types have podiums, with a height range of 24–40 m for Building Type 1, and 8–24 m for Building Types 2 and 3. The floor height was set to be 4 m across all the buildings.

Figure 2. The study site, highlighted in red, and its surrounding urban context.

The urban form generation process included nine major steps (Figure 3): (a) adding movable points on the site boundary; (b) connecting the points with offset to create streets; (c) creating parcels and allocating green spaces (occupancy range: 1–3 parcels); (d) allocating building types, with Type 1 in red (occupancy range: 1–3 parcels), Type 2 in blue (occupancy range: 1–5 parcels), and Type 3 in yellow (occupancy range: remaining parcels); (e) randomly selecting one out of the four corner points of each parcel, and scaling down the parcel to be the footprint of the main building; (f) scaling up the parcel based on the point selected to create the footprint of the associated podium; (g) extruding the footprint of the main buildings; (h) extruding the footprint of the podiums; And (i) identifying extra green spaces in addition to the green spaces on the ground allocated at Step (c), including unbuilt lands within the parcels if the area is greater than 1000 m^2 (otherwise turned into a plaza) as well as the podium roofs if the area is greater than 500 m^2 (otherwise turned into a roof deck).

As introduced in the Research Methodology section, four performance objectives were integrated into the evolutionary algorithm (Figure 4), through the following calculation and measurement: (a) The Floor Area Ratio (FAR) = the total floor area/total site area. (b) The Green Area Ratio (GAR) = the total area of the green spaces/total site area. The green spaces included the green corridor (5997 m^2) alongside the canal, the dedicated parcels for green spaces, as well as the areas of the podium roofs (if >500 m^2). (c) Daylight Potential is measured by the ratio of the building floors with direct sun exposure through the facades to all the building floors. The analysis ran from 9 a.m. to 5 p.m. on the Winter Solstice in Guangzhou, with 2 h of minimal direct sun exposure. An algorithm was developed to contour the buildings every two floors and then connect the midpoint on each exploded segment of the buildings to the points in the sky presenting the sun's locations during the analysis period. (d) The construction cost, measured by the total construction cost of all the buildings in USD in millions. The price per square meter for each building was determined based on the building height and the quadratic regression model ($y = 0.0809x^2 - 16.13x + 2062.6$, $R^2 = 0.32$), and then was multiplied by the GFA of each respective building to estimate the total construction cost.

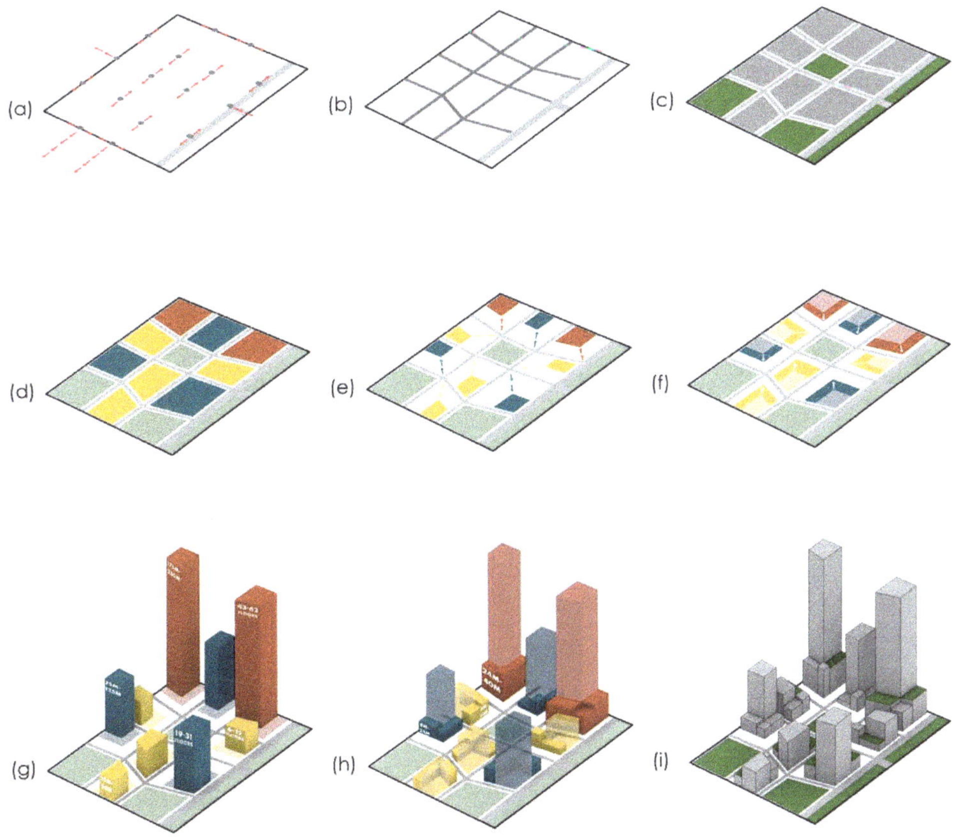

Figure 3. The step-by-step process for generating the urban form: (**a**) street node generation; (**b**) street network generation; (**c**) green space allocation; (**d**) building allocation with different typologies; (**e**) tower footprint generation; (**f**) podium footprint generation; (**g**) tower volume generation; (**h**) podium volume generation; (**i**) urban massing generation.

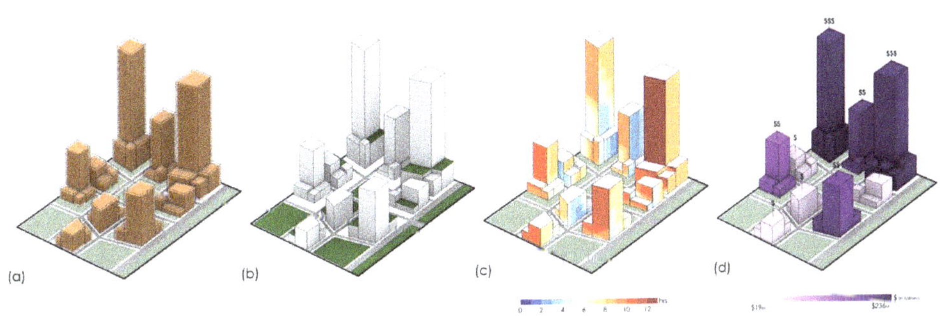

Figure 4. A visualization of the quantification of the 4 performance objectives in the first-round evaluation process (from left to right): (**a**) Floor Area Ratio (FAR); (**b**) Green Area Ratio (GAR); (**c**) Daylight Potential; (**d**) construction cost.

5. Analysis and Selection

5.1. Result Analysis

The algorithm ran a total population of 7500 (design solutions) via Wallacei comprised of 100 generations with 75 solutions each. Each solution included four performance objective values, so there were 30,000 values in total. Each performance objective (fitness function) was analyzed separately across four key metrics: the standard deviation, fitness values (performance values), the standard deviation trendline, and the mean value trendline (Figure 5). The simulation was proven to be successful in improving the mean values for Performance Objective 1 (FAR) and Performance Objective 2 (GAR) based on the mean value trendline. The variation of solutions fluctuated throughout and did not converge toward an optimal result for Performance Objective 3 (Daylight Potential) and Performance Objective 4 (construction cost).

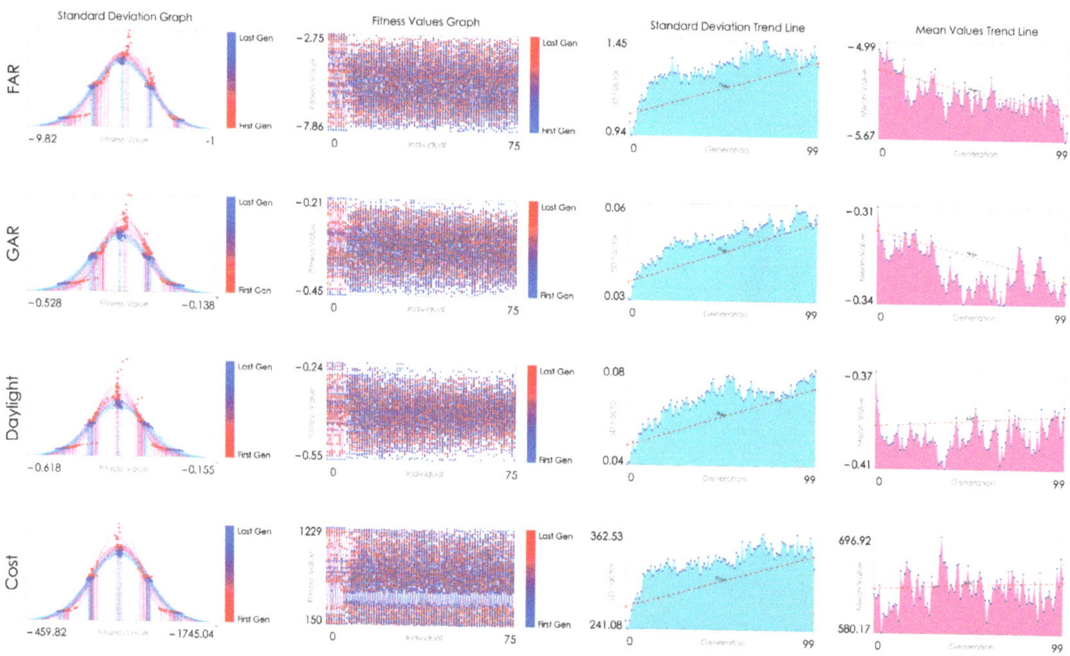

Figure 5. The results of the multi-objective evolutionary algorithm. Each performance objective (fitness function) was analyzed against four metrics (from left to right): the standard deviation, fitness values (performance values), the standard deviation trendline, and the mean value trendline.

5.2. Evaluation and Selection

As introduced in the Research Methodology section, a two-round selection process was conducted. In the first round, 809 pareto front solutions out of the entire population (e.g., 7500 iterations) were identified. Then, these selected solutions were grouped into 20 clusters via a K-means clustering algorithm. Specifically, the solution closest to the center of each cluster was picked, so a total of 20 solutions were selected. In addition, the optimal solution for each performance objective, e.g., the highest Floor Area Ratio (FAR), the highest Green Area Ratio (GAR), the greatest Daylight Potential, and the lowest construction cost, were selected, totaling fou solutions. Finally, two more special solutions, the "Average Solution" and "Ideal Solution", were selected. The "Ideal Solu-tion" happened to be the same as the optimal solution for the Daylight Potential, so, the number of the selected solutions was reduced to 25 from 7500 (Figure 6).

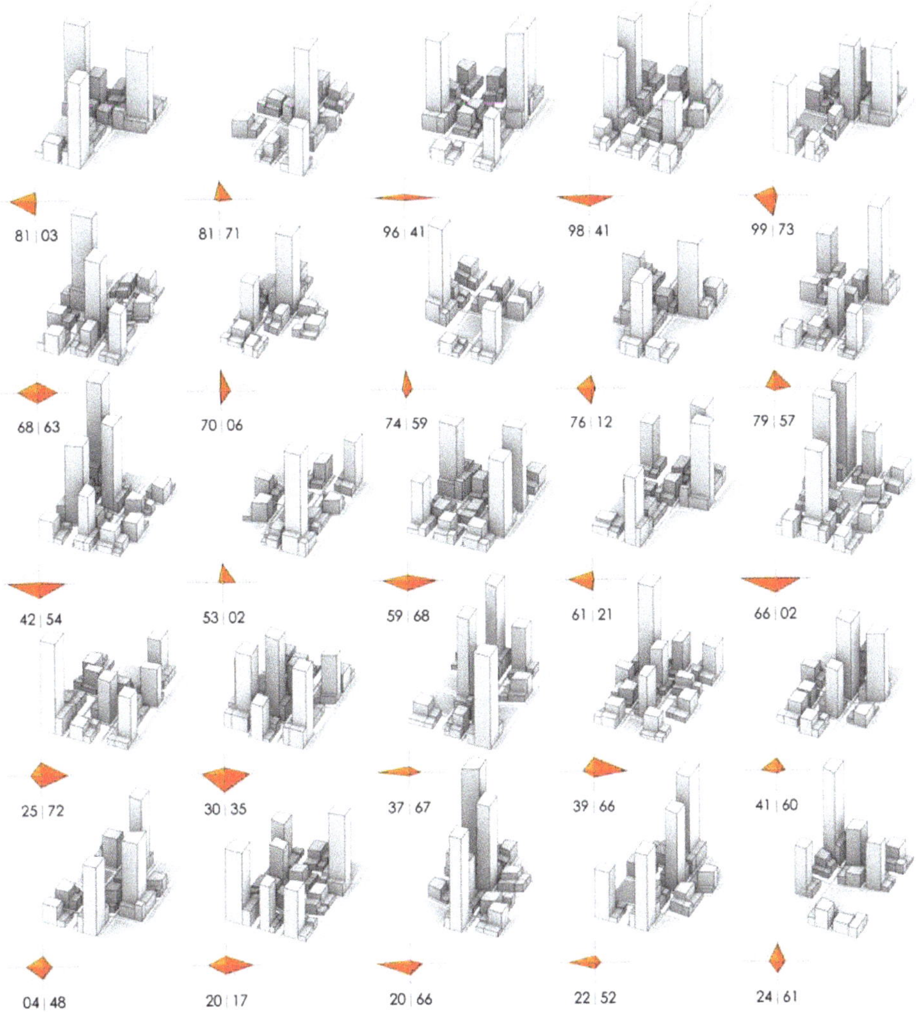

Figure 6. A 3D aerial view render of the selected 25 solutions and their associated radar charts.

The result of each performance objective for each solution, including the absolute value and its ranking among all the 7500 solutions, was exported, analyzed, and visualized via Wallacei (Figure 7). Lower values represent better performance on Wallacei. "Solution (Gen. 68 | Ind. 43)" was found to be the "Average Solution" (with the closet equal weighting between the four performance objectives), presenting 5.33 FAR (ranking # 3741/7500), 32% Green Area Ratio (ranking # 4229/7500), 39% Daylight Potential (ranking# 3802/7500), and a construction cost of USD 714.48 million (ranking # 4171/7500). "Solution (Gen. 53 | Ind. 2)" was found to be the "Ideal Solution" (the closest optimal result for the four performance objectives), presenting 3.75 FAR (ranking # 6517/7500), 36% Green Area Ratio (ranking # 2551/7500), 55% Daylight Potential (ranking# 1/7500), and a construction cost of USD 193.17 million (ranking # 764/7500).

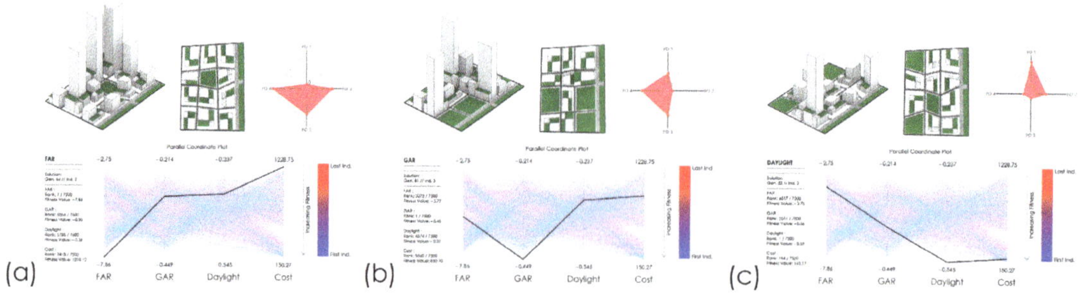

Figure 7. Examples of the design (3D model and site plan) and performance analysis (radar chart and parallel coordinate plot chart) (from left to right): (**a**) the optimal solution for the Floor Area Ratio (FAR); (**b**) the optimal solution for the Green Area Ratio (GAR); and (**c**) the "Ideal Solution", the same as the optimal solution for the Daylight Potential.

5.3. Second-Round Evaluation

In order to offer a more comprehensive understanding of the overall performance of the design proposals produced by the algorithms, an additional four performance objectives were added to further evaluate the 25 selected solutions (Figure 8) through the following calculations and measurements: (a) Heat Stress was measured by the ratio of the number of "heat stress" hours to the total hours reflecting the Universal Thermal Climate Index (UTCI) on one of the hottest days in Guangzhou. Six categories are defined to reflect the UTCI conditions: No Thermal Stress ($9 \leq$ UTCI < 26), Slight Heat Stress ($26 \leq$ UTCI < 28), Moderate Heat Stress ($28 \leq$ UTCI < 32), Strong Heat Stress ($32 \leq$ UTCI < 38), Very Strong Heat Stress ($38 \leq$ UTCI < 46), and Extreme Heat Stress ($46 <$ UTCI). The "heat stress" defined in this study included the three strongest Heat Stress levels: Strong Heat Stress, Very Strong Heat Stress, and Extreme Heat Stress. 26 August to 1 September was found to be the hottest week in Guangzhou. The study ran the UTCI simulation on 26 August for the full day and calculated the "heat stress" hours. (b) Green Space Accessibility (GSA) was measured by the walking distances between the green spaces and the surrounding neighborhoods. Two corner points on the north boundary of the site were set as the starting points for walking towards the green spaces within the site. The average length of the pedestrian paths from the starting points to the middle points of the designated green spaces was used for comparison. (c) The View Interest was measured by the ratio of the building floors with a direct visual connection with the canal to all the building floors. The visual connection to the river was excluded in the quantification of the View Interest. The river is in the south, so optimizing the view to the river would lead to the production of similar solutions to optimizing the Daylight Potential. It is important to set up competing objectives in the multi-objective evolutionary algorithm. An algorithm was developed to contour the buildings every two floors and to connect the midpoint on each exploded segment of the buildings to the middle line of the portion of the canal adjacent to the site. (d) The Shadow Impact was measured by the change of the direct sun hours on the facades of the surrounding buildings immediately outside the site boundary with and without the proposed urban development.

The study conducted simulations on these four additional performance objectives for the 25 selected solutions through various Rhino/Grasshopper plugins. Then, the simulation results were combined with the results of the previous four performance objectives for the final evaluation. A ranking criterion was developed to evaluate the overall performance of each solution. Specifically, the average value of each performance objective was set as a threshold, and if the performance value was better than the average, one point will be gained. Performing better than the average did not mean that the value was greater than the average. The higher value the better for the performance objectives FAR, GAR, Daylight, and View Interest; and the lower value the better for the performance objectives construction cost, Heat Stress (UTCI), Green Space Accessibility, and Shadow Impact. For example, the average FAR among all the 25 solutions was 5.43, so a solution with a FAR greater than 5.43 received one point, otherwise it received no point. The final ranking of the solutions was generated based on the total points (Figure 9).

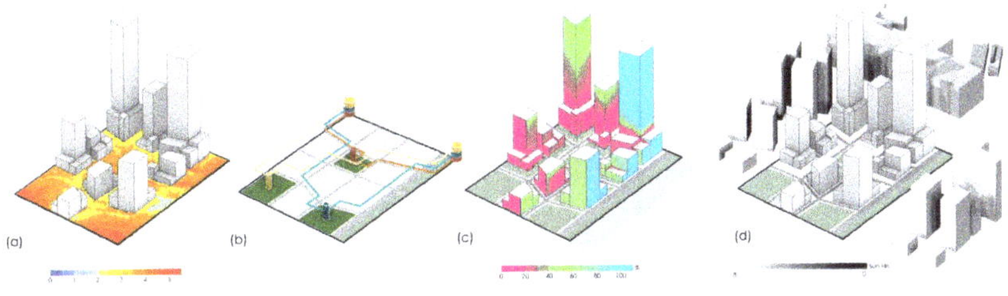

Figure 8. A visualization of the quantification of the 4 performance objectives in the second-round evaluation process (from left to right): (**a**) Heat Stress; (**b**) Green Space Accessibility (GSA); (**c**) View Interest; (**d**) Shadow Impact.

	First-Round Performance Objectives				Second-Round Performance Objectives				
Gen. | Ind.	FAR	GAR	Daylight	Cost	UTCI	GSA	View	Shadow	Points
41 | 60	5.05	0.38	0.47	603.72	0.37	229.97	0.47	0.02	6
70 | 06	3.04	0.37	0.37	150.27	0.38	225.18	0.43	0.02	5
53 | 02	3.75	0.36	0.55	193.17	0.38	270.62	0.44	0.01	5
81 | 71	3.79	0.36	0.51	197.58	0.38	269.92	0.44	0.02	5
20 | 17	5.69	0.29	0.41	671.57	0.34	153.80	0.43	0.03	5
74 | 59	3.49	0.39	0.41	180.97	0.38	322.85	0.44	0.03	4
79 | 57	4.11	0.34	0.46	268.52	0.38	272.61	0.47	0.02	4
39 | 66	4.70	0.25	0.43	337.93	0.37	377.99	0.37	0.03	4
25 | 72	4.74	0.29	0.38	341.41	0.36	160.82	0.41	0.03	4
61 | 21	5.99	0.44	0.39	809.39	0.37	328.38	0.48	0.03	4
22 | 52	6.30	0.40	0.43	886.40	0.37	277.82	0.45	0.05	4
30 | 35	6.33	0.29	0.32	773.57	0.36	224.90	0.35	0.03	4
37 | 67	6.47	0.36	0.46	1042.01	0.37	271.77	0.48	0.05	4
96 | 41	6.85	0.28	0.48	985.90	0.36	364.51	0.46	0.05	4
66 | 02	7.86	0.30	0.36	1218.12	0.36	236.25	0.43	0.07	4
24 | 61	3.97	0.39	0.38	268.44	0.38	318.16	0.45	0.04	3
76 | 12	4.95	0.41	0.35	573.72	0.37	278.78	0.36	0.03	3
04 | 48	5.33	0.38	0.38	647.25	0.38	250.79	0.38	0.03	3
81 | 03	5.77	0.45	0.37	850.70	0.37	329.49	0.42	0.06	3
99 | 73	5.93	0.41	0.27	702.67	0.38	320.42	0.38	0.04	3
59 | 68	6.44	0.28	0.39	820.95	0.36	153.63	0.36	0.05	3
20 | 66	6.91	0.36	0.39	1127.39	0.37	262.92	0.40	0.06	3
98 | 41	7.35	0.30	0.40	1114.21	0.35	236.25	0.38	0.07	3
42 | 54	7.36	0.30	0.36	1077.61	0.36	153.29	0.37	0.07	3
68 | 63	5.33	0.32	0.39	714.48	0.36	160.05	0.39	0.04	2

Figure 9. The final ranking of the selected 25 solutions, highlighting the values of the performance objectives performing better than the averages.

Following this approach, no solution received eight or seven points. The top-ranked solutions included one solution with six points and four solutions with five points each (Figure 10). "Solution (Gen. 41 | Ind. 60)" was found to have the best overall performance, with six objectives performing better than the averages, including the Green Space Ratio (GAR), the Daylight Potential, the construction cost, Green Space Accessibility (GSA), View Interest, and Shadow Impact.

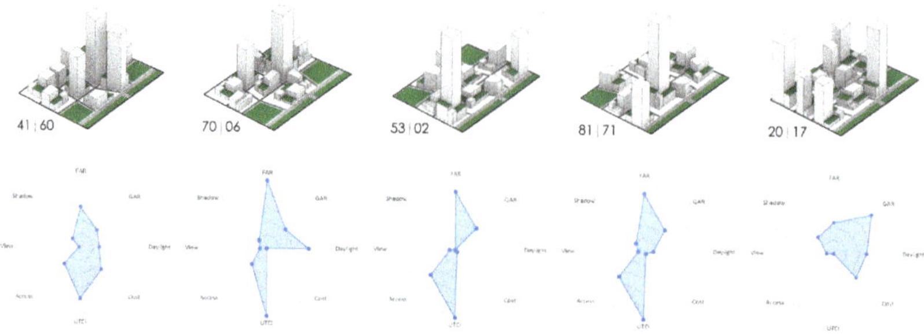

Figure 10. The top 5 best overall performing solutions and the associated radar charts representing the performance results.

Linear relationships between some performance indicators were found as predicted, e.g., the higher the FAR the greater the construction cost, but some relationships were dynamic, which offered unique insights for urban designers. For example, "Solution (Gen. 20 | Ind. 66)" had the fourth highest FAR (6.91) with an above-average GAR (36%), meaning that we could still obtain large green spaces in an ultra-dense urban development; "Solution (Gen. 66 | Ind. 02)" had a below-average GAR (29%) with the third greatest GSA (153.8), meaning that we could still make the green space very much accessible even though the size of the green space is limited.

6. Discussion

This study introduced a generative urban design approach to address complex and interdisciplinary challenges in the context of high-rise urban development. Specifically, multi-objective evolutionary algorithms were used to generate and evaluate how morphological variation evolved in response to conflicting performance objectives across environmental, social, and economic aspects. One of the major limitations of multi-objective evolutionary algorithms in urban design is the very limited number of objectives that can be applied in the algorithms due to the capacity of the tools that are currently available, as well as the simulation time; therefore, many studies have used subjective indicators to expand the scope of the performance matrix. The two-round, quantitative evaluation process with two different sets of performance objectives (totaling eight performance objectives) presented in this study offers an innovative approach to expanding the number of quantifiable performance objectives, filtering a large solution set (totaling 7500 urban forms with 30,000 performance values), and understanding each solution's performance environmentally, socially, and economically. Particularly, integrating construction cost into the algorithms as one performance objective is a unique contribution to the early stage of urban design. There are other important indicators involved in urban design that are hard to quantify, e.g., cultural and aesthetic considerations. How to integrate such indicators into a generative urban design workflow, either through algorithms or other analytical approaches, should be explored in future research.

It is important to highlight that the generative urban design workflow developed in this study was not intended to give a definitive answer on which solution is the best. As discussed above, the urban design process is complex, so it requires input from a variety of stakeholders who often represent conflicting requirements and interests. The ranking criterion proposed in the study treated each of the eight performance objectives equally, but future research may apply different weights to the metrics representing the priorities of certain stakeholders. Using the performance objectives in this study as an example, developers might give more weight to the construction cost, urban designers might give more weight to Shadow Impact, and landscape architects might give more weight to the Green Area Ratio and accessibility. Also, different weights might be given based on the project's location and climatic environment. In this study, for example, more weight could be given to Heat Stress and Shadow Impact, as Guangzhou has a subtropical monsoon climate with high temperatures and humidity in summer.

Author Contributions: Conceptualization, P.D.; methodology and analysis, P.D. and G.L.; software and visualization, G.L. and E.R.; writing—original draft preparation, P.D.; writing—review and editing, P.D., G.L. and E.R. All authors have read and agreed to the published version of the manuscript.

Funding: This research received no external funding.

Data Availability Statement: The original contributions presented in this study are included in the article. Further inquiries can be directed to the corresponding author.

Conflicts of Interest: The authors declare no conflicts of interest.

References

1. Du, P.; Al-Kodmany, K.; Ali, M.M. (Eds.) *The Routledge Handbook on Greening High-Density Cities: Climate, Society and Health*, 1st ed.; Routledge: New York, NY, USA, 2024.
2. Du, P.; Wood, A.; Stephens, B. Empirical Operational Energy Analysis of Downtown High-Rise vs. Suburban Low-Rise Lifestyles: A Chicago Case Study. *Energies* **2016**, *9*, 445. [CrossRef]
3. Nagy, D.; Villaggi, L.; Benjamin, D. Generative Urban Design: Integrating Financial and Energy Goals for Automated Neighborhood Layout. In Proceedings of the Symposium for Architecture and Urban Design, Delft, The Netherlands, 4–7 June 2018. [CrossRef]
4. Besserud, K.; Hussey, T. Urban Design, Urban Simulation, and the Need for Computational Tools. *IBM J. Res. Dev.* **2011**, *55*, 2:1–2:17. [CrossRef]
5. Sun, Y. Generative Methods for Urban Design and Rapid Solution Space Exploration. Master's Thesis, Cornell University, Ithaca, NY, USA, 2021. [CrossRef]
6. Wilson, L.; Danforth, J.; Davila, C.C.; Harvey, D. How to Generate a Thousand Master Plans: A Framework for Computational Urban Design. In Proceedings of the Symposium on Simulation for Architecture and Urban Design, SIMAUD '19, Atlanta, GA, USA, 7–9 April 2019; Society for Computer Simulation International: San Diego, CA, USA, 2019; pp. 1–8.
7. Jiang, F.; Ma, J.; Webster, C.J.; Chiaradia, A.J.F.; Zhou, Y.; Zhao, Z.; Zhang, X. Generative Urban Design: A Systematic Review on Problem Formulation, Design Generation, and Decision-Making. *Prog. Plan.* **2024**, *180*, 100795. [CrossRef]
8. Newton, D.W. Identifying Correlations between Depression and Urban Morphology through Generative Deep Learning. *Int. J. Archit. Comput.* **2023**, *21*, 136–157. [CrossRef]
9. Barr, J.M. The Economics of Skyscraper Height. In *Building the Skyline: The Birth and Growth of Manhattan's Skyscrapers*; Oxford University Press: Oxford, UK, 2016; pp. 141–183. [CrossRef]
10. Chau, K.-W.; Wong, S.K.; Yau, Y.; Yeung, A.K.C. Determining Optimal Building Height. *Urban Stud.* **2007**, *44*, 591–607. [CrossRef]
11. Picken, D.H.; Ilozor, B.D. Height and Construction Costs of Buildings in Hong Kong. *Constr. Manag. Econ.* **2003**, *21*, 107–111. [CrossRef]
12. Ashworth, A.; Perera, S. The relationship between construction price and height. *Chart. Surv. Build. Quant. Surv. Q.* **1978**, *5*, 68–71.
13. Picken, D.; Ilozor, B. The Relationship between Building Height and Construction Costs. In *Design Economics for the Built Environment*; John Wiley & Sons, Ltd.: Hoboken, NJ, USA, 2015; pp. 47–60. [CrossRef]
14. Blackman, I.Q.; Picken, D.H. Height and Construction Costs of Residential High-Rise Buildings in Shanghai. *J. Constr. Eng. Manag.* **2010**, *136*, 1169–1180. [CrossRef]
15. Showkatbakhsh, M.; Makki, M. Multi-Objective Optimisation of Urban Form: A Framework for Selecting the Optimal Solution. *Buildings* **2022**, *12*, 1473. [CrossRef]
16. Balling, R.J.; Taber, J.T.; Brown, M.R.; Day, K. Multiobjective Urban Planning Using Genetic Algorithm. *J. Urban Plan. Dev.* **1999**, *125*, 86–99. Available online: https://trid.trb.org/view/503961 (accessed on 1 July 2024). [CrossRef]
17. Koenig, R.; Miao, Y.; Aichinger, A.; Knecht, K.; Konieva, K. Integrating Urban Analysis, Generative Design, and Evolutionary Optimization for Solving Urban Design Problems. *Environ. Plan. B Urban Anal. City Sci.* **2020**, *47*, 997–1013. [CrossRef]

18. Choi, J.; Nguyen, P.C.T.; Makki, M. The Design of Social and Cultural Orientated Urban Tissues through Evolutionary Processes. *Int. J. Archit. Comput.* **2020**, *19*, 331–359. [CrossRef]
19. Dogan, T.; Samaranayake, S.; Saraf, N. Urbano: A New Tool to Promote Mobility-Aware Urban Design, Active Transportation Modeling and Access Analysis for Amenities and Public Transport. In Proceedings of the Symposium on Simulation for Architecture and Urban Design, SIMAUD '18, Delft, The Netherlands, 4–7 June 2018; Society for Computer Simulation International: San Diego, CA, USA, 2018; pp. 1–8.
20. Dogan, T.; Yang, Y.; Samaranayake, S.; Saraf, N. Urbano: A Tool to Promote Active Mobility Modeling and Amenity Analysis in Urban Design. *Technol. Archit. Des.* **2020**, *4*, 92–105. [CrossRef]
21. Hong Kong Architectural Services Department. Building Works Tender Price Index Report. Available online: https://www.archsd.gov.hk/en/reports/building-works-tender-price-index.html# (accessed on 15 February 2024).
22. May, E.; Du, P.; Martine, V. Environmental Justice: A Case Study into the Heat Vulnerable Neighborhoods of Philadelphia. 2023. Available online: https://www.acsa-arch.org/chapter/environmental-justice-a-case-study-into-the-heat-vulnerable-neighborhoods-of-philadelphia/ (accessed on 1 July 2024).

buildings

MDPI

Article

A Grammar-Based Optimization Approach for Designing Urban Fabrics and Locating Amenities for 15-Minute Cities

Fernando T. Lima [1,2,*], Nathan C. Brown [3] and Jose P. Duarte [2]

1 O'More College of Architecture and Design, Belmont University, Nashville, TN 37212, USA
2 Stuckeman Center for Design Computing, The Pennsylvania State University, University Park, State College, PA 16802, USA; jxp400@psu.edu
3 Department of Architectural Engineering, The Pennsylvania State University, University Park, State College, PA 16802, USA; ncb5048@psu.edu
* Correspondence: fernando.lima@belmont.edu

Abstract: Providing pedestrian accessibility to urban services is a big challenge and a key factor in creating more walkable urban areas. Moreover, it is a critical aspect of climate-resilient urban planning as it is broadly assumed that neighborhoods with greater walkability discourage automobile use and reduce CO_2 emissions. The idea of 15-minute cities, defined as urban environments where most places that residents need to access are within a 15-minute walk, is gaining increasing attention worldwide. Because aspects of urban performance are increasingly quantifiable, generative, and data-driven design approaches can explore broader sets of potential solutions, while optimization can help identify designs with desired properties. This work demonstrates and tests a new approach that combines shape grammars, a formal method for shape generation that facilitates the elaboration of complex patterns and meaningful solutions, with multi-objective optimization. The goal was to optimize the design of urban fabric layouts and the location of amenities to provide 15-minute neighborhood configurations that minimize infrastructure cost (as estimated by cumulative street length) and the number of amenities, while maximizing pedestrian accessibility to urban services (as assessed by overall integration and the average distance from all plots to nearest amenities).

Keywords: urban design optimization; walkability; generative design; shape grammars; 15-minute cities

1. Introduction

Automobile dependency combined with urban population growth has resulted in significant challenges in cities worldwide. Increasing CO_2 emissions, noise pollution, traffic jams, and social segregation are among the most harmful and severe impacts in contemporary urban life [1–3]. Cities are home to 56% of the world's population [4] and are accountable for more than 70% of global CO_2 emissions, while fuel combustion in automobiles creates up to 75% of urban air pollution [5]. Thus, designing areas where most daily needs can be accessed by either walking or cycling has been progressively adopted as a necessary premise regarding the current urban planning agenda.

The concept of 15-minute cities relies on locating urban services within walkable and bikeable distances from all places in a neighborhood. It envisions more sustainable communities by promoting proximity, diversity, density, and ubiquity, requiring a social, environmental, and urban reconfiguration of communities. The ultimate goal is to obtain neighborhoods with short perimeter access to essential social functions such as living, working, supplying, caring, learning, and enjoying [6].

The term 15-minute city was first coined in 2016 by Carlos Moreno [7]. Its main logic can be found in similar approaches that rely on "Spatio-temporal" metrics, such as the 15-minute walkable neighborhood proposed by Weng et al. [8] and the 20-Minute City model, explored by Da Silva et al. [9]. Cities such as Paris, Melbourne, Portland, and

Bogotá have implemented urban models of neighborhoods that can be traversed in a few minutes [6].

At the same time, the advent of novel technologies within the urban design framework has brought a whole new set of possibilities for assessing the efficiency and performance of cities. As the 15-minute city concept relies on measurable features for evaluating its performance, computational approaches can contribute to the improvement of proposals within this scope. This work focuses on exploring computation to support decision-making concerning the layout of urban fabrics and the location of amenities in a neighborhood. These two critical features of pedestrian accessibility [8], essential for promoting 15-minute urban areas, are used to generate fabrics that minimize physical and topological distances to basic urban needs while reducing the street network length, a proxy for infrastructure costs.

The proposed grammar-based approach to the generation of urban fabrics falls within the category of discursive grammars [10,11], which are grammars that can generate designs that are both syntactically and semantically adequate, meaning that they match the design context, or in other words, they satisfy user and site requirements. Frameworks that implement a discursive grammar typically include a generative system to produce candidate solutions, an evaluation system to rate and rank the solutions, and a search mechanism to browse the design space in search of the most adequate solutions.

In previous work [12], we have shown a grammar-based optimization approach to generate fabrics that simultaneously minimize the physical distance between locations in a neighborhood, considering Physical Proximity Index [13,14] and its street network length. In sequence [15], we have investigated the generation of optimal urban grid layouts, considering both the Physical Proximity Index and topological measures from Space Syntax [16] like integration and connectivity. These works, however, focused just on generating and evaluating fabric layouts without addressing the location and distribution of facilities or desirable secondary goals between them, such as minimizing the distances between schools and parks, maximizing the distance between bars and schools, and prioritizing the location of retail in corners.

The work described in this article is focused on the design of new urban fabrics for 15-minute cities, that is, cities in which most amenities and other targets sought by people in their daily life are within a walking distance. In the assessment of walkability, we considered both dimensional (length) and topological (number of turns) measurements of physical and psychological distance, as proposed in Space Syntax and other urban planning and design literature concerned with walkability and accessibility. Shape grammars provide a suitable basis to develop a generative design system for this purpose because of their ability to generate solutions with dimensional and topological variation, thus yielding diverse solutions in potentially wide design spaces. They are thus used as the backbone of the generative system. Dimensional and topological measurements of walkability are encoded into the evaluation system. The coupling of shape grammars with space syntax was first proposed by [17] for the mass customization of housing but is applied to urban design in the present work. The use of optimization in the design of housing was introduced by [18], and the current work extends such research by utilizing optimization to find the best solutions in urban design.

For the application of the framework combining a grammar-based generative system with evaluation and optimization to the retrofitting of existing urban fabrics see, for instance, [19].

This paper aims to address the above limitations in the proposed grammar-based optimization approach. By developing and improving the design method, this research seeks to support proposals for 15-minute urban areas. We are interested in gaining a better understanding of how to maximize the overall integration of a given urban fabric while minimizing its street network length and maximizing the location of amenities, simultaneously considering physical and topological proximities and attending to a set of desirable secondary goals. To this end, we present an approach using a generative system combining (1) a parametric urban grammar interpreter, (2) a set of computational tools to assess conflicting metrics such as Physical Proximity Index, Space Syntax integration, and street network length, with (3) multi-objective optimization algorithms. The goal of the demonstrated method is to find urban arrangements that maximize proximity (from both physical and topological viewpoints) while minimizing the infrastructure cost and additional secondary goals. A shape grammar depicting the urban fabric of Chicago (one of the top 10 cities in the world with the largest carbon footprints) is used to (1) to ground the research on an existing, recognizable grid type to avoid generating designs that could be perceived as inappropriate, (2) address an orthogonal grid, paradigmatic of American cities, and (3) compare the performance of the obtained fabrics with an existing area.

In the proposed approach, design flexibility exists on three levels. In general terms, it exists in the selection of rules to generate different urban fabric types—orthogonal, organic, etc. The current research is focused on orthogonal grids, namely, Chicago's grid, but for other grid types see [12,15,20]. In the case of the Chicago grammar, flexibility exists in how the rules are applied to generate regular or irregular grids [12,20]; for instance, in the subdivision of large blocks into small blocks, as shown in Section 2. Flexibility also exists in how the amenities are distributed within the selected grid. In the current approach it is used a two-step sequential optimization, with the grid layout being optimized first and the amenity distribution later. This decreases the complexity of the optimization process, but it may constitute a limitation, as theoretically it may miss designs with a higher overall score in which the grid layout might not be optimal. To overcome this limitation, concurrent optimization can be explored in the future.

2. Materials and Methods

2.1. Parametric Urban Grammar Interpreter: Chicago Urban Grammar

Shape grammars are systems of simple visual rules used to transform one shape into another [20,21], which include an initial state, a set of instructions, and a termination condition. They support the exploration of designs within an existing design language, by providing the means to describe their generation, or the production of new design languages. Although their use is still more prominent in architecture, shape grammars have been increasingly used at the urban design scale [12,22–26], due to software and hardware improvements that have supported their use for flexible urban planning in recent years.

This work builds on the parametric implementation of the Chicago Urban Grammar developed previously [12]. The logic behind Chicago's urban fabric was encoded into a shape grammar and then translated into a parametric system to support the generation of urban fabrics within the design language defined by the grammar.

The Chicago Urban Grammar consists of two sets of rules. The first set defines the compositional structure of Chicago's street grid (Rules 1–6) and the orientation of blocks (Rules 7 and 8). It is used to create a series of perpendicular (vertical and horizontal) axes, creating a square grid of half a mile (roughly 800 m). These axes define the city's major streets and establish proper subdivision operations resulting in standard blocks of 330 by 660 feet (approximately 100 m $\times$ 200 m) that can be oriented vertically or horizontally. Thus, this first set of rules can create urban plans that vary from four east-west blocks by eight north-south blocks to eight east-west blocks by four north-south blocks and all the combinations in between.

The second set of rules (Rules 10–18) defines alternatives for block arrangements by inserting alleys and dividing blocks into parcels that can then be further divided into plots. Additional rules can be utilized for deleting street segments when needed. Different applications of these rules result in various urban fabrics within the Chicago language. The resulting fabric layouts can be evaluated and optimized according to several performance-related metrics. In the context of this work, we consider Physical Proximity, integration, and street network length as metrics. Figure 1 depicts the two sets of rules and some example fabrics within the Chicago Urban Grammar.

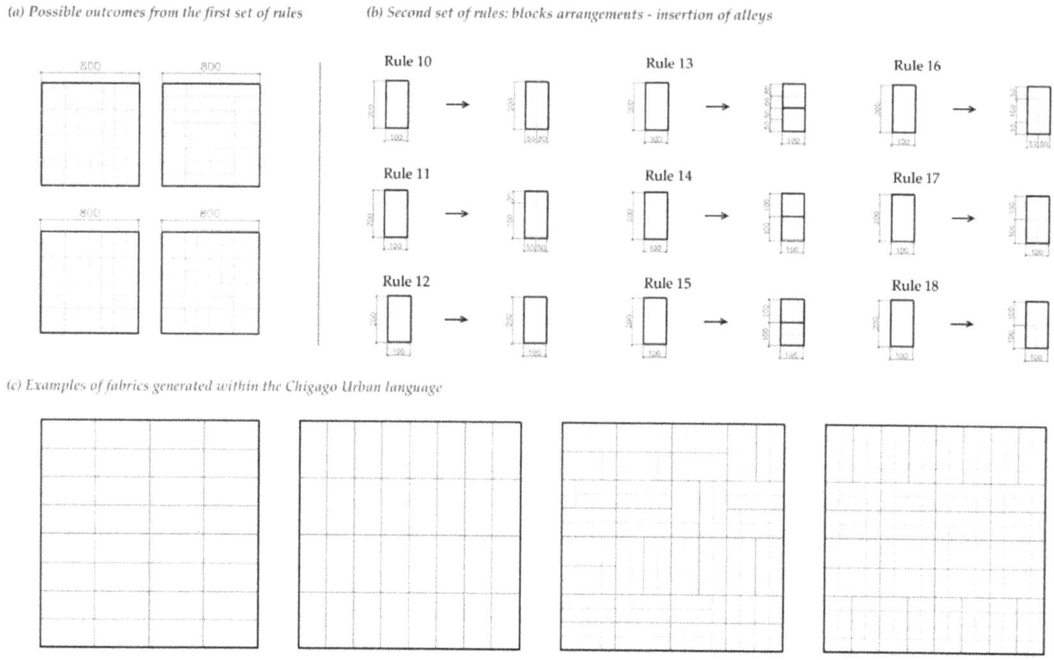

Figure 1. The basic logic of the Chicago Urban Grammar: example of rules for grid generation and urban fabrics within the grammar, obtained through different rules combinations.

2.2. *Physical Proximity—A Metric for Pedestrian Accessibility*

Several scholars have assumed that pedestrian accessibility, as estimated by the proximity to everyday urban services, has substantial implications for the walkability of an urban area [27–32]. As the likelihood of walking trips decreases with greater distances [33], a robust body of research has estimated pedestrian accessibility by calculating distances to facilities, focusing on grid layout and the location of amenities.

This work employs the Physical Proximity Calculator (PPC), a CityMetrics tool developed by Lima [13], to measure pedestrian accessibility. The PPC computes the smallest distance between two or more points of interest, assigning a Physical Proximity Index (PPI) on a scale of 0–1. For example, a 400 m or less distance between two spots corresponds to a PPI of 1. The index decreases as the distance approaches 1600 m and is 0 when the distance becomes greater than 1600 m. Our experiments aim to estimate and optimize pedestrian accessibility. We use the PPC to calculate the PPIs between each block and identify possible locations for amenities considering six categories of urban services: educational (elementary and middle schools), food (bars and restaurants), retail (stores), entertainment (theaters), recreation (parks), and "others" (banks, gyms, etc.). Other categories can be considered in the proposed methodology, depending on the researcher's goals. Figure 2 depicts the PPC procedure considered in this work.

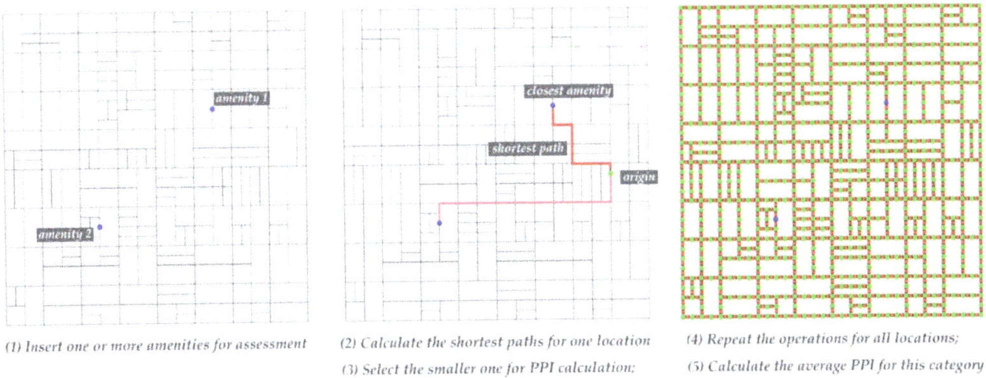

Figure 2. The main steps to calculate the PPI of a given fabric, considering only one type of amenity (e.g., educational). The used tool calculates the shortest path that connects an amenity and a dwelling and repeats that for all assigned locations. The same operations should be performed simultaneously or consecutively for all categories considered.

2.3. Integration

Integration is a syntactic measure from Space Syntax, a set of techniques introduced by Hillier and Hanson [16] to model and analyze spatial configurations and understand social dynamics. Space Syntax advocates that spaces can be decomposed into system components, such as maps and graphs, that permit to assess different features (syntactic measures) related to these spaces. Integration captures how close one space is to everything else in an urban system, referring to the to-movement potential [34–36]. In other words, it is a normalized measure of distance from any place to all others in a system, similarly to how closeness centrality works in network science [35]. In theory, it shows the cognitive complexity of reaching a street, allowing one to predict its pedestrian importance by assessing how close one street segment is in relation to all others in a system.

The importance of coupling integration with Physical Proximity to assess pedestrian accessibility relies on the fact that, in addition to being physically accessible (providing smaller physical distances to basic urban amenities), a neighborhood should also provide smaller topological distances between its locations; that is, fewer street turns, aiming at more connected and legible fabrics [15]. Legibility, considered by Kevin Lynch [37] a major quality of a city, is a key feature that integration helps to assess, so we incorporate it in this study.

2.4. Street Network Length as a Proxy for Pedestrian Accessibility and Infrastructure Cost

It is hypothesized that pedestrian accessibility increases as fabrics become denser as the likelihood of connecting destinations through shorter routes rises within more extensive street networks. Sevtsuk et al. [27] concluded that when various dimensions for plots, block length, and street width are mixed, fabrics with smaller blocks provide higher pedestrian accessibility than the ones with larger blocks. Accordingly, Zhao et al. [38] confirmed that orthogonal street grids with high street density present greater accessibility, while Stahle et al. [39] explored the distribution of land uses while addressing physical proximity and Space Syntax. Feng and Peponis [40,41], in turn, explored different street network and grid typologies.

At the same time, as assessed by the network street length, infrastructure cost proportionally rises with the density of fabrics, defining a trade-off to be addressed toward less CO_2 emitting and more climate-resilient cities—which is a primary goal of 15-minute cities. According to the American Road and Transportation Builders Association [42], the cost of constructing new roads can range from \$2.8 million to \$11.2 million per mile, depending on the location, the number of lanes, and the type of road. Therefore, the street network length is the third performance metric we explore in our proposed approach.

2.5. Computational Optimization in Urban Contexts

Computational optimization methods and tools to support design tasks have improved substantially in recent years. Their use on the urban scale is still limited compared to architecture due greater complexity and computational requirements. However, many scholars have explored optimization as a tool to discover potentially improved urban designs from several viewpoints [14,15,20,43,44]. In our work, we understand optimization as a resource for finding possibilities or even directions for further exploration and refinement instead of a deterministic approach to choose a single, perfect solution.

The proposed approach encompasses both single and multi-criteria optimization, depending on the goals and context. It uses several tools encoding different algorithms to generate optimized urban fabrics (multi-criteria) and locate amenities while attending to PPI, integration, and certain secondary goals (single criteria supported by a weighting equation).

In summary, we considered three Grasshopper optimization tools (Radical [45], Opossum [46], and Wallacei [47]) and their potential to implement the four following algorithms: (i) Radical—SBPLX; (ii) Opossum—RBFOpt; (iii) Opossum—CMAES, and; (iv) Wallacei—NSGA-2. The algorithms have varying strengths and weaknesses for different problem types. The algorithm SBPLX is a variant of the numerical Nelder-Mead simplex optimization method, RBFOpt builds and iteratively refines a surrogate model to reduce the number of simulations required during optimization, and CMAES and NSGA-2 are evolutionary strategies that create progressively better generations of possible design outcomes. As explained in below, these algorithms were utilized in two different stages of the proposed approach.

2.6. A Grammar-Based Optimization Approach for 15-Minute Fabrics—Exploring Pedestrian Accessibility and Infrastructure Cost

As mentioned above, the proposed approach combined a parametric interpreter for generating urban fabrics within a specific language and a set of computational tools developed to assess their performance within a multi-objective optimization framework. The approach included two main optimization stages. The first drove the generation of fabrics belonging to the Chicago Urban Grammar, maximizing their integration while minimizing street network length. The second stage focused on selecting an optimal fabric and finding optimal locations for different amenities while considering desired secondary goals. To this end, a workflow of eight steps was established, including the two optimization stages, as depicted in Figure 3: (1) the generation of a base grid and its parcels using the Chicago Urban Grammar rules 1 to 8; (2) the subdivision of the base grid parcels into blocks by inserting additional streets and alleys, according to the rules 10–18; (3) the assessment of the generated fabrics considering integration and street network length; (4) the multi-optimization step addressing the trade-off between the overall fabric integration and street network length (using Wallacei X); (5) the selection of one optimal fabric among the best solutions found (Pareto-frontier); (6) the elaboration of an algorithm informing the types and number of amenities to be located on the fabric and a collection of secondary goals (established by the users) to drive this location; (7) the setting of an equation that weights the performance criteria and secondary goals into a single fitness function for assessing the amenities distribution; and (8) the optimization step (testing SBPLX, RBFOpt, and CMAES) locating all amenities while maximizing pedestrian accessibility and attending to additional desired secondary goals.

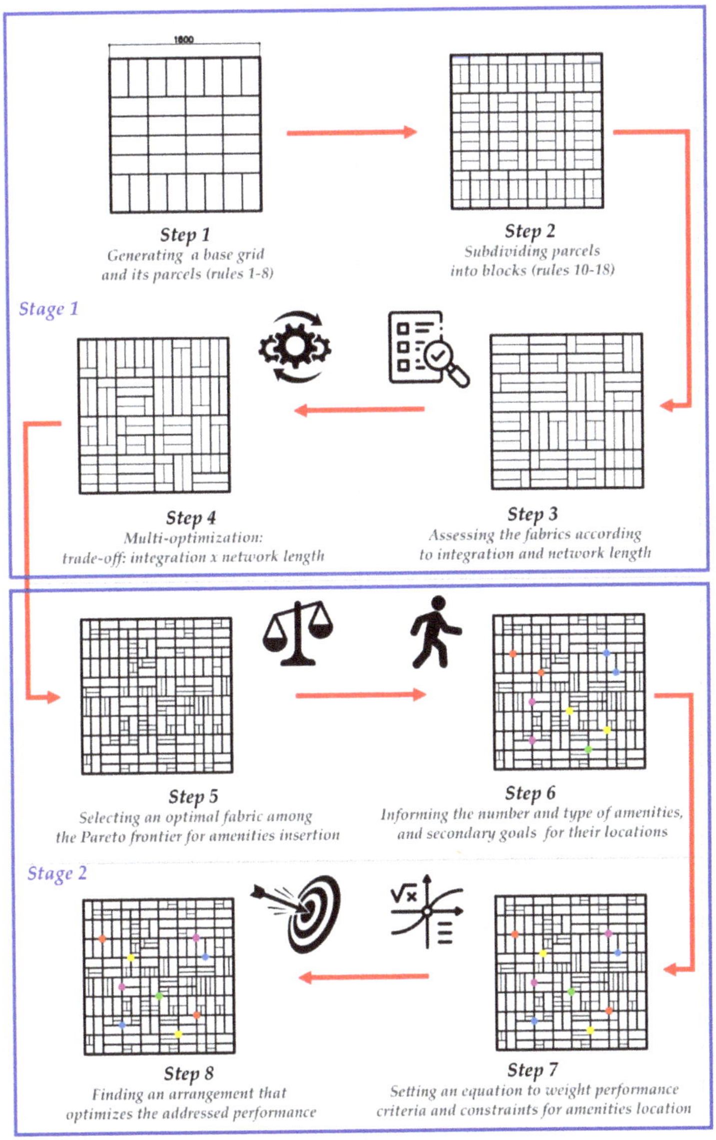

Figure 3. A summarized description of the proposed workflow. The generation of the base grid and its parcels is followed by a subdivision of the parcels into blocks. These configurations are assessed and their performances are set as fitness functions in multi-optimization procedures to determine amenities optimal positions.

3. Case Study

We explored a case study addressing an existing neighborhood in Chicago to test our grammar-based optimization approach. The main goal was to generate an alternative neighborhood configuration that maximized pedestrian accessibility, minimized the infrastructure cost, and attended to a set of desirable secondary goals. Considering the idea of 15-minute cities and the Chicago Urban Grammar framework, we opted for a one-mile square sample area. We did not consider the influence of nearby amenities located outside the studied fabric, although we recognize that they play an important role. The guidelines for determining the sort and number of amenities were extracted from the actual neighborhood features. At the end of the experiment, the existing site and the proposed one were compared.

3.1. *Area of Study*

The city of Chicago can be divided according to three different categories: wards, neighborhoods, and community areas. According to the City of Chicago website [48], a ward is a political district whose boundaries change after each United States census to reflect population shifts within the city and ensure that each ward has approximately the same population and representativity. Similarly, Chicago neighborhood names and boundaries can change over time. Different people may have different perceptions of specific neighborhood names and locations, and the city government does not recognize neighborhood boundaries for any official purposes. Community area boundaries, in turn, do not vary over time, meaning that information about the city can be consistently collected and analyzed over long periods. Chicago is divided into seventy-seven community areas, amongst which we selected Portage Park, a walkable-residential neighborhood located on the northwest side of the city.

According to the Chicago Metropolitan Agency for Planning [49], Portage Park has a population of 63,020 inhabitants, a population density of 15,834 hab/sq mile, and 100% of its population in areas of high walkability. In addition to that, it has a relatively large territorial area, allowing us to explore a one-mile square for the experiment. The idea was to test our approach's potential in a highly walkable area and search for improvements within this context. Figure 4 illustrates the location of Portage Park, its boundaries, fabric, and the area we are addressing in the experiment.

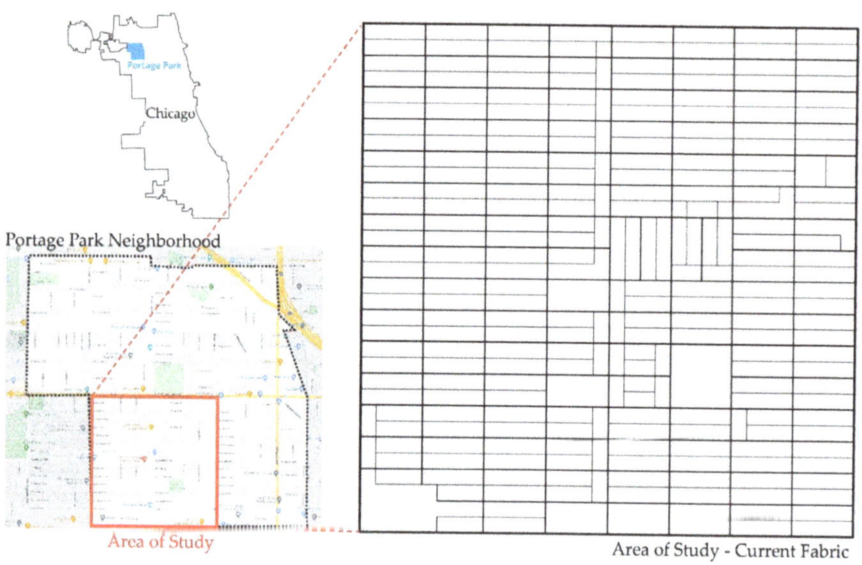

Figure 4. Portage Park location, fabric, and the selected area for the experiment.

3.2. *Experiment*

The experiment tests the grammar-based optimization approach by implementing two subsequent stages. The first consisted of generating an alternative fabric for the study area, maximizing its overall integration while minimizing its infrastructure cost. The second stage was to identify optimal locations for amenities within the fabric selected from the first stage, seeking to simultaneously: (1) maximize the average Physical Proximity between all the sites in the area and their nearest amenities, considering six categories (education, retail, food, entertainment, recreation, and others), (2) insert the amenities in street segments with higher integration, thereby maximizing their overall topological proximity, and (3) comply with a set of desired secondary goals between them (minimize the distances between food and retail, schools and recreation, retail and other, entertainment and food, and maximize the distances between schools and bars).

Additionally, it was established that retail should be on the corners and education should not so that we could ensure more accessibility to the former and more privacy to the latter. The criteria for choosing the number of amenities for each category followed the current Portage Park ratio for education, recreation, and entertainment [48]. A minimum number of four facilities for retail, food, and other categories was also established. Due to the 15-minute city concept and the scope of the study, which aimed at the distributing amenities within a walkable perimeter, it was decided to consider only amenities within the limits of the study area. Nevertheless, it is acknowledged that nearby facilities may serve the site and improve pedestrian accessibility. Consequently, the distribution of amenities was expected to be concentrated in the middle of the neighborhood. Future implementations considering a larger area may result in the distribution of amenities in a larger and continuous fabric. Table 1 summarizes the amenities categories, their numbers, and secondary goals.

Table 1. Amenities considered in this study: categories, number, and secondary goals.

Amenity Category	Number of Amenities	Secondary Goals
education	2	not on corners
retail	4	on corners
food	4	minimize distances to retail [1]
others	4	minimize distances to retail
recreation	3	minimize distances to education
entertainment	1	minimize distances to Food

[1] One specific food amenity was considered a bar and set to maximize its distances to education.

Different optimization techniques were considered in the two experimental stages. To first find a fabric layout that addressed the trade-off between integration and infrastructure cost, Wallacei X [47] was used to employ NSGA-II, a common multi-objective evolutionary solver [50]. This tool enables the analysis and selection of one specific fabric from a set of equivalently efficient solutions (the Pareto Frontier).

The second stage consisted of: (1) proposing the amenities distribution visually, aiming to provide a balanced configuration based on their perceptions, and (2) using three different optimization algorithms (CMAES, RBFOpt, and SPBLX) to find optimal amenities distributions while considering several metrics and secondary goals. To accurately weigh the importance of multiple viewpoints, the following equation was developed to define a performance metric incorporating all viewpoints and secondary goals as a single fitness function to be optimized:

$$\text{Performance metric} = 5a + 2.5b + 0.5c + 0.5d + 0.5e + 0.5f + 0.5g \quad (1)$$

In the above scalarized objective function, a is the normalized average PPI of the neighborhood, b is the normalized average integration of the street segments where amenities are located, c is the normalized average distance between food and retail, d is the normalized average distance between others and retail, e is the normalized average distance between recreation and education, f is the normalized average distance between entertainment and food, and g is the normalized average distance between schools and food amenity number 4 (bar).

Finally, to guide the calculations and better identify the potential locations of the study area, we subdivided each street segment into smaller segments of 50 m. The resulting points of these operations were set as possible locations for the amenities. Figure 5 illustrates this subdivision.

Figure 5. The subdivision of a street network into segments of 50 m. Each point represents an approximate access point to amenities.

4. Results

4.1. Optimization Step 1 Results—Fabric Layout Definition

Stage 1 optimization procedures lasted 25 h and 15 min on an Intel (R) Core (TM) i9-9900X, 3.50 GHz computer, with 128 GB of installed RAM. It led to 25,000 alternative configurations (500 generations with 50 individuals) for fabric layout. The Pareto frontier consisted of 50 solutions sorted into 31 clusters of designs with the same performance and different configurations. The clustering of designs is due to how the grammatical rules are established and translated through a parametric interpreter, as well as how the objectives are calculated. Because the rules add and take away discrete streets on a rectangular grid, many of the solutions snap to similar performance results. Nevertheless, the optimization algorithm demonstrates an ability to improve solutions over generations. Considering the integration vs. total streets length trade-off, most of the solutions found performed better than the existing fabric. Among the Pareto Frontier solutions, the one that provided the best average integration was selected to proceed to stage 2. This design provided an integration improvement of 0.305 (0.926–0.621) compared to the existing layout, and it decreased the total street length by 14,650 m (=68,400 − 53,750 m), meaning a 27.4 million US dollar cost reduction, following estimate guidelines by the American Road and Transportation Builders Association [42]. Figure 6 illustrates all step 1 solutions and highlights three fabrics from the Pareto Frontier: the one with the highest integration (A), the one with the smallest street length (C), and one solution in between them (B). Figure 7 compares the existing fabric and solution A, the fabric set as input to step 2.

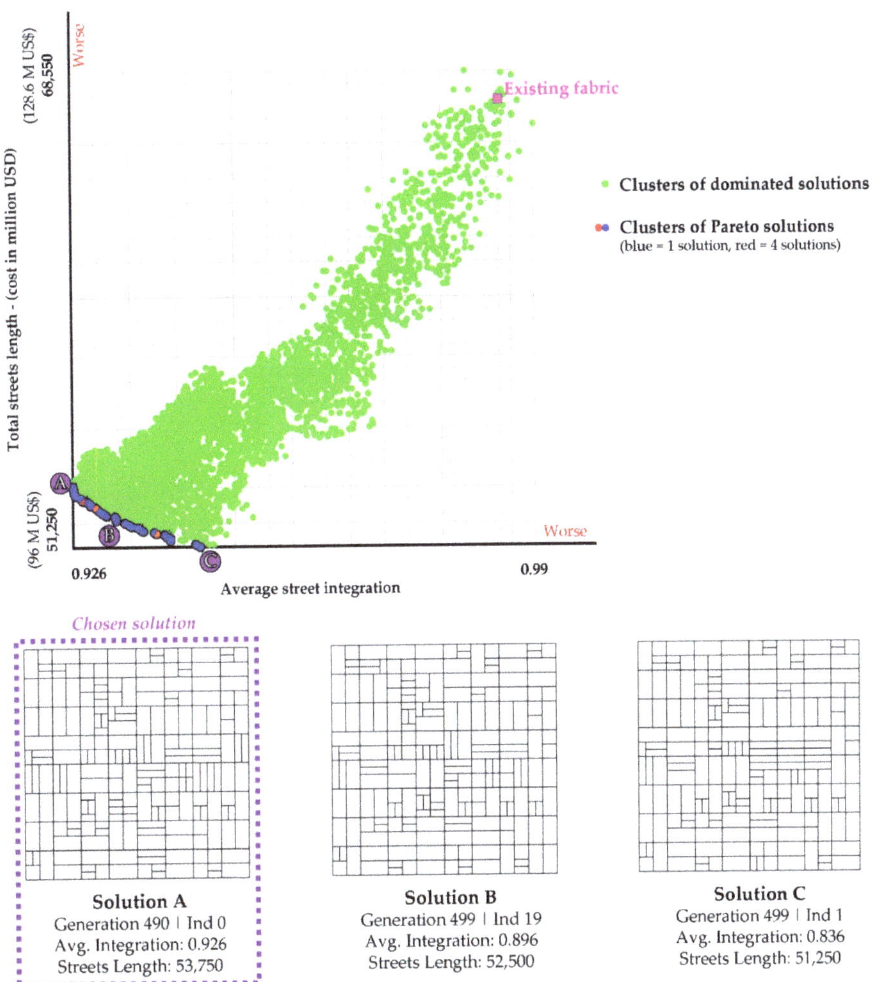

Figure 6. All optimization step 1 solutions and three illustrating fabrics within the Pareto Frontier: the one with the highest integration (**A**), the one with the smallest street length (**C**), and one solution in between them (**B**).

4.2. Optimization Step 2 Results—Distribution of Amenities

As indicated above, optimization step 2 resulted in four different amenities distributions, shown in Figure 8. The distribution manually proposed by the authors, which consisted of an attempt to visually balance the location of amenities within the fabric, had a performance metric of 7.51. This distribution was used as the starting configuration for the SBPLX optimization procedures, as the CMAES and RBFOpt algorithms start with a randomized solution. The best distribution found with this procedure after 4 h returned a configuration with slightly better performance: 7.52. The RBFOpt procedure yielded a configuration with a performance metric of 8.19 (after 8 h), and the CMAES procedure found a configuration with a performance metric of 8.24 (after 8 h). The ranking of solutions according to performance metrics did not correspond to the ranking of solutions considering PPI and integration. Table 2 presents the obtained solutions and their performance.

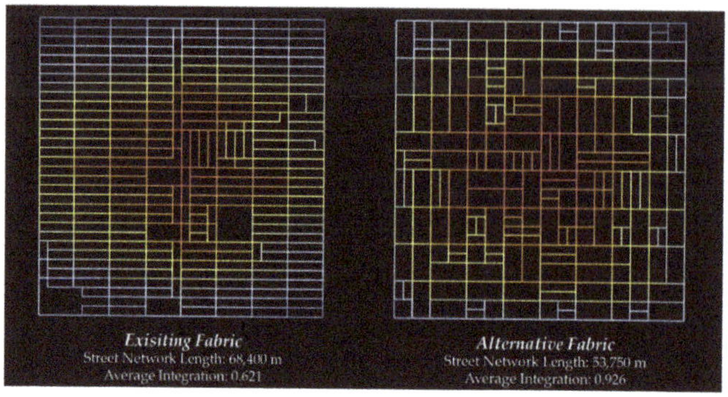

Figure 7. A comparison between the existing fabric and the one obtained in step 1 in terms of integration, with higher integrated streets shown in red and less integrated ones in blue.

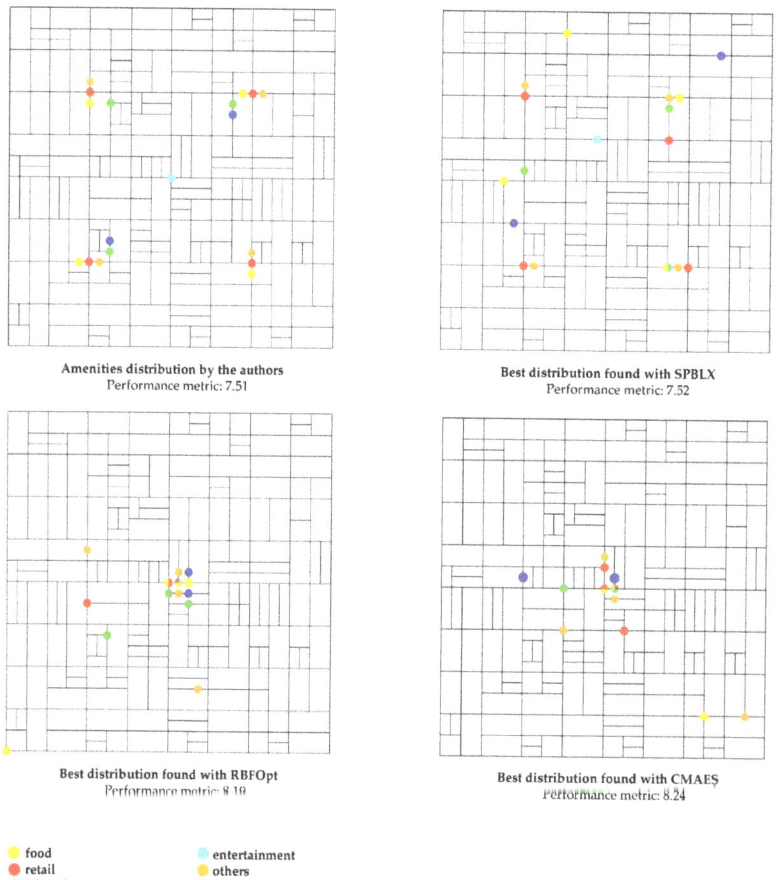

Figure 8. The distribution of amenities manually elaborated by the authors and the best distributions found with different optimization algorithms.

Table 2. The performance of the solution manually proposed by the authors and of the best solutions found by the different optimization algorithms according to different viewpoints.

Procedure	Performance Metric	Average PPI	Average Integration
By Authors	7.51	0.84	0.79
SPBLX (Radical)	7.52	0.83	0.80
RBFOpt (Opossum)	8.19	0.72	0.86
CMAES (Opossum)	8.24	0.74	0.87

5. Discussion

Orthogonal grids like the Chicago grid can be extended indefinitely. The proposed approach could be applied to an entire city, but this possibility greatly increased computational and design complexity. As our main goal was to verify the affordances of the approach, it was decided to focus on a one-mile square area, encompassing four super-blocks. This allowed us to study the relationships between super-blocks and the entire area, and super-blocks and smaller blocks, while facilitating comparison with the actual area. In addition, if the entire city was considered, the increase in scale would require the use of other metrics and the consideration of other amenities, proper of such a large scale. Nonetheless, the application of the proposed approach to the entire city could, in fact, mimic its application to the one-mile area, with the relationship between the city and the super-blocks being "analogous" to the relationship between the super-blocks and the blocks, in a hierarchical fashion. In this fashion, the city would be considered as a polycentric entity with semi-autonomous neighborhoods. This would have clear advantages, such as decreasing the need for movements between neighborhoods and facilitating intermediate lockdowns [51] with positive impacts on CO2 emissions and on limiting the spread of infectious diseases [32], respectively. Ultimately, the choice of how to apply the proposed approach would depend on a variety of factors, including, the preferred urban design theory.

Our goal was not to validate shape grammars but to create a framework to support designers in their quest for producing urban designs that are as adequate as possible to the design context, while matching specific design objectives, which in the current case is walkability. It just happens that this goal cannot be achieved by shape grammars alone. This refers to the discussion between strong AI, like rule-based design systems of which shape grammars are an example, and weak AI, like search and learning. There is not enough domain and contextual knowledge to rely solely on shape grammars. Therefore, a hybrid system is proposed. The first work to propose such a system was by Cagan and Mitchell [52], and the concept was formalized by Duarte [10], where a more in depth discussion on the subject can be found.

Stage 1 results demonstrated a broad spectrum of possibilities to improve the overall integration of the existing fabric while minimizing its total street network length. The vast majority of the 25,000 solutions obtained performed better considering this trade-off than the existing fabric. The application of multi-criteria optimization in this stage enables a designer to view and choose from 50 different solutions found on the Pareto front. The chosen fabric (the one that provided the highest average integration) allowed for a 27.4 million dollars less expensive configuration while delivering a more integrated network within the Chicago urban language. Moreover, if the Pareto frontier solution that provided the less expensive fabric was chosen, this would mean a 32.3 million dollars cost decrease.

Step 2 results reflect an even more difficult problem. The distribution of amenities should deliver less than 15-minute distances (an average PPI of 0.6) between all locations and the closest amenities. In addition, we were seeking configurations that located urban services in street segments with higher integrations while complying with a set of secondary goals. The distribution proposed by the authors intended to respond to these conditions using their perception and expertise as designers. This solution provided the lowest performance metric value but the highest average PPI. The best solution for SPBLX started with the distribution proposed by the authors as input and could not significantly increase its performance (actually, it improved the integration and decreased the PPI) before getting stuck in a locally optimal solution.

Both the RBFOpt and the CMAES best solutions provided significantly better performance metric scores (8.19 and 8.84, respectively). However, these solutions delivered worse PPI performances (yet desirable for 15-minute neighborhoods as they are >0.6), probably due to their trend to centralize the amenities.

6. Conclusions

This study builds on the research thread of grammar-based optimization approaches to address the generation of urban fabrics in early design. The presented approach is intended to parametrically implement a shape grammar that codifies the rules for designing urban fabrics within the Chicago language while attending to multiple viewpoints related to the 15-minute cities concept.

In summary, this study contributes to 15-minute city design methods by exploiting the potential of coupling shape grammars and computational optimization to filter solutions at the urban design scale. A grammar-based optimization approach was used to find 15-minute neighborhood configurations that minimized infrastructure cost (as estimated by cumulative street length) and the number of facilities while maximizing pedestrian accessibility to urban services (as assessed by the fabrics overall integration and the average distance from all plots to nearest facilities). As the 15-minute city concept envisions more sustainable communities by promoting proximity, we believe the presented approach is an initial step toward computational methods to support its proposition.

Despite achieving meaningful findings, the authors acknowledge some limitations of the conducted study. The Chicago Urban Grammar does not address irregular or non-orthogonal urban block patterns, and the influence of nearby amenities located outside the studied fabric was not considered. In addition, other critical aspects related to the 15-minute city concept, like density or the number of street intersections, should be incorporated as objective functions or in the proposed performance metric equation. We also aim to explore shape grammars for irregular urban block patterns. Therefore, there is a wider spectrum of opportunities for future work and improvement of the presented approach.

Author Contributions: Methodology, F.T.L., N.C.B. and J.P.D.; analysis, F.T.L., N.C.B. and J.P.D.; investigation, F.T.L.; supervision, N.C.B. and J.P.D.; writing—original draft preparation, F.T.L.; writing—review and editing, N.C.B. and J.P.D. All authors have read and agreed to the published version of the manuscript.

Funding: This study was financed in part by the Stuckeman Center for Design Computing and the Stuckeman School of Architecture and Landscape Architecture, The Pennsylvania State University, United States, by the Brazilian Coordination of Superior Level Staff Improvement—Coordenação de Aperfeiçoamento de Pessoal de Nível Superior—Brasil (CAPES) Finance Code 001, and by FAPEMIG, the research supporting foundation of Minas Gerais state, Brazil.

Institutional Review Board Statement: Not applicable.

Informed Consent Statement: Not applicable.

Data Availability Statement: Not applicable.

Conflicts of Interest: The authors declare no conflict of interest.

References

1. Appleyard, D. Livable Streets: Protected Neighborhoods? *Ann. Am. Acad. Polit. Soc. Sci.* **1980**, *451*, 106–117. [CrossRef]
2. Gössling, S. Why cities need to take road space from cars—And how this could be done. *J. Urban Des.* **2020**, *25*, 443–448. [CrossRef]
3. Brown, L. *Redesigning Cities for People: Planning Cities for People*; Earth Policy Institute: Washington, DC, USA, 2001.
4. United Nations. *World Urbanization Prospects*; United Nations: New York, NY, USA, 2018.
5. Global Fuel Economy. Available online: http://www.globalfueleconomy.org/ (accessed on 4 October 2016).

6. Nicoli, T. 15-Minute City: A Vision for the Future? | Urban Design Lab. Available online: https://urbandesignlab.in/15-minute-city-a-vision-for-the-future/ (accessed on 25 April 2022).
7. Moreno, C.; Allam, Z.; Chabaud, D.; Gall, C.; Pratlong, F. Introducing the "15-minute city": Sustainability, resilience and place identity in future post-pandemic cities. *Smart Cities* **2021**, *4*, 93–111. [CrossRef]
8. Weng, M.; Ding, N.; Li, J.; Jin, X.; Xiao, H.; He, Z.; Su, S. The 15-minute walkable neighborhoods: Measurement, social inequalities and implications for building healthy communities in urban China. *J. Transp. Health* **2019**, *13*, 259–273. [CrossRef]
9. Capasso Da Silva, D.; King, D.A.; Lemar, S. Accessibility in practice: 20-minute city as a sustainability planning goal. *Sustainability* **2020**, *12*, 129. [CrossRef]
10. Duarte, J.P. Customizing Mass Housing: A Discursive Grammar for Siza's Malagueira Houses. Ph.D. Thesis, Massachusetts Institute of Technology, Cambridge, MA, USA, 2001.
11. Duarte, J.P. A discursive grammar for customizing mass housing: The case of Siza's houses at Malagueira. *Autom. Constr.* **2005**, *14*, 265–275. [CrossRef]
12. Lima, F.; Brown, N.; Duarte, J.P. A grammar-based optimization approach for walkable urban fabrics considering pedestrian accessibility and infrastructure cost. *Environ. Plan. B Urban Anal. City Sci.* **2021**, *498*, 1489–1506. [CrossRef]
13. Lima, F. Urban Metrics: (Para)metric System for Analysis and Optimization of Urban Configurations. Ph.D. Thesis, Federal University of Rio de Janeiro, Rio de Janeiro, Brazil, 2017.
14. Lima, F.; Montenegro, N.; Paraizo, R.; Kós, J. Urbanmetrics: An algorithmic-(para)metric methodology for analysis and optimization of urban configurations. In *Planning Support Science for Smarter Urban Futures*; Lecture Notes in Geoinformation and Cartography; Geertman, S., Allan, A., Pettit, C., Stillwell, J., Eds.; Springer International Publishing: Cham, Switzerland, 2017; pp. 47–64. ISBN 978-3-319-57819-4.
15. Lima, F.T.; Brown, N.; Duarte, J.P. Optimizing urban grid layouts using proximity metrics. In *Artificial Intelligence in Urban Planning and Design*; Elsevier: Amsterdam, The Netherlands, 2022.
16. Hillier, B.; Hanson, J. *The Social Logic of Space*; Cambridge University Press: Cambridge, UK, 1984.
17. Heitor, T.; Duarte, J.P.; Marques, R.P. Combining Grammars and Space Syntax. *IJAC Int. J. Archit. Comput.* **2004**, *2*, 491–515.
18. Granadeiro, V.; Pina, L.; Duarte, J.P.; Correia, J.R.; Leal, V.M.S. A general indirect representation for optimization of generative design systems by genetic algorithms: Application to a shape grammar-based design system. *Autom. Constr.* **2013**, *35*, 374–382. [CrossRef]
19. Noronha, M.; Celani, G.; Duarte, J.P. Computationally evaluating street retrofitting interventions. *Nexus Netw. J.* **2022**, *24*, 481–502. [CrossRef]
20. Lima, F. Urban design optimization: Generative approaches towards urban fabrics with improved transit accessibility and walkability—Generative approaches towards urban fabrics with improved transit accessibility and walkability. In Proceedings of the PROJECTIONS—26th CAADRIA Conference, Hong Kong, China, 29 March–1 April 2021; Volume 2, pp. 719–728.
21. Stiny, G. Introduction to shape and shape grammars. *Environ. Plan. B Plan. Des.* **1980**, *7*, 343–351. [CrossRef]
22. Teeling, C. Algorithmic design: Generating urban form. *Urban Des. Stud.* **1996**, *2*, 89–100.
23. Mayall, K.; Hall, G.B. Landscape grammar 1: Spatial grammar theory and landscape planning. *Environ. Plan. B Plan. Des.* **2005**, *32*, 895–920. [CrossRef]
24. Duarte, J.P.; Rocha, J.; Soares, G. Unveiling the structure of the Marrakech Medina: A shape grammar and an interpreter for generating urban form. *Artif. Intell. Eng. Des. Anal. Manuf.* **2007**, *21*, 317–349. [CrossRef]
25. Duarte, J.P.; Beirão, J. Towards a methodology for flexible urban design: Designing with urban patterns and shape grammars. *Environ. Plan. B Plan. Des.* **2011**, *38*, 879–902. [CrossRef]
26. Beirão, J. *CItyMaker: Designing Grammars for Urban Design*; TU Delft: Delft, The Netherlands, 2012.
27. Brewster, M.; Hurtado, D.; Olson, S.; Yen, J. Walkscore.com: A new methodology to explore associations between neighborhood resources, race, and health. In Proceedings of the 137st APHA Annual Meeting and Exposition 2009, Philadelphia, PA, USA, 10 November 2009.
28. Carr, L.; Dunsigen, I.; Marcus, B. Validation of walk score for estimating access to walkable amenities. *Br. J. Sports Med.* **2011**, *45*, 1144–1158. [CrossRef]
29. Dogan, T.; Yang, Y.; Samaranayake, S.; Saraf, N. Urbano: A tool to promote active mobility modeling and amenity analysis in urban design. *Technol. Archit. Des.* **2020**, *4*, 92–105. [CrossRef]
30. Lima, F.; Paraízo, R.C.; Kós, J.R. Algorithms-aided sustainable urban design: Geometric and parametric tools for transit-oriented development. In *Handbook of Research on Visual Computing and Emerging Geometrical Design Tools*; Amoruso, G., Ed.; IGI Global: Hershey, PA, USA, 2016; pp. 875–897.
31. Nourian, P.; Rezvani, S.; Sariyildiz, I.; Van der Hoeven, F. Configurbanist: Urban configuration analysis for walking and cycling via easiest paths. In Proceedings of the eCAADe 2015 33rd Annual Conference, Vienna, Austria, 16–18 September 2015.
32. Lima, F.T.; Brown, N.C.; Duarte, J.P. Understanding the impact of walkability, population density, and population size on COVID-19 spread: A pilot study of the early contagion in the United States. *Entropy* **2021**, *23*, 1512. [CrossRef]
33. Sevtsuk, A.; Kalvo, R.; Ekmekci, O. Pedestrian accessibility in grid layouts: The role of block, plot and street dimensions. *Urban Morphol.* **2016**, *20*, 89–106.
34. Hillier, B.; Penn, A.; Hanson, J.; Grajewski, T.; Xu, J. Natural Movement: Or, configuration and attraction in urban pedestrian movement. *Environ. Plan. B Plan. Des.* **1993**, *20*, 29–66. [CrossRef]

35. Hillier, B. A theory of the city as object: Or, how spatial laws mediate the social construction of urban space. *Urban Des. Int.* **2002**, *7*, 153–179. [CrossRef]
36. Wasserman, S.; Faust, K. *Social Network Analysis: Methods and Applications*; Structural Analysis in the Social Sciences; Cambridge University Press: Cambridge, UK, 1994; ISBN 978-0-521-38707-1.
37. Lynch, K. *The Image of the City*, Illustrated ed.; The MIT Press: Cambridge, MA, USA, 1960; ISBN 978-0-262-62001-7.
38. Zhao, P.; Yen, Y.; Bailey, E.; Sohail, M.T. Analysis of urban drivable and walkable street networks of the ASEAN smart cities network. *ISPRS Int. J. Geo-Inf.* **2019**, *8*, 459. [CrossRef]
39. Ståhle, A.; Marcus, L.; Karlström, A. Place syntax—Geographic accessibility with axial lines in GIS. In Proceedings of the Fifth International Space Syntax Symposium, Delft, The Netherlands, 13–17 June 2005; pp. 131–144.
40. Feng, C.; Peponis, J. The definition of syntactic types: The generation, analysis, and sorting of universes of superblock designs. *Environ. Plan. B Urban Anal. City Sci.* **2020**, *47*, 1031–1046. [CrossRef]
41. Feng, C.; Peponis, J. Pathways to creating differentiated grids: Types, benefits and costs. *Environ. Plan. B Urban Anal. City Sci.* **2021**, *49*, 535–548. [CrossRef]
42. ARTBA. *How Much Does It Cost to Build a Mile of Road*; American Road & Transportation Builders Association Web Site: Washington, DC, USA, 2021.
43. Navarro-Mateu, D.; Makki, M.; Cocho-Bermejo, A. Urban-Tissue Optimization through Evolutionary Computation. *Mathematics* **2018**, *6*, 189. [CrossRef]
44. Makki, M.; Showkatbakhsh, M.; Tabony, A.; Weinstock, M. Evolutionary algorithms for generating urban morphology: Variations and multiple objectives. *Int. J. Archit. Comput.* **2019**, *17*, 5–35. [CrossRef]
45. Brown, N.; Jusiega, V.; Mueller, C. Implementing data-driven parametric building design with a flexible toolbox. *Autom. Constr.* **2020**, *118*, 103252. [CrossRef]
46. Wortmann, T. Opossum: Introducing and evaluating a model-based optimization tool for grasshopper. In Proceedings of the Protocols, Flows and Glitches, 22nd CAADRIA Conference, Suzhou, China, 5–8 April 2017; pp. 283–292.
47. Makki, M.; Weinstock, M.; Showkatbajhsh, M. Wallacei. 2020. Available online: https://www.wallacei.com/ (accessed on 30 June 2022).
48. Chicago Ward, Community Area and Neighborhood Maps. Available online: https://www.chicago.gov/content/city/en/depts/dgs/supp_info/citywide_maps.html (accessed on 13 May 2022).
49. Chicago Metropolitan Agency for Planning CMAP Community Data Snapshot | Portage Park. Available online: https://www.cmap.illinois.gov/search?q=portage%20park (accessed on 17 May 2022).
50. Deb, K.; Agrawal, S.; Pratap, A.; Meyarivan, T. A fast elitist non-dominated sorting genetic algorithm for multi-objective optimization: NSGA-II. In *Lecture Notes in Computer Sciences, Proceedings of the Parallel Problem Solving from Nature PPSN VI, Paris, France, 18–20 September 2000*; Schoenauer, M., Deb, K., Rudolph, G., Yao, X., Lutton, E., Merelo, J.J., Schwefel, H.-P., Eds.; Springer: Berlin/Heidelberg, Germany, 2000; pp. 849–858.
51. Carpio-Pinedo, J.; Pozo Menéndez, E.; Lamíquiz Daudén, F.J.; Higueras García, E. When a city must be a tree: Rethinking the spatial approach to fighting epidemics based on the notion of 'intermediate confinement'. *Urban Des. Int.* **2021**, *26*, 332–347. [CrossRef]
52. Cagan, J.; Mitchell, W.J. Optimally directed shape generation by shape annealing. *Environ. Plan. B Plan. Des.* **1993**, *20*, 5–12. [CrossRef]

buildings

MDPI

Article

Coastal Waterfront Vibrancy: An Exploration from the Perspective of Quantitative Urban Morphology

Lung Shih Huang, Yun Han * and Yu Ye *

College of Architecture and Urban Planning, Joint International Research Lab of Eco-Urban Design, Tongji University, Shanghai 200092, China
* Correspondence: yunhan1124@163.com (Y.H.); yye@tongji.edu.cn (Y.Y.); Tel.: +86-199-2127-6317 (Y.H.); +86-182-1726-8257 (Y.Y.)

Abstract: Contemporary urban design, requiring a deep understanding of urban form and its performance, has recently shifted its focus on the vibrancy of waterfronts in coastal cities. Based on analytical methods of quantitative urban morphology, this study aims to explore the common morphological features of waterfronts with high urban vibrancy. We selected vibrant waterfront cases from different countries as the benchmark and collected the multi-sourced urban data. The quantitative analysis extracts the common morphological characteristics of vibrant waterfront by calculating the range of those indicators in different cases. The results indicate that those successful waterfronts comprise compact street networks and are mostly dominated by building types favorable for urban vibrancy. They possess high development intensity and mixed functions. Consequently, the compact urban form and dense-mixed land use are recommended for developing vital waterfronts. Moreover, considering the problematic waterfront area of the Jinshatan area in Yantai, quantitative urban morphology methods can be adopted to develop precise urban design guidance for vibrancy-oriented design practice. This study, thus, provides comprehensive insights for shaping the vibrancy of the waterfronts in coastal cities.

Keywords: coastal waterfront; urban vibrancy; quantitative urban morphology; multi-sourced urban data

1. Background

1.1. The Significance of a Vibrant Waterfront in Coastal Cities

The vibrant waterfronts are often the core areas of coastal cities, attracting diverse activities and representing the image of the city. Thus, they tend to have important historical value and potential for tourism development, benefiting economic development to some extent. The mid-20th century witnessed a decline in coastal manufacturing industries due to the transfer of international labor division, with industries shifting from developed to developing countries. However, some coastal cities, such as Baltimore's Inner Harbor [1], Boston's Quincy Market [2], and Sydney's Darling Harbor, have revitalized their waterfront spaces through urban regeneration. In this context, there is a broad consensus on integrating waterfront areas into coastal zone management through urban design [3], revitalizing the waterfront while ensuring sustainable development, and creating livable spaces [4]. Accordingly, quantitative urban morphology has garnered considerable attention from researchers and practitioners as an effective urban design tool.

1.2. Urban Vibrancy as an Important Indicator of a Living Waterfront Space

Many coastal waterfronts need to be shaped into livable places for human life. Urban vibrancy denotes the intangible spatial quality created by diverse social activities [5], thus serving as a key indicator of the livability of the waterfronts. The various amenities create the conditions for attracting social activities continuously throughout the day [6].

Numerous empirical studies have demonstrated that diverse social activities can promote social interactions and enhance social cohesion [7]. Additionally, creating urban vibrancy serves to enhance the gateway function of waterfront areas in coastal cities [8]. Moreover, vibrancy intensity characterizes the concentration of social activities, which is an essential indicator of the efficiency of waterfront space utilization [9]. Consequently, urban vibrancy is a key indicator of spatial quality, related to the livability, attractiveness, and sustainability of waterfronts [10].

1.3. Research Objectives and Significance

According to the above introduction, it is obvious that enhancing the vibrancy of human habitation is an effective way to achieve holistic and sustainable development. Although many studies have proved the correlation between urban morphological characteristics and urban vibrancy [5,6], it still lacks a systematic understanding of the common characteristics of those urban form features of the vibrant waterfronts. In line with this gap, there is still a lack of an operational path to guide the arrangement of urban form indicators that is aimed at developing vibrancy.

In response, this study targets the human-settled coastal waterfronts that are faced with the urgent need to revive vibrancy. A series of quantitative morphological indicators is demonstrated according to the literature review in Section 2. Based on those indicators, this study selects the benchmark and problematic cases and proposes a systematic path to support the vibrancy-oriented urban design practice for coastal waterfronts in Section 3. Then, the morphological indicators summarize the common characteristics of vibrant waterfronts and show specific guidance for the problematic area through the quantitative urban morphology methods in Section 4. Finally, the contribution and deficiencies of this study are further discussed and concluded in Section 5.

2. Literature Review

2.1. The Vibrancy of Coastal Waterfronts and Their Relationship with Urban Form

In urban design practice, how to enhance spatial quality through intervention in urban form has always been a central concern [11]. Therefore, creating vibrancy from the perspective of spatial form has always been the focus of the coastal waterfront.

The relationship between spatial form and urban vibrancy has been demonstrated in many previous studies. From the qualitative perspective, researchers pointed out that urban vibrancy consists of an integrated urban spatial quality of social activity and is associated with many spatial forms tightly [5,6,12,13]. According to Jacob's classical urban design theory, the vibrancy-related urban form features include four aspects, the mix of building ages, multiple building functions, building density, and street accessibility. Among them, the concentration of buildings will facilitate social interaction, and the diverse functions and chronology of buildings will attract a diverse range of people to gather in public spaces. Moreover, a high level of spatial pedestrian accessibility will promote walking trips for city dwellers and enhance social interaction activities in public spaces [12,13]. In terms of the quantitative aspects, researchers further verified the contribution of spatial elements to vibrancy, such as locations [7], small blocks [5,12,14,15], dense street networks, high intensity, and mixed land use [16]. Some researchers also proved the effectiveness of the accessibility to certain functions, such as small catering facilities [17] and subway stations [18]. In general, there is a broad consensus on the spatial characteristics that are conducive to vibrancy, such as street intersection density, building density, block size, and land use diversity.

2.2. Quantitative Urban Morphology and Its New Potentials

With the progress in urban analytics methodology over the past decades, the methods to describe, classify, and represent the spatial and socio-economic performance of urban form gradually become mature, known as quantitative urban morphology [19]. Urban morphology was defined by Conzen's theory originally, including street configuration, development intensity, typology, and land use aspects [20]. Quantitative urban morphology uses new analytical tools to quantify the urban form and extract the indicators from the urban morphology.

Recently, the emergence of multi-sourced urban data has opened up new opportunities for quantitative urban morphology as an approach to urban design analysis. First, geospatial data make it possible to refine the measurement of urban spatial form on a large scale [21]. Second, the functional service facilities data help to quantify urban land use features in detail, such as the diversity and density of different urban facilities [17,22]. More importantly, the quantitative urban analytics tools make it possible for automatic urban analysis [16,23]. For example, based on the space syntax, the accessibility of streets could be predicted accurately through the spatial analysis network plugin on the software of ArcGIS [21]. Moreover, Spacematrix, invented by Berghauser Pont, is a useful technique that can classify various building types quantitatively with a set of morphological indicators, e.g., GSI and average floors [24]. Overall, the urban morphology methods assisted by big data and new analytical tools will bring new insights into the urban design of developing vibrancy under the specific environmental context.

2.3. Research Gap and the Purpose of This Study

Based on the above literature review, two research gaps could be concluded. First, although many empirical studies have been conducted to discover the elements of urban form that are closely related to vibrancy, few have proposed a systematic path to guide the formation of vibrancy from an urban morphology perspective. Second, although the correlation coefficients between key morphological features and urban vibrancy have been analyzed, detailed thresholds of these key elements of urban form are often missed. An understanding of the valid range of thresholds would help to provide precise guidance on urban design practice.

To address the above gaps, the quantitative morphological indicators could be applied to guide the vibrancy-making process of waterfronts in coastal cities. First, the successful benchmark waterfronts in different coastal cities are selected, and the common urban form characteristics of those benchmarks are summarized. Second, this study takes the problematic coastal waterfront area as an example, exploring the specific guidance of spatial elements from the perspective of quantitative urban morphology.

3. Research Method

Firstly, waterfronts famous for their vibrancy were selected as the benchmark cases and the Jinshatan waterfront area in Yantai was selected as a comparative case. Secondly, we identified the urban morphological indicators in three dimensions, including street network configuration, land use, typology, and development intensity. Detailed sources for the three dimensions can be found in Section 3.2. Subsequently, we summarized the overall range of values for each morphological indicator of those benchmark cases. Then, the urban form indicators of benchmark cases were compared to the problematic case. The analysis results would finally help us to propose the targeted urban form guidance and control strategy for a problematic case. The analysis framework is shown in Figure 1.

3.1. The Case Selection and the Research Scope Definition

3.1.1. The Benchmark Cases: The Vibrant Coastal Waterfront Areas

The successful waterfront areas were selected according to the following specific criteria: (a) they feature important coastlines near city centers or urban areas; (b) they contain abundant good water and green resources; (c) they are known internationally for their urban vibrancy, providing abundant leisure lives that attract urban dwellers; and (d) they have long histories, sufficient development, and well-complete urban form.

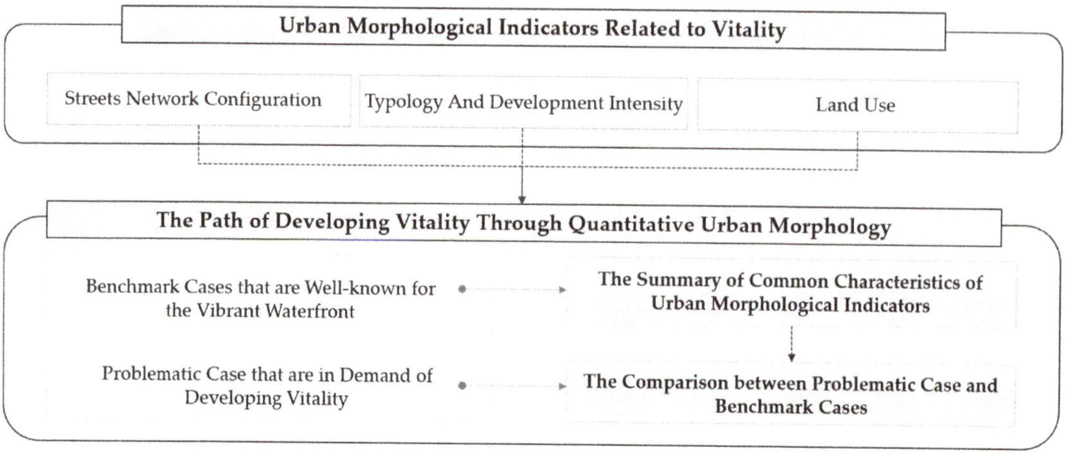

Figure 1. The analytical framework.

Accordingly, through the world's leading travel website, tripadvisor.com (accessed on 12 September 2022), we found some famous vibrant waterfronts located in the USA and Europe. Then, we selected the study areas conforming to the conditions above as benchmark cases [25]. The study area of each case is from the coastline to the hinterland one kilometer away (a ten-minute walking distance), with an area of 3 to 5 square kilometers (Figure 2).

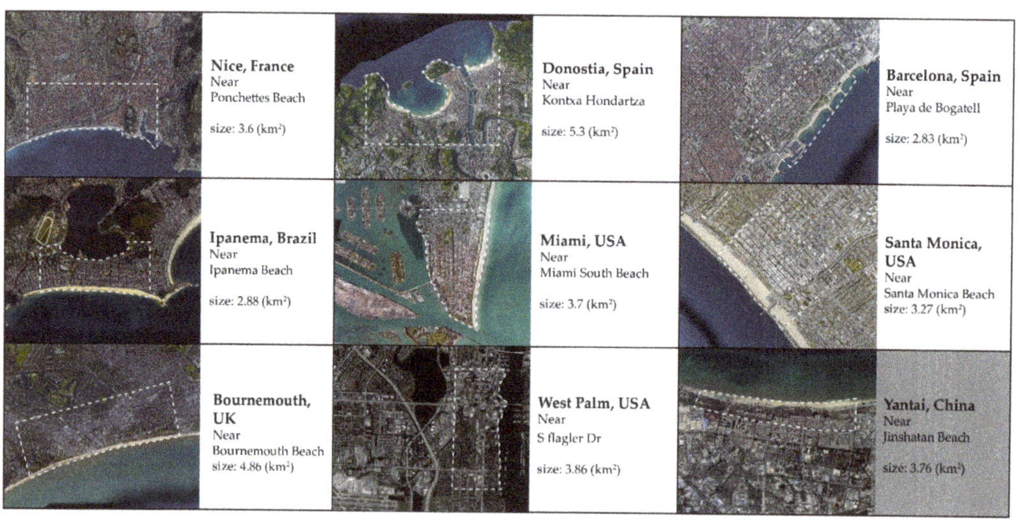

Figure 2. The waterfront areas selected by the vibrant coastal cases and the Jinshatan area.

The values of the indicators from these cases could guide the form of other areas with similar environmental backgrounds but lower vibrancy. Although the cases we choose come from different places of the world and even have different urban morphology in some aspects, their many important features above are the same and are all famous for their urban vibrancy. Consequently, we assume that the famous waterfront urban areas might have common ranges of indicators in urban morphology. Their common ranges might be able to form advice for the Jinshatan area, a waterfront area with a mixed function but lacking urban vibrancy.

3.1.2. The Problematic Examples: A Coastal Waterfront Area in Yantai

Jinshatan is a provincial tourist resort located in the Yantai Economic and Technological Development Zone, in the coastal belt of northern China. It has more than 10 km of sandy beaches. It is also a comprehensive area dominated by the tourism leisure industry and is adjacent to a residential area. The Jinshatan area has numerous geographical and environmental advantages. However, based on the relevant reviews on the website of Tripadvisor.com [25], it can be found that, compared to other famous waterfront areas in China, this area has a lower level of tourism prosperity.

3.2. Morphological Indicators Aimed at Developing Vibrancy of the Coastal Waterfronts

Starting from the "Anglo-Germanic Historico-geographers" and "town-plan analysis" research methods proposed by Conzen [20], the urban form can be disassembled into town planning, typology models, and land use models. Town planning can be divided into street organizations, plots, blocks, and buildings. According to the definitions mentioned above, to summarize the relevant features, the quantitative urban morphology can be divided into three dimensions, as shown in Table 1, including street network configuration, typology and development intensity, and land use. The meaning of each indicator has also been explained below.

After disassembling the urban form into three types, according to the classical theory of urban vibrancy and urban form [5,6,26–28], the indicators affecting urban vibrancy can be extracted from each dimension. Specifically, road density, road area density, street accessibility, block size, and road intersection density are extracted from street network configuration. Generally, compact road networks, short blocks, and high-accessibility areas contribute to urban vibrancy [29–35]. From the dimension of typology and development intensity, building types, floor area ratio, ground space index, open space ratio, and average building floors can be extracted. Usually, areas with high development intensity may contribute to urban vibrancy [36–41]. Furthermore, several kinds of building are conducive to urban vibrancy. In addition, the average mixed-use index, the proportion of functions, and the density of facilities are extracted from the land use dimension. Empirical studies show that the higher the degree of mixing and density of facilities in an area, the more conducive it is to the vibrancy of the city [42–44]. The function proportion also affects urban vibrancy [45]. Table 1 lists urban morphological types, indicators, definitions, and formulas in detail.

The data utilized in this study were obtained from open-source data websites, such as Openstreetmap.org, and Python programing language was used to request Pois from Google Map API. The data were integrated and calculated using ArcGIS software. The betweenness indicator was developed using the sDNA, an ArcGIS plug-in.

To provide insight into the urban design that is often practiced based on block scale, we have used the block as the calculation unit for each indicator, except the indicator of betweenness. There are two reasons for this consideration. First, the block is widely regarded as an operation spatial unit when designers intervene in the urban form [46]. Second, the effectiveness of block has been verified by many empirical studies [14,15,34]. As to the indicators of road density, road area density, and road intersection density, we calculated their values by the road within the walking distance of 300 m [47] for each block. However, the road segment has been used as the calculation unit for the indicator of betweenness. That is because this indicator mainly reflects the street network structure; thus, we want to know which roads have higher values and have more opportunities to be the regional or city centers through the betweenness.

Table 1. The morphological indicators.

Morphology	Indicators	Definition	Formulas
Streets Network Configuration	Road density [29,30]	The total length of the street network divided by the area	R_d = L/A R_d: The road density, L: Total length of roads in the area (km), A: The land area
	Road area density [29,30]	The sum of the street network area divided by the area	R_ad = R_a/A R_ad: The road area density, R_a: Total road area A: The land area
	Street accessibility [31–33]	The number of times each street segment x is traversed by the "shortest" path between any two other street segments y and z within a particular analysis radius	$Betweeness(x) = \sum_{y \epsilon N} \sum_{Z \epsilon Ry} P(z)OD(y,z,x)$
	Block size [34]	The average side length of the blocks	B_s = B_p/4 B_s: The average block size, B_p: The perimeter of a block
	Road intersection density [35]	The sum of the road intersections divided by the area	R_id = R_i/A R_id: The road intersection density in the area, R_i: Total number of intersections, A: The land area
Typology and Development intensity	Building types [36,37]	The most simplified classification of buildings according to floors and plan configurations	Using the Space Matrix technique: 0 < GSI ≤ 0.2: point, 0.2 < GSI ≤ 0.35: plate, GSI > 0.35: enclosed 0 < Floor (average) ≤ 3: low-rise, 3 < Floor (average) ≤ 6: multi-rise, 6 < Floor (average) ≤ 11: small-high rise, Floor (average) > 11: high-rise Ex: 0 < GSI ≤ 0.2 and 0 < Floor (average) ≤ 3: Low-rise point buildings
	FAR [38–40]	Floor-to-area ratio	FAR = F/A FAR: Floor-to-area ratio, F: Total floor area, A: The land area
	GSI [39,40]	Ground space index	GSI = B/A GSI: Ground space index, B: Total footprint of buildings in the area, A: The land area
	OSR [41]	Open space ratio	OSR = (1−GSI)/FAR FAR: Floor-to-area ratio, GSI: Ground space index
	Average floor [40]	The average number of floors	FAR/GSI FAR: Floor-to-area ratio, GSI: Ground space index
Land use	Function proportion [45]	The proportion of various functions	A, B, C, D, E, F, G, H, I, J, K, and M are the number of different POIs Function ratio = A:B:C:D:E:F:G:H:I:J:K
	Facility density [45]	The total number of facilities divided by the area	F_d = P/A F_d: Facility density, P: The sum of POIs in an area, A: The land area
	Average mixed-use index (MXI) [42–44]	The entropy of the various functional POIs divided by the area	$S(p_1, p_2, \ldots, p_n) = -K \sum_{i=1}^{n} p_i log_2 p_i$ MXI = S/A, A: The land area

4. Analysis

4.1. The Common Features of Urban Morphology in Benchmark Cases

This section illustrates the common characteristics of indicators in successful coastal cases. The following examines the quantitative analysis results to discuss and interpret its meaning and common features.

4.1.1. Street Network Configuration

Road density and road area density reflect the abundance of roads in an area; the higher the value, the more roads and coverage of roads in the area. Additionally, accessible roads can service various areas and pave the way for developing urban vibrancy [29,30].

There are complete and compact street networks near the shorelines and water and green resources to form the urban areas. Therefore, the study areas have good road density (14–33.5/km^2) and road area density (7–21%). The compact street network enables the blocks with water and green resources to have good road density and road area density. The blocks in or near the regional centers and compact roads have higher road density and road area density values. This is shown in Figures 3 and 4.

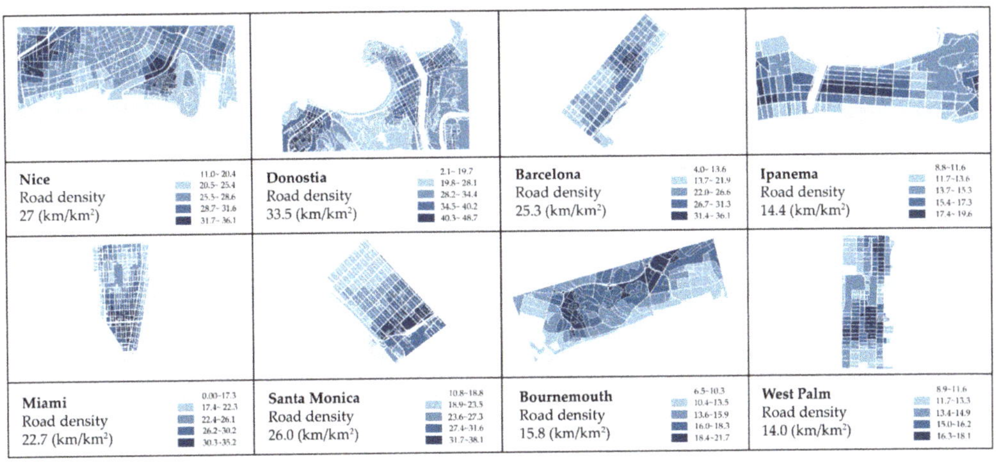

Figure 3. The visualization results of road density in successful coastal cases.

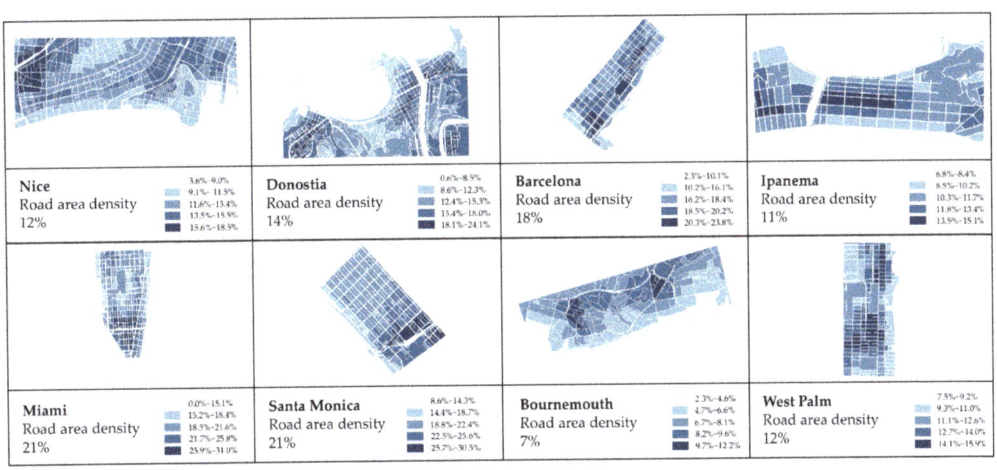

Figure 4. The visualization results of road area density in successful coastal cases.

The betweenness of a segment can be obtained using sDNA [32,33], which represents the selection level of the road segment [31]. We utilized BTA400, which means that the measurement is calculated on a pedestrian scale, 400 m, an acceptable 5-min walking distance in the United States [48], and used BTA 5600, which means that the measurement is calculated on a driving scale, 5600 m, that is above the driving scale of 5000 m and is a multiple of 400 m. Generally, areas with high pedestrian choice become regional service centers more easily, and areas with high vehicle choice become city centers more easily [48]. The degree of choice is positively related to urban vibrancy, and streets with high betweenness usually have high urban vibrancy.

- BTA400:

In the benchmark cases, in Figure 5, because of the compact street networks, the study areas have good average values of BTA400, 116.9–1424.9, and there are some segments with high pedestrian betweenness in the study areas. Some are parallel or perpendicular and close to the coastline, and some in the depth of the hinterlands are parallel or perpendicular to the coastline. Segments with high pedestrian betweenness serve as regional service centers [48].

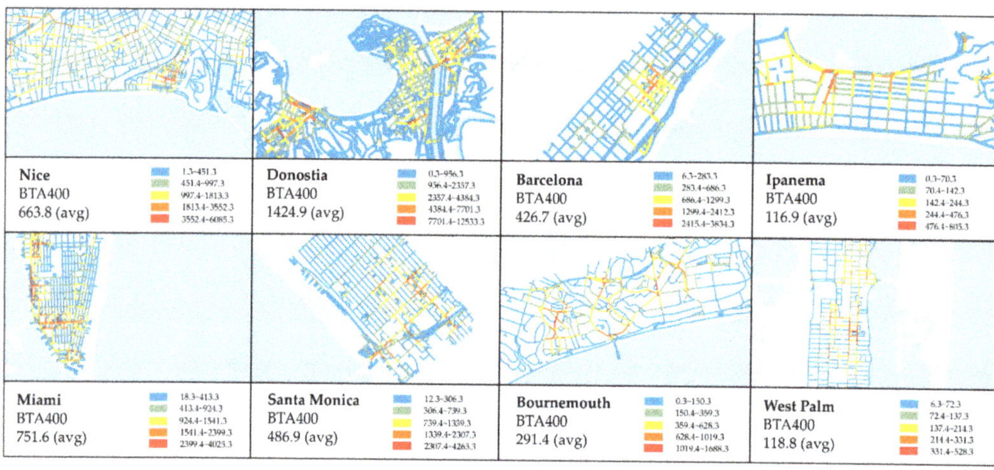

Figure 5. The analytical results of BTA 400 in successful coastal cases.

- BTA5600:

As shown in Figure 6, the study areas have good average values of BTA5600, between 6269 and 100,762.2. These successful cases reveal that there are some roads with high betweenness for vehicles in the study areas. Some of them are parallel or perpendicular and close to the coastline, and some are in the depth of the hinterlands and perpendicular or parallel to the coastline. These roads are also usually wider for vehicles and the parts of the city centers.

Road intersection density and average block size can reflect the sizes of the blocks and the ability to turn [34,35]. It is easier to boost urban vibrancy when people can turn and change routes conveniently. Generally, smaller blocks can improve urban vibrancy. Nevertheless, they are not favorable for cars, demonstrating that suitable street blocks are beneficial to urban vibrancy.

Near the shorelines, there are complete, short, and small blocks forming the urban areas. Therefore, the whole areas have a high road intersection density (86.3–533.4/km^2) and low block sizes (73.98–85.8 m). The blocks with water and green resources still have a good road intersection density because of the compact block organization. The value of road intersection density gradually increases toward the hinterland until it reaches the highest level in the regional center, as shown in Figure 7.

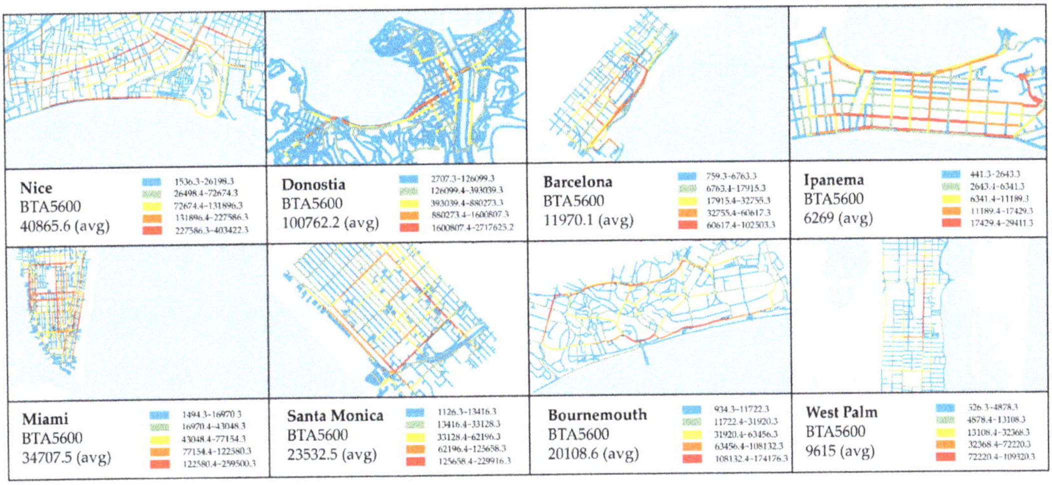

Figure 6. The analytical results of BTA 5600 in successful coastal cases.

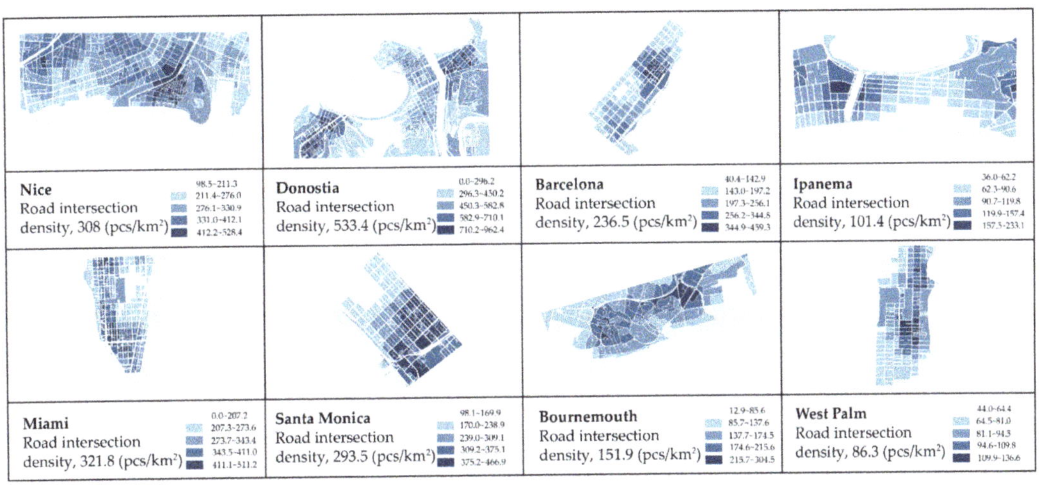

Figure 7. The visualization results of road intersection density in successful coastal cases.

The blocks of beaches, mountains, and green spaces form large block sizes, and other compact and short blocks form a completely urban area near these large-sized blocks. These short blocks have square, rectangular, and irregular shapes, as envisaged in Figure 8.

4.1.2. Typology and Development Intensity

The spatial matrix method can be used to quickly learn the dominant building types and their composition methods in each block in the coastal area [36,37]. According to the previous understanding of building types and urban vibrancy [24], as highlighted in Table 2, building types can be categorized into low, medium, and high vibrancy forms, depending on their level of urban vibrancy.

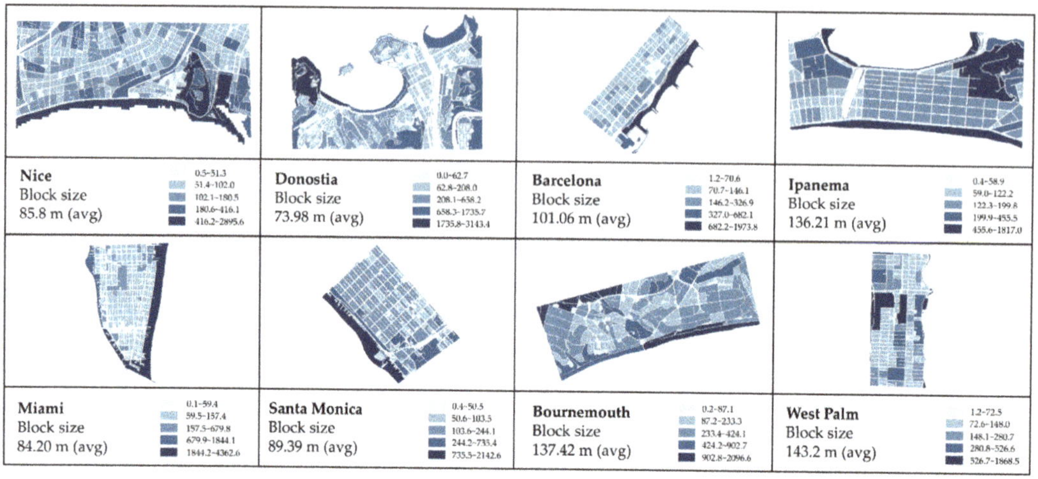

Figure 8. The visualization results of average block size in successful coastal cases.

Table 2. The urban vibrancy levels of different building types [24].

Space Matrix	**High urban vibrancy**	Multi-rise plate buildings, multi-rise enclosed buildings, and (small) high-rise enclosed buildings
	Medium urban vibrancy	Multi-rise-point buildings, (small) high-rise-point buildings, and (small) high-rise plate buildings
	Low urban vibrancy	Low-rise-point buildings, low-rise plate buildings, and low-rise enclosed buildings

Figure 9 provides the number of blocks dominated by various building types. Consequently, the dominant building types in the whole district can be known.

The successful coastal cases can be categorized into the following forms according to the dominant building types in the regions in Figure 9. These are dominated by:

- Multi-rise and small-high-rise enclosure buildings (vibrancy: high): Nice and Donostia.
- Multi-rise enclosed buildings (vibrancy: high): Barcelona.
- Small-high-rise row houses (vibrancy: high): Ipanema.
- Low-rise and multi-rise row houses (vibrancy: high, low): Miami and Santa Monica.
- Low-rise-point buildings (vibrancy: low): Bournemouth and West Palm Beach.

The building types of cases can be categorized into three forms. The first form is dominated by building types conducive to forming urban vibrancy, such as multi-rise and small-high-rise enclosed buildings and small-high-rise row houses. In the second form, the multi-rise townhouses conducive to urban vibrancy are distributed near the coastline and support the public spaces, while the low-rise townhouses that are not conducive to urban vibrancy are distributed in the hinterland. The third form is dominated in the whole area by the low-rise-point buildings that are not conducive to the vibrancy of the city, but the regional centers have many high-rise row houses that are beneficial to the urban vibrancy to support the regional vibrancy. In Figure-Ground [49,50] (Figure 10), the compositions of the building masses are mainly open blocks. The townhouses or enclosed buildings in the blocks, side by side, form continuous street interfaces, which are advantageous for the formation of urban vibrancy.

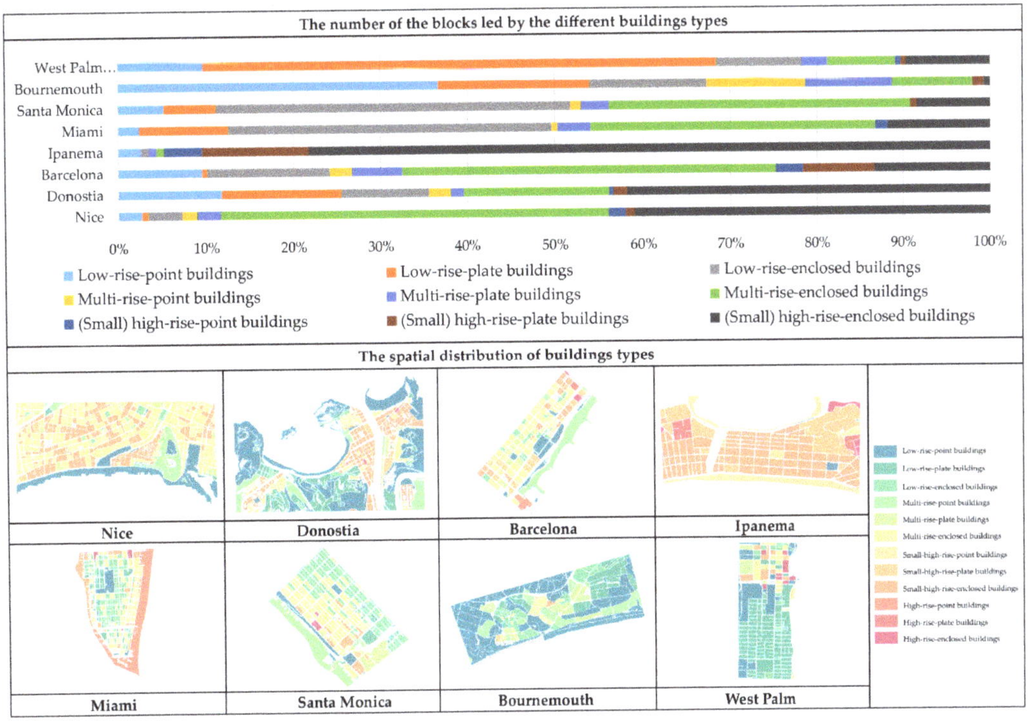

Figure 9. The number of blocks led by the different building types and the distribution of building types.

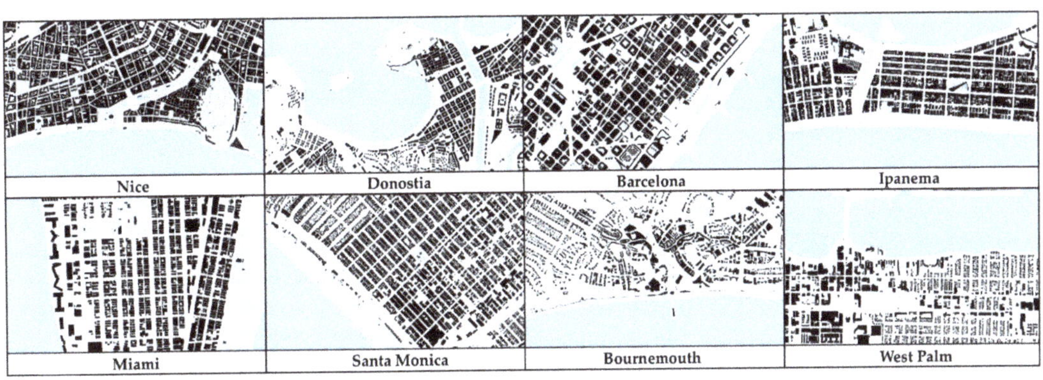

Figure 10. The Figure-Ground in successful coastal cases.

The FAR reflects development intensity, and areas with a higher FAR can usually support better urban vibrancy [38–40]. Nevertheless, a higher FAR can also result in a decline in the quality of life; thus, an appropriate FAR range is required. The GSI can be used to determine the development intensity of a site and reflects the coverage of the building area of a region [39,40]. An excessively high coverage will adversely affect fire protection, sanitation, and greening, whereas an excessively low coverage will be unfavorable for urban vibrancy. The open space ratio (OSR) index represents the ratio of vacant land, which means lands without coverage to floor area [41]. The higher the index, the more vacant land available per capita. However, a high index suggests the availability of more vacant land, which is not conducive to the formation of urban vibrancy. Average floors can reflect the height of the houses in the blocks and affect the development intensity to a certain extent.

Because of the compact urban areas in the study areas, they have high development density, with high FAR (63%–310%), high GSI (19-43%), high average floors (3.3–8.2), and low OSR (0.2–1.26). In these cases, the distribution of the blocks with high development intensity can be categorized into three forms, as shown in Figures 11–13. The first form is that the blocks with high development intensity (with high FAR, GSI, and average floors) are all over and form the urban areas near the coastline. The second form is that the blocks with high development intensity (with high FAR, GSI, and average floors) are mainly close to the shoreline to support the services near these huge public spaces, while the development intensity (FAR, GSI, and average floors) of the blocks in the hinterland are low. The third form is that the blocks with high development density (with high FAR, GSI, and average floors) are concentrated in the centers of the areas that are near the coastline with a walkable distance, while the development intensity of the blocks in the remaining areas is low.

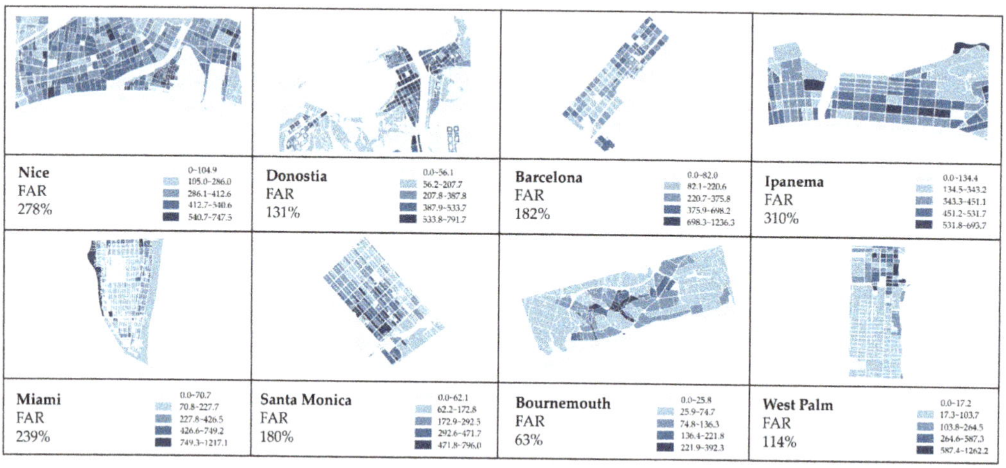

Figure 11. The distribution of FAR in successful coastal cases.

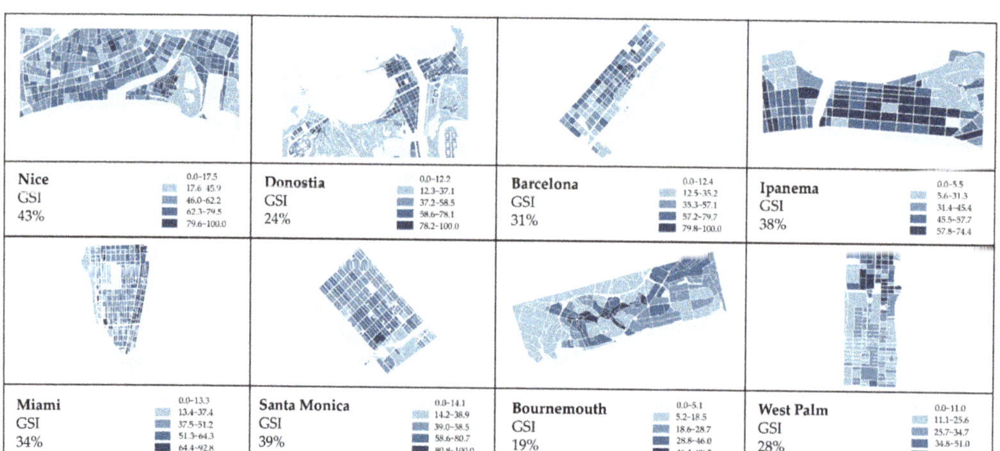

Figure 12. The distribution of GSI in successful coastal cases.

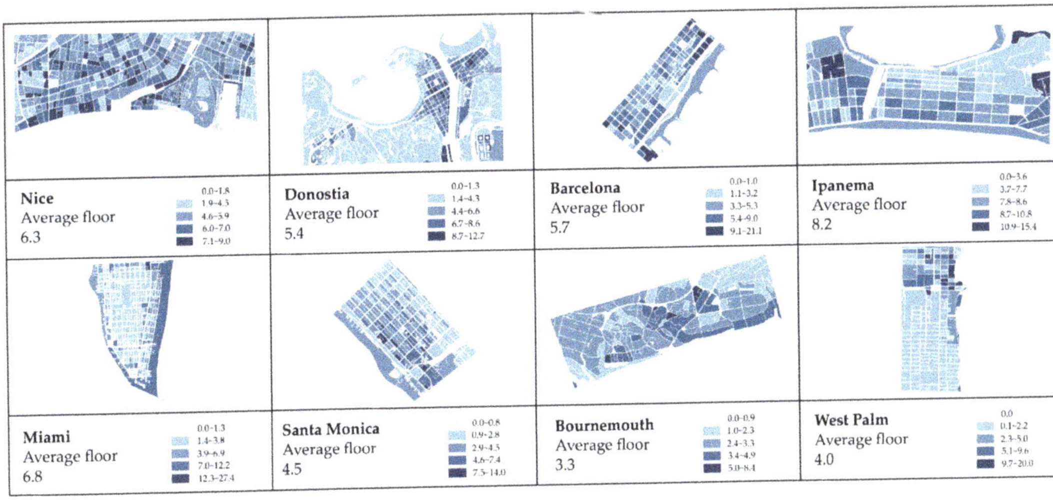

Figure 13. The distribution of average floors in successful coastal cases.

In Figure 14, the blocks of coastal beaches, green spaces, and other resources have the highest OSR. However, the blocks near the coastline and in the hinterland are not high. This reveals that the blocks of the urban areas near the coastal resources are all compactly developed. There is a marked contrast between the low OSR of the blocks near the coastline and natural resources and the high OSR of the blocks with natural resources.

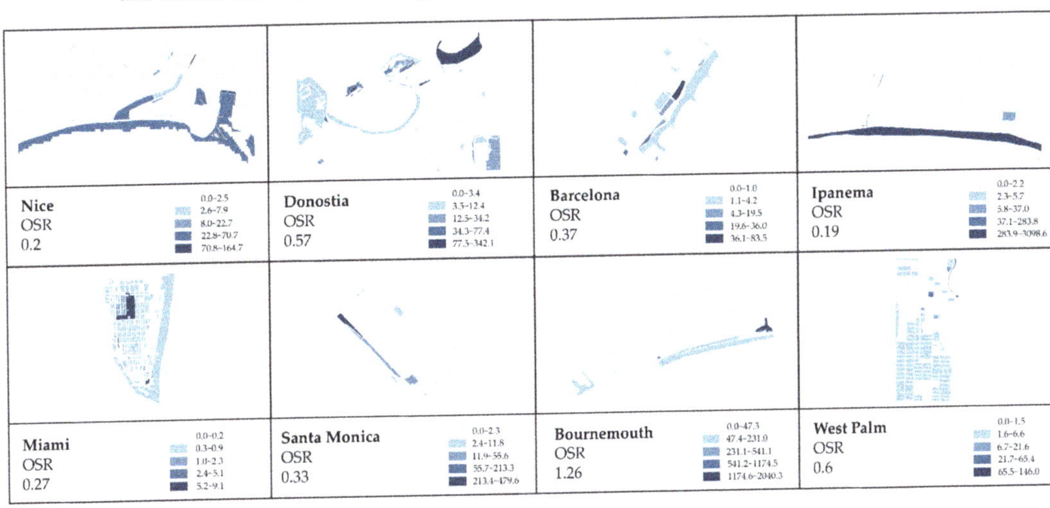

Figure 14. The distribution of OSR in successful coastal cases.

4.1.3. Land Use

The facility density reflects the number of facilities in an area [45]. Generally, the higher the facility density, the more convenient and favorable it is for urban vibrancy. The function proportion illustrates the relationship between the various facility functions and is an important factor affecting vibrancy [45]. The MXI reflects the functional mixing situation of the region [42–44]. Generally, the higher the MXI, the more favorable it is to urban vibrancy.

The study areas have high land use with a high facility density (416.4–2687.4/km^2) and a high MXI (30.6–127.9/km^2). As shown in Figures 15 and 16, the land use of the cases can be categorized into three forms. The first form is that the blocks with high land use (with a high MXI and facility density) are all over and form an urban area near the coastline. The second form is that the blocks with high land use (with a high MXI and facility density) are mainly close to the shoreline, while the low land use (with a low MXI and facility density) blocks are in the hinterland. The third form is that the blocks with high land use (with a high MXI and facility density) are concentrated in the center of the area, and the values of the remaining areas are low. Furthermore, there is a high proportion of life, shopping, financial, medical, lodging, and food services in the successful coastal cases (Figure 15).

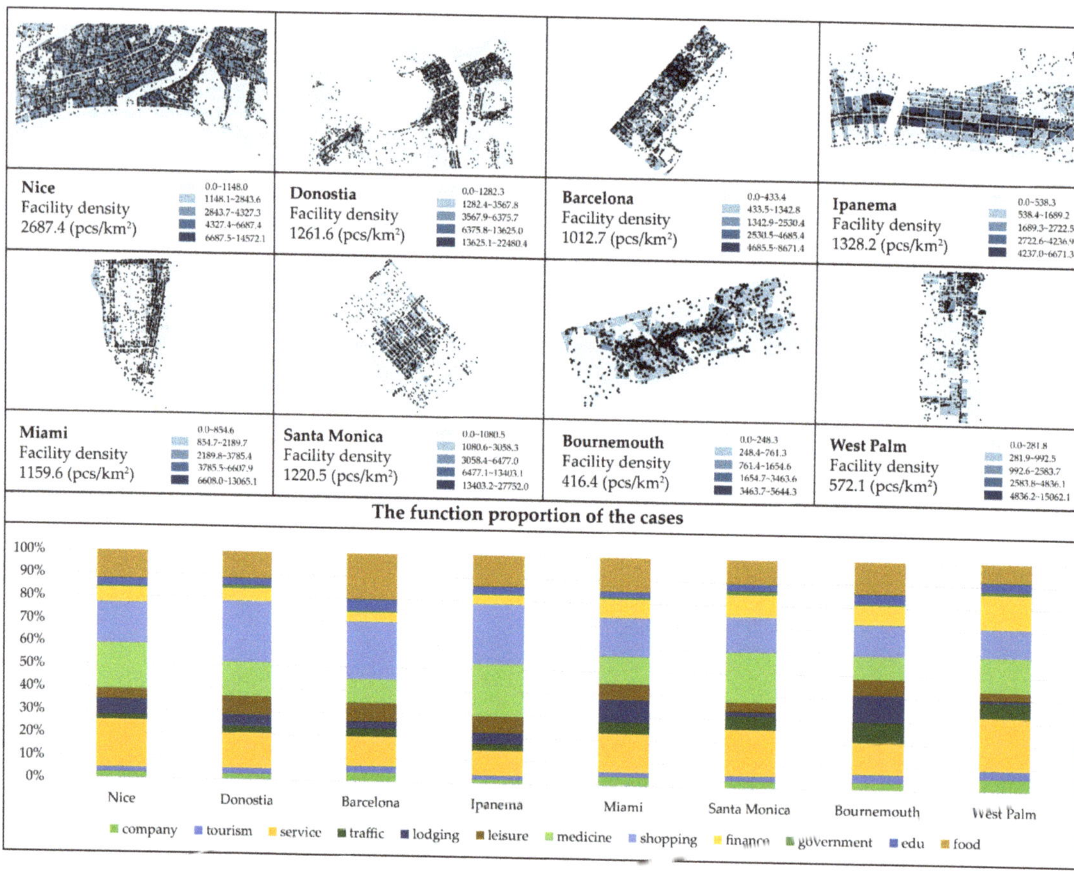

Figure 15. The distribution of facility density and the function proportion in successful coastal cases.

4.1.1. The Summary of the Common Range of Each Morphological Indicator in the Benchmark Cases

The quantitative urban form analysis of the benchmark cases could aid in summarizing the common range of morphological indicators of creating a successful waterfront area. Table 3 presents the values of those indicators in different cases, including the minimum, maximum, average values, and standard deviation. To be specific, in the street network configuration aspect, the road density, road area density, road intersection density, and block size are expected to be set between 14 and 33.5, 7 and 21, 86.3 and 533.3, and 73.9 and 143.2, respectively. Second, the FAR, GSI, OSR, and average floor are expected to be set between 0.63 and 3.1, 19 and 43, 3.3 and 8.2, and 30.6 and 127.9, respectively. Finally, the MXI and facility density are expected to be set between 30.6 and 127.9, and 416.4 and 2687.4, respectively.

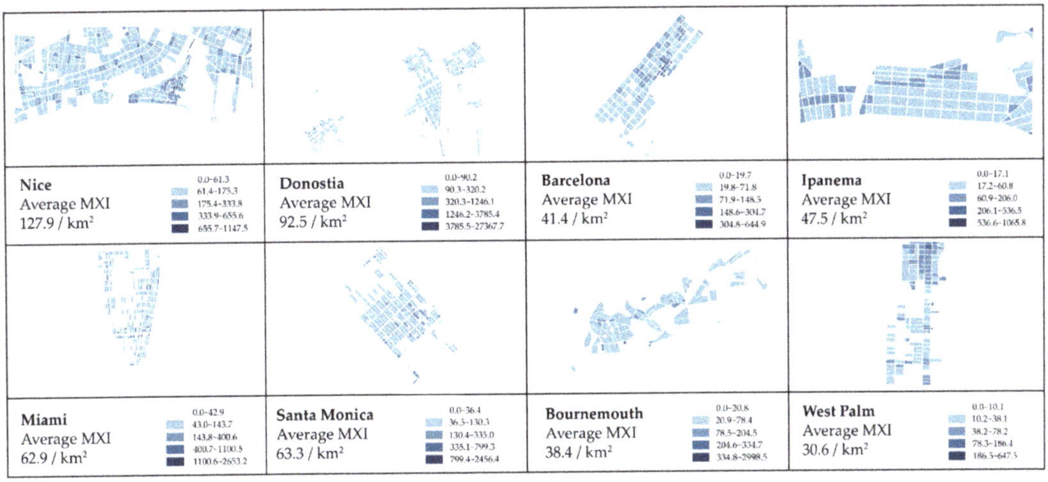

Figure 16. The distribution of MXI in the successful coastal cases.

Table 3. Quantitative analysis summary.

	Street Network Configuration				**Typology and Development Intensity**						**Land Use**	
Indicators	Road density (km/km^2)	Road area density (km^2/km^2)	Road intersection density (pcs/km^2)	Block average size (m)	BTA 400	BTA 5600	FAR (%)	GSI (%)	OSR (km^2/km^2)	Average floor	MXI (/km^2)	Facility density (pcs/km^2)
Minimum value	14.0	7.0	86.3	73.9	116.9	6259	63	19.0	0.2	3.3	30.6	416.4
Maximum value	33.5	21.0	533.4	143.2	1424.9	100,762.2	310	43.0	1.3	8.2	127.9	2687.4
Average value	22.3	14.5	254.1	106.4	535.1	30,978.8	183	32.0	0.5	5.5	63.1	1207.3
Standard Deviation	6.5	4.7	136.60	26.2	399.5	28,686.4	78.4	7.5	0.3	1.4	30.5	640.0

Through the comparison of those values between the benchmark and problematic cases, precise guidance for the problematic area with low urban vibrancy can be obtained. That is demonstrated in detail in the next section.

4.2. Design Applications of Theoretical Insights: The Suggestions for the Jinshatan Area as an Example

Jinshatan is a waterfront tourism resort located in Yantai. Built over the last two decades, it comprises several tourist and leisure facilities along the waterfront area. Compared to the benchmark cases that have a long history, Jinshatan lacks urban vibrancy. As shown in Figure 17, this research explores the reason for this gap between Jinshatan and the benchmark cases through the analysis of the urban morphological indicators. The quantitative analysis results offer some operational guidance on the urban form.

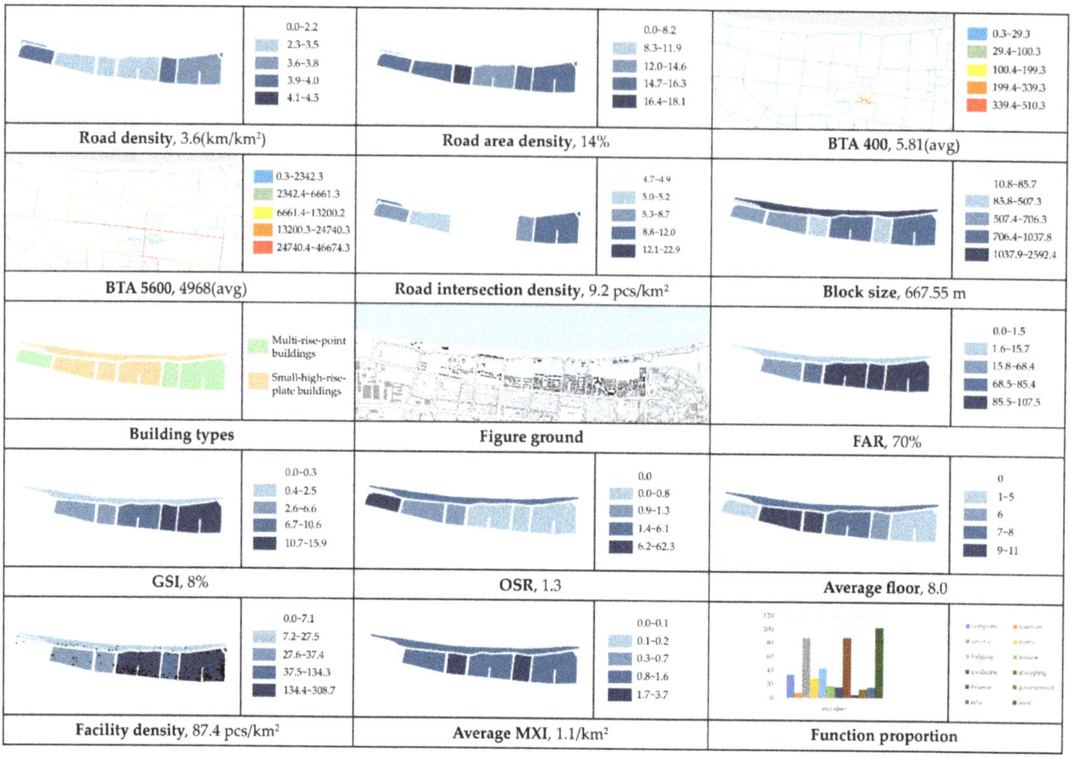

Figure 17. The analysis result of the Jinshatan area, Yantai.

- Suggestions for Jinshatan's street network configuration

As shown in Figure 17, Jinshatan has a morphological structure with a sparse street network and huge blocks. It would be ideal to reform it into an urban area with a compact street network and shorter blocks to increase its accessibility and develop some regional or city centers. Specifically, through the comparison between Jinshatan and the successful cases highlighted in Figure 18, the road density could be changed from 3.6 to 14–33.5, and the road area density could be changed from 14% to 21%. Additionally, road intersection density could be increased from 9.2 to 86.3–533.4, and the average block size may be reduced from 667.5 to 73.9–143.2. It is recommended that the BTA400 be promoted from 5.1 to 116.9–1424 and the BTA5600 from 4968 to 6259–100,762.2.

- Suggestions for the typology and the development intensity of Jinshatan, Yantai

As shown in Figure 17, Jinshatan features multi-rise-point buildings and small-high-rise plate buildings that contribute to medium urban vibrancy. The analysis recommends the addition of building types that are conducive to urban vibrancy, including multi-rise and small-high-rise enclosed buildings and multi-rise and small-high-rise row houses. Additionally, these buildings can be arranged at the Jinshatan waterfront evenly or concentrated in the center of the hinterland, near the coastline. Furthermore, the gated communities could also be changed to open blocks to form continuous street interfaces.

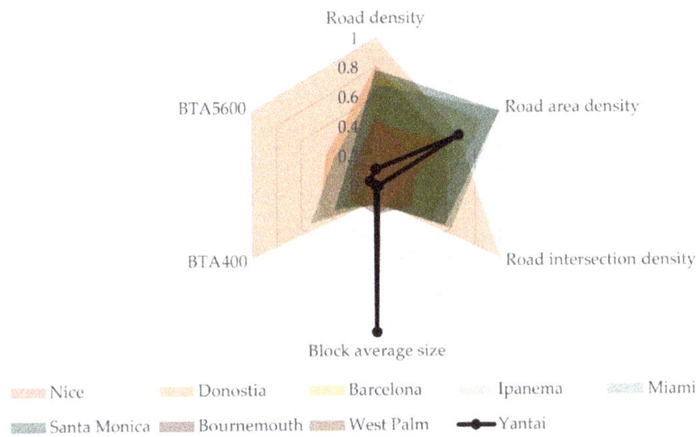

Figure 18. The suggestions for the street network configuration.

The development density could be improved to form a compact urban area that can support urban vibrancy. As to the specific guidance of indicator range, in Figure 19, it is recommended that the FAR for Jinshatan be changed from 70 to 63–310, the GSI from 8 to 19–43, and the OSR from 1.3 to 0.19–1.26. The following strategies relating to development intensity are recommended: (1) the development intensity can be increased evenly throughout the waterfront area, (2) the high development intensity blocks should be arranged close to the shoreline and the low development intensity blocks in the hinterland, and (3) the high development intensity blocks should be arranged close to the regional centers, with the remaining area retaining low development intensity.

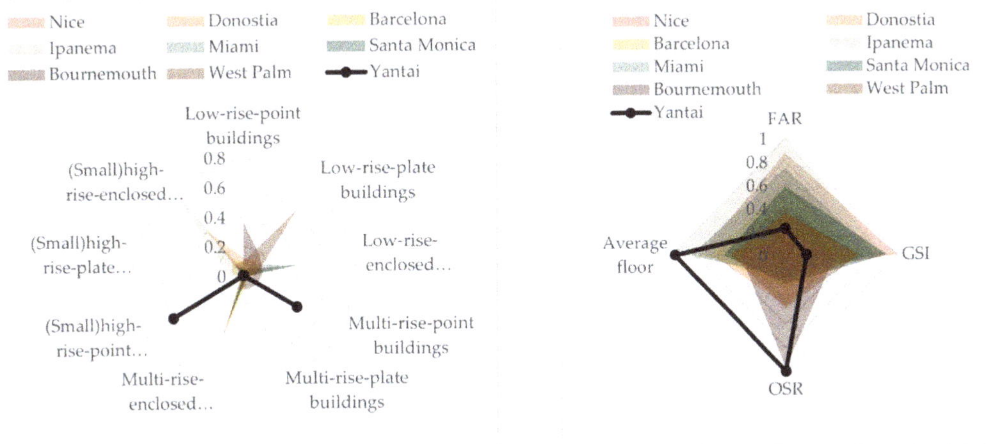

Figure 19. The suggestions for the typology and the development intensity.

- Suggestions for Jinshatan's land use

As shown in Figure 17, the Jinshatan waterfront area lacks dense functional facilities, with a low level of function mixture. It is recommended that the overall number of facilities be increased, raising the facilities' density from 87.4 to 416.4–2687.4. The MXI could also be increased from 1.1 to 30.6–127.9. Additionally, as shown in Figure 20, Jinshatan mainly provides dining, living, and shopping services. The proportion of finance and medical services should be improved. In terms of the spatial distribution of land use, the following strategies could be adopted: 1) high land use blocks in the whole area, 2) arrangement of high land use blocks close to the shoreline, with low land use blocks in the hinterland, and 3) arrangement of high land use blocks near the regional centers, while that of the low land use blocks in the remaining area.

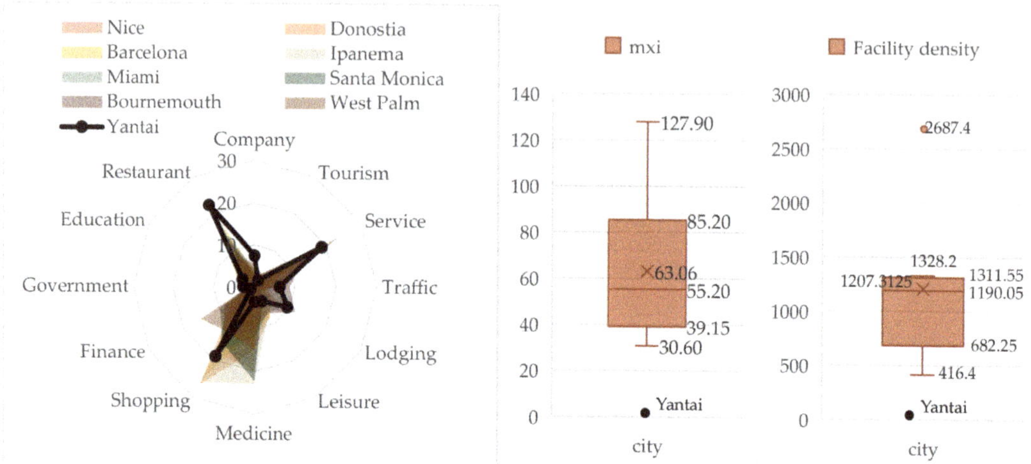

Figure 20. The suggestions for land use.

5. Discussions and Conclusions

5.1. Summary and Contribution of This Research

This study offers twofold contributions. It summarizes the common characteristics of successful waterfronts with a high degree of vibrancy. The features were extracted by the analysis of the morphological indicators related to vibrancy from representative benchmark cases, which were selected from travel websites and under strict requirements [25]. In addition, the comparative analysis of Section 4.2. shows that, despite the differences in the benchmark cases, they have greater differences from the problem case. Therefore, the benchmark cases could reflect the common urban form features to some extent of the vibrant waterfronts. To be specific, it can be found in the illustrations in Sections 4.1 and 4.2, that the vibrant waterfronts have compact street networks and building types that are conducive to public activities. They also possess high development intensity and mixed-dense land use. Thus, compact urban areas are recommended for developing vital waterfronts. These conclusions are consistent with the classical Jacob's urban theory. Most importantly, this study gives a specific range of relevant morphological indicators from the quantitative urban morphology, which will tell us what a reasonable level of indicators should be. For example, based on the results of this study, the road density for a vibrant waterfront is suggested to be set in the range of 14.0–33.5 km/km^2 (Figure 3), while the floor area ratio should be set in the range of 0.63–3.1 (Table 3).

This study also provides an operational approach for developing the urban vibrancy of the waterfronts based on the quantitative urban morphological analysis. Comparing the indicators between Jinshatan (the problematic case) and the famous coastal waterfront cases, the following insights were obtained. In terms of the streets, blocks, and development intensity, the average block size and OSR should be reduced in Jinshatan, while the road density, road area density, FAR, and GSI should be increased correspondingly. As for the building types, an appropriate increase is needed in the proportion of small high-rise enclosed buildings, and multi-rise and small high-rise row houses. For land use, there is a large gap in the functional density and mixture between Jinshatan and the other case areas. Both should be enhanced by introducing multiple and complex functional services. Additionally, the proportion of medicine and leisure facilities could be increased appropriately. Overall, this study provides a comprehensive and operational path to enhance the vibrancy of the coastal waterfront from the perspective of quantitative urban morphology.

5.2. Research Limitations and Future Perspective

This study has several limitations. Firstly, for the methods of case selection, due to the difficulty of using quantitative methods, the benchmarks were selected by a qualitative approach based on the predefined conditions (Section 3.1.1). More cases should be included in the future, and quantitative sampling methods should be attempted to bring more robust results. Secondly, human-scale morphological indicators, such as street interfaces, were also identified as important factors influencing the vibrancy of public space. In addition, more features, such as pedestrian roads and the availability of public spaces, may be related to urban vibrancy. We may take them into account in the future. Thirdly, the results of this study also provide some clues to discovering the classification of urban morphology, for which different types of coastal waterfronts may have different paths of vibrancy enhancement. Under the constraints of costs and benefits, their paths need to be further explored. Fourthly, although current study mainly focuses on urban morphological indicators, we would involve the indicators from social dimensions in our future study to achieve a more systematic consideration.

Author Contributions: Conceptualization, Y.Y.; methodology, Y.Y. and Y.H.; software, L.S.H.; validation, Y.Y. and Y.H.; formal analysis, L.S.H. and Y.H.; investigation, L.S.H. and Y.H.; resources, Y.H.; data curation, L.S.H. and Y.H.; writing—original draft preparation, L.S.H. and Y.H.; writing—review and editing, Y.Y. and Y.H.; visualization, L.S.H.; supervision, Y.Y.; project administration, Y.Y.; funding acquisition, Y.Y. All authors have read and agreed to the published version of the manuscript.

Funding: This research was funded by National Natural Science Foundation of China, grant number 52078343 and 51838002, Natural Science Foundation of Shanghai, grant number 20ZR1462200, and Fundamental Research Funds for the Central Universities, grant number 22120210540.

Data Availability Statement: Not applicable.

Acknowledgments: We would like to thanks Changyu Chen, Xiaoyu Chen and Chengcheng Huang for their support on data preparation.

Conflicts of Interest: The authors declare no conflict of interest.

References

1. Del Rio, V. From Downtown to the Inner Harbor: Baltimore's Sustainable Revitalization—Part 2: The Inner Harbor Plan (1967 to 2005). *Focus* **2018**, *14*, 62–76.
2. Quincy, J., Jr. *Quincy's Market: A Boston Landmark*; Northeastern University Press: Boston, MA, USA, 2019; ISBN 13-978-1555535520.
3. Vallega, A. Urban waterfront facing integrated coastal management. *Ocean. Coast. Manag.* **2001**, *44*, 379–410. [CrossRef]
4. Sue, B. *The Routledge Companion to Urban Regeneration*; Routledge: London, UK, 2013.
5. Gehl, J.K.J. *Life Between Buildings: Using Public Space*; Island Press: Washington, DC, USA, 2011; ISBN 10-978-1597268271.
6. Jacobs, J. *The Death and Life of Great American Cities*; Vintage Books: New York, NY, USA, 1961.
7. Mouratidis, K.; Poortinga, W. Built environment, urban vitality and social cohesion: Do vibrant neighborhoods foster strong communities? *Landsc. Urban Plan.* **2020**, *204*, 103951. [CrossRef]
8. Gravagnuolo, A.; Angrisano, M. Assessment of Urban Attractiveness of Port Cities in Southern Italy—A Case Study of Torre Annunziata. *Sustainability* **2013**, *5*, 3906. [CrossRef]
9. Forward, C.N. Waterfront Land Use in the Six Australian State Capitals. *Ann. Assoc. Am. Geogr.* **1970**, *60*, 517–532. [CrossRef]
10. Liu, S.; Lai, S.-Q.; Liu, C.; Jiang, L. What influenced the vitality of the waterfront open space? A case study of Huangpu River in Shanghai, China. *Cities* **2021**, *114*, 103197. [CrossRef]
11. Tonkiss, F. *Cities by Design: The Social Life of Urban Form*; John Wiley & Sons: New York, NY, USA, 2014; ISBN 978-0-745-64898-9.
12. Buchanan, P.J.A.R. What city? A plea for place in the public realm. *Archit. Rev.* **1988**, *184*, 31–41.
13. Maas, P.R. Towards a Theory of Urban Vitality. Ph.D. Thesis, The University of British Columblia, Vancouver, BC, Canada, 1984.
14. Gan, X.; Huang, L.; Wang, H.; Mou, Y.; Wang, D.; Hu, A. Optimal Block Size for Improving Urban Vitality: An Exploratory Analysis with Multiple Vitality Indicators. *J. Urban Plan. Dev.* **2021**, *147*, 04021027. [CrossRef]

15. Long, Y.; Huang, C.C. Does block size matter? The impact of urban design on economic vitality for Chinese cities. *Environ. Plan. B Urban Anal. City Sci.* **2017**, *46*, 406–422. [CrossRef]
16. Jiang, Y.; Han, Y.; Liu, M.; Ye, Y. Street vitality and built environment features: A data-informed approach from fourteen Chinese cities. *Sustain. Cities Soc.* **2022**, *79*, 103724. [CrossRef]
17. Ye, Y.; Li, D.; Liu, X. How block density and typology affect urban vitality: An exploratory analysis in Shenzhen, China. *Urban Geogr.* **2018**, *39*, 631–652. [CrossRef]
18. Xiao, L.; Lo, S.; Liu, J.; Zhou, J.; Li, Q. Nonlinear and synergistic effects of TOD on urban vibrancy: Applying local explanations for gradient boosting decision tree. *Sustain. Cities Soc.* **2021**, *72*, 103063. [CrossRef]
19. Ye, Y.; Yeh, A.; Zhuang, Y.; van Nes, A.; Liu, J. "Form Syntax" as a contribution to geodesign: A morphological tool for urbanity-making in urban design. *Urban Des. Int.* **2017**, *22*, 73–90. [CrossRef]
20. Conzen, M.R.G. Alnwick, Northumberland: A Study in Town-Plan Analysis. *Trans. Pap. (Inst. Br. Geogr.)* **1960**, *27*, iii-122. [CrossRef]
21. Zhang, L.; Ye, Y.; Zeng, W.; Chiaradia, A. A Systematic Measurement of Street Quality through Multi-Sourced Urban Data: A Human-Oriented Analysis. *Int. J. Environ. Res. Public Health* **2019**, *16*, 1782. [CrossRef]
22. Guo, X.; Chen, H.; Yang, X. An Evaluation of Street Dynamic Vitality and Its Influential Factors Based on Multi-Source Big Data. *ISPRS Int. J. Geo-Inf.* **2021**, *10*, 143. [CrossRef]
23. Kandt, J.; Batty, M. Smart cities, big data and urban policy: Towards urban analytics for the long run. *Cities* **2021**, *109*, 102992. [CrossRef]
24. Ye, Y.; Nes, A. Quantitative tools in urban morphology: Combining space syntax, spacematrix and mixed-use index in a GIS framework. *Urban Morphol.* **2014**, *18*, 97–118. [CrossRef]
25. TripAdvisor. Available online: https://www.tripadvisor.com/TravelersChoice (accessed on 12 September 2022).
26. Trancik, R. *Finding Lost Space: Theories of Urban Design*; John Wiley & Sons: New York, NY, USA, 1991.
27. Katz, P.; Scully, V.; Bressi, T.W. *The New Urbanism: Toward an Architecture of Community*; McGraw-Hill: New York, NY, USA, 1994.
28. Montgomery, J. Making a city: Urbanity, vitality and urban design. *J. Urban Des.* **1998**, *3*, 93–116. [CrossRef]
29. Pengyao, Y.E.; Xiaohong, C. Study of Urban Optimal Density of Road Network Based on Transportation Efficiency. *China J. Highw. Transp.* **2008**, *21*, 94–98.
30. Fei, S. New Research on Urban Road Network Mode and Density Values. In Proceedings of the 11th International Conference of Chinese Transportation Professionals, ICCTP 2011, Nanjing, China, 14–17 August 2011.
31. Hillier, B.; Hanson, J. *The Social Logic of Space*; Cambridge University Press: Cambridge, UK, 1989.
32. Cooper, C.; Chiaradia, A.J. sDNA: How and why we reinvented Spatial Network Analysis for health, economics and active modes of transport. In Proceedings of the 23rd GIS Research UK (GISRUK) Conference, Leeds, UK, 15–17 April 2015.
33. Cooper, C.H.V. Spatial localization of closeness and betweenness measures: A self-contradictory but useful form of network analysis. *Int. J. Geogr. Inf. Sci.* **2015**, *29*, 1293–1309. [CrossRef]
34. Zhang, L.; Menendez, M. Modeling and Evaluating the Impact of City Block Size on Traffic Performance. *J. Urban Plan. Dev.* **2020**, *146*, 04020021. [CrossRef]
35. Grasser, G.; van Dyck, D.; Titze, S.; Stronegger, W.J. A European perspective on GIS-based walkability and active modes of transport. *Eur. J. Public Health* **2016**, *27*, 145–151. [CrossRef] [PubMed]
36. Steadman, P. Density and Built Form: Integrating 'Spacemate' with the Work of Martin and March. *Environ. Plan. B Plan. Des.* **2014**, *41*, 341–358. [CrossRef]
37. Berghauser Pont, M.H.P. *Spacematrix: Space, Density and Urban Form*; Nai010 Publishers: Rotterdam, The Netherlands, 2021.
38. Ng, E. *Designing High-Density Cities: For Social and Environmental Sustainability*; Routledge: London, UK, 2009.
39. Dovey, K.; Pafka, E. The urban density assemblage: Modelling multiple measures. *Urban Des. Int.* **2014**, *19*, 66–76. [CrossRef]
40. Boyko, C.T.; Cooper, R. Clarifying and re-conceptualising density. *Prog. Plan.* **2011**, *76*, 1–61. [CrossRef]
41. Yi, X.; Huang, M.; Zhao, B.; Zhao, B. Investigation and Research on the Open Space Ratio of Residential Plots in Typical Cities of China. *City Plan. Rev.* **2022**, *46*, 30–43.
42. Yue, Y.; Zhuang, Y.; Yeh, A.G.O.; Xie, J.-Y.; Ma, C.-L.; Li, Q.-Q. Measurements of POI-based mixed use and their relationships with neighbourhood vibrancy. *Int. J. Geogr. Inf. Sci.* **2017**, *31*, 658–675. [CrossRef]
43. Van den Hoek, J.W. The MXI (Mixed-use Index) as Tool for Urban Planning and Analysis. In Proceedings of the Corporations and Cities: Envisioning Corporate Real Estate in the Urban Future, Delft, The Netherlands, 26–28 May 2008.
44. Jost, L. Entropy and diversity. *Oikos* **2006**, *113*, 363–375. [CrossRef]
45. Lu, C.; Pang, M.; Zhang, Y.; Li, H.; Lu, C.; Tang, X.; Cheng, W. Mapping Urban Spatial Structure Based on POI (Point of Interest) Data: A Case Study of the Central City of Lanzhou, China. *ISPRS Int. J. Geo-Inf.* **2020**, *9*, 92. [CrossRef]
46. Zhang, A.; Li, W.; Wu, J.; Lin, J.; Chu, J.; Xia, C. How can the urban landscape affect urban vitality at the street block level? A case study of 15 metropolises in China. *Environ. Plan. B Urban Anal. City Sci.* **2020**, *48*, 1245–1262. [CrossRef]
47. Yang, Y.; Diez-Roux, A.V. Walking Distance by Trip Purpose and Population Subgroups. *Am. J. Prev. Med.* **2012**, *43*, 11–19. [CrossRef] [PubMed]
48. Chiaradia, A.; Hillier, B.; Schwander, C.; Wedderburn, M. Compositional and urban form effects on centres in Greater London, UK. *Proc. Inst. Civ. Eng. Urban Des. Plan.* **2012**, *165*, 21–42. [CrossRef]

49. Rowe, C.; Slutzky, R. Transparency: Literal and Phenomenal. *Perspecta* **1963**, *8*, 45–54. [CrossRef]
50. Hebbert, M. Figure-ground: History and practice of a planning technique. *Town Plan. Rev.* **2016**, *87*, 705–728. [CrossRef]

buildings

MDPI

Article

Developing Urban Heat Mitigation Strategies for a Historic Area Using a High-Fidelity Parametric Numerical Simulation: A Case Study in Singapore

Wei Zhu, Liqing Zhang , Shuo-Jun Mei and Chao Yuan *

Department of Architecture, National University of Singapore, Singapore 117566, Singapore
* Correspondence: akiyuan@nus.edu.sg

Abstract: The coexistence of developed areas and historic buildings is an important topic in urban planning. Our study focuses on this topic from the perspective of urban microclimate. A multi-physics CFD simulation is applied to model urban microclimate with anthropogenic heat and buoyancy effects. First, we clarified the impact of new development, i.e., high-rises on pedestrian-level air flow by comparing city structures in Case A (the past, 1960s) and Case B (the current, 2020s). The results showed an average wind speed decrease of 43% over time. Second, we assessed air temperature increments from anthropogenic heat emitted from Case C (high-rises), Case D (historic buildings), and Case E (both). We found that the mean air temperature increased by 0.16 °C for Case C, 0.52 °C for Case D, and 0.87 °C for Case E, respectively. Third, we developed heat mitigation strategies based on the assessment results in the previous steps. The integration of open spaces and building porosity, which create wind corridors together, can promote outdoor ventilation and heat dispersion in the study area. Compared with Case E, the three mitigation cases improve outdoor thermal environment, with mean temperature reductions of 33%, 25%, and 21%, respectively. Finally, we developed new mitigation strategies by considering the constraints in this special region, where modernity and history coexist. Our practical mitigation strategies will aid urban planning and support conservation efforts not only in Singapore, but also in other tropical and subtropical cities.

Keywords: historic area; urban heat island; mitigation strategies; anthropogenic heat; wind corridor

1. Introduction

Many cities have historic areas in their centers where historic buildings and a modern city context (i.e., high-rise buildings) coexist [1,2]. These historic areas are an important part of local tourism and cultural conservation [3,4]. Due to the economic and cultural significance of historic buildings, current measures for their conservation aim to keep their original form and allow occupancy of these buildings [5–7]. However, city centers tend to experience Urban Heat Island effect (UHI) due to factors arising from modern urban development such as high density, high-rise buildings and anthropogenic heat release [8–12]. The UHI phenomenon in city centers can lead to negative health impacts [13] for users of the historic area, especially tourists and occupants of the historic buildings. Therefore, a comprehensive method to improve the thermal environment of historic areas in city centers is needed that balances the users' needs and architectural conservation. This study contributes to a better understanding of scientific and practical mitigation strategies for urban planning in historic areas in the process of urban development. Chinatown in Singapore is a typical area where historic shophouses and modern urban development coexist. Shophouses with two–three stories are the primary building form in Chinatown [14,15], whereas skyscrapers are dominant in other areas of the Central Business District (CBD) (Figure 1). Meanwhile, Singapore, one of densest cities worldwide, tends to suffer intense UHI effects [11,16–19]. Additionally, Singapore is in a tropical climate zone, and relies on air-conditioning (AC) for year-round thermal comfort [20]. The effect of anthropogenic heat from AC can be significant [10,12,21], especially in Chinatown, which is the densest city area of Singapore.

Figure 1. Downtown area in Singapore, where modernity and history coexist.

There are two main challenges regarding the thermal environment in Singapore's Chinatown area. The first challenge is that the existing shophouses were designed to adapt to the functions and ambient environment of the past, without considering the use of AC systems which cause anthropogenic heat accumulation in the area [15]. The second one is the negative effect of surrounding high-rise buildings that have been constructed over the past 40 years. On the one hand, a large amount of AC-related anthropogenic heat emissions from skyscrapers can lead to localized temperature increases [10,11,22]. On the other hand, the blockage of airflow may result in anthropogenic heat not being readily removed [11]. Therefore, understanding heat dispersion within and around historic regions helps to improve future urban and architectural design.

Urban heat mitigation strategies can be implemented in newly built areas in or near existing historic regions in the process of urban development. Various mitigation strategies regarding urban morphology and anthropogenic heat have been proposed in previous studies. These strategies generally rely on modifying urban morphology to enhance airflow, and thus decrease the surrounding air temperature [23]. Some studies have proposed the recommended geometries of street canyons, one of the basic urban forms in a city [24,25]. In a high-density urban context, the effects of high-rise buildings are one of the main issues to explore. An earlier study from Priyadarsini et al. [17] suggested reducing the number of high-rise buildings to achieve a better thermal environment in street canyons. However, this does not appear possible in practice because a high-rise and high-density environment is an inevitable trend in Singapore due to the limited land availability [19]. This means that limiting the amount or the height of high-rise buildings in street canyons may not be an appropriate strategy. More practical suggestions have been provided by some later studies, focusing on the geometry of individual buildings. One useful strategy was proposed by Adelia et al. [22], which emphasized the importance of emission positions of outdoor AC units and building porosity in street canyons in Singapore residential areas. This study also indicated the positive impact of uniformity of building height, which concurred with studies by Erell & Williamson [26] and Kondo et al. [27]. Another strategy is from a study by Yuan et al. [28] who proposed using gaps between buildings with long façades in order to improve the wind environment on urban streets. These studies tend to focus on modifying building geometry to mitigate the localized UHI effect.

However, the above literature mainly focused on the localized effects of urban morphology and anthropogenic heat, such as in street canyons, while the effects of surrounding neighborhoods are overlooked. Therefore, there is a clear need to better understand the effects of urban morphology of both the historic area itself and the surrounding areas on the microclimate at the targeted historic area. Therefore, this study aims, firstly, to develop urban heat mitigation strategies for a historic area—Chinatown, Singapore—and secondly, to analyze the integrated effects of urban morphology and anthropogenic heat from both the historic shophouses and the surrounding high-rise buildings in the area. In this study, we conducted a high-fidelity CFD simulation to model the microclimate with anthropogenic heat and buoyancy effects and quantitatively investigate wind environment and air temperature increment in the Chinatown area.

The research structure is organized as follows. The design of scenarios for the numerical simulation and data analysis method are described in Section 2. In Section 3, the simulation results of all the cases are presented. Section 4 presents discussion on the effects of urban morphology and practical implications for historical areas. The main conclusions of this research and future study are presented in Section 5.

2. Methodology

This study comprised three steps. First, we compared the pedestrian-level air flow of the historic area in the 1960s and 2020s, considering the different urban and building forms of the Singapore CBD area in the past and current scenarios. In the second step, we compared the air temperature increment when involving the effects of anthropogenic heat due to air conditioning (AC) from both the surrounding high-rise buildings and the existing shophouses in the Chinatown area. The third step was to develop mitigation strategies based on the evaluation of results from the previous two steps. The mitigation strategies were developed from three aspects: focusing on the Chinatown area; focusing on the surrounding high-rise buildings; and integrating both strategies.

2.1. Simulation Scenarios Description

Eight parametric cases were simulated using CFD simulation: Five cases to investigate the impact of surrounding new development and anthropogenic heat due to AC on air flow and air temperature increment at the historic area as the first and second steps, and three cases to develop and evaluate mitigation strategies as the third step. As Table 1 and Figure 2a show, the first two cases represent the urban form in the study area in the 1960s (Case A) and 2020s (Case B). In these two cases, the effect of anthropogenic heat due to ACs was not considered because they were not widely used in the 1960s. Therefore, the impact of new high-rise buildings on the wind environment can be investigated by cross-comparing Cases A and B. The differences in anthropogenic heat dispersion among three cases at the second step (Cases C, D, and E) were mainly focused on anthropogenic heat emissions from ACs (Table 1 & Figure 2b). Specifically, at Case C, we only considered the anthropogenic heat due to ACs from the surrounding high-rise buildings. Case D only considers anthropogenic heat from shophouses in the historic area of Chinatown. In Case E, anthropogenic heat emission in both surrounding high-rise buildings and shophouses in the historic area were applied for simulation to represent the real scenario. Cases F, G, and H represent the proposed mitigation strategies (Table 1 & Figure 2c). The principle of the proposed mitigation strategies was to create a ventilation corridor by increasing porosity. In Case F, horizontal porosity was created by removing some shophouses. Case G focuses on the modification of the surrounding high-rise buildings to increase both vertical and horizontal porosity. The specific modifying methods in Cases F and G were based on the analysis of previous evaluation steps. Case H integrated the previous two methods into one case.

Table 1. Description of Case A-H and simulation aim of step 1–3.

Cases	Description	Aim
Case A	Represents urban geometry of the study area in 1960s	Step 1: To clarity the impact of new development on pedestrian-level air flow
Case B	Represents urban geometry of the study area in 2020s	
Case C	The urban geometry of 2020s and anthropogenic heat from the surrounding high-rise buildings are applied	
Case D	The urban geometry of 2020s and anthropogenic heat from the historic shophouses are applied	Step 2: To assess the air temperature increment caused by anthropogenic heat emitted
Case E	The urban geometry of 2020s and anthropogenic heat from both high-rise buildings and the historic shophouses are applied	
Case F	The mitigation strategy is to create horizontal porosity and open spaces by removing some shophouses	
Case G	The mitigation strategy is to create both vertical and horizontal porosity by modifying the geometry of the surrounding high-rise buildings	Step 3: To develop the mitigation strategies based on the assessment from previous two steps
Case H	The mitigation strategies in Case F and G are integrated in this case	

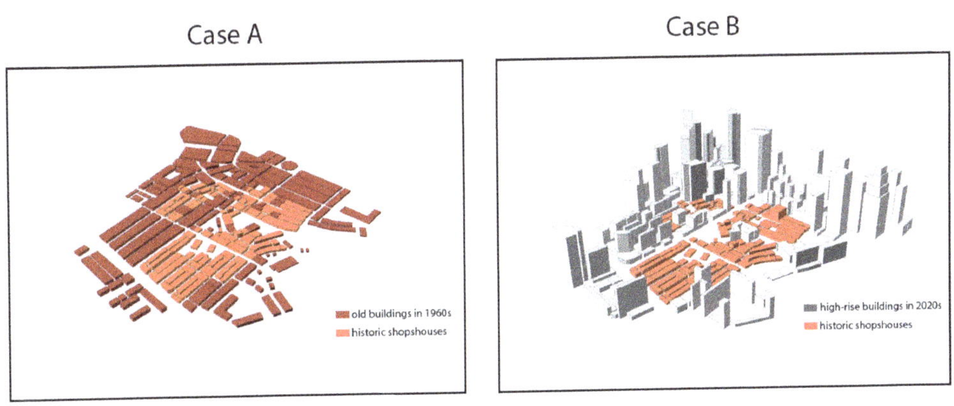

(a)

Figure 2. *Cont.*

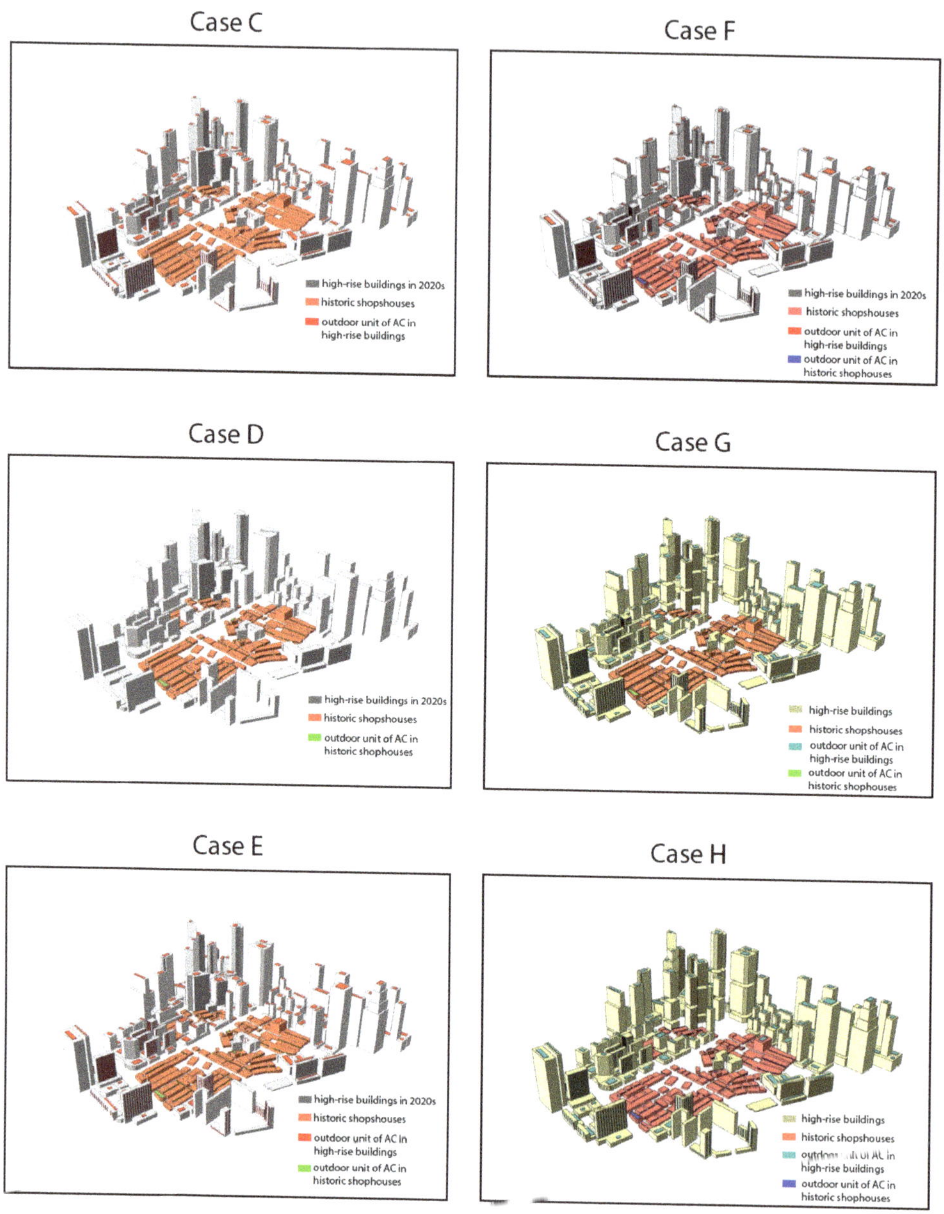

(b) (c)

Figure 2. (**a**) Simulation scenarios of Case A and B to represent the urban geometry in the past (1960s) and the current (2020s). (**b**) (**Left**) Simulation scenarios of Case C–E to represent the impact of anthropogenic heat emitted from high-rise buildings, historic shophouses, and both. (**c**) (**Right**) Simulation scenarios of Case F–H to represent the developed mitigation strategies focusing on the historic shophouses, high-rise buildings, and both.

2.2. CFD Simulation Setup

2.2.1. Computational Domain and Meshing

As shown in Figure 3, a simplified district model which represents the study area, was created and placed in the center of the computational domain. As suggested by Tominaga et al. [29], considering that maximum building height in the study area was approximately 33 m (Case A) and 280 m (Cases B–H), the domain heights were set as five times these heights, i.e., 165 m and 1400 m, respectively. The side dimensions were set at 15 times these heights, i.e., 495 m and 4200 m, respectively, which are the minimum dimensions for the inlet. Simulations were conducted with NE (northeast, i.e., the prevailing wind direction of Singapore [30], as the incoming wind direction. Figure 4 shows the applied mesh type is unstructured poly-hexcore mesh. We assigned the maximum size with 100 m for the less important area in the domain and 2 m for the surrounding high-rise buildings. In the targeted area with shophouses, we applied the maximum size as 1 m. In addition, we added four boundary layers with 1 m height for building walls. Taking Case B as an example, the average value of wall y^+ of targeted buildings is around 250, which is well within the logarithm region.

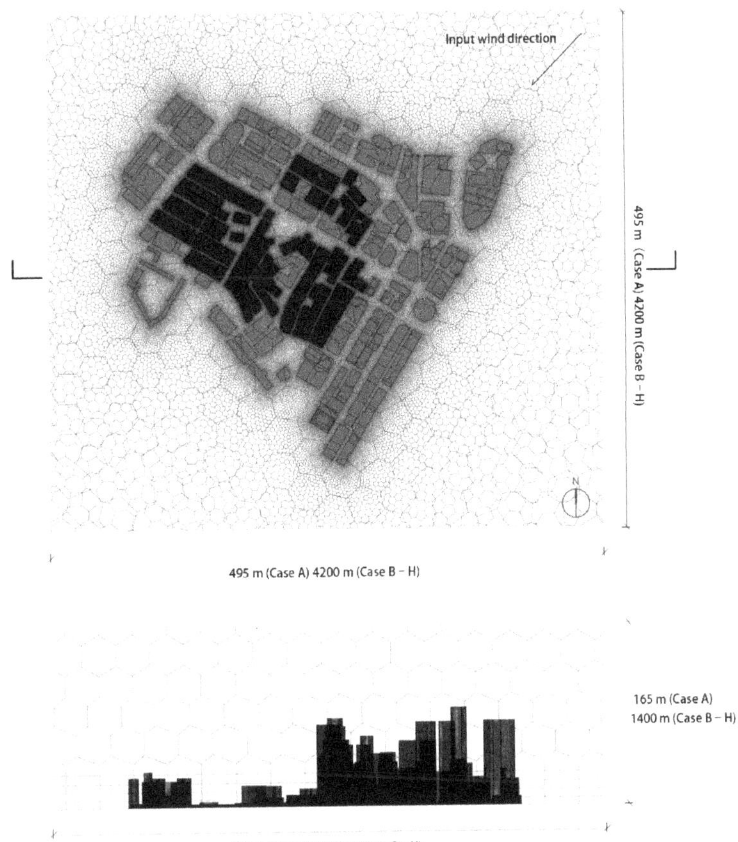

Figure 3. Horizontal and vertical cross-sections of the computational domain of Case B as an example to present domain size, meshing, and input wind direction.

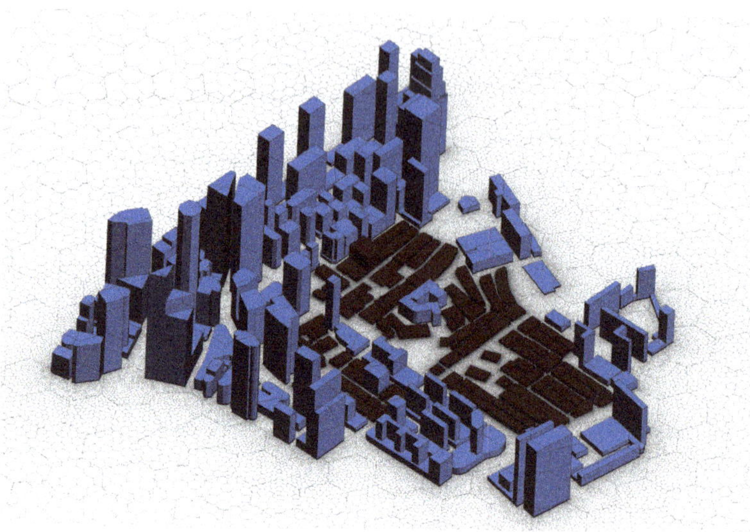

Figure 4. Meshing for Case B as an example (Buildings highlighted as red are the historic shophouses conserved at the Chinatown area).

2.2.2. Boundary Conditions

Table 2 summarized the CFD simulation settings. The bottom of the computational domain and building surfaces were set as a non-slip plane. Top side was set as symmetric plane. Regarding incoming wind speed, the annual average incoming wind speed, i.e., 7.6 m/s at 300 m above the ground, was applied in this study as a common scenario. Additionally, we used a power law equation for the incoming wind speed profile. Horizontal inflow velocity at the inlet of the computational domain reproduces an Atmospheric Boundary Layer (ABL) velocity profile, where the turbulence originates only from the friction and shear [31].

$$U_{(z)} = \frac{u^*_{ABL}}{\kappa} \ln\left(\frac{z + z_0}{z_0}\right). \quad (1)$$

$$k_{(z)} = \frac{{u^*_{ABL}}^2}{\sqrt{C_\mu}}. \quad (2)$$

$$\varepsilon_{(z)} = \frac{{u^*_{ABL}}^3}{\kappa(z + z_0)}. \quad (3)$$

$$\omega_{(z)} = \varepsilon_{(z)} \frac{C_\mu}{k(z)}. \quad (4)$$

In Equations (1)–(4), u^*_{ABL} is the atmospheric boundary layer friction velocity, z is the height above the ground, $z_0 = 0.5$, which is the aerodynamic roughness length, $\kappa = 0.42$ and $C_\mu = 0.09$, which are constants of the turbulence model, U is the horizontal inflow velocity, k is the turbulence kinetic energy and ε and ω are the turbulence dissipation rate and specific dissipation rate respectively.

This study applied the SST k-ω model as the turbulence model, which has been validated by our team in a previous study [22]. In the validation, we compared the CFD simulation results and wind tunnel data from a study worked by Allegrini et al. [24]. All the simulation settings, i.e., domain configurations and boundary conditions, are the same as the ones at the wind tunnel experiment. The cross-comparison between simulation result using SST k-ω model and wind tunnel experiment shows there is a good agreement between SST k-ω and wind tunnel data, as Figure 5 shown. The validation results suggest the SST k-ω model can reproduce the mixed convection, caused by input air flow and buoyancy effect. Therefore, SST k-ω model was used in the simulation in this study.

Table 2. CFD simulation settings.

Turbulence Model	SST k-ω
Computational grid type	Unstructured poly-hexcore meshes
Blockage ratio	<5%
Grid expansion ratio	1.2
Density	Boussinesq
Solving algorithms	SIMPLE
Input wind profile	Power law equation
Inflow boundary condition	Operating temperature: 27 °C
Incoming wind speed	Power-law profile with the reference wind speed of 7.6 m/s at 300 m above the ground
Incoming wind directions	northeast (the prevailing wind direction of Singapore)
Heat flow specification method	Mass flow rate inlet: normal to boundary direction (Heat emission temperature: 40 °C)
Other boundary conditions	Outflow: pressure outlet Bottom and buildings: Wall Top: Symmetry
Convergence criteria	1E-6 for all variables

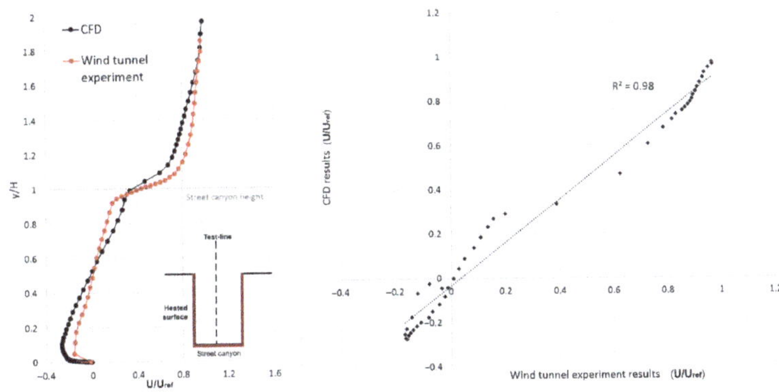

Figure 5. Validation results [22].

The locations and geometries of AC condensers were set according to the real situation of the buildings in the study area. The buildings include two types of AC condensers, i.e., air-cooling and water-cooling. The air-cooling type was utilized in shophouses in Chinatown, residential buildings, and some office buildings in the surrounding area. The water-cooling type was the dominant type in retails, hotels, and office buildings in the surrounding area. Examples of locations and geometries of heat sources which represented the AC condensers in simulations are shown in Figure 6. AC condensers in the study area were set either on exterior walls or on the rooftop for both shophouses and surrounding high-rise buildings. Therefore, we modelled the AC condensers as vertical or horizontal stripes on exterior walls or blocks on the rooftop according to the real locations of the AC condensers in individual buildings.

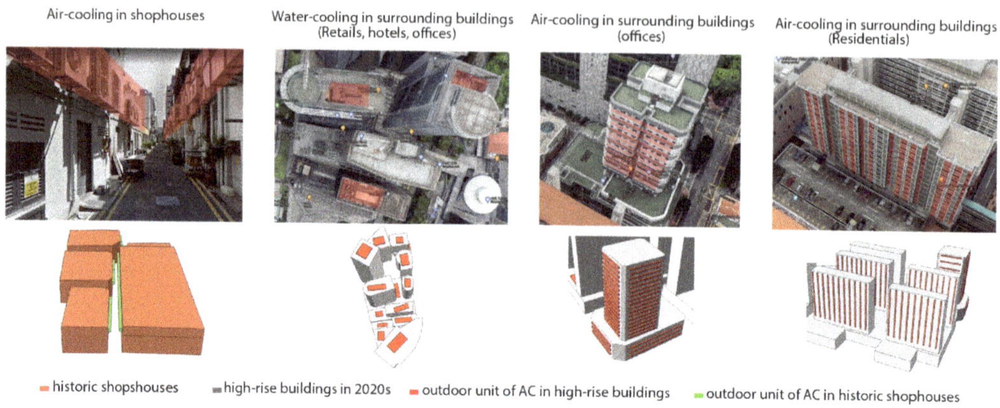

Figure 6. Examples of locations and geometries of heat sources regarding different AC types and building functions (AC—air conditioning).

Additionally, we need to apply the heat emission from AC condensers in the simulations to represent the impacts of anthropogenic heat from AC [10,32]. We applied the following equation to calculate the mass flow rate (m_a) of heat emission from AC condensers [33]:

$$Q = m_a c_a \Delta T. \tag{5}$$

Therefore, the mass flow rate (m_a) is:

$$m_a = \frac{Q}{c_a \Delta T} \tag{6}$$

where Q is the anthropogenic heat from the AC condenser of each building, c_a is the specific heat of air, and 1005 J/kg K. ΔT is the increase in air temperature from the background air temperature near the AC condensers. Accordingly, to determine the input amount of mass flow rate (m_a), we also needed to determine the increase in air temperature from background air temperature near the condensers of AC, ΔT. Adelia et al. [22] and Yuan et al. [11] set m_a as 0.245 kg/s, according to the study from Bojic et al. [34], and c_a was set to 1.005 kj/kg.K, resulting in a ΔT of 13 °C. This value is in agreement with the 10–13 °C range of temperature increase near the condenser unit in Singapore measured by Bruelisauer et al. [35]. This study applied a ΔT of 13 °C for calculating the m_a in our simulations.

In addition to the values of c_a and ΔT, we need to estimate anthropogenic heat from the AC operation of each building (Q). The anthropogenic heat from the AC condenser (Q) consists of the cooling loads (Q_c) of buildings and the energy needed (E_c) for AC operation [36]. Therefore, Q can be calculated as follows:

$$\begin{gathered} Q = Q_c + E_c, \\ \text{where } E_c = Q_c/COP,\ Q_c = E_c.COP \end{gathered} \tag{7}$$

$$\text{Thus, } Q = Q_c\left(\frac{COP+1}{COP}\right) \tag{8}$$

In this study, we calculated the anthropogenic heat emission for the different types individually, including retail buildings with air-cooling (shophouses); office and residential buildings with air-cooling; and retail, hotel, and office buildings with water-cooling. One of the differences between the air-cooling and water-cooling types is the ratio of sensible heat to the dispersed heat. The air-cooling type is considered as 100% sensible heat whereas the water-cooling type which applies to the cooling towers is normally approximately 75% of latent heat due to the evaporation effect and 25% of sensible heat [37,38]. Therefore, for the water-cooling type, the anthropogenic heat from one condenser unit is calculated as follows [36]:

$$Q = 0.25Q_c\left(\frac{COP + 1}{COP}\right) \tag{9}$$

For commercial buildings, according to Version 3 of the Code for Environmental Sustainability of Buildings, issued in 2012, from the Building and Construction Authority (BCA) of Singapore, [39] the coefficient of performance (COP) of 4.2–4.29 (4.2 for multi-function areas, such as the lobby on the first floor and 4.29 for single function, such as offices above) is recommended for non-residential buildings when at full installed capacity. In this study, COP of 4.2 was applied. In addition, the design peak cooling load (Q_c) was in the range of 100–180 W/m^2 for office buildings, 120–260 W/m^2 for hotels, and 250–350 W/m^2 for retail buildings, according to the cooling load data obtained from the energy audit results during the operating hours, complied by the Building Construction Authority (BCA) and National Environment Agency (NEA). The dataset on the measured cooling load showed that the mean cooling load per air-conditioned floor area of office buildings (from 58 projects), hotels (from 32 projects), and retail buildings (from 28 projects), ranged from 54–100 W/m^2, 40–98 W/m^2, and 93–195 W/m^2, respectively with mean values of 74 W/m^2, 61 W/m^2, and 130 W/m^2, respectively [40]. Therefore, taking office building as an example, the average anthropogenic heat emission can be calculated as follows:

$$\overline{Q} = \overline{Q_c}\left(\frac{4.2 + 1}{4.2}\right) \tag{10}$$

In addition, Quah and Roth [10] measured and calculated the total anthropogenic heat flux of $\overline{Q}$ in commercial buildings in Singapore and showed that it ranged from 40–120 W/m^2, with the largest mean hourly flux of 113 W/m^2.

By understanding the average anthropogenic heat flux of $\overline{Q}$, we can calculate the sensible heat Q as:

$$Q = \overline{Q}A, \tag{11}$$

where A is the gross floor area of the involved buildings.

In terms of residential buildings, the COP of the AC operation was set to 3.34 according to the energy label and tick rating of ACs operating in the apartments in Singapore [41]. Additionally, Q_c is 2.5 kW per apartment in residential buildings, assuming one-split-AC operating in full-load capacity in each apartment unit. Q_c values were found from Energy Efficient (E_2) Singapore [42] and were calculated based on the total air-conditioned area [22]. Therefore, we can produce the result that Q_A which represents anthropogenic heat from of each unit was 3.25 kW. Therefore, the total anthropogenic heat (Q) was calculated using equation as:

$$Q = Q_A N \tag{12}$$

where Q_A is 3.25 KW and N is the number of units in the residential buildings, and was calculated as:

$$N = \frac{(1 - R_p)A_{GF}}{A_h}, \tag{13}$$

where A_h is assumed as 80 m^2 which is the typical area of one apartment unit in Singapore; and R_p is the public area ratio which was assumed to be 20%. We calculated the gross floor area (A_{GF}) as:

$$A_{GF} = A_F N_F, \tag{14}$$

where A_F is the floor area of one building and N_F is the number of floors of the building.

Table A1 at Appendix A shows a summary of the equations to calculate m_a for different types of AC and building functions. By estimating the anthropogenic heat from the AC condenser (Q) with the above method, which has been applied in previous studies [11,12,22], we are able to calculate the m_a for simulation input.

2.3. Data Analysis

The simulation's outputs, including air temperature increment and wind speed, were exported for further comparison. Specifically, we selected 18 test lines where major visitors' activities occurred: 13 test lines at the streets with both roadways and footways and 5 test lines at the roads with only footways (Figure 7). The test lines were selected from the middle of the streets, and average value within the analysis points from the selected test lines for each road were reported as a single value to represent their temperature increment and wind speed. The single temperature increment and wind speed values in each case scenarios were compared for further evaluation.

Figure 7. Locations of test lines for data analysis (RTL represents the test line on streets with both roadway and footway and FTL represents the test line on streets with only footway.).

3. Results and Analysis

In this section, we presented the simulation results of proposed eight cases. Figure 8 shows the wind speed contours of Case A and Case B. Figure 9a presents the air temperature increment contours of Case C, D, and E. Figure 9b presents the air temperature increment contours of Case F, G, and H. The quantitative data, including wind speed and air temperature increment in each test line and the average value, are presented at Tables A2 and A3 at Appendix B.

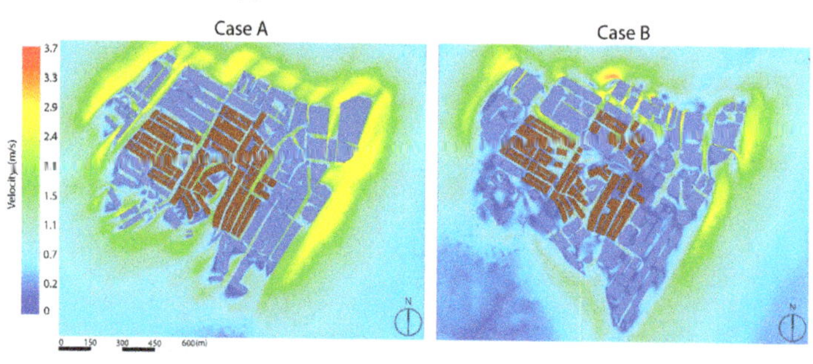

Figure 8. Wind speed contours of Case A (**left**) and Case B (**right**) at the study area (buildings highlighted as red are the historic shophouses conserved at the Chinatown area).

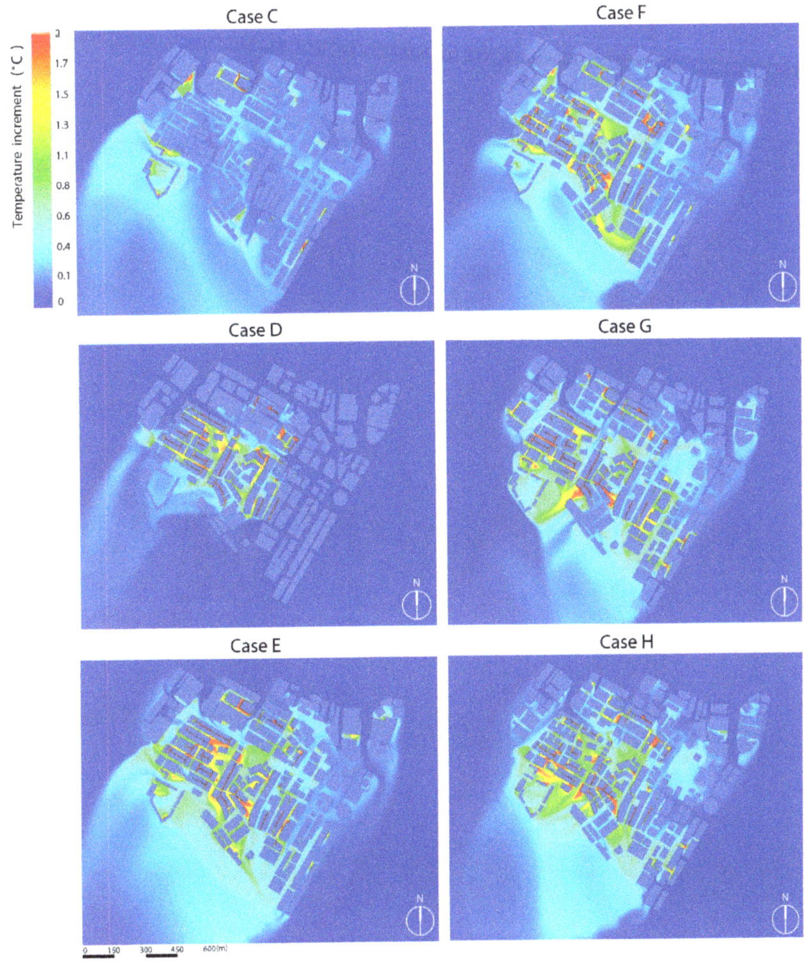

(**a**) (**b**)

Figure 9. (**a**) (**Left**). Air Temperature increment contours of Case C–E. (**b**) (**Right**) Air Temperature increment contours of Case F–H.

3.1. Comparison of Wind Speed between Cases A and B

To evaluate the impacts of newly built high-rise buildings on the wind speed in the Chinatown area, the CFD simulation was taken to obtain the wind speed at Case A (representing the 1960s scenario) and Case B (representing the current scenario). The wind environment results show that considerably more areas of Chinatown experienced extremely low wind speed (nearly 0 m/s) at the pedestrian level in Case B compared with Case A (Figure 8). Figure 10 shows the wind speed at the test streets showed differences from 84% to −80%, with an average of −43%, from 1.11 m/s decreased to 0.63 m/s for the tested values of wind speed, compared with Case B and Case A. Additionally, it can be noted that in Case B, only three tested streets (RTL1, RTL13 and FTL4) had higher wind speed and the other 15 streets had lower wind speed than the tested values of wind speed in Case A. This result clearly showed that the surrounding high-rise buildings caused reductions in wind speed in the Chinatown area. The only three streets with increased wind speed were observed at the locations closer to the wind inlet. The lower wind speed was likely to have been due to the high-rise buildings on the leeward side (wind shadow), especially in the areas farther from the wind corridor and the streets which are perpendicular to the prevailing wind direction. For example, at the test line on RTL7, which was the furthest test line from the oncoming wind, the wind speed decreased by 43% from 0.79 m/s to 0.45 m/s. The wind speed at the streets which are perpendicular to the prevailing wind direction, such as RTL9, RTL10, RTL11, and RTL12, also decreased significantly, i.e., by 47–80%.

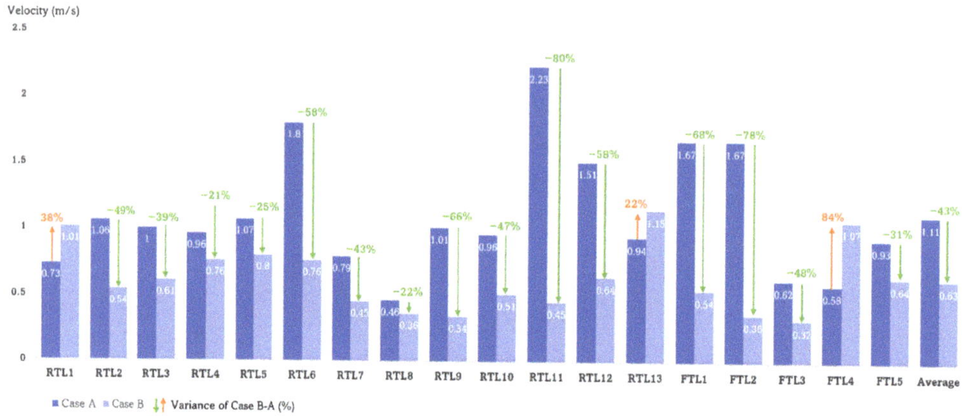

Figure 10. Wind speed of Case A & B and the variance of Case (B−A) % of each test line and the average value. Variance of Case (B−A) was calculated with the equation Case $(B-A) = \frac{v_{CaseB} - v_{CaseA}}{v_{CaseA}}\%$ (v_{CaseA} is the wind velocity of each test line in Case A and v_{CaseB} is the wind velocity of each test line in Case B).

3.2. Comparison of Air Temperature Increments among Cases C, D, and E

Cases C and D assumes that the shophouses in the Chinatown area and the high-rise buildings in the surrounding area did not use AC, respectively. Case E represents the real scenario of the study area, with AC in use. In this way, we are able to compare the effects of anthropogenic heat in different areas on thermal environment in analyzing the Chinatown area. Figure 11a,b show the temperature increment in each test line of the three cases, comparing Case C with Case E and Case D with Case E. In the real scenario, Case E, the temperature increment at the pedestrian level at all tested streets increases by 0.87 °C on average, ranging from 0.13 °C at RTL13 to 1.45 °C at FTL1. The temperature increments of Cases C and D were compared with that of Case E to evaluate the effects of anthropogenic heat. Specifically, compared with Case E, the mean temperature increments of both of Cases C and D decreased, by averages of 82% and 41%, respectively. The result indicates that anthropogenic heat due to AC in both shophouses in Chinatown and high-rise buildings in the surrounding area significantly influenced the outdoor thermal environment in the Chinatown area. In addition, the result shows that the anthropogenic heat due to AC in shophouses influenced the outdoor air temperature more than those in high-rise buildings, most likely due to the types (i.e., water-cooling) and locations (i.e., rooftop) of the AC condensers in the skyscrapers.

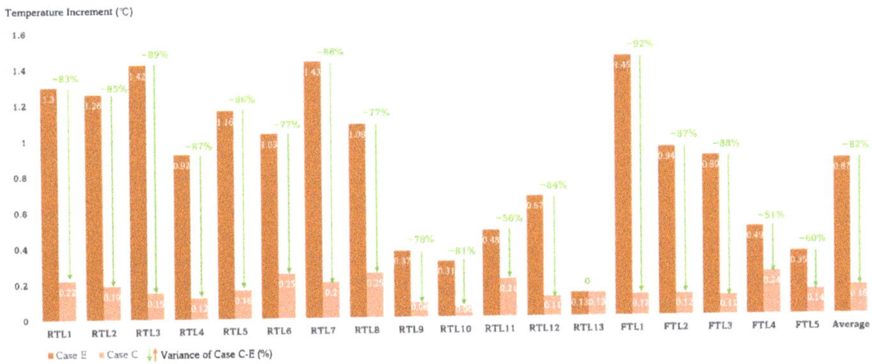

(a)

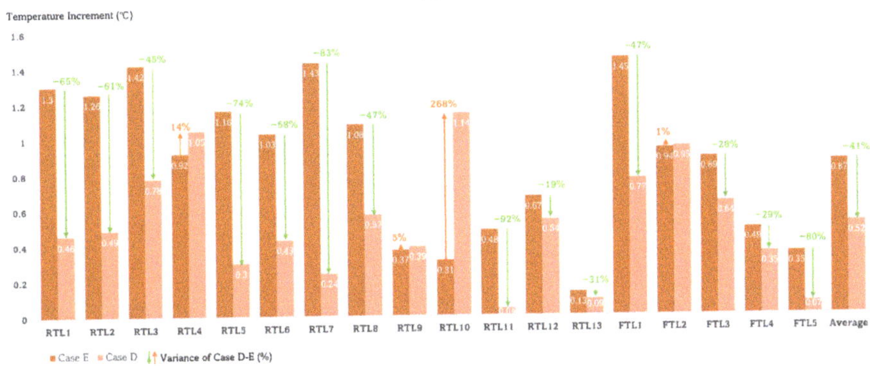

(b)

Figure 11. (**a**). Temperature increment of Case C & E and the variance of Case (C–E) % of each test line and the average value, Variance of Case (C–E) was calculated with the equation Case $(C - E) = \frac{T_{CaseC} - T_{CaseE}}{T_{CaseE}}\%$ (T_{CaseC} is the temperature increment of each test line in Case C and T_{CaseE} is the temperature increment of each test line in Case E). (**b**). Temperature increment of Case D & E and the variance of Case (D–E) % of each test line and the average value, Variance of Case (D–E) was calculated with the equation Case $(D - E) = \frac{T_{CaseD} - T_{CaseE}}{T_{CaseE}}\%$ (T_{CaseD} is the temperature increment of each test line in Case D and T_{CaseE} is the temperature increment of each test line in Case E).

3.3. Urban Heat Mitigation Strategies Evaluation among Case F, G, and H

According to the previous evaluation and analysis, the detailed mitigation strategies are described as Figure 12a,b. The main principle is to create porosity both vertically and horizontally to enhance ventilation at the pedestrian level in the Chinatown area. For Case F (Figure 12a), we focused on the Chinatown area. The principle is to remove some shophouses to allow the ventilation into the historic area, and we also consider the locations and amount of the removed shophouses carefully to retain the texture of the existing area simultaneously. In most cases, we removed one shophouse at a time instead of removing several adjacent shophouses, especially for the rows of shophouses closer to the wind inlet and parallel to the wind direction. Generally, the width of one shophouse is approximately 6 m [14]. One or more of 6 m porosity have little influence on the original texture of the historical area. Additionally, a group of (i.e., two or more) shophouses at the end of a row of shophouses where heat emission accumulated are removed to create open spaces for heat diffusion. Although two or more shophouses may be removed, it would have little effect on the texture of Chinatown because these buildings are at the end of a row of shophouses. In some rare cases, we remove a few adjacent shophouses in the middle of a row of shophouses which closer to the wind outlet. In terms of Case G (Figure 12b), both vertical and pedestrian-level porosity was applied accordingly for the high-rise buildings in the surrounding area. Specifically, in place where airflow is noticeably blocked and thus not able to entry Chinatown, buildings were removed and substituted by one or more super high-rise buildings. This is to create more open spaces at the wind gateway while maintain the gross floor areas for the original sites. Moreover, some buildings were redesigned to generate porosity to allow ventilation. Vertical porosity at different heights is applied in the building towers, which enables the incoming wind to be diverted into different levels and promotes vertical flow within the street canyon. Meanwhile, pedestrian-level porosity was applied in podium structures. Study shows that wind environment at the pedestrian level is significantly influenced by building morphology at podium layer [43]. Large podium volume tends to reduce the air volume and block the air flow, which is not desirable for pedestrians [44]. Thus, we diminished the large podium structures in surrounding buildings by random breaks and encourage wind penetration by providing ventilation corridors. Case H was simulated as the scenario including mitigation strategies in both areas.

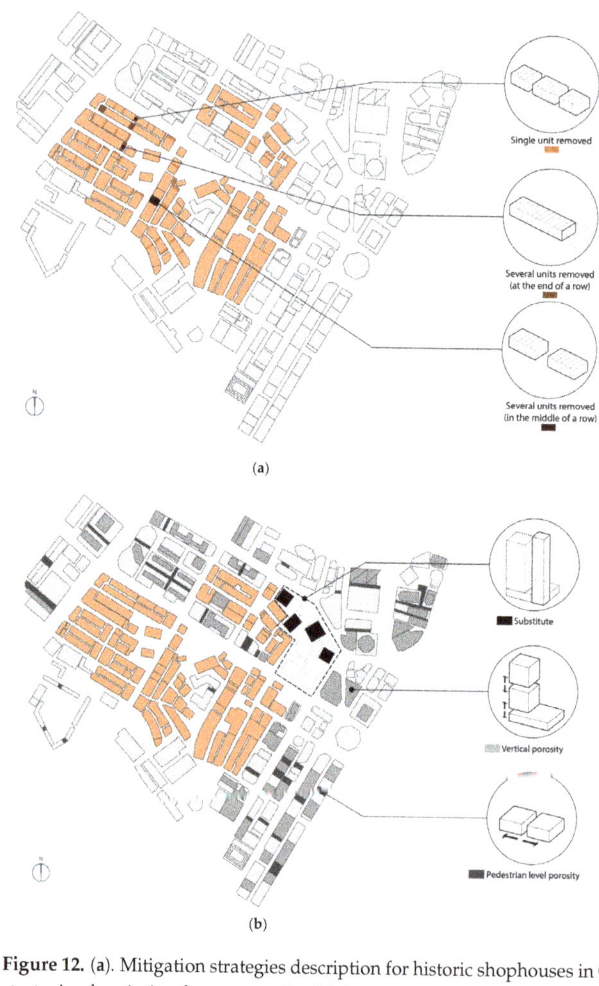

Figure 12. (**a**). Mitigation strategies description for historic shophouses in Chinatown. (**b**). Mitigation strategies description for surrounding high-rise buildings.

Comparing the three mitigated cases with the real scenario, Case E, the results showed that all the three cases with mitigation strategies had decreased temperature increments, with average decreases of −33% for Case F, −25% for Case G, and −21% for Case H, respectively (Figure 13a–c).

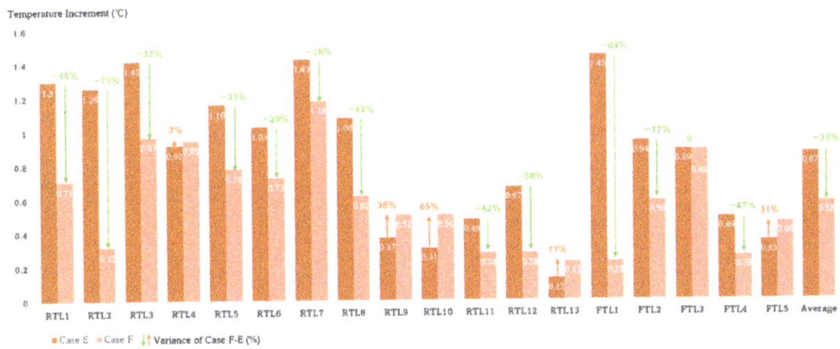

(**a**)

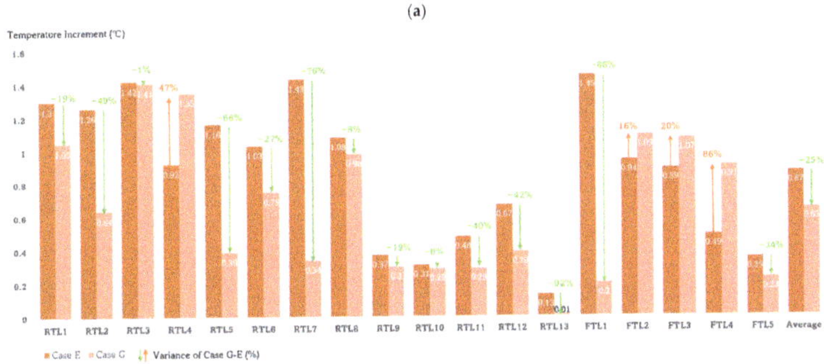

(**b**)

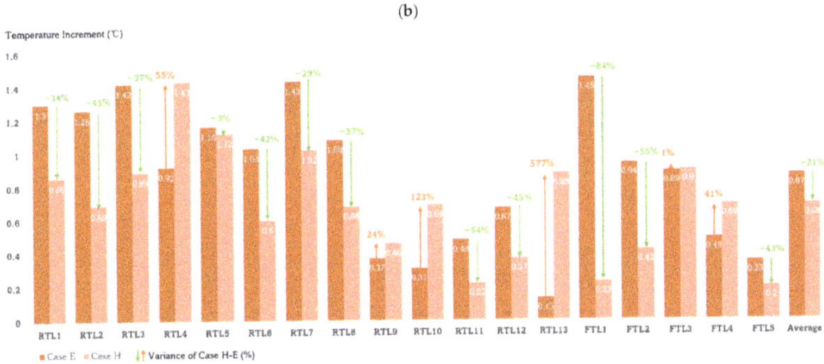

(**c**)

Figure 13. (**a**). Temperature increment of Case F & E and the variance of Case (F–E) % of each test line and the average value, Variance of Case (F–E) was calculated with the equation Case $(F - E) = \frac{T_{CaseF} - T_{CaseE}}{T_{CaseE}}\%$ (T_{CaseF} is the temperature increment of each test line in Case D and T_{CaseE} is the temperature increment of each test line in Case E). (**b**). Temperature increment of Case G & E and the variance of Case (G–E) % of each test line and the average value, Variance of Case (G–E) was calculated with the equation Case $(G - E) = \frac{T_{CaseG} - T_{CaseE}}{T_{CaseG}}\%$ (T_{CaseG} is the temperature increment of each test line in Case D and T_{CaseE} is the temperature increment of each test line in Case E). (**c**). Temperature increment of Case H & E and the variance of Case (H–E) % of each test line and the average value, Variance of Case (H–E) was calculated with the equation Case $(H - E) = \frac{T_{CaseH} - T_{CaseE}}{T_{CaseE}}\%$ (T_{CaseH} is the temperature increment of each test line in Case D and T_{CaseE} is the temperature increment of each test line in Case E).

This indicated that these strategies were effective to reduce outdoor air temperature in the Chinatown area. Although the mean value of all the three cases showed decreased temperature increments, some test streets with increased values can offer more information to select locations of porosity more effectively. More specifically, in Case F (Figure 13a), if the test streets are far from the wind inlet, such as RTL4 which is closed to the surrounding area, the porosity method applied to shophouses in Chinatown is unsatisfactory because the temperature increment of RTL4 increased by 3% compared with that of RTL4 in Case E. Additionally, the other test lines (i.e., RTL9, RTL10, RTL13, FTL3, and FTL5) with increased temperature increments were parallel to the prevailing wind direction. This means that the strategy to remove some shophouses to increase pedestrian-level porosity should consider the locations of existing shophouses, either near the wind inlet or perpendicular to the prevailing wind direction. For Case G (Figure 13b), although the mean value of the decreasing temperature increment in Chinatown was lower than the value of Case F, most test streets (14 of 18) still showed decreases ranging from 1% (RTL3) to 92% (RTL13). Similar to Case F, RTL4, which was far from the wind inlet, showed an unsatisfactory result because the temperature increment increased by 47% compared with the value of Case E. All the other test streets with increased values were those with only footway (FTL2, FRL3, and FTL4), which were narrower than the streets with both roadways and footways. These results indicated that the mitigation strategies for the surrounding area allowed more wind to enter into Chinatown for lowering the temperature increment of the entire area. However, some streets may be too narrow (i.e., streets with only footway) to adequately allow ventilation, resulting in unsatisfactory results locally. Case H (Figure 13c), which integrated both mitigation strategies in shophouses in Chinatown and high-rise buildings in the surrounding area, however, was less effective at decreasing the temperature increment in Chinatown than Cases F and G. The decreased temperature increment of Case H, 21–25% was slightly lower than that of Case G, but much lower than that of Case F (21–33% on average). There may be two reasons for this result. Firstly, it could be that anthropogenic heat from high-rise buildings in the surrounding area was dispersed into Chinatown via the ventilation corridor. Evidence of this was shown from RTL4, in Case F, which showed only a 3% temperature increment compared with Case E; however, for Cases G and H, when the ventilation corridor of the surrounding area was created, this value increased by 47% in Case G and 55% in Case H. Secondly, the wind environment was influenced by both the wind corridor in the surrounding area and the porosity in Chinatown; however, they may have opposing or negligible effects. As Table A2 in Appendix B shows, the wind speed in streets, RTL 3 and RTL 5 increased in both Cases F and G whereas it decreased in Case H, compared with the values in Case E.

4. Discussion

4.1. Effect of Anthropogenic Heat and Urban Morphology on the Urban Heat Environment in the Singapore Chinatown Area

In this study, the unsatisfactory urban heat environment at the pedestrian-level in Chinatown was due to the combined effect of heat emission from AC in both Chinatown and its surrounding areas, and that low wind speed was most likely due to blockages from surrounding high-rise buildings and the building geometry of Chinatown. Firstly, the high-rise buildings in the surrounding area can block the strong prevailing wind [13,45]. The wind environment simulation results of Cases A and B support this finding. However, this does not mean that high-rise buildings should be limited during the urban development process. In contrast, high-rise buildings are among the appropriate solutions for densely populated regions, such as Singapore [46]. Therefore, carefully designed high-rise buildings should be handled by related professionals in order to mitigate the negative effects for localized and existing wind environments. Secondly, the geometry of the Singapore Chinatown (i.e., the shophouses as a row) could be a barrier because it would obstruct the anthropogenic heat from dissipating and cause it to aggregate. Although the historic regions have restrictions for development and renovation, actions may be taken to simultaneously conserve existing historic buildings and enhance the outdoor thermal environment. The mitigation strategies on urban geometry in the Singapore Chinatown developed in this study considered both aspects. Third, urban heat mitigation strategies should consider the effect of anthropogenic heat, including the locations, types, and potential heat emissions. For example, in this study, the water-cooling type of AC on rooftops of high-rise buildings limited the negative effects on the pedestrian-level thermal environment in the Chinatown area. In contrast, although the shophouses in Chinatown dispersed much less anthropogenic heat, compared with high-rise buildings in surrounding areas, the effect was larger, which was probably due to the ineffective AC type and unplanned locations of AC condensers.

4.2. Practical Implications for Proposed Mitigation Strategies in Historic Areas

Based on our analysis, it is more effective to modify urban morphology at Chinatown area to mitigate the urban heat as the existing building geometry was not designed for its current functions and macro- and micro- climate conditions. However, these shophouses are historic buildings which need to be handled carefully with regard to any demolition or reconstruction actions. Although urban heat mitigation strategies in the surrounding area appeared to not be as effective as those in the Chinatown area, the results still showed great potential for improvements in the outdoor thermal environment. For architects, urban designers, and related professionals, it can be easier to consider mitigation strategies when starting new buildings. Thus, considering existing historic areas in advance is of great importance for these professionals. Additionally, a combination of mitigation strategies should be applied based on different situations. In this study, we suggest pedestrian-level porosity by removing some existing shophouses near the prevailing wind inlet and those perpendicular to the wind direction. Thus, if a redevelopment plan is undertaken in the historic area, these kinds of buildings should be considered as a priority for the benefit of outdoor wind and heat environment of the entire area. In terms of newly-built high-rises in surrounding areas, the principle was to create a ventilation corridor in the surrounding area to allow wind to penetrate Chinatown more easily, further influencing outdoor thermal performance. We suggest both vertical and horizontal porosity to create the ventilation corridor. In addition, the orientation of the wind corridor should be matched to the prevailing wind direction. We also noted from this study that separate mitigation strategies for historic area and surrounding areas may have opposing or negligible effects. This highlights the need for professionals to carefully design the locations of porosity to avoid anthropogenic heat in the surrounding area dispersing to the historic areas.

5. Conclusions

This study evaluated the impact of surrounding high density urban areas on the microclimate at historical areas, and developed the corresponding urban heat mitigation strategies. We investigated the impact of high-rise buildings on pedestrian-level air flow and assessed air temperature increment caused by anthropogenic heat. Multi-physics CFD simulation was applied to model the microclimate with anthropogenic heat and buoyancy effects. A total of eight parametric cases were designed in three steps. The following are specific research outputs to support urban and architectural design in historic area in Singapore and other high density tropical cities.

- The average wind speed decreases of 43%, from 1.11 m/s in Case A to 0.63 m/s in Case B. This indicates the great impact of new development, i.e., high-rise buildings, on pedestrian-level air flow of the historic area.
- The mean air temperature increased by 0.16 °C for Case C, 0.52 °C for Case D and 0.87 °C for Case E, respectively. This indicates that the anthropogenic heat emission from surrounding high-rise buildings had less effect than that from historic shophouses in Chinatown.
- The integration of open spaces and building porosity, which create wind corridors together, can promote outdoor natural ventilation and heat dispersion at the study

area. Compared with Case E, the three mitigation cases improve outdoor thermal environment, with mean temperature reduction of 33%, 25%, and 21%, respectively.
- To retain the urban texture of the area, the locations and number of removed shophouses should include either one shophouse closer to the wind inlet and perpendicular to the wind direction as a priority, or a number of shophouses at the end or in the middle of the row where heat emission accumulates, in some cases.

6. Limitations and Future works

This study focuses on a common scenario with annual average wind speed. In future studies, we will consider various scenarios, e.g., the worse scenario, utilizing the lowest wind speed for simulation. Different wind directions will be utilized for further analysis in order to investigate other scenarios in addition to the prevailing wind direction. In addition, the effect of heat emission from vehicular traffic on the thermal environment will also be discussed. Moreover, urban heat plume impact will be included to make the simulation results more realistic. In terms of developing urban heat mitigation strategies, the contribution of urban greenery needs to be studied. Last but not least, quantitative analysis is needed to evaluate the performance of urban heat mitigation strategies, so as to give more practical suggestions to improve urban design in historic areas.

Author Contributions: Conceptualization, W.Z. and C.Y.; methodology, W.Z. and C.Y; software, W.Z., L.Z. and S.-J.M.; formal analysis, W.Z.; writing—original draft preparation, W.Z.; writing—review and editing, C.Y. and S.-J.M.; visualization, W.Z. and L.Z; supervision, C.Y.; project administration, C.Y.; funding acquisition, C.Y. All authors have read and agreed to the published version of the manuscript.

Funding: This research was funded by Singapore Ministry of Education, NUS Presidential Young Professorship, grant number R-295-000-172-133.

Data Availability Statement: Not applicable.

Acknowledgments: The authors would like to thank National University of Singapore for Wei ZHU's Ph.D. research scholarship.

Conflicts of Interest: The authors declare no conflict of interest.

Nomenclature

Abbreviations	
ABL	Atmospheric boundary layer
AC	Air-conditioning
BCA	Building and construction authority
CBD	Central Business District
CFD	Computational fluid dynamics
COP	Coefficient of performance
NE	Northeast
NEA	National environment agency
UHI	Urban heat island
Symbols	
A	Gross floor area of buildings (m^2)
A_h	Typical area of one unit of residential buildings (m^2)
A_{GF}	Gross floor area (m^2)
c_a	Specific heat of air ($J \cdot kg^{-1} \cdot K^{-1}$)
C_μ	Model constant
E_c	Cooling energy (kw)
E_2	Energy Efficient Singapore
m_a	Input mass flow rate of heat emissions ($kg \cdot s^{-2}$)
N	Number of units in residential buildings
N_F	Floor number of residential buildings

Q	Total anthropogenic heat (kw)
Q_A	Anthropogenic heat from each unit of residential buildings (kw)
Q_c	Cooling loads generated in the indoor spaces (kw)
R_p	Public area ratio
u^*_{ABL}	Atmospheric boundary layer friction velocity (m/s)
z	Hight above ground (m)
z_0	Aerodynamic roughness length (m)
ω	Specific dissipation (s^{-1})
κ	Turbulent kinetic energy ($m^2 \cdot s^{-2}$)
ε	TKE dissipation rate ($m^2 \cdot s^{-3}$)

Appendix A

Table A1. Equations to calculate m_a for buildings with different AC types and functions.

AC Type	Building Function	Equation
Air cooling	Shophouses (retail)	$m_a = \frac{130 \times \left(\frac{4.2+1}{4.2}\right) A}{C_a \Delta T}$
	Residential buildings	$m_a = \frac{1000 \times 2.5 \times \left(\frac{3.34+1}{3.34}\right) \times \frac{(1-R_p) A_{GF}}{A_h}}{C_a \Delta T}$
Water cooling	Retail	$m_a = \frac{0.25 \times 130 \times \left(\frac{4.2+1}{4.2}\right) A}{C_a \Delta T}$
	Office	$m_a = \frac{0.25 \times 74 \times \left(\frac{4.2+1}{4.2}\right) A}{C_a \Delta T}$
	Hotel	$m_a = \frac{0.25 \times 61 \times \left(\frac{4.2+1}{4.2}\right) A}{C_a \Delta T}$

Appendix B

Table A2. Wind speed of all test streets and average value.

	RTL1	RTL2	RTL3	RTL4	RTL5	RTL6	RTL7	RTL8	RTL9	RTL10	RTL11	RTL12	RTL13	FTL1	FTL2	FTL3	FTL4	FTL5	Ave
Case A	0.73	1.06	1	0.96	1.07	1.8	0.79	0.46	1.01	0.96	2.23	1.51	0.94	1.67	1.67	0.62	0.58	0.93	1.11
Case B	1.01	0.54	0.61	0.76	0.8	0.76	0.45	0.36	0.34	0.51	0.45	0.64	1.15	0.54	0.36	0.32	1.07	0.64	0.63
Case B-A	38%	−49%	−39%	−21%	−25%	−58%	−43%	−22%	−66%	−47%	−80%	−58%	22%	−68%	−78%	−48%	84%	−31%	−43%
Case C	1.39	0.92	1.7	2.26	1.75	0.69	0.47	0.52	0.84	1.92	1.16	1.16	2.34	1.21	1.13	1.94	2.23	0.9	1.36
Case D	1.39	1.47	0.99	1.43	1	0.91	2.82	0.57	0.88	1.1	1.08	0.69	1.57	0.43	1.36	0.81	0.96	1.51	1.17
Case E	1.62	1.62	1.52	2.38	1.44	1.19	0.39	0.33	1.52	1.86	1.27	1.31	3.04	1.36	1.17	0.62	2.09	2.25	1.47
Case C-E	−14%	−21%	12%	−5%	22%	42%	21%	58%	45%	3%	−9%	−12%	−23%	−11%	−3%	213%	7%	−60%	−8%
Case D-E	−14%	27%	35%	40%	−31%	−24%	623%	73%	−42%	−41%	−15%	−47%	−48%	−68%	16%	31%	−54%	−33%	−21%
Case F	1.9	1.8	2.08	2.49	1.64	1.07	0.58	0.42	1.13	1.81	2.35	2.23	2.21	2.28	1.39	0.47	2.44	0.76	1.61
Case G	1.61	1.38	1.77	1.89	2.25	2.06	0.68	0.82	1.24	1.69	1.96	1.5	3.84	1.86	0.77	0.89	1.06	1.31	1.59
Case H	1.22	1.62	1.16	1.62	0.79	1.98	0.77	1.18	1.23	1.46	2.32	1.88	1.74	2.39	0.97	0.79	1.26	1.29	1.43
Case F-E	17%	55%	37%	5%	14%	−10%	49%	27%	−26%	−3%	85%	70%	−27%	68%	19%	−24%	17%	−66%	10%
Case G-E	−1%	19%	16%	−21%	56%	73%	74%	148%	−18%	−9%	54%	15%	26%	37%	−34%	44%	−49%	−42%	8%
Case H-E	−25%	40%	−24%	−32%	−45%	66%	97%	258%	−19%	−22%	83%	44%	43%	76%	−17%	27%	−40%	−43%	−3%

Table A3. Temperature increment at all test streets and average value.

	RTL1	RTL2	RTL3	RTL4	RTL5	RTL6	RTL7	RTL8	RTL9	RTL10	RTL11	RTL12	RTL13	FTL1	FTL2	FTL3	FTL4	FTL5	Ave
Case C	0.22	0.19	0.15	0.12	0.16	0.25	0.2	0.25	0.08	0.06	0.21	0.11	0.13	0.12	0.12	0.11	0.24	0.14	0.16
Case D	0.46	0.49	0.78	1.05	0.3	0.43	0.24	0.57	0.39	1.14	0.04	0.54	0.09	0.77	0.95	0.64	0.35	0.07	0.52
Case E	1.3	1.26	1.42	0.92	1.16	1.03	1.43	1.08	0.37	0.31	0.48	0.67	0.13	1.45	0.94	0.89	0.49	0.35	0.87
Case C-E	−83%	−85%	−89%	−87%	−86%	−77%	−86%	−77%	−78%	−81%	−56%	−84%	0	−92%	−87%	−88%	−51%	−60%	−82%
Case D-E	−65%	−61%	−45%	14%	−74%	−58%	−83%	−47%	5%	268%	−92%	−19%	−31%	−47%	1%	−28%	−29%	−80%	−41%
Case F	0.71	0.32	0.97	0.95	0.78	0.73	1.18	0.62	0.51	0.51	0.28	0.28	0.23	0.23	0.59	0.89	0.26	0.46	0.58
Case G	1.05	0.64	1.41	1.35	0.39	0.75	0.34	0.98	0.3	0.29	0.29	0.39	0.01	0.2	1.09	1.07	0.91	0.23	0.65
Case H	0.86	0.69	0.89	1.43	1.12	0.6	1.02	0.68	0.46	0.69	0.22	0.37	0.88	0.23	0.42	0.9	0.69	0.2	0.69
Case F-E	−45%	−75%	−32%	3%	−33%	−29%	−18%	−43%	38%	65%	−42%	−58%	77%	−84%	−37%	0	−47%	31%	−33%
Case G-E	−19%	−49%	−1%	47%	−66%	−27%	−76%	−9%	−19%	−6%	−40%	−42%	−92%	−86%	16%	20%	86%	−34%	−25%
Case H-E	−34%	−45%	−37%	55%	−3%	−42%	−29%	−37%	24%	123%	−54%	−45%	577%	−84%	−55%	1%	41%	−43%	−21%

References

1. Sotoudeh, H.; Abdullah, W.M.Z.W. Evaluation of Fitness of Design in Urban Historical Context: From the Perspectives of Residents. *Front. Archit. Res.* **2013**, *2*, 85–93. [CrossRef]
2. Egusquiza, A.; Brostrom, T.; Izkara, J.L. Incremental Decision Making for Historic Urban Areas' Energy Retrofitting: EFFESUS DSS. *J. Cult. Herit.* **2022**, *54*, 68–78. [CrossRef]
3. Aigwi, I.E.; Ingham, J.; Phipps, R.; Filippova, O. Identifying Parameters for a Performance-Based Framework: Towards Prioritising Underutilised Historical Buildings for Adaptive Reuse in New Zealand. *Cities* **2020**, *102*, 102756. [CrossRef]
4. Gremmelspacher, J.M.; Campamà Pizarro, R.; van Jaarsveld, M.; Davidsson, H.; Johansson, D. Historical Building Renovation and PV Optimisation towards NetZEB in Sweden. *Sol. Energy* **2021**, *223*, 248–260. [CrossRef]

5. Bedate, A.; Herrero, L.C.; Sanz, J.Á. Economic Valuation of the Cultural Heritage: Application to Four Case Studies in Spain. *J. Cult. Herit.* **2004**, *5*, 101–111. [CrossRef]
6. Pedersen, A. *Managing Tourism at World Heritage Sites: A Practical Manual for World Heritage Site Managers*; UNESCO: Paris, France, 2002.
7. Webb, A.L. Energy Retrofits in Historic and Traditional Buildings: A Review of Problems and Methods. *Renew. Sustain. Energy Rev.* **2017**, *77*, 748–759. [CrossRef]
8. Enteria, N.; Cuartero-Enteria, O.; Santamouris, M.; Eicker, U. Morphology of Buildings and Cities in Hot and Humid Regions. In *Urban Heat Island (UHI) Mitigation: Hot and Humid Regions*; Enteria, N., Santamouris, M., Eicker, U., Eds.; Springer: Singapore, 2021; pp. 1–13, ISBN 978-981-334-050-3.
9. Giridharan, R.; Lau, S.S.Y.; Ganesan, S.; Givoni, B. Urban Design Factors Influencing Heat Island Intensity in High-Rise High-Density Environments of Hong Kong. *Build. Environ.* **2007**, *42*, 3669–3684. [CrossRef]
10. Quah, A.K.L.; Roth, M. Diurnal and Weekly Variation of Anthropogenic Heat Emissions in a Tropical City, Singapore. *Atmos. Environ.* **2012**, *46*, 92–103. [CrossRef]
11. Yuan, C.; Adelia, A.S.; Mei, S.; He, W.; Li, X.-X.; Norford, L. Mitigating Intensity of Urban Heat Island by Better Understanding on Urban Morphology and Anthropogenic Heat Dispersion. *Build. Environ.* **2020**, *176*, 106876. [CrossRef]
12. Yuan, C.; Zhu, R.; Tong, S.; Mei, S.; Zhu, W. Impact of Anthropogenic Heat from Air-Conditioning on Air Temperature of Naturally Ventilated Apartments at High-Density Tropical Cities. *Energy Build.* **2022**, *268*, 112171. [CrossRef]
13. Lan, H.; Lau, K.K.-L.; Shi, Y.; Ren, C. Improved Urban Heat Island Mitigation Using Bioclimatic Redevelopment along an Urban Waterfront at Victoria Dockside, Hong Kong. *Sustain. Cities Soc.* **2021**, *74*, 103172. [CrossRef]
14. Lee, S.L. The Conservation of Singapore's Chinatown. *J. Archit. Conserv.* **1997**, *3*, 43–57. [CrossRef]
15. Bruelisauer, M.; Berthold, S. *Reclaiming Backlanes: Design Vision for Increasing Building Performance and Reprogramming Common Spaces*; World Scientific: Singapore, 2015; ISBN 981-4740-87-X.
16. Kardinal Jusuf, S.; Wong, N.H.; Hagen, E.; Anggoro, R.; Hong, Y. The Influence of Land Use on the Urban Heat Island in Singapore. *Habitat Int.* **2007**, *31*, 232–242. [CrossRef]
17. Priyadarsini, R.; Hien, W.N.; Wai David, C.K. Microclimatic Modeling of the Urban Thermal Environment of Singapore to Mitigate Urban Heat Island. *Sol. Energy* **2008**, *82*, 727–745. [CrossRef]
18. Roth, M.; Chow, W.T.L. A Historical Review and Assessment of Urban Heat Island Research in Singapore. *Singap. J. Trop. Geogr.* **2012**, *33*, 381–397. [CrossRef]
19. Aflaki, A.; Mirnezhad, M.; Ghaffarianhoseini, A.; Ghaffarianhoseini, A.; Omrany, H.; Wang, Z.-H.; Akbari, H. Urban Heat Island Mitigation Strategies: A State-of-the-Art Review on Kuala Lumpur, Singapore and Hong Kong. *Cities* **2017**, *62*, 131–145. [CrossRef]
20. Liddle, B.; Loi, T.S.A.; Owen, A.D.; Tao, J. Evaluating Consumption and Cost Savings from New Air-Conditioner Purchases: The Case of Singapore. *Energy Policy* **2020**, *145*, 111722. [CrossRef]
21. Santos, L.G.R.; Singh, V.K.; Mughal, M.O.; Riegelbauer, E.; Fonseca, J.A.; Norford, L.K.; Nevat, I. *Building Anthropogenic Heat Flux in Singapore*; ETH Zurich: Zurich, Switzerland, 2020.
22. Adelia, A.S.; Yuan, C.; Liu, L.; Shan, R.Q. Effects of Urban Morphology on Anthropogenic Heat Dispersion in Tropical High-Density Residential Areas. *Energy Build.* **2019**, *186*, 368–383. [CrossRef]
23. Oke, T.R.; Mills, G.; Christen, A.; Voogt, J.A. *Urban Climates*; Cambridge University Press: Cambridge, UK, 2017; ISBN 978-1-139-01647-6.
24. Allegrini, J.; Dorer, V.; Carmeliet, J. Buoyant Flows in Street Canyons: Validation of CFD Simulations with Wind Tunnel Measurements. *Build. Environ.* **2014**, *72*, 63–74. [CrossRef]
25. Andreou, E. The Effect of Urban Layout, Street Geometry and Orientation on Shading Conditions in Urban Canyons in the Mediterranean. *Renew. Energy* **2014**, *63*, 587–596. [CrossRef]
26. Erell, E.; Williamson, T. Intra-urban Differences in Canopy Layer Air Temperature at a Mid-latitude City. *Int. J. Climatol. A J. R. Meteorol. Soc.* **2007**, *27*, 1243–1255. [CrossRef]
27. Kondo, A.; Ueno, M.; Kaga, A.; Yamaguchi, K. The Influence of Urban Canopy Configuration on Urban Albedo. *Bound.-Layer Meteorol.* **2001**, *100*, 225–242. [CrossRef]
28. Yuan, C.; Ng, E.; Norford, L.K. Improving Air Quality in High-Density Cities by Understanding the Relationship between Air Pollutant Dispersion and Urban Morphologies. *Build. Environ.* **2014**, *71*, 245–258. [CrossRef] [PubMed]
29. Tominaga, Y.; Mochida, A.; Yoshie, R.; Kataoka, H.; Nozu, T.; Yoshikawa, M.; Shirasawa, T. AIJ Guidelines for Practical Applications of CFD to Pedestrian Wind Environment around Buildings. *J. Wind. Eng. Ind. Aerodyn.* **2008**, *96*, 1749–1761. [CrossRef]
30. Meteorological Service Singapore. *Climate of Singapore*; Meteorological Service Singapore: Singapore, 2020.
31. Richards, P.J.; Norris, S.E. Appropriate Boundary Conditions for Computational Wind Engineering Models Revisited. *J. Wind. Eng. Ind. Aerodyn.* **2011**, *99*, 257–266. [CrossRef]
32. Parker, D.S.; Sherwin, J.R.; Hibbs, B. Development of High Efficiency Air Conditioner Condenser Fans. *Trans.-Am. Soc. Heat. Refrig. Air Cond. Eng.* **2005**, *111*, 511.
33. Hsieh, C.-M.; Aramaki, T.; Hanaki, K. Managing Heat Rejected from Air Conditioning Systems to Save Energy and Improve the Microclimates of Residential Buildings. *Comput. Environ. Urban Syst.* **2011**, *35*, 358–367. [CrossRef]

34. Bojic, M.; Yik, F.; Lee, M. Influence of Air-Conditioning Exhaust on Exterior Recessed Space. *Build. Res. Inf.* **2003**, *31*, 24–33. [CrossRef]
35. Bruelisauer, M.; Meggers, F.; Saber, E.; Li, C.; Leibundgut, H. Stuck in a Stack—Temperature Measurements of the Microclimate around Split Type Condensing Units in a High Rise Building in Singapore. *Energy Build.* **2014**, *71*, 28–37. [CrossRef]
36. Kikegawa, Y.; Genchi, Y.; Kondo, H.; Hanaki, K. Impacts of City-Block-Scale Countermeasures against Urban Heat-Island Phenomena upon a Building's Energy-Consumption for Air-Conditioning. *Appl. Energy* **2006**, *83*, 649–668. [CrossRef]
37. Stanford, H.W., III. *HVAC Water Chillers and Cooling Towers: Fundamentals, Application, and Operation*; CRC Press: Boca Raton, FL, USA, 2011; ISBN 1-4398-6202-8.
38. Jones, W.P. The Rejection of Heat from Condensers and Cooling Towers. In *Air Conditioning Engineering*; Routledge: London, UK, 2007; pp. 325–339.
39. Building and Construction Authority. *Code for Environmental Sustainability of Buildings*; BCA: Singapore, 2012.
40. Building and Construction Authority; National Environment Agency. *Benchmark Your Building's Cooling Load*; BCA: Singapore, 2018.
41. National Environment Agency. *Energy Label and Tick Rating Prior to 1 September 2014*; NEA: Singapore, 2014.
42. National Environment Agency; Energy Market Authority. *Energy Efficient Singapore*; NEA: Singapore, 2021.
43. Yuan, C.; Ng, E. Building Porosity for Better Urban Ventilation in High-Density Cities–A Computational Parametric Study. *Build. Environ.* **2012**, *50*, 176–189. [CrossRef] [PubMed]
44. Ng, E. *Designing High-Density Cities: For Social and Environmental Sustainability*; Earthscan: London, UK, 2009; ISBN 978-1-84977-444-4.
45. Juan, Y.-H.; Yang, A.-S.; Wen, C.-Y.; Lee, Y.-T.; Wang, P.-C. Optimization Procedures for Enhancement of City Breathability Using Arcade Design in a Realistic High-Rise Urban Area. *Build. Environ.* **2017**, *121*, 247–261. [CrossRef]
46. Yeh, A.G.; Yuen, B. Introduction: High-rise living in Asian cities. In *High-Rise Living in Asian Cities*; Springer: Dordrecht, The Netherlands, 2011; pp. 1–8, ISBN 978-90-481-9737-8.

buildings

MDPI

Article

Multi-Objective Optimisation of Urban Form: A Framework for Selecting the Optimal Solution

Milad Showkatbakhsh [1] and Mohammed Makki [2,*]

[1] Architectural Association, London WC1B 3ES, UK
[2] School of Architecture, University of Technology Sydney, Sydney, NSW 2007, Australia
* Correspondence: mohammed.makki@uts.edu.au

Abstract: The complexity associated with the design of urban tissues is driven by the multitude of design goals that influence urban development and growth. This complexity is amplified by the design goals being inherently conflicting, necessitating preference-based decisions within the design process—an approach that results in predetermined design solutions driven by personal biases. The utility of population-based optimisation algorithms addresses this by allowing for the examination of multiple conflicting objectives within the same design problem, negating the need for trade-off decisions between the design goals. The application of these algorithms is associated with three primary steps. The first is the formulation of the design problem, the second is the application of the algorithm, and the third is selecting the most optimal solution from the algorithm's output. This paper examines the third step in this process, in which various methods are employed to facilitate data-driven selection mechanisms that are both objective as well as subjective in their formulation. The selection mechanisms are demonstrated on a speculative urban tissue that examines the potential of inhabiting interstitial spaces, through various morphological interventions, within the urban fabric. The results present a scalable and adaptable framework that assists designers employing multi-objective evolutionary algorithms (MOEAs) to select the optimal solution from their generated populations, a challenge commonly associated with the application of MOEAs in design.

Keywords: MOEA; generative design; selection; evolutionary computation; urban design; optimisation

1. Introduction

The application of multi-objective evolutionary algorithms (MOEAs) in design, specifically urban design, has gained momentum over the last decade. The adoption of MOEAs to solve complex design problems has enabled designers to address one of the primary challenges for the design of urban spaces: the ability to simultaneously address multiple design objectives that are in conflict with one another without necessitating trade-off decisions between the design goals. Although the method itself is firmly rooted in computer science and continues to grow as a field [1–6], its translation to design and its widespread application is associated with the emergence of various software that do not necessitate coding skills to implement and run the algorithm [7–10]. In doing so, access to MOEAs by designers (both for developers as well as users) has been more streamlined than any other point in history, allowing for innovative approaches to responding to, and designing for, design problems comprising multiple conflicting objectives. In urban design, the value of MOEAs is amplified, as it requires a performance-based metric approach to the design problem, allowing designers to integrate climatic, environmental, and urban metrics as key drivers for the algorithm [11–13]. This data-driven approach to solving the complexity associated with the urban fabric informs all stages of the algorithmic process; however, key to this is the coupling of the user-driven analysis of the data and the way in which it informs the results outputted by the optimisation run.

Through employing MOEAs in design, three key stages to a successful optimisation run can be identified, each equally significant. The first stage focuses on how the design problem is formulated; this highlights the relationship between three key metrics: the parameters informing the morphology (genes), the morphological characteristics being generated (phenotype), and the fitness functions being optimised for. The way in which these metrics inform each other plays a vital role in a successful optimisation run. The second stage pertains to the algorithm chosen to optimise the design problem; there exists a large number of MOEAs that have been developed over the past few decades, each claiming greater efficiency and success to complex problems [14]. As such, the chosen algorithm plays a key role in achieving optimised results. The third and final stage is the methods and metrics used to select a solution (or solutions) from the algorithm's output. Conflicts among the fitness functions driving the MOEA prevent a single optimal solution; rather, what is usually generated, is a number of optimised trade-off solutions [15]. These trade-off solutions form part of the pareto front (the most optimal solutions generated by the MOEA) and, depending on the size of the generation in the algorithm, the pareto front is composed of a varied number of solutions [16]—the larger the number, the more challenging it is for the user to select the 'optimal solution'. The process of selection plays a key role in how the designer rationalises the algorithm's output—one that is directly correlated with the complexity of the design problem. Design problems with greater diversity in output (driven by greater conflict between fitness functions) demand a rigorous approach to analysis for the purpose of selection—one that is frequently overlooked in the design process.

The research presented in this paper focuses on this third stage of the optimisation process. Through the presentation of a speculative design experiment, evolving an imaginative urban superblock modelled after a high-density urban tissue, this paper presents an adaptable framework for selecting a single solution from the output of an MOEA. The framework employs objective and subjective methods of analysis, both of which are informed by the data generated from the algorithm as well as additional metrics informed by the user. The primary focus of the presented research is the selection mechanisms and processes associated with the application of MOEAs in design, in which the key contribution is a scalable model for designers and users of MOEAs to adapt in their work.

2. Materials and Methods

2.1. Selection Strategy

The advantage of employing an MOEA is primarily due to the algorithm's ability to solve multiple conflicting objectives without the requirement for the designer to make trade-off decisions; however, in doing so, the algorithm presents a primary challenge to the designer. Optimising for multiple conflicting objectives naturally leads to multiple optimal solutions, where there is no single 'best' solution for the designer (in this case referred to as the decision maker (DM)) to select [1]. There are three points in the optimisation process from where the DM can define a preference for the 'best' solution [17]. The first is before running the algorithm; i.e., during the formulation of the design problem (referred to as priori), in which the DM can skew the algorithm to optimise for one fitness function over the other, primarily through pre-defining weights to the fitness functions. Applying preferences priori transforms the multi-objective problem into a single objective one [18]. The second point of intervention for preference is during the algorithm itself (referred to as progressive), in which the DM intervenes at different points in the algorithm to direct the optimisation process towards a preferred solution. The third point of intervention is after the algorithmic run (referred to as posteriori), in which the DM chooses a solution from the outputted solution set (this solution is usually selected from within the pareto front). Most MOEAs are classified as posteriori, providing the DM with a set of non-dominated solutions to choose from. The following sections outline a model for designers to choose the 'best' solution posteriori.

Posteriori selection strategies are primarily driven by quantitative methods of analysis, mainly due to the volume of data generated by the algorithm and the associated population size needed for the algorithm to successfully converge towards optima solutions. Conversely, qualitative selection strategies (if not translated numerically) necessitate a solution set that is small enough for the visual analysis and identification of qualitative conditions not easily represented through quantitative representations. The presented framework combines objective and subjective trade-offs [19] in a two-stage quantitative and qualitative selection process: the first, a quantitative approach that utilises the fitness functions as the primary metric for selection, and the second, a quantitative and qualitative approach that utilises user-driven 'phenotype indicators' for selection. Stage 1 aims to ensure that the entirety of the pareto front is adequately represented while minimising the need to analyse each solution on the front. To do so, various methods are employed to extract solutions from the pareto front, including clustering (for example through k-means or hierarchical clustering methods), fittest solution for each function, solution closest to the 'Utopia' point, and the solution with the most equal weight between all functions. From the filtered list of pareto front solutions, Stage 2 requires the DM to analyse the solutions based on additional metrics they find most relevant to the problem being investigated, ranking solutions according to their performance in these user-driven metrics as well as implementing a qualitative selection process through a visual analysis of the filtered solution set (Figure 1).

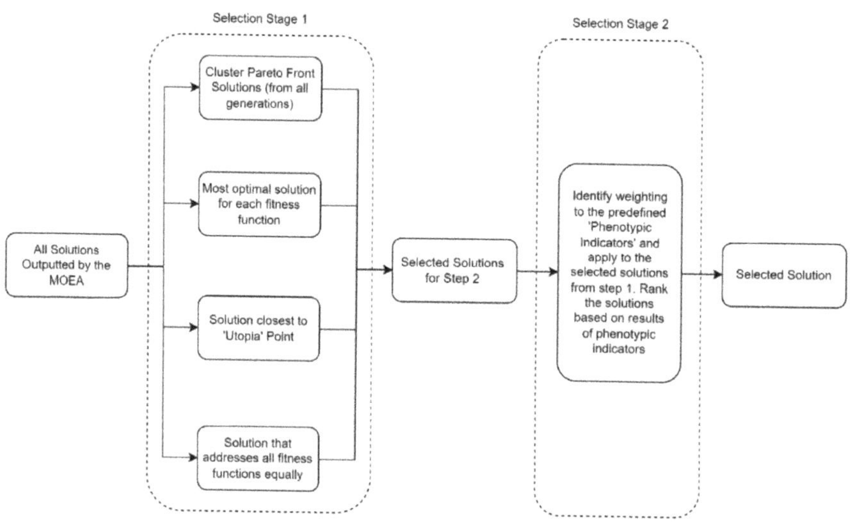

Figure 1. Pseudo code of both stages of the selection framework.

2.2. Selection Stage 1

Stage 1 of the posteriori selection process aims to filter the solutions outputted by the algorithm to a manageable size. As designers, there is a vested interest in the phenotypic characteristics of the solutions evolved by the MOEA, thus highlighting a significant advantage to the visual and comparative analysis of the solutions. However, due to the complexity of the problem being investigated and the size of the solution set, conducting this analysis on the entirety of the pareto front may not be feasible. Therefore, the selected subset of solutions from the pareto front must retain the variation and diversity of solutions in the pareto front. To do so, the methods employed in Stage 1 use objective metrics, primarily driven by the fitness values of each solution to facilitate selection. Solutions are selected based on the following four steps.

2.2.1. Clustering the Pareto Front

The first step selects a subset of solutions from the pareto front through employing a clustering algorithm (the type of clustering algorithm used should respond to the manner by which solutions are distributed across the pareto front [20]). The K-value (i.e., the number of clusters) is to be determined by the size of the pareto front, ensuring that each cluster represents a small number of solutions to maintain minimal variation between solutions within a cluster. Doing so ensures that the centre of each cluster (which in this case will be the solution selected) is an accurate representative of all solutions within the cluster. A general guide would be to limit the size of each cluster to between 5 and 15 solutions (Figure 2).

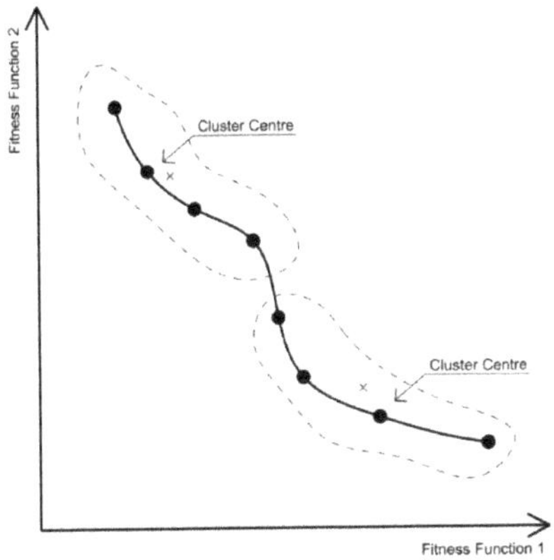

Figure 2. Example of clustering solutions along the pareto front. The 'representative' solution is the closest one to the cluster centre (represented in the figure by 'x').

2.2.2. Fittest Solutions (Outliers)

In most cases, clustering the pareto front and choosing the cluster centre for further analysis will minimise the chance for the outlier fittest solutions to be selected. Therefore, the second step selects the fittest solutions for each fitness function for inclusion in the selected pool. As these outlier solutions are less likely to be selected by the DM (since they perform well for one fitness function but poorly for all the others), their inclusion in the selection pool is critical to provide the DM with an indication of the 'extremes' generated by the algorithm and the morphological characteristics associated with these extremes (Figure 3).

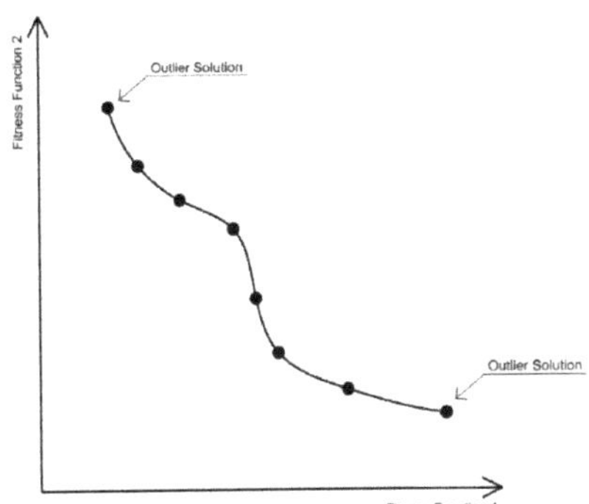

Figure 3. Example of the fittest solutions (or 'outliers') for each fitness function.

2.2.3. Global Compromise

Yu [21] and Zeleny [22] proposed that when the ideal (or 'Utopia') solution is not accessible, the next best solution would be the closest one to the ideal solution. In this case, the ideal solution is considered the non-existent optimal solution in the objective space, and thus the selected solution is the one nearest to this point based on a Euclidean measure between the two. This 'global compromise' approach is usually considered by decision makers as the solution that best represents all fitness functions and is the closest to the optimal solution; however, this does not automatically translate to it being the ideal solution for the problem being investigated (Figure 4).

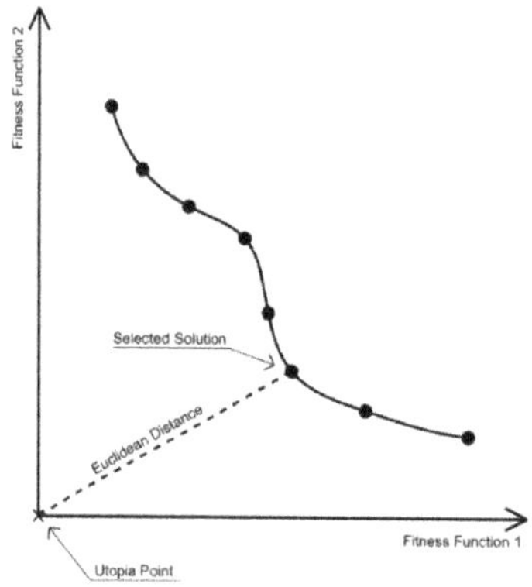

Figure 4. Example of the 'global compromise' solution. This solution is the closest (Euclidean) solution to the 'Utopia' solution.

2.2.4. Equal Fitness for All Functions

The final step is to select the solution that holds equal weight for all fitness functions. The 'Utopia' solution in the previous step does not correlate to all fitness functions being equally weighted. Where Step 2 selects the outlier solutions, this final step selects the opposite by ensuring each fitness function is optimised equally. It should be noted that this solution may not be located on the pareto front, and may be considered to be a 'poor-performing' solution; however, it is the solution that presents the least (if any) trade-offs between fitness functions (Figure 5).

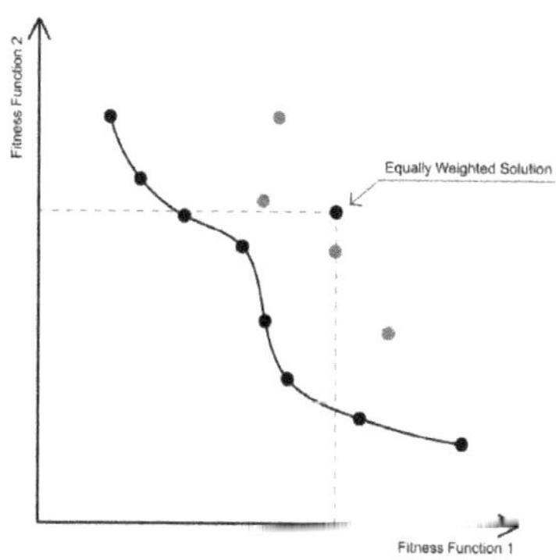

Figure 5. Example of the solution with the most equal weights among the fitness ranks; i.e., the solution with the least trade-offs between fitness functions. This solution may not be located on the pareto front.

2.3. Selection Stage 1

Depending on the number of fitness functions of the design problem, along with the number of clusters used in Step 1 above, Stage 1 of the selection process drastically reduces the solution set to one that is more manageable for evaluation (primarily visual) by the DM.

Where Stage 1 employed objective metrics for selection, Stage 2 employs a subjective (yet still metric) approach to selection.

The primary challenge with complex urban problems is the number of fitness functions that can be optimised in the MOEA. The greater the number of fitness functions, the more complex the fitness landscape, and thus the ability for the algorithm to optimise the solution set towards a meaningful pareto front [16]. In most cases, this requires the DM to limit the number of fitness functions used in the algorithm despite there being a comparatively high number of fitness measures that are relevant to the problem. In this context, Stage 2 of the selection process engages with the relevant metrics that were not included in the algorithmic run, utilising them to further analyse the solution pool from Stage 1 and rank the solutions based on how they perform against these added metrics.

Moreover, as these added metrics are subjective to the DM, their impact on analysis and ranking of the solution pool is equally so. Considering the number of added metrics being analysed for, the DM applies preference through weighting, giving some more priority than others. Despite being a subjective approach, this process remains data driven, requiring the designer to rationalise the analytic and selection metrics through a quantifiable (and reproducible) process (and thus avoid choosing a solution based on visual preference alone). Finally, this approach removes the requirement to limit the number of metrics being used for analysing the filtered solution set, giving greater freedom to the DM as well as providing them with an adaptive model that can be modified to the problem at hand. In this context, Stage 2 of the selection framework comprises two key steps: the first is to identify the metrics for analysis, and the second is to apply weighting to these metrics (Figure 6).

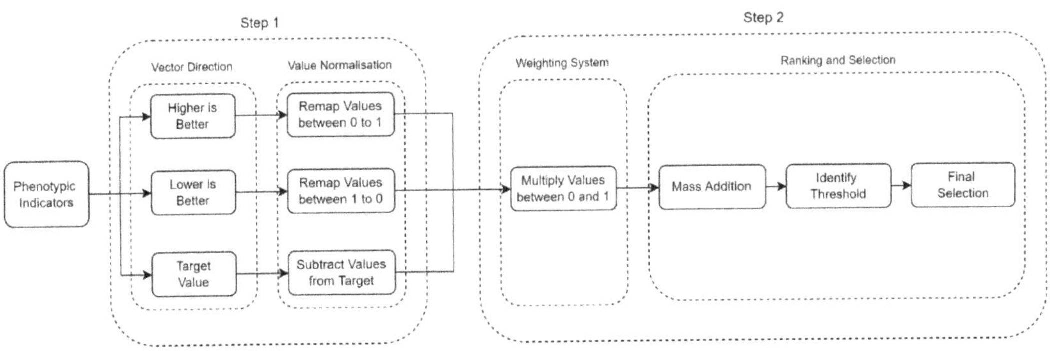

Figure 6. Pseudo code for Stage 2 of the selection framework.

The following sections demonstrate the application of Stages 1 and 2 of the described selection framework on a speculative high-density urban block. The presented experiment utilises the abstracted morphological and behavioural principles of homeostasis [23] to inform the evolutionary design processes to evolve an urban superblock with a degree of formal and behavioural adaptation to environmental changes.

3. Case Study

The selection framework presented above is demonstrated through the evolutionary optimization of a speculative urban tissue that examines morphological interventions in the interstitial spaces (the spaces between buildings) of an urban block. The morphological interventions explored in the experiment are driven by Herbert Simon's argument that the urban fabric's internal components and external conditions must conform to one another for the city to fulfil its purpose. As such, the case study proposes urban scenarios that connect urban morphologies to dynamic changes and is presented as an extension to radical urbanism through an exploration and habitation of interstitial spaces. The proposed morphological interventions within the urban fabric's interstitial spaces aim to serve as an explorative study that offer an alternative perspective to urban habitation. The use of an MOEA, and the explorative nature of the evolutionary process, plays a critical role in this explorative study—one that is driven by environmental, micro-climatic, and societal considerations.

3.1. *The Design Problem*

Rapidly changing environmental and climatic conditions, coupled with the growing numbers of urbanised populations, have stressed the ability of existing cities to cope with these sudden and highly impactful changes. Crossing the critical threshold of stability in a city [24] is when its population grows beyond its maximum capacity, thus straining its resources and ecological demand, transforming the city into one that is highly sensitive to changes in its environment. Although this is a scenario that has repeated itself multiple times across different geographic locations and periods, its occurrence in modern-day cities carries dire impacts as the current rate of change to environmental conditions is an unprecedented one.

The adaptation of cities that have approached their critical threshold is highly contingent on the rate of change in the environment; historically, the rate of environmental and climatic changes allowed for cities to evolve in response to these changes. However, the rate of environmental fluctuations observed in the 20th century, as well as predicted throughout the 21st century, coupled with the exponential rate of population growth (including the migration of people from rural settlements to urbanised ones) highlights the necessity to re-evaluate the city's ability to maintain a balanced relationship between the internal processes that govern the city's growth and development and its environment. Traditionally, the evolution of cities was driven by the adaptation of their spatial distributions in response to a narrow range of change in its environmental context. However, changes in environmental and urban conditions witnessed throughout the last century are occurring at a more frequent and rapid rate, within a broader range and higher intensities.

The vulnerability of urban settlements to environmental disasters (natural or otherwise), has made it necessary to re-evaluate urban morphological configurations. The presented experiment addresses three key elements as part of the superblock's urban morphology: elevated public networks, the spatial distribution of public space, and the ecological aspects of the city. These three elements are explained briefly in the following section. Future research that focuses on the formulation of the design problem will present the following in greater detail.

3.1.1. Elevating the Flow of the City

The morphology of urban areas and the efficiency by which they grow and occupy their environment has gained significant attention [25]. Current urban morphologies developed through lateral growth and centralised nodes of activities have two conflicting objectives: first, to be as compact as possible—centralisation; and second, to be as dispersed as possible—decentralisation [26]. In recent years, the dramatic increase in urbanites has placed tremendous demand on the spatial distribution and resource management of existing urban settings. Although its impact naturally leads to verticality, this has been implemented within the scale of a single building, yet the city's flow continues to grow laterally at ground level. As a result, the programmes that are dependent on this circulatory system are being distributed at the street level, while buildings continue to develop as separate entities vertically. This led to Harvey Wiley Corbett, primarily known for his skyscraper designs, to be one of the first figures to suggest in the early 20th century the integration between multi-level street networks and mixed-use skyscrapers [27]. Patterns of settlements in many urban areas are transforming from dispersed fabrics to centralised entities with integrated infrastructures [28], mostly in the form of segregated and standalone mixed-use buildings. However, there are cities where the application of higher-level connections has been proven to benefit the urban context. In Hong Kong, traffic congestion, accompanied by air and noise pollution, are the reasons for the incorporation of high-level connections within an urban block. While in Minneapolis and Calgary, their application was in response to the region's severe climatic conditions, leading to an 18-km network of higher-level connections throughout the urban fabric. In addition to the climatic advantages, Corbett et al. [29] argue that such spatial relationships allow for greater and more efficient circulation paths across the urban morphology.

Elevating the flow of the city is one of the urban objectives considered for the presented case study. The spatial proliferation of different urban activities across the urban morphology leads to the emergence of interstitial spaces at different heights with different spatial attributes.

3.1.2. Spatial Distribution of Public Space

The physical and social structures of a city have a reciprocal influence on one another as they continue to develop [26]. Interactions between individuals occur at different spatial scales and locations within a city. However, these networks of interactions are not constrained to their physical structures and surpass the current physical attributes of cities. Public spaces across the city are examples of such areas, where the spatial structure facilitates the social interactions of its inhabitants. The majority of public spaces accessible to the public are located at the street level, while the network of interactions goes beyond a singular level. According to Frans Dieleman and Michael Wegner [30], one of the consequences of urban sprawl is the loss of open space; therefore, by allowing the vertical development and distribution of public space, the morphology and its organisational structure would transform the conventional spatial distribution of an urban sprawl. Although an urban patch may be constrained to its geographical boundaries at the ground level, thus limiting the development of public spaces, the vertical distribution of such spaces bypasses this constraint. Such spaces have great potential to be considered not as confined areas but as a network of spaces connected through higher-level connections.

The distribution of public spaces on multiple levels is another urban objective that is addressed through the presented case study. However, rather than emerging as a by-product of the spatial distribution of buildings, the experiment designates a distinct identity to the interstitial spaces across different levels, allowing for their propagation throughout the urban morphology.

3.1.3. Ecology and Spatial Qualities of the City

Changing environmental and climatic conditions, coupled with a growing demographic, has challenged cities' ecological capabilities to adapt to these fluctuations. The strains of energy consumption have substantially influenced cities' internal environmental and ecological contexts. Processes of urban development, more particularly, urban morphological configurations and reconfigurations, influence the city ecology across a range of spatial domains—from interstitial spaces to the regional environment. As such, ecological systems are dynamic and continuously adapt to other components, such as social and biophysical ones [31]. Therefore, an urban ecological model is integral to the adaptability and flexibility of an urban area. In contrast to many of the planned cities of the 20th century, evolving cities have been closely coupled to their immediate territories, with distinct morphologies, integrated infrastructure, and urban cultures that have evolved in response to the specific ecological and climatic conditions of the region. As these cities grow and develop in complexity, they have become less dependent on their immediate surroundings by obtaining the required energy demands from their local territories [23]. The exchange of energy between the built structures of the urban fabric and the surrounding environment is influenced by the spatial distribution of the urban morphology. It also influences the microecology of interstitial spaces across a range of temporal and spatial scales in the tissue. Therefore, morphological relationships across the urban fabric, such as volume-to-surface ratios [32], surface resolutions, proximities of the buildings, and the spatial relationships between the different components, play a crucial role in adaptation to different environmental conditions and the ecology of a city at large.

Enhancing ecological interactions, more particularly, the energy exchange expressed via intricate spatial distribution of spaces across the urban area, is another urban objective in the presented case study, examining the energy exchange between the built environment of the urban fabric and the interstitial spaces evolved as a result of the spatial distribution of buildings.

3.2. Experiment Setup

3.2.1. Spatial Distribution of Public Space

The base phenotype for the presented experiment imagines a high-density urban block (Figure 7) as the basic geometric component. The phenotype comprises a 5 × 10 grid layout that spans over 375 m in width and 470 m in length. The orthogonal grid structure is distorted in several locations to enable the phenotype to break free from the grid should the algorithm favour such a change. By allowing the nonorthogonal formation of the grid, it enables the building footprints to cluster in different proximities, leading to an increase in variation of the interstitial spaces throughout the urban tissue. The generated footprints were divided into two distinct categories, namely, low-rise building blocks and tower blocks, according to their areas. The morphology comprises rectangular footprints that are distributed amongst the building blocks, with heights varying from 5 to 25 stories.

Figure 7. The primitive phenotype for the design problem.

The low-rise building blocks develop alleys and central courtyards. The tower building blocks comprise two high-rise buildings and an open corner at the street level to serve as the open plaza for each tower. A network of open spaces, comprising the courtyards and open plazas, is formed, which is distributed across an extended network of streets and alleys with different widths and lengths at the ground level. A set of transformations control the attributes of this network and consists of the building heights and footprints as well as the size of the street networks and alleyways.

Tower blocks across the urban tissue cluster into five groups based on their proximities. The shorter tower of each group is designated as the connection point of the skyways that locally connect the towers of the cluster. As the rate of the towers grows in the phenotype, the frequency of the skyways increases and enables circulatory paths to emerge between the towers in close proximity.

The phenotype is constructed to enable elevated connections to emerge across the urban fabric for two primary reasons: first, to extend the circulation beyond the street level, and second, to provide a blueprint for further development of the city should the population expand to its maximum capacity. In addition to the skyways in each tower cluster, a set of higher-level connections that stretches over the entire urban fabric are designated in the phenotype. To ensure that additional higher-level connections across the elevation cover the entirety of the urban area, but which will not overshadow the ground level excessively, the phenotype comprises a set of genes that allows higher-level connections to emerge across the urban fabric at two different elevations. The higher-level connections across the urban area are achieved by enabling spatial units to emerge across the tissue should simulation favour such a change. These spatial units facilitate the morphological intervention of the urban form.

The mid-level elevated circulatory networks connect the towers in the urban tissue while developing a point of contact with the low-rise building blocks. The high-level elevated connections grow from the tallest towers within each cluster and circulate around the urban fabric and create a ring of horizontal habitable spaces. The transformations within the genome of the phenotype control the length, distance, and surface area of these elevated connections, to ensure adequate solar gain on the ground level, adequate solar gain on the elevated surfaces, and their accessibility across the urban fabric. Through interactions between the buildings (low-rise and towers) and the elevated connections, various interstitial spaces with unique spatial qualities emerge across the tissue. The benefit of the morphological interventions (through adding spatial units, voxels) of the elevated connections for the urban area is threefold: first, to elevate the flow of the city; second, to enable habitable spaces to emerge in the interstitial spaces between towers; and third, to develop a balance between horizontality and verticality, to attain a higher density. The chromosomes that define the phenotype's construction are presented in Figure 8.

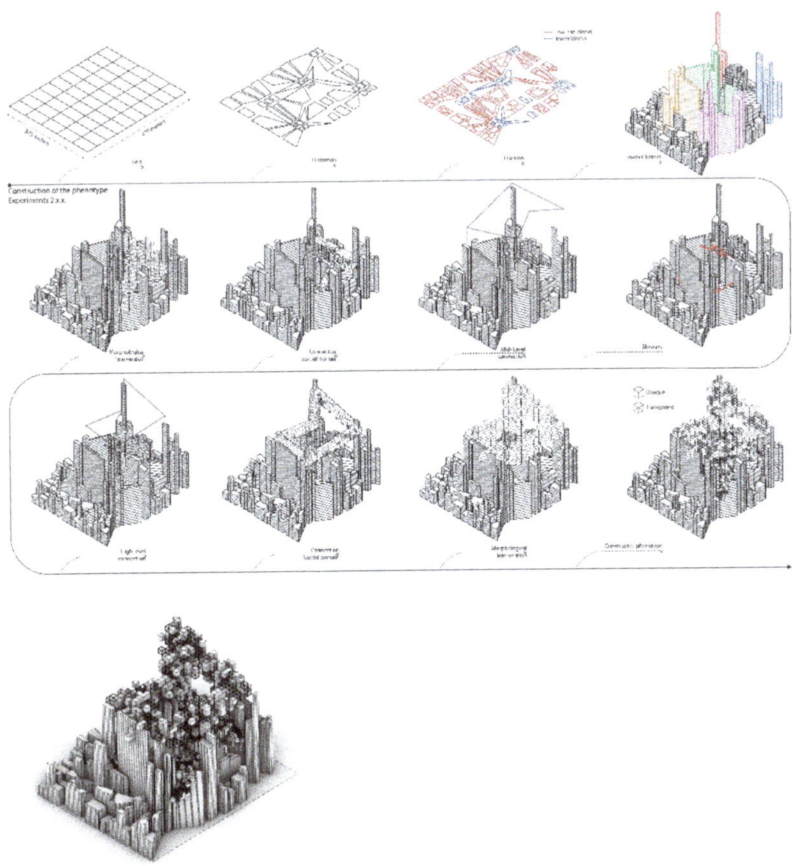

Figure 8. Step-by-step process for the phenotype's construction; each step presents the chromosome applied and the morphological impact of its application.

3.2.2. Fitness Functions

A multitude of metrics have been identified to optimise the developed phenotype. Four of these metrics have been selected as the fitness functions within the MOEA (Figure 9), while the remainder will be utilised for the purposes of analysis and selection (presented in Stage 2 of the selection framework).

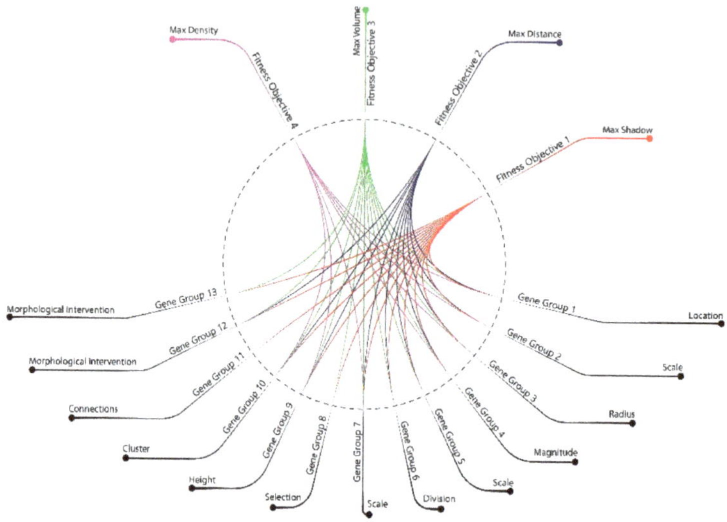

Figure 9. The relationship between the fitness functions and chromosomes defining the design problem.

The first fitness function concerns the solar gain on two categories of spaces across the urban tissue: the first is the generated courtyard areas in the low-rise building blocks, and the second is the outer surfaces of the mid-level elevated connections. By controlling the solar gain on the courtyard areas, the algorithm attempts to explore various distributions of low-rise and tower blocks across the urban tissue. This naturally leads to the emergence of unique interstitial spaces across the ground level. Simultaneously, the second part of this objective (to control the solar gain on the mid-level elevated connections) forces the simulation to explore the spatial distributions of mid-level elevated surfaces in relation to the urban morphology. It creates elevated surfaces across different levels of the urban area that can be utilised for outdoor urban activities (Figure 10).

The second fitness function increases the length of the mid-level and high-level elevated connections and the distance between them. As previously discussed, the phenotype is created to facilitate the development of elevated connections across the urban fabric for two primary reasons: to extend the circulation beyond the street level and to provide a blueprint for further development. Thus, the fitness function is defined to increase the length of these elevated connections (both mid-level and high-level) as well as to encourage their distribution in elevation. By doing so, solutions with elevated connections that are stretched horizontally and vertically across the urban tissue are favoured. As a result of this spatial distribution, interstitial spaces with unique spatial qualities emerge at different levels across the tissue (Figure 11).

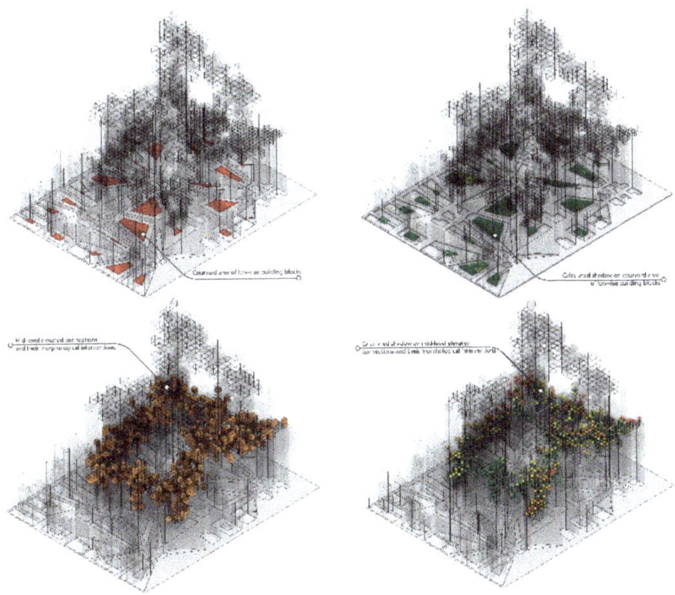

Figure 10. Fitness Function 1: Solar gain calculation on the different morphological characteristics of the superblock.

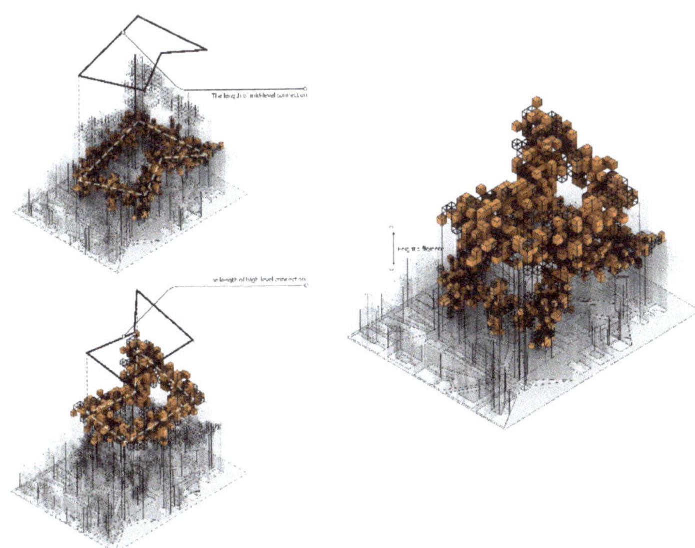

Figure 11. Fitness Function 2: Length and distribution of the spatial interventions throughout the superblock.

The third fitness function increases the volume of the elevated connections and the surface-to-volume ratio. This fitness function focuses on the formation of elevated connections throughout the urban tissue. It is defined to increase their volume as well as the ratio of their surface area to their volume. By increasing the volume of the elevated connections, the simulation generates solutions with volumetric connections that can be considered as extensions to the current urban morphology and be utilised as habitable spaces. It provides an alternative urban morphology for increasing density horizontally and vertically. The increase in the ratio of the surface area to volume contributes to self-shading mechanisms and produces shaded elevated outdoor surfaces (Figure 12).

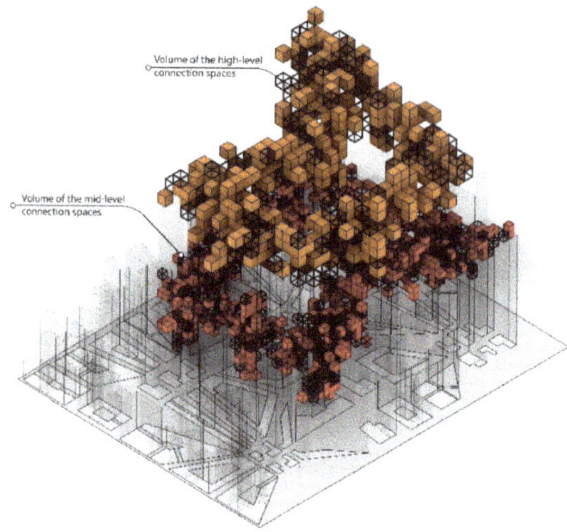

Figure 12. Fitness Function 3: Calculation of the volumetric mass of the spatial interventions throughout the superblock.

Finally, the fourth fitness function increases the built volume of the towers. This fitness function drives the algorithm to evolve larger tower blocks which are associated with an increase in density. Due to the elevated connections being attached to the tower blocks, increasing the volume enhances the structural performance of the buildings to which the elevated connections are connected (Figure 13).

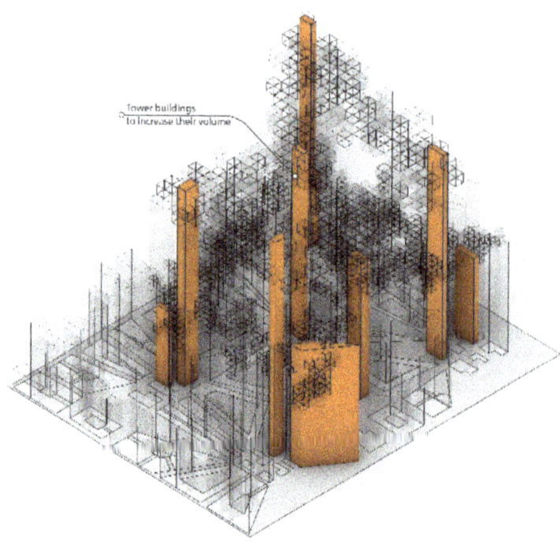

Figure 13. Fitness Function 4: Calculation of the volumetric mass of the larger towers, in this case the primary structural supports for the spatial interventions.

3.2.3. Algorithm Settings

The design problem presented above was run in the software Wallacei [9], which utilises the NSGA-2 MOEA developed by Deb et al. [33]. The algorithm was run on a consumer-grade laptop with the following settings (Table 1) and specifications: Intel Core i7, 2.4 GHz processor, and 32 GB Ram.

Table 1. Simulation and algorithm settings.

Simulation Size		Algorithm Settings	
Generation Size	50	Mutation Rate	1/n (n = no. of var.)
Generation Count	100	Crossover Probability	0.9
Population Size	5000	Mutation Distribution Index	20
No. of Chromosome	13	Crossover Distribution Index	20
No. of Variables	88,445	Runtime	17:46:48

4. Results

4.1. Algorithm Results

Although the presented research focuses on the selection mechanisms employed on the algorithm's output, an overview of the algorithm's results is presented. As can be observed from Figure 14, all four fitness functions exhibit varying degrees of convergence towards local optima, with Fitness Functions 1 and 3 (shadows and volumetric connections) displaying a more evident improvement in fitness throughout the evolved generations. In contrast, the other two objectives (skyways and built volume) are comparatively less so: although presenting improved fitness overall, they are slower in converging towards an optimum. As discussed in previous sections, the formulation of the design problem is key, and although outside the scope of this paper, the successful optimisation observed in the algorithm's output is a result of multiple iterations of reformulating the design problem, constantly striving to find a balanced relationship between the chromosomes, phenotype, and fitness functions.

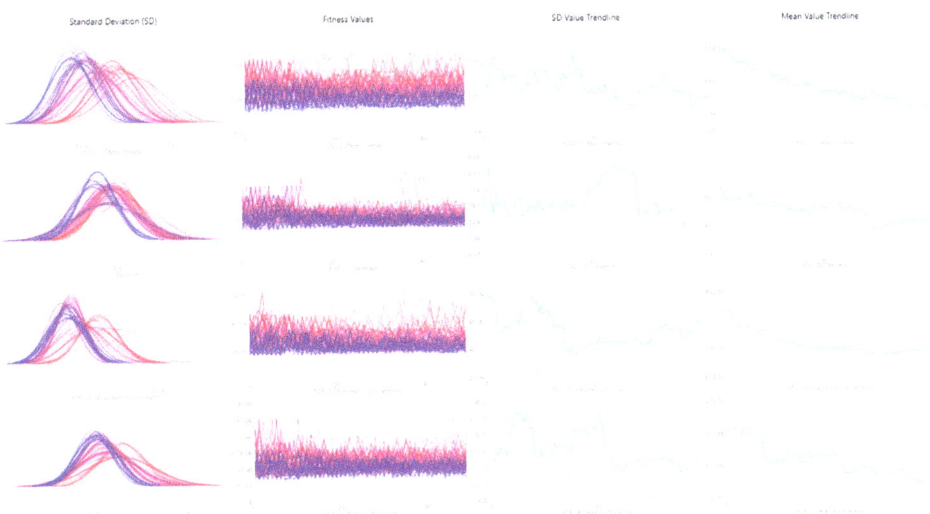

Figure 14. The results of the evolutionary algorithm. Each fitness function was analysed separately across four key metrics (from **left to right**): standard deviation, fitness values, standard deviation trendline, and mean value trendline.

4.2. Selection

As discussed in the Materials and Methods section, the proposed selection framework comprises 2 stages: an objective Stage 1, which filters the population down to a manageable solution pool, and a subjective Stage 2, which ranks the solution pool based on weighted indicators. The results from these two stages are presented below.

4.2.1. Selection Stage 1

Stage 1 filters down the population using four steps. Step 1 applies a clustering algorithm on the pareto front of the entire population. In the case of the presented experiment, the MOEA resulted in 120 pareto front solutions. A hierarchical (average linkage) clustering algorithm was employed, with a K-value of 20 and an average cluster size comprising 8 solutions (Figure 15). The solution closest to the centre of each cluster was selected and added to the solution pool (Figure 16).

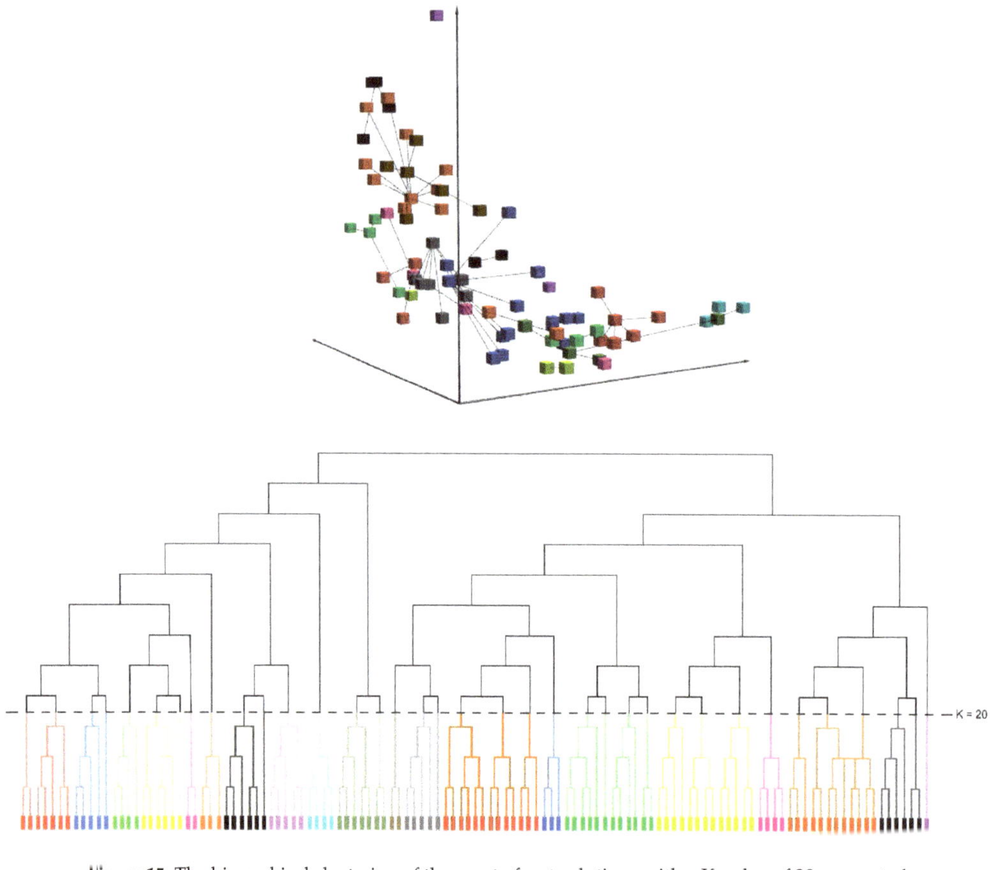

Figure 15. The hierarchical clustering of the pareto front solutions with a K-value of 20, presented through the objective space and dendrogram. As the problem comprised of four fitness functions, it is difficult to read the cluster distribution in the objective space (which is 3-dimensional). In this case the dendrogram provides a clearer understanding of the relationship between solutions, both within and between clusters.

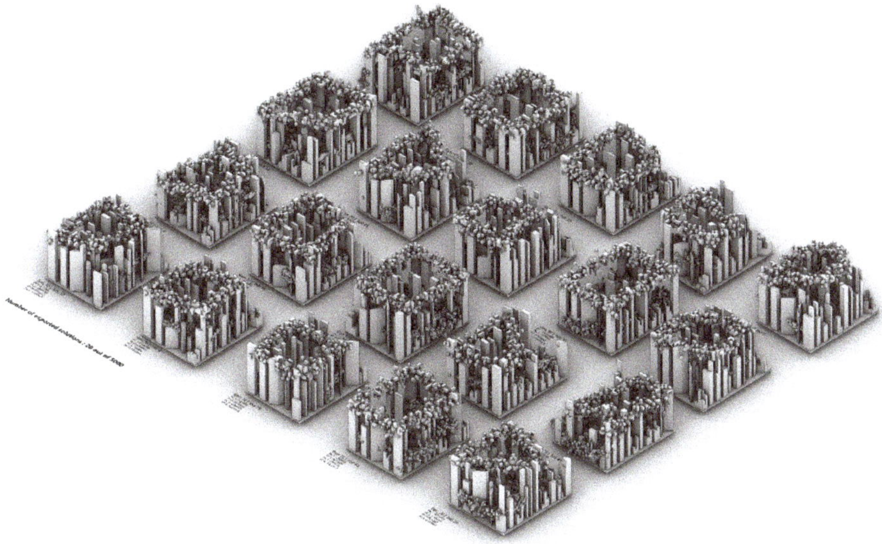

Figure 16. The phenotypes of the cluster centres extracted from the clustering algorithm.

Step 2 selects the 'outliers' in the pareto front; i.e., the fittest solution for each fitness function (Figure 17). As the fitness functions in the design problem are conflicting, the fittest solution for one function usually exhibits poor fitness for other functions.

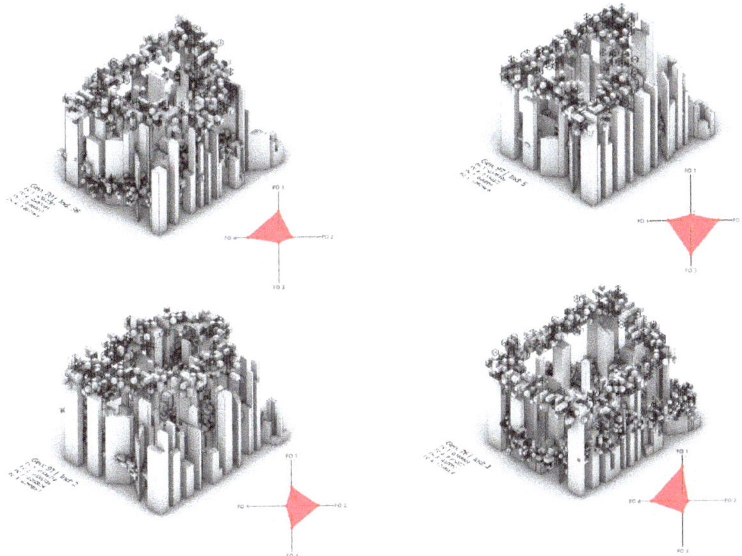

Figure 17. The fittest solution for each fitness function (outliers) alongside each solution's respective diamond chart.

Step 3 selects the solution closest to the 'Utopia' point (Figure 18). Identifying this solution requires a simple Euclidean calculation between the 'Utopia' point (x), which in this case has a vectorial value of 0,0,0,0, and the normalised vectors of all the solutions in the pareto front (y). The solution with the shortest Euclidean distance is closest to the 'Utopia' point. An example of this calculation is provided below:

$$\text{distance}(y,x) = \text{sqrt}((y1 - x1)^2 + (y2 - x2)^2 + (y3 - x3)^2 + (y4 - x4)^2)$$

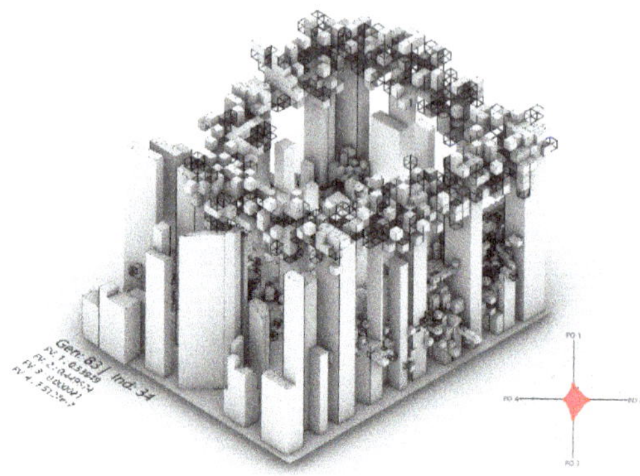

Figure 18. The 'global compromise' solution (solution closest to the 'Utopia' point) alongside its diamond chart.

Finally, Step 4 selects the solution that exhibits an equal weighting (or the closest equal weighting) between all four fitness functions (Figure 19). Identifying this solution requires the normalisation of the fitness values for all fitness functions and selecting the solution with the same normalised fitness value for all functions. Plotting this solution on the parallel coordinate plot and on the diamond chart should present a straight line and an equilateral diamond, respectively. If there exists no solution with an exactly equal weighting between the fitness functions, such as the example presented here, the closest solution to the 'ideal' equally weighted solution is selected. As described previously, and unlike the three steps outlined above, this solution may not fall on the pareto front (which is the case in the selected solution).

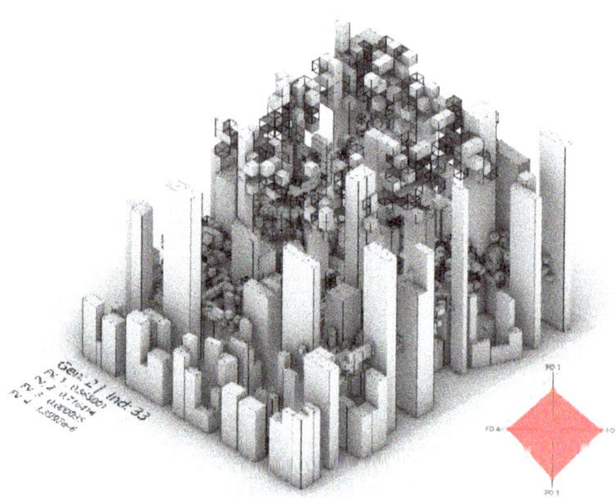

Figure 19. The solution with equal weighting for all fitness functions. Although this solution minimises trade-offs between fitness functions, its performance is significantly poor when compared to the other selected solutions.

Through the selection mechanism of Stage 1, the solutions selected from the population have been filtered down to a solution pool comprising 26 solutions. Most importantly, however, these 26 solutions represent and maintain the diversity generated in the pareto front while simultaneously allowing for a visual analysis of the results.

4.2.2. Selection Stage 2

In contrast to the objective selection mechanisms presented in Stage 1, the second stage of the selection framework requires a high degree of subjective decisions by the DM, who in this case is the person with the greatest expertise on the problem being investigated. However, selection based on purely visual analysis negates the data-driven approach that has so far been utilised in the MOEA process. Therefore, Stage 2 requires the DM to identify relevant 'phenotypic indicators' that evaluate each solution in the filtered solution pool. These indicators provide two key advantages to the DM: first, they provide a metric approach to analyse the solutions and identify differences between them that go beyond the metrics used in the fitness functions (optimised by the algorithm); and second, through using a numeric approach to evaluate the solutions, weighting can be applied to the metrics used, giving the DM the ability to dictate priority of metrics for selection, driven by the parameters that define the design problem. In both steps above, the DM can respond to external demands imposed on the project by additional stakeholders that are not directly involved in the problem's formulation.

The metrics identified to be relevant to the problem being investigated are listed in Table 2. Note that it is critical for the DM to normalise all the metrics as well as identify the direction of each metric (i.e., is it better to minimise the value, maximise the value, or drive it towards a specific target value). Once the metrics have been calculated for each solution in the solution pool, Step 2 applies various weights to the calculated metrics; this step is critical, as it provides added agency to the DM to select solutions based on rationalised and quantified subjective decisions. The final step is to add the resulting values for all metrics per solution; this provides the DM with a single number that represents the rank of each solution based on the results of the 'phenotypic indicators'.

Table 2. Each solution carried forward from the first stage of selection was evaluated according to the phenotypic indicators and weights listed below.

Phenotypic Indicator	Vector Direction	Weighting
Urban FSI (w/o morphological interventions)	Positive	0.25
Urban FSI (w/morphological interventions)	Positive	0.5
Surface to volume ratio (w/o morphological interventions)	Positive	0.25
Surface to volume ratio (w/morphological interventions)	Positive	0.5
Surface Coverage Ratio	Target: 35%	0.25
Average Height	Target: 20 Floors	0.25
Open Space Ratio	Target: 60%	0.5
Mid-Level Private Spatial Units Volume	Positive	0.25
Mid-Level Semi-Private Spatial Units Volume	Positive	0.25
Mid-Level Walking Length	Positive	0.25
High Level Private Spatial Units Volume	Positive	0.25
High Level Semi-Private Spatial Units Volume	Positive	0.25
High Level Walking Length	Positive	0.25
Urban Access	Negative	0.75
Local Open Space	Positive	1.00
Low-rise Building Volume	Positive	0.25
High-rise Building Volume	Negative	0.25
Length of Skyways	Positive	0.75

The phenotypic indicators, along with the defined weighting for each indicator, were applied to the solution pool from Stage 1. The results and ranking are presented in Figure 20. The three top solutions, with a ranking above a certain threshold (in this case 3.80), were selected for selection by the DM (Figure 21).

Phenotypic Incicators	Selection Pool Stage 1																								
	{2;33}	{70;36}	{70;37}	{72;3}	{72;16}	{75;44}	{76;3}	{76;36}	{77;3}	{77;19}	{77;27}	{83;34}	{86;38}	{87;21}	{90;12}	{91;2}	{91;35}	{92;19}	{93;16}	{96;2}	{96;38}	{97;5}	{97;10}	{97;19}	{97;31}	{99;2}
Urban FSI (w/o morph. interventions)	0.12	0.13	0.22	0.21	0.22	0.14	0.12	0.15	0.14	0.19	0.17	0.17	0.18	0.13	0.16	0.22	0.21	0.23	0.16	0.22	0.13	0.23	0.15	0.13	0.17	0.22
Urban FSI (w/ morph. interventions)	0.18	0.31	0.44	0.39	0.43	0.35	0.30	0.34	0.32	0.38	0.37	0.38	0.36	0.34	0.35	0.43	0.42	0.43	0.38	0.39	0.32	0.44	0.35	0.31	0.31	0.43
Surface to volume ratio (w/o morph. Interv.)	0.06	0.09	0.04	0.04	0.03	0.10	0.10	0.07	0.07	0.05	0.06	0.03	0.04	0.08	0.05	0.03	0.03	0.03	0.06	0.04	0.07	0.03	0.07	0.11	0.08	0.03
Surface to volume ratio (w/ morph. Interv.)	0.10	0.26	0.13	0.09	0.10	0.28	0.28	0.22	0.20	0.14	0.19	0.13	0.10	0.27	0.17	0.10	0.10	0.05	0.21	0.04	0.25	0.06	0.20	0.28	0.16	0.10
Surface Coverage Ratio	0.21	0.23	0.12	0.15	0.12	0.24	0.23	0.23	0.24	0.25	0.18	0.18	0.18	0.22	0.20	0.21	0.23	0.18	0.20	0.20	0.23	0.19	0.15	0.23	0.23	0.21
Average Height	0.24	0.08	0.09	0.09	0.08	0.04	0.06	0.05	0.03	0.09	0.09	0.02	0.10	0.13	0.05	0.09	0.12	0.10	0.05	0.10	0.07	0.07	0.07	0.05	0.06	0.09
Open Space Ratio	0.35	0.28	0.46	0.42	0.46	0.29	0.28	0.28	0.29	0.31	0.22	0.22	0.22	0.27	0.25	0.35	0.33	0.39	0.25	0.37	0.28	0.38	0.18	0.32	0.33	0.35
Mid Level Private Spatial Units Volume	0.10	0.15	0.14	0.07	0.09	0.17	0.15	0.17	0.12	0.12	0.15	0.13	0.10	0.20	0.14	0.11	0.11	0.05	0.18	0.04	0.17	0.07	0.14	0.14	0.10	0.11
Mid Level Semi Private Spatial Units Volume	0.12	0.20	0.19	0.11	0.13	0.22	0.20	0.22	0.17	0.16	0.20	0.17	0.14	0.25	0.18	0.16	0.15	0.09	0.23	0.07	0.22	0.10	0.19	0.19	0.14	0.16
Mid Level Walking Length	0.04	0.12	0.06	0.03	0.05	0.10	0.09	0.07	0.12	0.06	0.10	0.07	0.07	0.08	0.08	0.04	0.14	0.03	0.10	0.01	0.08	0.02	0.07	0.07	0.04	0.04
High Level Private Spatial Units Volume	0.03	0.16	0.11	0.12	0.15	0.18	0.18	0.13	0.18	0.13	0.17	0.18	0.13	0.21	0.16	0.11	0.15	0.13	0.16	0.12	0.23	0.15	0.18	0.17	0.08	0.11
High Level Semi Private Spatial Units Volume	0.01	0.20	0.14	0.15	0.18	0.22	0.22	0.17	0.22	0.17	0.19	0.22	0.15	0.18	0.20	0.14	0.19	0.17	0.20	0.15	0.20	0.18	0.23	0.22	0.11	0.14
High Level Walking Length	0.00	0.09	0.05	0.07	0.08	0.11	0.12	0.07	0.12	0.09	0.12	0.10	0.07	0.12	0.10	0.06	0.08	0.06	0.12	0.06	0.13	0.09	0.11	0.09	0.05	0.06
Urban Access	0.41	0.30	0.19	0.27	0.21	0.35	0.31	0.30	0.30	0.29	0.45	0.26	0.35	0.29	0.46	0.24	0.27	0.28	0.40	0.32	0.40	0.29	0.44	0.37	0.30	0.24
Local Open Space	0.31	0.32	0.08	0.02	0.10	0.36	0.38	0.35	0.37	0.16	0.57	0.33	0.33	0.49	0.39	0.11	0.19	0.03	0.30	0.07	0.45	0.03	0.52	0.33	0.15	0.11
Low-rise Building Volume	0.08	0.11	0.00	0.02	0.01	0.12	0.12	0.10	0.11	0.06	0.10	0.12	0.09	0.11	0.13	0.04	0.05	0.02	0.12	0.03	0.11	0.02	0.14	0.10	0.07	0.04
High-rise Building Volume	0.13	0.12	0.03	0.04	0.03	0.11	0.13	0.10	0.11	0.06	0.08	0.08	0.07	0.11	0.09	0.03	0.04	0.02	0.09	0.03	0.12	0.02	0.10	0.12	0.08	0.03
Length of Skyways	0.40	0.53	0.51	0.60	0.59	0.32	0.23	0.47	0.31	0.47	0.39	0.30	0.51	0.38	0.36	0.71	0.59	0.59	0.28	0.50	0.31	0.60	0.45	0.40	0.40	0.71
Score	**2.89**	**3.68**	**3.00**	**2.88**	**3.07**	**3.69**	**3.47**	**3.48**	**3.41**	**3.18**	**3.81**	**3.08**	**3.20**	**3.86**	**3.52**	**3.18**	**3.40**	**2.89**	**3.46**	**2.76**	**3.79**	**2.98**	**3.73**	**3.65**	**2.86**	**3.18**

Figure 20. The results of each solution and its applied phenotypic indicator. The solutions above a specific threshold are highlighted as the top-performing solutions.

Figure 21. The phenotypes of the selected three solutions from Figure 20.

5. Discussion and Conclusions

The application of MOEAs in design, specifically urban design, requires the decision maker to critically review all aspects associated with the algorithmic process. Although the formulation of the design problem, and the chosen algorithm to run the simulation, are critical for generating successful results, the method for selecting the 'best solution' from a population of solutions that are all non-dominated (i.e., improving one solution will make another one worse) is considerably consequential. The selection method presented in this paper proposes a framework for designers to evaluate and filter a larger solution set through both an objective and subjective process, each equally significant.

To demonstrate the proposed selection framework, a design problem comprised of a speculative urban tissue was developed to respond to contemporary urban and environmental objectives; that is, to evolve urban configurations comprised of intricate spatial and morphological organisations within the urban fabric's interstitial spaces. The conflicting nature of the fitness functions driving the experiment resulted in wide-ranging diversity amongst the optimised solutions, demanding a critical and thorough selection process that filters through this variation by both quantitative and qualitative modes of analysis.

The key objective of the proposed selection framework is to maintain diversity in the filtered solution set while delivering a pool of solutions that is feasible for visual analysis. In Stage 1 of the selection framework, four key selection mechanisms are applied. Clustering the pareto front (with a high K-value) ensures that all solutions are represented in the filtered solution pool, while selecting the fittest solutions for each objective ensures the DM incorporates the outliers of each fitness function in the solution pool. Moreover, the outlier solutions provide the DM with the domain of the pareto front. Selecting the solution closest to 'Utopia' provides the DM with the result that is in the nearest proximity to the ideal (non-existent) solution to the problem, while the final mechanism for selecting the solution with the most equal weight delivers the result that applies no trade-offs to the fitness functions. These four selection mechanisms provide the DM with an objective and comprehensive representation of the optimised results.

Although Stage 2 aims to apply subjective decisions for analysis and selection, filtering the solution set to a manageable size in Stage 1 equips the DM with the ability to visually inspect the results of Stage 2. As discussed, the subjective approach in Stage 2 requires the DM to identify additional metrics (phenotypic indicators) to further analyse the results. Although not essential, there is added value to weighting the phenotypic indicators according to their importance to the problem being investigated. This added value increases as the number of phenotypic indicators. Once the solutions are evaluated and ranked, the DM may choose to visually inspect the top solutions and apply an additional layer of analysis to those top solutions, or choose the best solution based on the ranked results. The results in the case study presented herein intentionally avoided selecting the best solution, instead proposing the top solutions for the DM to choose from. The DM's role in visually inspecting solutions and making subjective decisions remains critical to the design problem; however, this process is significantly strengthened when complemented with a numerically rationalised objective and subjective evaluation process.

Although the primary mechanism for maintaining diversity in Stage 1 is through clustering the pareto front, the clustering algorithm is applied on the fitness values of the solutions. Despite the algorithm successfully grouping solutions based on the Euclidean proximity of their fitness values, this does not necessarily translate to phenotypic similarity. It is possible for two solutions to present fitness similarity while simultaneously maintaining phenotype diversity [34]; therefore, future work should focus on integrating (or replacing) fitness clustering with genomic clustering for an accurate correlation between the clusters and morphological characteristics of the selected solutions. It is critical to highlight that the presented framework is highly adaptable to the problem at hand. Although the four steps of selection in Stage 1 aim to maintain diversity, the DM may employ additional selection mechanisms to add solutions to the filtered solution set. Such selection mechanisms include the Knee-Point solution, located on a 'bulge' or 'knee' of the pareto front [35], or the average ranked solution (the solution with a mean ranking between fitness functions) [36]. In Stage 2, the phenotypic indicators most relevant to the DM should be applied for evaluation; these indicators will widely differ based on the problem being investigated, for example metrics that specifically address liveability and quality of life (such as walkability, privacy, access to daylight and ventilation, etc.). Fundamental to this process is the user's ability to numerically quantify the urban condition being analysed. Although the presented indicators do not consider opportunities for temporal growth or adaptive response of the generated solutions, integrating metrics related to these modes of analyses for the purposes of selection is made possible through favouring variation over optimisation; however, a priori approach to adaptive growth remains the most suitable approach.

Author Contributions: Methodology, M.S., M.M.; Case Study, M.S.; analysis, M.S., M.M.; writing—original draft preparation, M.S., M.M.; writing—review and editing, M.S., M.M. All authors have read and agreed to the published version of the manuscript.

Funding: This research received no external funding.

Institutional Review Board Statement: Not applicable.

Informed Consent Statement: Not applicable.

Data Availability Statement: Not applicable.

Conflicts of Interest: The authors declare no conflict of interest.

References

1. Deb, K. *Multi-Objective Optimization Using Evolutionary Algorithms*, 1st ed.; Wiley-Interscience Series in Systems and Optimization; John Wiley & Sons: Chichester, UK; New York, NY, USA, 2001; ISBN 978-0-471-87339-6.
2. Fonseca, C.M.; Fleming, P.J. Genetic Algorithms for Multiobjective Optimization: Formulation, Discussion and Generalization. In *Proceedings of the Fifth International Conference on Genetic Algorithms*; Morgan Kauffman Publishers: San Mateo, CA, USA, 1993; Volume 93, pp. 416–423.
3. Goldberg, D.E. *Genetic Algorithms in Search, Optimization, and Machine Learning*, 13th ed.; Addison Wesley: Reading, MI, USA, 1989; ISBN 978-0-201-15767-3.
4. Schaffer, J.D. Multiple Objective Optimization with Vector Evaluated Genetic Algorithms. Ph.D. Thesis, Vanderbilt University, Nashville, TN, USA, 1984.
5. Seada, H.; Deb, K. U-NSGA-III: A Unified Evolutionary Optimization Procedure for Single, Multiple, and Many Objectives: Proof-of-Principle Results. In *Evolutionary Multi-Criterion Optimization*; Springer: Cham, Switzerland, 2015; pp. 34–49.
6. Zitzler, E.; Laumanns, M.; Thiele, L. *SPEA2: Improving the Strength Pareto Evolutionary Algorithm*; ETH Library: Zurich, Switzerland, 2001.
7. Abdulmawla, A.; Bielik, M.; Bus, P.; Dennemark, M.; Fuchkina, E.; Miao, Y.; Knecht, K.; Konig, R.; Schneider, S. Decoding Spaces Toolbox 2017. Available online: https://toolbox.decodingspaces.net/ (accessed on 20 February 2022).
8. Harding, J. Biomorpher. Grasshopp. 3D. 2016. Available online: https://www.grasshopper3d.com/group/biomorpher?overrideMobileRedirect=1 (accessed on 20 February 2022).
9. Makki, M.; Showkatbakhsh, M.; Song, Y. Wallacei: An Evolutionary and Analytic Engine for Grasshopper 3D. Available online: https://www.wallacei.com/ (accessed on 20 February 2022).
10. Vierlinger, R. Octopus. Available online: https://www.food4rhino.com/app/octopus (accessed on 23 July 2018).
11. Choi, J.; Nguyen, P.C.T.; Makki, M. The Design of Social and Cultural Orientated Urban Tissues through Evolutionary Processes. *Int. J. Archit. Comput.* **2021**, *19*, 331–359. [CrossRef]
12. Koenig, R.; Miao, Y.; Aichinger, A.; Knecht, K.; Konieva, K. Integrating Urban Analysis, Generative Design, and Evolutionary Optimization for Solving Urban Design Problems. *Environ. Plan. B Urban Anal. City Sci.* **2020**, *47*, 997–1013. [CrossRef]
13. Makki, M.; Showkatbakhsh, M.; Tabony, A.; Weinstock, M. Evolutionary Algorithms for Generating Urban Morphology: Variations and Multiple Objectives. *Int. J. Archit. Comput.* **2018**, *17*, 1–31. [CrossRef]
14. Huang, W.; Zhang, Y.; Li, L. Survey on Multi-Objective Evolutionary Algorithms. *J. Phys. Conf. Ser.* **2019**, *1288*, 012057. [CrossRef]
15. von Lücken, C.; Barán, B.; Brizuela, C. A Survey on Multi-Objective Evolutionary Algorithms for Many-Objective Problems. *Comput. Optim. Appl.* **2014**, *58*, 707–756. [CrossRef]
16. Luke, S. *Essentials of Metaheuristics*, 2nd ed.; Lulu: Morrisville, NC, USA, 2013; ISBN 978-1-300-54962-8.
17. Van Veldhuizen, D.A.; Lamont, G.B. Multiobjective Evolutionary Algorithms: Analyzing the State-of-the-Art. *Evol. Comput.* **2000**, *8*, 125–147. [CrossRef] [PubMed]
18. Branke, J.; Deb, K.; Dierolf, H.; Osswald, M. Finding Knees in Multi-Objective Optimization. In *Parallel Problem Solving from Nature-PPSN VIII*; Yao, X., Burke, E.K., Lozano, J.A., Smith, J., Merelo-Guervós, J.J., Bullinaria, J.A., Rowe, J.E., Tiňo, P., Kabán, A., Schwefel, H.-P., Eds.; Springer: Berlin/Heidelberg, Germany, 2004; pp. 722–731.
19. Eskelinen, P.; Miettinen, K. Trade-off Analysis Approach for Interactive Nonlinear Multiobjective Optimization. *Spectr* **2012**, *34*, 803–816. [CrossRef]
20. Kaushik, M.; Mathur, B. Comparative Study of K-Means and Hierarchical Clustering Techniques. *Int. J. Softw. Hardw. Res. Eng.* **2014**, *2*, 93–98.
21. Yu, P.-L. A Class of Solutions for Group Decision Problems. *Manag. Sci.* **1973**, *19*, 936–946. [CrossRef]
22. Zeleny, M. Compromise Programming. In *Multiobjective Optimization: Behavioral and Computational Considerations*; Springer: Berlin/Heidelberg, Germany, 1973.
23. Showkatbakhsh, M.; Makki, M. Application of Homeostatic Principles within Evolutionary Design Processes: Adaptive Urban Tissues. *J. Comput. Des. Eng.* **2018**, *7*, 1–17. [CrossRef]
24. Weinstock, M. *The Architecture of Emergence: The Evolution of Form in Nature and Civilisation*; Wiley: Chichester, UK, 2010; ISBN 978-0-470-06632-4.

25. Batty, M.; Longley, P.A. *Fractal Cities: A Geometry of Form and Function*; Academic Press: Cambridge, MA, USA, 1994; ISBN 0-12-455570-5.
26. Batty, M. *The New Science of Cities*; The MIT Press: Cambridge, MA, USA, 2013; ISBN 978-0-262-01952-1.
27. Goodman, A.E. *A Brief History of the Future: The United States in a Changing World Order*; Routledge: London, UK, 2019; ISBN 0-429-03472-5.
28. Kern, L. Reshaping the Boundaries of Public and Private Life: Gender, Condominium Development, and the Neoliberalization of Urban Living. *Urban Geogr.* **2007**, *28*, 657–681. [CrossRef]
29. Corbett, M.J.; Xie, F.; Levinson, D. Evolution of the Second-Story City: The Minneapolis Skyway System. Environ. *Plan. B Plan. Des.* **2009**, *36*, 711–724. [CrossRef]
30. Dieleman, F.; Wegener, M. Compact City and Urban Sprawl. *Built Environ.* **2004**, *30*, 308–323. [CrossRef]
31. Pickett, S. Ecology of the City: A Perspective from Science. In *Urban Design Ecologies*; John Wiley & Sons, Ltd.: Hoboken, NJ, USA, 2012; pp. 162–173.
32. Allen, J.A. The Influence of Physical Conditions in the Genesis of Species. *Radic. Rev.* **1877**, *1*, 108–140.
33. Deb, K.; Agrawal, S.; Pratap, A.; Meyarivan, T. A Fast Elitist Non-Dominated Sorting Genetic Algorithm for Multi-Objective Optimization: NSGA-II. In *International Conference on Parallel Problem Solving From Nature*; Springer: Paris, France, 2000; pp. 849–858.
34. Makki, M.; Navarro-Mateu, D.; Showkatbakhsh, M. Decoding the Architectural Genome: Multi-Objective Evolutionary Algorithms in Design. *Technol. Des.* **2022**, *6*, 68–79. [CrossRef]
35. Das, I. On Characterizing the "Knee" of the Pareto Curve Based on Normal-Boundary Intersection. *Struct. Optim.* **1999**, *18*, 107–115. [CrossRef]
36. Makki, M.; Showkatbakhsh, M.; Song, Y. *Wallacei Primer 1.0*. 2019. Available online: https://www.wallacei.com/ (accessed on 20 February 2022).

buildings

MDPI

Article

Urban Microclimate, Outdoor Thermal Comfort, and Socio-Economic Mapping: A Case Study of Philadelphia, PA

Farzad Hashemi [1,*], Ute Poerschke [1], Lisa D. Iulo [1] and Guangqing Chi [2]

[1] Department of Architecture, The Pennsylvania State University, University Park, PA 16802, USA
[2] Department of Agricultural Economics, Sociology, and Education, Population Research Institute, and Social Science Research Institute, The Pennsylvania State University, University Park, PA 16802, USA
* Correspondence: fxh99@psu.edu

Abstract: Urban areas are often warmer than rural areas due to the phenomenon known as the "urban heat island" (UHI) effect, which can cause discomfort for those engaging in outdoor activities and can have a disproportionate impact on low-income communities, people of color, and the elderly. The intensity of the UHI effect is influenced by a variety of factors, including urban morphology, which can vary from one area to another. To investigate the relationship between outdoor thermal comfort and urban morphology in different urban blocks with varying social vulnerability status, this study developed a geographic information system (GIS)-based workflow that combined the "local climate zone" (LCZ) classification system and an urban microclimate assessment tool called ENVI-met. To demonstrate the effectiveness of this methodology, the study selected two different urban blocks in Philadelphia, Pennsylvania–with high and low social vulnerability indices (SVI)–to compare their microclimate conditions in association with urban morphological characteristics such as green coverage area, sky view factor (SVF), albedo, and street height to width (H/W) ratio. The results of the study showed that there was a strong correlation between tree and grass coverage and outdoor air and mean radiant temperature during hot seasons and extremely hot days, which in turn affected simulated predicted mean vote (PMV). The effects of greenery were more significant in the block associated with a low SVI, where nearly 50% of the site was covered by trees and grass, compared to only 0.02% of the other block associated with a high SVI. Furthermore, the investigation discovered that reduced SVF, along with increased albedo and H/W ratio, had a beneficial impact on the microclimate at the pedestrian level within the two studied urban blocks. This study provided an effective and easy-to-implement method for tackling the inequity issue of outdoor thermal comfort and urban morphology at fine geographic scales.

Keywords: urban heat island; outdoor thermal comfort; predicted mean vote (PMV); geographic information system (GIS); local climate zones (LCZ); social vulnerability index (SVI)

1. Introduction

In 2017, human-induced warming was about 1 °C above pre-industrial levels and increasing at a rate of 0.2 °C per decade. If the current rate continues, there is a strong possibility of a further rise of 1.5 °C in global temperatures between 2030 and 2052 [1]. With 55% of the world's population living in urban areas in 2017, the UN world urbanization prospect [2] estimates the proportion of the population living in urban areas will increase to 68% (i.e., 2 of 3 people will live in urban areas) by 2050. As a result of urbanization, most cities worldwide have experienced a significant conversion of natural landscapes to built environments that are associated with low vegetation index, the emergence of dark surfaces with low albedo, increased heat waste from both buildings and traffic, and reduced sky view [3,4]. As a result of these land use conversions, air temperatures in urban areas are typically higher than their surrounding rural areas, which is termed the "urban heat island" (UHI) effect [5,6], one of the most evident anthropogenic impacts on urban microclimates.

Increases in urban air temperature have a severe impact on building energy uses and may endanger life and increase mortality rates.

UHIs have been observed globally in cities regardless of their locations and size [7–10]. The UHI effect and its direct consequence (i.e., heat stress in cities) cause severe discomfort in outdoor activities, especially for groups of people who are known to be socially vulnerable [11]. From 2008 to 2017, extreme heat was associated with higher all-cause mortality in the contiguous U.S., with a greater increase noted among older adults, men, and non-Hispanic Black individuals [12]. More than 3800 death were recorded to be caused by heat-related illness (HRI), and almost 23,000 people were hospitalized for HRI between 2016–2020 across the U.S. [13]. Overall, heat island hotspots and mortality rates are found to be greater in urban blocks with a socially disadvantaged population [14–16]. According to a recent study by Hsu et al. [17], the mean and dispersion of summer daytime Surface UHI intensity are worse for both people of color and the poor compared to white and wealthier populations in nearly all major U.S. cities.

Due to the significant impacts of UHIs on human health and comfort, several studies have been conducted to evaluate these impacts during past decades in cities located in different climate zones. Field observation and numerical models are the two main methods used in these studies to investigate the impacts of the induced temperature on comfort sensations. A field study was carried out in outdoor urban spaces of Singapore with a focus on the human thermal sensation of urban dwellers from August 2010 to May 2011 [18]. Solar radiation showed the most significant effect on human thermal sensation, and to improve outdoor thermal comfort, a combination of lower-density urban areas with higher building height was found to be an effective solution. During the summers and winters of 2008 and 2009, a field survey was conducted in two hot and arid climate cities-Marrakech, Morocco and Phoenix, Arizona. The survey included structured interviews using a standard questionnaire, observations of human activities, and microclimatic monitoring [19]. They found that the socioeconomic status and cultural background of attendees showed a significant impact on their thermal sensation and comfort expectation. Noticeable differences between the two cultures in the type of outdoor activities were also found. Field measurement and surveys were employed by several other studies in cities such as Rome, Italy [20], Wuhan and Xi'an, China [21,22], Belo Horizonte, Brazil [23], and Tempe, Arizona [24], to name a few.

Due to the differences in urban fabrics in the examined cities, as well as the overall efforts required to conduct field observations/surveys, this method imposes substantial limitations in terms of the applicability of results and the required time and cost to accomplish similar studies. Hence, numerical models have become of significant interest to planners, architects, and environmental assessors to simulate urban microclimate and perform comparative analyses based on different scenarios; these methods often use spatial overlay and simulation methods to combine both physical/climatic and socioeconomic data [25,26]. Urban microclimate modeling tools, especially simulation with Computational Fluid Dynamics (CFD), are used both to simulate UHI intensities and to estimate how effective it would be to apply different mitigation measures. The yearly number of studies on the topic has increased rapidly since 1998 [27]. In recent years, many studies have been conducted on modeling urban microclimates using tools such as OpenFOAM, FLUENT, STAR-CCM+, PHOENICS, and ENVI-met. Among these CFD-based models, ENVI-met was found to be the most popular due to a balance of sophistication, user-friendliness, and lower computational costs [28]. From whole microclimate simulation at a neighborhood scale to evaluation of hourly outdoor thermal comfort, ENVI-met has been validated by several studies [29–36]. Forouzandeh [37] compared the ENVI-met V4 outputs with values measured experimentally under various seasonal conditions inside a courtyard located in Hannover, Germany. The ENVI-met model was proved to be accurate in predicting microclimate variables in medium-narrow courtyards with low root mean square error (RMSE) values for temperature, humidity, and wind speed. RMSE values ranged between 0.73 °C for air temperature, 3.34% for humidity, and 0.01 m/s for wind speed.

ENVI-met's capacity to estimate Physiological Equivalent Temperature (PET), a thermal index, was evaluated [38] by comparing it to field measurements for various points in a street canyon in Port Said, Egypt, throughout the summer and winter seasons. Except for the limited air velocity correlation, the results presented a strong agreement, particularly with respect to the final results of the PET, with an index of agreement value ranging from 0.81 to 0.95.

Existing literature adopted urban blocks or master plan developments, normally only one in each study, for urban microclimate and/or outdoor thermal sensations studies. Assessment of microclimate conditions in multiple urban blocks of a city for a meaningful comparison of impacts caused by urban parameters and the effectiveness of heat mitigation efforts remains limited. In addition, the areas being studied are typically modeled with added details for urban parameters and building properties, which often involves a time-consuming and resource-intensive process of collecting real data, including manpower and funding. To overcome these disadvantages, a GIS-based methodology was developed and used in this study to first identify urban blocks with different socioeconomic statuses in U.S. cities and then model urban microclimate inside the blocks creating a set of generic inputs eliminating the need for field data collection. The proposed workflow is replicable and can be applied by researchers and practitioners, from novices to experts, to better address urban microclimate conditions in their designs and urban masterplans.

2. Methodology and Case Study

This study investigates outdoor thermal comfort levels in two Philadelphia, PA neighborhoods, one with a high social vulnerability index (SVI) and one with a low index. In 2021, Philadelphia was the city with the highest poverty rate (with 22.8% of people who live below the poverty level) among the United States' most populated cities [39]. Moreover, since 2010, Philadelphia experienced a variety of extreme weather events, including the snowiest winter, the two warmest summers, the wettest day, and the two wettest years on record [40]. Philadelphia's architectural landscape is mainly composed of narrow, attached rowhouses that have a one-to-four-story structure. The city experienced rapid growth during the early 20th century resulting in the construction of numerous high-rise buildings and skyscrapers. The city's street grid layout was established in the late 17th century, and its urban structure has undergone several changes. Urban renewal projects in the 1950s and 1960s led to the demolition of several historic buildings and the construction of modernist structures [41].

According to the U.S. Department of Energy's Building America Program, Philadelphia is located in a Mixed-Humid region, which is characterized by receiving over 20 inches (50 cm) of yearly rainfall, having around 5400 heating degrees days (based on a temperature of 65 °F), and experiencing an average monthly outdoor temperature below 45 °F (7 °C) during winter [42]. The proposed methodology in this research consists of two primary steps, which are outlined below.

2.1. Step 1: Exploring Social Vulnerability Index to Identify Neighborhoods

Using U.S. Census data, the SVI is gathered by the Agency for Toxic Substances & Disease Registry (ATSDR) at the Centers for Disease Control and Prevention (CDC) to identify the United States communities that will likely need support before, during, and after a public health emergency [43]. Typical census tracts incorporate 1500 to 8000 people, with an optimum size of 4000 people [44], and they are designed to be demographically homogeneous. The CDC/ATSDR SVI ranks U.S. census tracts according to 15 individual variables, i.e., poverty, unemployment, income, education, age (65 or older and 17 or younger), disability, single-parent households, minority, language (speak English less than well), housing unit structure, mobile homes, crowding, lack of vehicle, and group quarter. These 15 variables are further categorized into four related themes including; 1-socioeconomic status, 2-household competition and disability, 3-minority status and language, and 4-housing type and transportation.

The CDC SVI data (2018, the latest version at the time of conducting this research) were used to determine the social vulnerability of every census tract in Philadelphia. Figure 1 shows the tasks to rank the tracts for the four vulnerability themes and the overall vulnerability of the tracts as the last task.

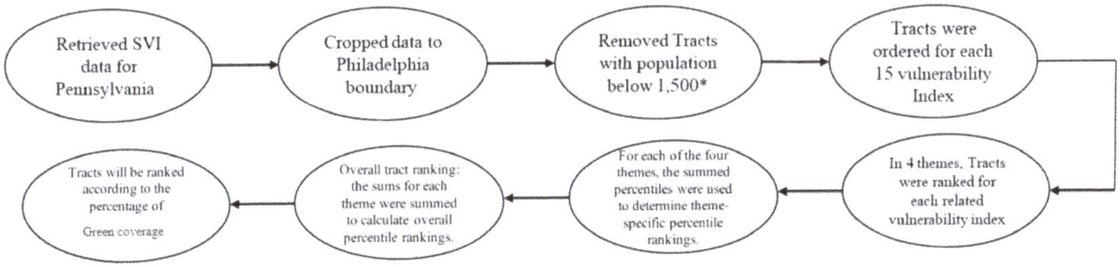

Figure 1. The workflow for ranking Philadelphia's census tract according to their SVI, Source: Authors.

Census tracts in Philadelphia are compared to one another in order to map and analyze relative vulnerability as a consequence of the four aforementioned themes called the overall vulnerability index across the city. Tract rankings are based on percentiles. Percentile ranking values range from 0 (dark green) to 1 (dark red), with higher values indicating greater vulnerability (Figure 2). Tracts for which there is no SVI data available are shown as black and white stripes.

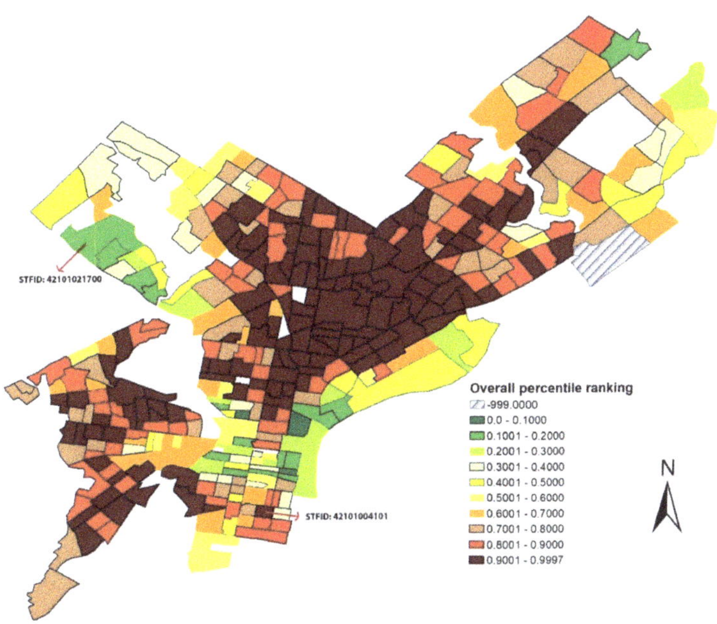

Figure 2. Maps show the range of vulnerability in Philadelphia, PA as a consequence of the four themes (overall ranking).

According to the CDC SVI data, census tracts in Philadelphia show significant differences in 15 socioeconomic characteristics, such as poverty, lack of vehicle access, housing type, education, living alone, and disability across the city. Tracts located in northwestern

Philadelphia are associated with a low percentile ranking for SVI, 0.1 to 0.2, while tracts in the south, central, and a big part of the north sides show a very high percentile ranking, more than 0.8 indicating greater vulnerability (Figure 2). Moreover, Figure 3 shows a high density of tree canopy in neighborhoods with a low vulnerable population and, on the other hand, a low density of trees in tracts with a higher vulnerability index. Accordingly, the process of selecting two urban blocks in Philadelphia, one with a low level of vulnerability and a high concentration of tree canopy and another with a high level of vulnerability and a low concentration of tree coverage, was achieved by overlaying the SVI and tree canopy coverage from GIS shapefiles.

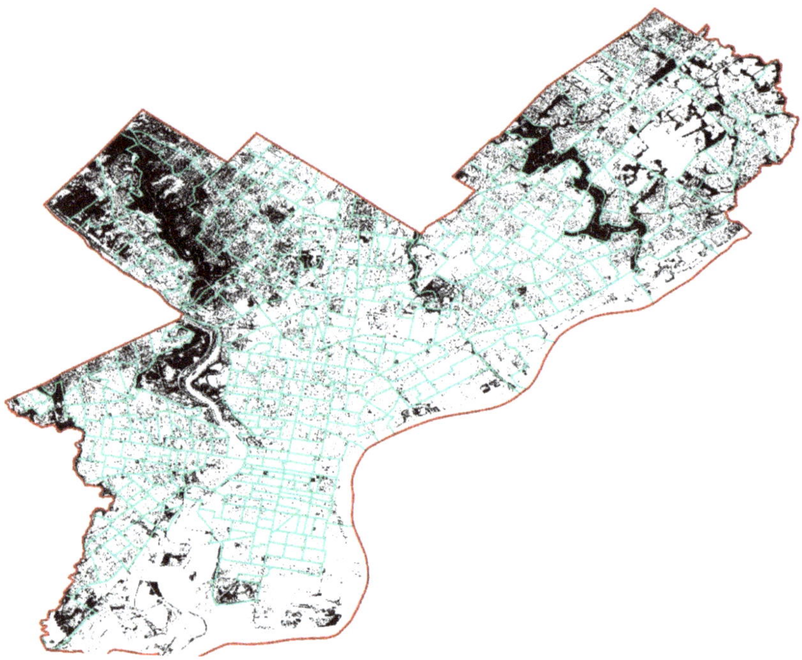

Figure 3. Tree canopy coverage in Philadelphia, PA.

The outcome of this stage involved the selection of two tracts out of 359 tracts in Philadelphia. One tract was chosen from the tracts with the lowest vulnerability index (STFID: 42101021700, see Figure 2) and high tree coverage, while the other was selected from the tracts with the highest vulnerability index (STFID: 42101004101, see Figure 2) and the lowest tree coverage to proceed with the next step: microclimate modeling. These two tracts and their snapshots are illustrated in Figure 4.

Figure 4. Two selected tracts for microclimate simulation; top left with low vulnerability index and bottom right with high vulnerability index, the purple highlights show the exact boundary of each census tract.

2.2. Step 2: Microclimate and Outdoor Thermal Comfort Simulation

The physical basis and spatial/temporal resolution of urban microclimate models differ greatly. ENVI-met [45], one of the most widely used tools for simulating urban microclimates, can be used to analyze the thermal comfort regime within the street canyon at fine resolutions, down to 0.5×0.5 m. The user-friendly tool aims at replicating the primary atmospheric processes that affect the microclimate on a physical basis (i.e., the fundamental laws of fluid dynamics and thermodynamics). BioMet is a plug-in for ENVI-met that calculates human thermal comfort indices from ENVI-met model output files. ENVI-met BIO-met can be used to assess the effects of the presence of vegetation in an urban space, local shading on the person, the position of the person in relation to buildings, and the albedo of paving materials.

ENVI-met BioMet is able to predict outdoor thermal comfort in terms of indices such as a predicted mean vote (PMV), physiological equivalent temperature (PET), universal thermal climate index (UTCI), and standard effective temperature (SET). In this research, the PMV index was adopted to evaluate the outdoor thermal comfort at the street level of the selected tracts in Philadelphia. The PMV is a thermal index developed by Fanger [46] to assess indoor thermal comfort. The PMV calculated by ENVI-met has been adapted for outdoor conditions, including solar radiation and wind speed. To perform a simulation through ENVI-met BioMet, four environmental factors, i.e., air temperature (T_a), mean radiant temperature (MRT), wind speed (UV), and the specific humidity (q), were extracted from the ENVI-met outputs. Moreover, two personal factors, i.e., metabolic rate (ISO 8996) and clothing index (ISO 9920) for four seasons were extracted from the literature [47] and were incorporated into the BIO-met model.

To simulate the PMV index for the selected tracts, two high-resolution 3-D models were created in ENVI-met (Figure 5), with green color representing grass coverage and trees in each model. As a main input to ENVI-met, the geometrical data of the models were obtained from Philadelphia GIS Shapefile, open-source data provided by Pennsylvania Spatial Data Access (PASDA) [48].

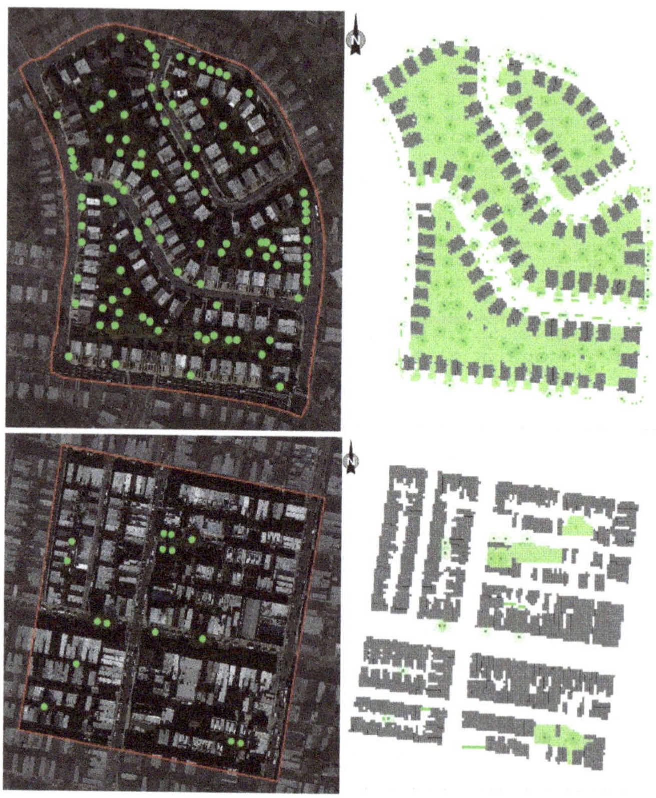

Figure 5. The exact boundary of selected blocks within the census tracts (**left images**) and ENVI-met models created for each block (**right images**).

In order to construct three-dimensional models of neighborhoods, Grasshopper 3-D and Meerkat, which are based on Rhinoceros software, were utilized. The process of using these tools to create urban 3-D models is explained in detail in [49,50]. The resulting raw 3-D geometries were then imported into Dragonfly, a plug-in created [51] for use in Grasshopper, which enables the creation of 3-D models using Envi-met (INX). To incorporate specific urban characteristics, such as construction materials, land cover, albedo, and anthropogenic heat generated by human activities, the two areas under investigation were classified based on the local climate zone (LCZ) system. The LCZs categorize landscapes based on a typical range of values for surface cover, urban structure, fabrics (radiative and thermal properties of construction materials), and anthropogenic heat flux [52]. The LCZ classification scheme consists of 10 built and 7 land cover types, and each of the 17 basic types is associated with typical value ranges for a set of key urban parameters. The definition and physical properties of all LCZs are provided in 17 illustrative datasheets. Accordingly, the tract with STFID of 42101021700 (top images in Figure 4, simply Block 1, hereafter) was classified as LCZ6: Open low-rise, where buildings are detached and small size with 1–3 stories tall. On the other hand, the tract with the STFID of 42101004101 (bottom images in Figure 4, simply Block 2, hereafter) was classified as LCZ 3: Compact low-rise in which buildings are attached or closely spaced with 1–3 stories tall. Land cover in LCZ 3 is mostly paved or hard-packed; LCZ6 is mostly covered by scattered trees and abundant plants (Figure 6). In order to balance the need for accurate simulation results with the computational time required, the 3-D models were constructed for only a portion of the selected tracts. Specifically, the models were created to cover an area with a minimum diameter of 400 m (or a radius of 200 m), which is the minimum size specified in the LCZ description.

LCZ	Definition	Aerial View
LCZ 3: Compact low-rise STFID: 42101004101 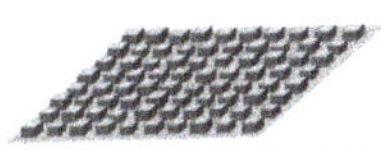	***Form:*** Attached or closely spaced buildings 1–3 stories tall. Buildings small and tightly packed along narrow streets. Sky view from street level significantly reduced. Heavy construction materials (stone, concrete, brick, tile); thick roofs and walls. Land cover mostly paved or hard-packed. Few or no trees. Moderate space heating/cooling demand. Low-moderate traffic flow. ***Function***: Residential (single-unit housing, high-density terrace/row housing); commercial (small retail shops). ***Location***: Old or densely populated cities, towns, or villages. Core (central or inner city); periphery (high-density sprawl).	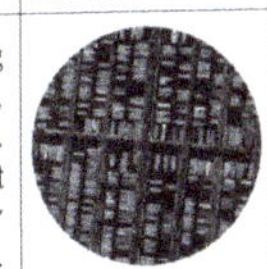
LCZ 6: Open low-rise STFID: 42101021700 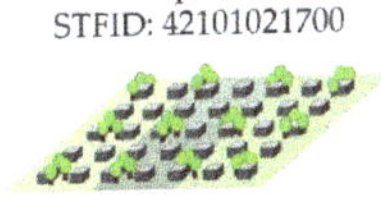	***Form:*** Small buildings 1–3 stories tall. Buildings detached or attached in rows, often in grid pattern. Sky view from street level slightly reduced. Construction materials vary (wood, brick, stone, tile). Scattered trees and abundant plant cover. Low space heating/cooling demand. Low traffic flow. ***Function***: Residential (single or multi-unit housing, low density terrace/row housing); commercial (small retail shops). ***Location***: City (medium density); periphery ("suburbs"). Commuter towns. Rural towns.	

Figure 6. LCZ description, data adapted from [52], aerial images from Google Earth (Philadelphia, PA).

Table 1 summarizes the details of each model for geometrical data and urban configuration. Although the buildings in the two models have almost the same height (6 to 7 m), other properties such as greenery area, site coverage by built elements, and pavement coverage are substantially different between the two models. Additionally, the canyon aspect ratio, height-to-width (H/W), for streets located within Block 1 is consistently 0.25; this ratio varies for streets in different directions within Block 2. Specifically, the H/W ratio in Block 2 ranges from 0.32 for primary East-West streets to 0.8 for secondary streets in the North-South and East-West directions.

In total, 180 h of medium forcing and parallel microclimate simulations were conducted for the two models. Each model was subject to six different weather scenarios, with four scenarios for each of the four seasons (winter, spring, summer, and fall), as well as two scenarios for extreme hot and cold days. The simulations were conducted separately for each scenario, using weather data obtained from the National Oceanic and Atmospheric Administration (NOAA) for the year 2021, which included hourly measurements of air temperature, relative humidity, wind speed, and wind direction and are listed in Table 2. To represent each season, average data for a month within each season (February, May, July, and October) were selected. The literature suggests beginning the simulation process during the early hours of the morning, specifically after sunrise, as this time is characterized by stable atmospheric conditions and minimal impact from the warming effects of the sun on the urban environment [33,53]. Therefore, the simulation for each case was started at 06:00 a.m. and lasted for 24 h to accurately reproduce the characteristics of the selected day.

Table 1. Model dimensions and site properties for two blocks.

Census Block	Model Dimensions	Site Properties							
		Site Area	Number of Buildings	Average Building Height	Site Coverage Ratio	Grass Coverage Ratio	Pavement Coverage Ratio	Road Coverage Ratio	H/W Ratio
Block 1	152 × 184 × 30 dx = 2 m dy = 2 m dz = 2 m	81,182 (m^2)	183	6 m	0.21	0.48	0.03	0.28	0.25
Block 2	132 × 137 × 30 dx = 2 m dy = 2 m dz = 2 m	53,026 (m^2)	302	7 m	0.42	0.02	0.18	0.22	Primary E-W: 0.32 Primary N-S: 0.5 Secondary N-S, E-W: 0.8

Table 2. Climatology data for 6 defined cases, source: NOAA.

Cases	T_{max} (°C)	T_{min} (°C)	T_{AVG} (°C)	Average Wind Speed (m/s)	Wind Direction
Winter Case	4.24	−1.84	1.23	4.35	NW
Coldest Day (2 February 2021)	0	−9.99	−4.99	7.07	NW
Summer Case	30	21.6	25.98	3.34	SW
Hottest Day (7 July 2021)	35.52	24.42	29.99	3.52	SW
Fall Case (24 October 2021)	17	7.8	12.47	3.7	SE
Spring Case (12 May 2021)	20.1	8.48	14.13	4.11	NW

3. Results and Discussion

To determine the climate variables and outdoor thermal comfort in various seasons, fine-level (2 m × 2 m) microclimate and the PMV index were simulated using the ENVI-met and BIO-met models, respectively. As it has been mentioned, 24 h microclimate simulation for six scenarios (winter, spring, summer, fall, extreme hot, and extreme cold) was run for the two models. It means, for example, the PMV inside one model was simulated 144 times to cover 24 h of the six scenarios. After conducting simulations for all six scenarios, the ENVI-met Leonardo tool was employed to visualize the results.

Figure 7 depicts the mean radiant temperature (MRT) simulated at a height of 1.5 m above the ground level at noon for the two blocks. MRT is considered the most important meteorological factor for evaluating outdoor thermal sensation under sunny conditions, regardless of the comfort index used [54]. The MRT range for both models is almost identical, with values exceeding 60 °C for streets that are fully exposed to solar radiation and constructed with asphalt. However, the effects of shading and evapotranspiration from trees result in a range of MRTs between 47 °C and 51 °C for areas below or close to trees (yellow to green areas), leading to an average MRT of 54.5 °C at street level in Block 1, while a very low number of trees in Block 2 led to an average simulated MRT of 61.5 °C. It should be noted that the average values are calculated for the entire street surfaces, including areas under trees and other surfaces made of asphalt and concrete pavement.

The influence of grass coverage on MRTs was negligible, with grassy areas exhibiting MRTs that were at most 2 °C lower than other surfaces in the two blocks. Additionally, the simulations revealed a minimum MRT of 35.41 °C and 38.47 °C and a maximum of 66.45 °C and 67.12 °C for Block 1 and Block 2, respectively.

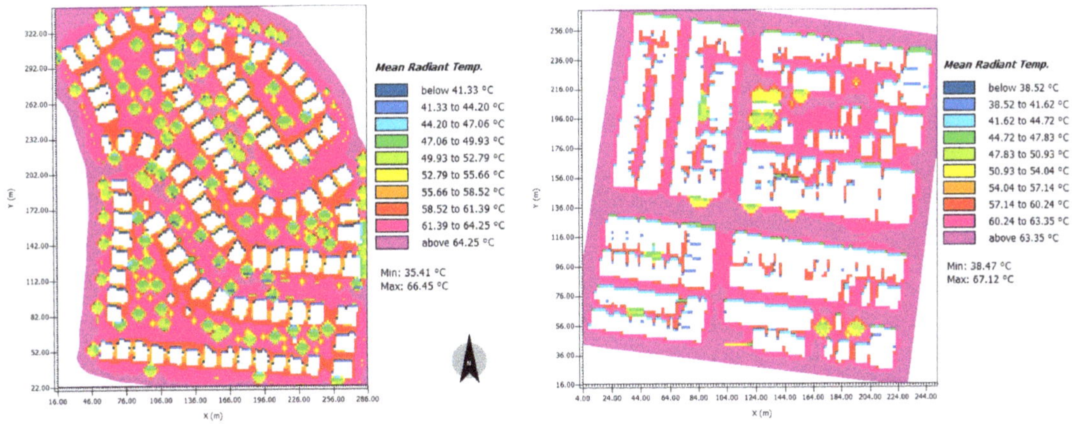

Figure 7. Simulated MRT at noon and 1.5 m above the ground level, summer case.

Figure 8 provides a visual comparison of surface temperatures for the summer case at noon. It should be noted that the dark blue areas represent the footprint of buildings and are excluded from the surface temperature analysis. These areas are not reflective of the surface material or vegetation cover and, therefore, would not provide accurate information on surface temperatures. However, the maps show significant differences in surface temperatures across various materials, such as asphalt, concrete pavement, and grass areas. Among these materials, the largest temperature difference is observed in asphalt, where the temperature exceeds 47 °C (purple and red areas). In contrast, the lowest temperature of 23 °C (blue areas) is recorded on the surface right below trees. Additionally, the surface temperature of grass areas was simulated at a range between 34.5 °C to 38 °C (yellow areas). Accordingly, Block 1, with almost 48% covered by grass, benefits from shading and cooling effects that mitigate heat buildup on surfaces compared to Block 2, with only 0.02% grass coverage but 40% covered by impervious surfaces such as asphalt and concrete pavement.

The output files from ENVI-met, along with two personal factors, i.e., metabolism rate and clothing index for each season were incorporated into ENVI-met BioMet to simulate the PMV indices. The PMV model is based on heat balance principles and connects six crucial elements of thermal comfort to the average human response on a specified scale. Normally, the PMV scale is defined on a seven-point thermal sensation scale ranging from cold (−3) to neutral or comfort (0) to hot (+3).

In Figure 9, the output of the ENVI-met model is shown as PMV index maps for the summer scenario at four different hours of the day: 10:00 am, 02:00 pm, 06:00 pm, and 10:00 pm. Maps in Figure 9 serve as examples to demonstrate the variations of comfort index across the models' area at different hours, and findings are explained in the following sections.

PMV values at 10:00 am, summer

According to the results of the simulations, the PMV values at street level and grass areas that were fully exposed to the sun were above 3, which indicates a hot environment and is represented by red to pink colors in the output maps of both models. For areas near or directly below trees, PMV values were recorded in a range from slightly warm (PMV = 1.5) to warm (PMV = 2.5), as indicated by the blue to yellow colors in the maps. For areas located 2 to 4 m to the north or northwest of the building geometries, the PMV values were recorded at the lowest value, below 2, which represents a slightly warm environment. However, no spot was found within the two models that had a PMV value lower than 1, which is the comfort range.

PMV values at 02:00 pm, summer

Based on the ENVI-met output maps at 2:00 pm, it can be observed that the entire areas of the two blocks showed PMV values higher than 2.5, which indicates a warm environment. These PMV values were recorded in a few spots located under trees or in areas 2 to 4 m to the north or northeast of the buildings, as shown by light to dark blue colors in the maps. However, spots located on grass areas, asphalt, and concrete pavement all showed PMV values higher than 4, which represents a very hot environment. This indicates that the shading effects of the trees were able to provide some relief from the heat, but other areas that were fully exposed to solar radiation experienced extremely hot conditions.

PMV values at 06:00 pm, summer

During the simulation at 6:00 pm, there were still no spots found in both Block 1 or Block 2 with PMV values within the comfort range. However, spots that were affected by trees or located at a distance of 2 to 10 m on the east side of the buildings showed PMV values between 1 and 2, indicating a slightly warm environment. Based on the results, the shading effect of trees and buildings had a significant impact on reducing the PMV index compared to spots on the street level, which had PMV values higher than 3, indicating a hot environment.

Upon comparing the maps of the two blocks, it was observed that Block 1, which had more trees, had more areas with blue color (light to dark), indicating a lower average of PMV values compared to Block 2.

PMV values at 10:00 pm, summer

The simulation results for 10:00 pm indicate that nearly all spots within Blocks 1 and 2 have PMV values ranging between −0.5 to +0.5, which is within the comfort range. This suggests that trees and buildings have no significant impact on the PMV values after sunset. It should be noted that this result is observed regardless of the location of the spot within the two blocks, whether it is influenced by trees or buildings. Overall, the PMV values at 10:00 pm suggest that the thermal comfort conditions have improved significantly within the urban environment of the two blocks.

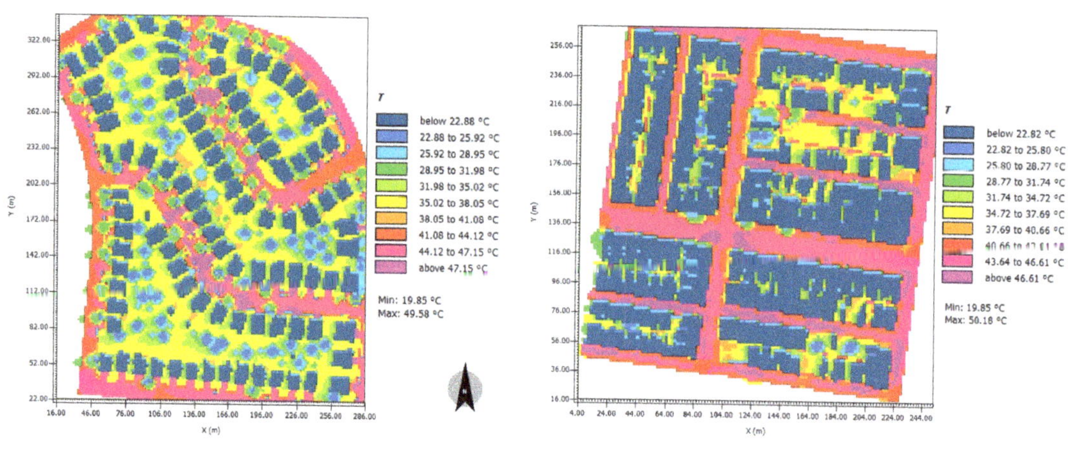

Figure 8. Simulated Surface temperature at noon, summer case.

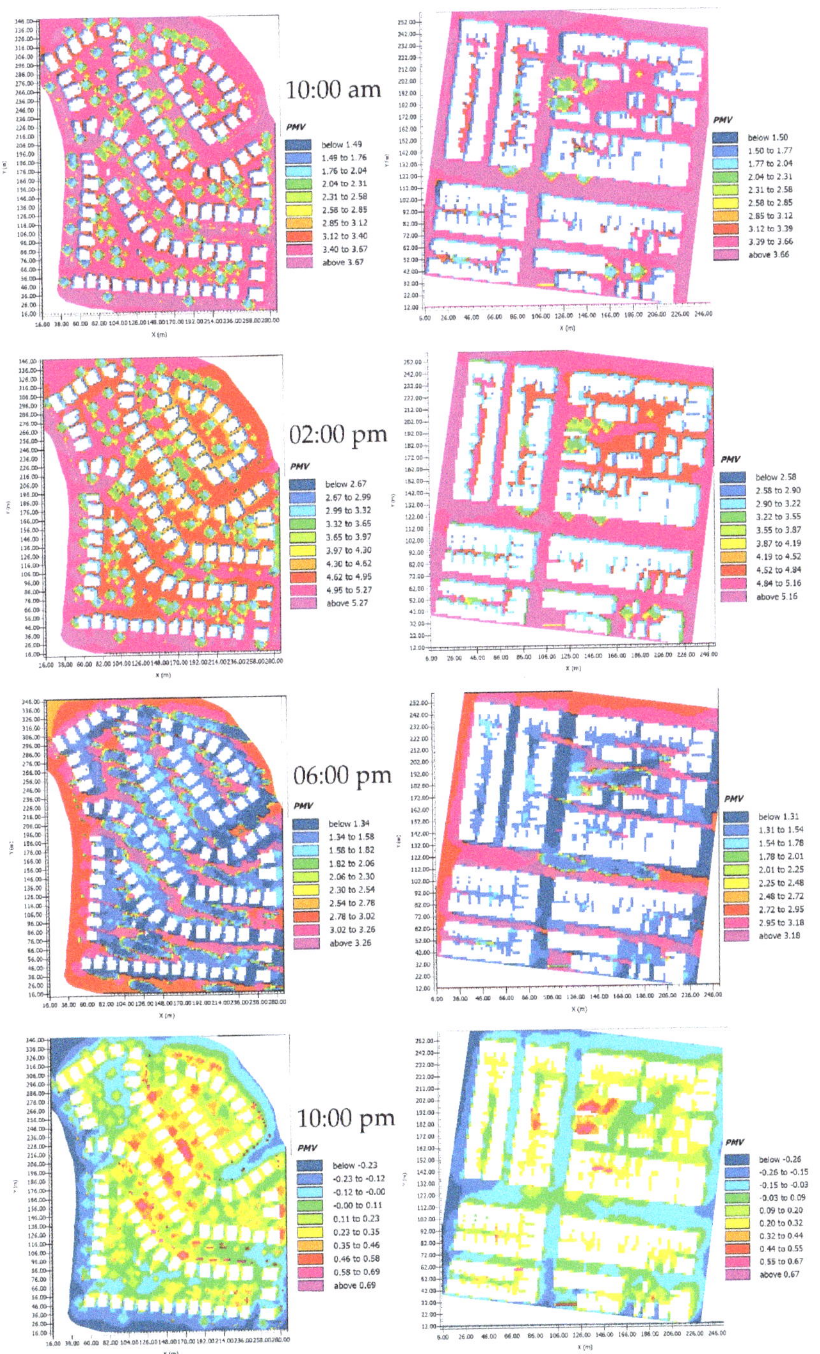

Figure 9. Simulated PMV inside the two models during the summer case for every 4 h starting at 10 am and ending by 10 pm.

The findings from the summer scenario and the other five weather scenarios demonstrate a strong correlation between tree and grass coverage and outdoor air temperature, mean radiant temperature, and surface temperature during hot seasons and extreme heat events. Areas with a higher concentration of tree canopies experienced a maximum reduction of 1.5 °C in outdoor air temperatures and up to 31 °C in mean radiant temperatures during the daytime of summer and extremely hot conditions. The effects of greenery were more significant in Block 1 (associated with a low SVI), where nearly 50% of the site was covered by trees and grass, compared to only 0.02% of Block 2 (associated with a high SVI). However, the benefits of tree and grass coverage were limited to their immediate surroundings, with areas without green coverage exhibiting significantly higher mean radiant temperatures.

Furthermore, the results showed a strong relationship between PMV and MRT, with a Pearson correlation analysis indicating a significant correlation of 0.987. A significant correlation of 0.859 was also found between air temperature and PMV. The study also found that a higher tree canopy cover in Block 1 was associated with lower outdoor heat exposure in that area, while more impervious surfaces such as streets and pavements in Block 2 were linked to higher heat exposure in that area.

To obtain a more accurate representation of the simulated climate variables and comfort index in the two models, the study included nine specific points for closer investigation, as shown in Figure 10. These points were strategically placed in primary and secondary streets, on asphalt surfaces, sidewalks, and under trees, with the street direction either running from North to South or East to West. The nine specific points were named as follows:

- North to South primary street, on asphalt (N-S)PA
- East to West primary street, on asphalt (E-W)PA
- North to South secondary street, on asphalt (N-S)SA
- East to West secondary street, on asphalt (E-W)SA
- North to South primary street, on pavement (N-S)PP
- East to West primary street, on pavement (E-W)PP
- North to South secondary street, on pavement (N-S)SP
- East to West secondary street, on pavement (E-W)SP
- Under Tress (TRS)

Figure 10. Location of nine points inside the two models.

By selecting points in different locations, the study was able to gather data on the varying effects of different surface types and orientations, which allowed for a more nuanced analysis of the simulated climate variables and their impact on human comfort. Figures 11 and 12 illustrate hourly MRTs simulated at each point inside Block 1 and 2 during the summer case, respectively. The simulation was programmed to take measurements every hour, starting at 6:00 am and ending at 6:00 am, 24 h overall.

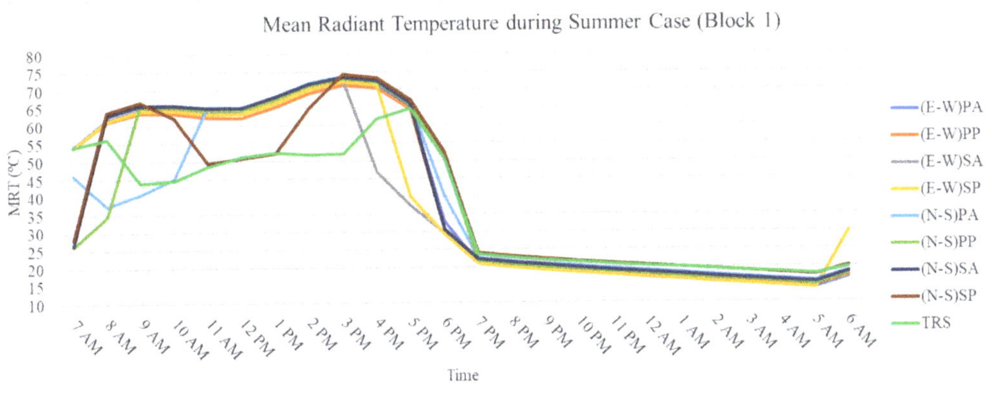

Figure 11. MRTs during summer case at nine points inside Block 1.

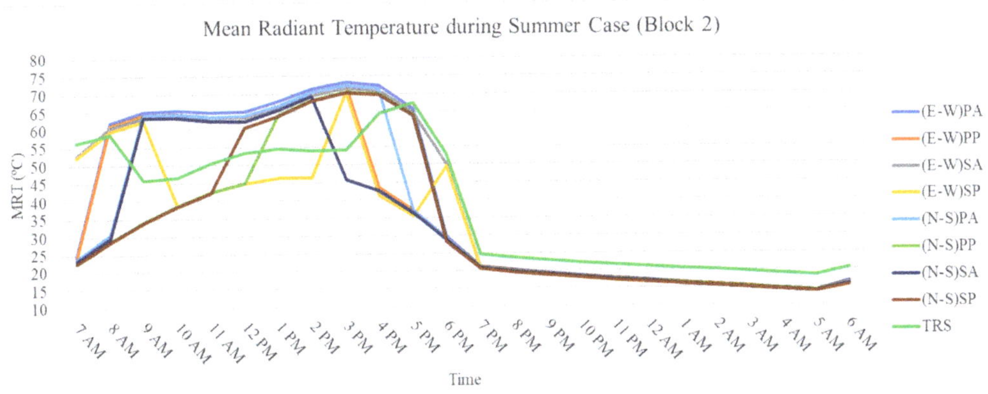

Figure 12. MRTs during summer case at nine points inside Block 2.

The points located under trees showed lower mean radiant temperatures (MRTs) throughout the daytime compared to other points. For further investigation of the impact of other factors on MRT and PMV, the other eight points were selected at places without tree and greenery effects. The urban morphology parameters, including Sky View Factor (SVF), Albedo, and H/W ratio, were recorded at each point and analyzed for their effect on MRT and PMV (Table 3). These parameters describe the physical characteristics of urban environments, and they can affect various aspects of the urban environment, including microclimate and human comfort.

Table 3. SVF, Albedo, and H/W ratio at each point inside two blocks.

	Blocks	Sky View Factor (SVF)	Surface Albedo	H/W Ratio
(E-W)PA	Block 1	0.84	0.2	0.25
	Block 2	0.78	0.2	0.32
(E-W)PP	Block 1	0.75	0.35	0.25
	Block 2	0.68	0.35	0.32
(E-W)SA	Block 1	0.84	0.2	0.25
	Block 2	0.62	0.2	0.8
(E-W)SP	Block 1	0.81	0.35	0.25
	Block 2	0.52	0.35	0.8
(N-S)PA	Block 1	0.61	0.2	0.25
	Block 2	0.70	0.2	0.5
(N-S)PP	Block 1	0.72	0.35	0.25
	Block 2	0.52	0.35	0.5
(N-S)SA	Block 1	0.66	0.2	0.25
	Block 2	0.61	0.2	0.8
(N-S)SP	Block 1	0.45	0.35	0.25
	Block 2	0.51	0.35	0.8

In Block 1, six points showed the approximately same amount of MRT during the daytime, while the other three points, including (N-S)PA, (N-S)SP, and TRS, showed lower MRT values at certain times between 7:00 am and 7:00 pm. For instance, the point located on the primary asphalt in the North-South direction, (N-S)PA, showed a value of 45 °C at 7:00 am, 37 °C at 8:00 am, 40 °C at 9:00 am, and 45 °C at 10:00 am. For the rest of the day, MRT recorded at this point was similar to that of other points, which was above 65 °C. The MRT at the point located on the pavement surface of the secondary street in the North-South direction, (N-S)SP, fluctuated between 65 °C at 9:00 am and 50 °C at noon, returning to the same amount as other points after 3:00 pm. The point located under a tree (TRS) had lower MRT throughout the daytime, with a maximum value of 20 °C at 9:00 am compared to the other points.

In Block 2, MRT recorded at two points was lower than other points, including the TRS point, during the morning. These two points were both in the North-South direction and located on both asphalt and pavement surfaces in a secondary street. These two points were well-shaded by the surrounding building, especially buildings on the right side that blocked east sun radiation.

Although the H/W ratio was constant (0.25 for all the points) inside Block 1, SVF and albedo showed a strong effect on MRTs and PMV. A Pearson correlation of 0.61 was found between SVF and MRTs inside Block 1, while it was found up to 0.75 inside Block 2. It means that points with higher SVF have more exposure to heat, resulting in higher MRTs and PMV. Conversely, the urban morphology parameter of albedo displayed a negative correlation of −0.60 with MRTs inside Block 1 and −0.55 inside Block 2, indicating a significant yet inverse relationship. Thus, it can be inferred that higher albedo values correspond to lower MRTs. In Model 2, where different streets had varying H/W ratios, the coefficient with MRTs was −0.39, suggesting that this parameter also has a notable impact on MRT values.

4. Conclusions

This research presented a method that integrates a climate classification system with a 3-D microclimate model to investigate the correlation between outdoor urban climate and the socioeconomic status of urban blocks, using two urban blocks in Philadelphia, Pennsylvania, USA, as the study case. The study involved creating and simulating microclimate models in two urban blocks, which were extracted from two census tracts with varying vulnerability indices. The results demonstrated that urban morphological factors such as vegetated areas, sky view factor (SVF), albedo, and height-to-width (H/W) ratio have an influence on air temperature, mean radiant temperature (MRT), and predicted mean vote (PMV).

The study revealed that trees have a significant cooling effect on outdoor air temperature, MRT, and PMV index during the summer day scenario. These findings are consistent with previous research [55–57] that highlighted the beneficial impacts of urban vegetation and roadside trees on the microclimate by reducing surface temperature and MRT, increasing evapotranspiration, providing shading, and reducing PMV. However, the study also found that the benefits of trees are limited to their immediate surroundings, with areas lacking green coverage exhibiting significantly higher MRTs. Consistent with the finding from [56], this study shows the effect of trees on air temperature decreases with increasing distance from the tree canopy.

The study also revealed that the impact of the H/W ratio and surface albedo on outdoor thermal comfort is significant, with higher H/W ratios and albedo leading to lower MRT and greater comfort levels, which is consistent with findings from [53,58,59]. Furthermore, the results presented in this research align with those of a study conducted by [60], which also indicated that factors related to urban morphology, such as building density and orientation, play a notable role in influencing microclimate parameters and thermal comfort.

In conclusion, it is recommended that the utilization of taller and denser buildings with higher albedo pavement surfaces should be considered to enhance thermal comfort conditions, particularly in urban blocks with limited vegetation coverage. The proposed workflow provides evidence to support incorporating urban climate knowledge into the planning and design of healthful and resilient urban environments. Overall, the study highlights the importance of considering urban morphology and vegetation coverage in the design of sustainable and livable urban environments.

This research could be extended in four directions. First, the research involved a 24-h simulation cycle, while it is recommended for future studies to extend the timeframe to 48-h or 72-h cycles to obtain a more comprehensive understanding. Second, the ENVI-met models were supplemented with generic data from the LCZ datasheet for building typology and properties. However, to improve accuracy, genuine information on building materials, A/C types, green roofs and walls, and blue (water) surfaces, if available within the studied blocks, can be added to the models. Third, the research solely concentrated on outdoor weather variables and comfort levels. However, researchers in this era are interested in indoor comfort, which can be investigated in future works. Last, the research concentrated on Philadelphia, PA, from the U.S. Northeast region. However, to gain a better understanding of the effects of urban properties on outdoor climate variables and thermal comfort inside urban blocks with varying social vulnerability indexes, this research should be expanded to incorporate more cities from other climate zones, including hot and dry or extremely cold.

Author Contributions: Conceptualization, F.H., U.P., L.D.I. and G.C.; methodology, F.H. and U.P. and G.C.; software, F.H.; writing—original draft preparation, F.H.; review and editing, U.P., L.D.I. and G.C. All authors have read and agreed to the published version of the manuscript.

Funding: This research was supported in part by the National Science Foundation (Award #2207436), the USDA National Institute of Food and Agriculture and Multistate Research Project (#PEN04623; Accession #1013257), the Eunice Kennedy Shriver National Institute of Child Health and Human Development (Award #P2C HD041025), and the 2021 Institute of Energy and the Environment (IEE) seed grant program at the Pennsylvania State University.

Data Availability Statement: The data presented in this study are available on request from the corresponding author.

Acknowledgments: The authors are grateful for the support from the Hamer Center for Community Design, a research arm of Stuckeman School, and the Department of Architecture at the Pennsylvania State University. Additionally, the authors acknowledge the contributions of Yosef Bodovski, a GIS research analyst at Penn State's Computational and Spatial Analysis (CSA) Core, for his aid with the GIS maps.

Conflicts of Interest: The authors declare no conflict of interest.

References

1. Masson-Delmotte, T.W.; Zhai, P.; Pörtner, H.-O.; Roberts, D.; Skea, J.; Shukla, P.R.; Pirani, A.; Moufouma-Okia, W.; Péan, C.; Pidcock, R.; et al. Global Warming of 1.5 °C. An IPCC Special Report on the Impacts of Global Warming of 1.5 °C above Pre-industrial Levels and Related Global Greenhouse Gas Emission Pathways, in the Context of Strengthening the Global Response to the Threat of Climate Change. 2018. Available online: https://www.ipcc.ch/sr15/ (accessed on 17 February 2022).
2. P. D. United Nations, Department of Economic & Social Affairs. *World Urbanization Prospects: The 2018 Revision*; United Nations: New York, NY, USA, 2019; Available online: https://population.un.org/wup/publications/Files/WUP2018-Report.pdf (accessed on 13 March 2022).
3. Taha, H. Urban climates and heat islands: Albedo, evapotranspiration, and anthropogenic heat. *Energy Build.* **1997**, *25*, 99–103. [CrossRef]
4. Unger, J. Connection between urban heat island and sky view factor approximated by a software tool on a 3D urban database. *Int. J. Environ. Pollut.* **2009**, *36*, 59. [CrossRef]
5. Landsberg, H.E. Urban energy fluxes. In *The Urban Climate*; University of Maryland: College Park, MD, USA, 1981; pp. 53–82.
6. Oke, T.R. The energetic basis of the urban heat island. *Q. J. R. Meteorol. Soc.* **1982**, *108*, 1–24. [CrossRef]
7. Taha, H. Characterization of Urban Heat and Exacerbation: Development of a Heat Island Index for California. *Climate* **2017**, *5*, 59. [CrossRef]
8. Santamouris, M. Science of the Total Environment Analyzing the heat island magnitude and characteristics in one hundred Asian and Australian cities and regions. *Sci. Total Environ.* **2015**, *512–513*, 582–598. [CrossRef]
9. Peng, S.; Piao, S.; Ciais, P.; Friedlingstein, P.; Ottle, C.; Bréon, F.-M.; Nan, H.; Zhou, L.; Myneni, R.B. Surface Urban Heat Island Across 419 Global Big Cities. *Environ. Sci. Technol.* **2011**, *46*, 696–703. [CrossRef] [PubMed]
10. Estoque, R.C.; Murayama, Y.; Myint, S.W. Science of the Total Environment Effects of landscape composition and pattern on land surface temperature: An urban heat island study in the megacities of Southeast Asia. *Sci. Total. Environ.* **2017**, *577*, 349–359. [CrossRef] [PubMed]
11. Ho, H.C.; Knudby, A.; Chi, G.; Aminipouri, M.; Lai, D.Y.-F. Spatiotemporal analysis of regional socio-economic vulnerability change associated with heat risks in Canada. *Appl. Geogr.* **2018**, *95*, 61–70. [CrossRef] [PubMed]
12. Khatana, S.A.M.; Werner, R.M.; Groeneveld, P.W. Association of Extreme Heat with All-Cause Mortality in the Contiguous US, 2008–2017. *JAMA Netw. Open* **2022**, *5*, e2212957. [CrossRef]
13. Centers for Disease Control and Prevention. National Environmental Public Health Tracking Network. 2023. Available online: www.cdc.gov/ephtracking (accessed on 10 May 2022).
14. Anderson, G.B.; Bell, M.L. Heat Waves in the United States: Mortality Risk during Heat Waves and Effect Modification by Heat Wave Characteristics in 43 U.S. Communities. *Environ. Health Perspect.* **2011**, *119*, 210–218. [CrossRef]
15. Heaton, M.J.; Sain, S.R.; Greasby, T.A.; Uejio, C.K.; Hayden, M.H.; Monaghan, A.J.; Boehnert, J.; Sampson, K.; Banerjee, D.; Nepal, V.; et al. Characterizing urban vulnerability to heat stress using a spatially varying coefficient model. *Spat. Spatio-Temporal Epidemiol.* **2014**, *8*, 23–33. [CrossRef] [PubMed]
16. Madrigano, J.; Ito, K.; Johnson, S.; Kinney, P.L.; Matte, T. A Case-Only Study of Vulnerability to Heat Wave—Related Mortality. *Environ. Health Perspect.* **2015**, *123*, 672–678. [CrossRef] [PubMed]
17. Hsu, A.; Manya, D. Intensity across Major US Cities. *Nat. Commun.* **2021**, *12*, 2721. [CrossRef] [PubMed]
18. Yang, W.; Wong, N.H.; Jusuf, S.K. Thermal comfort in outdoor urban spaces in Singapore. *Build. Environ.* **2013**, *59*, 426–435. [CrossRef]
19. Aljawabra, F.; Nikolopoulou, M. Thermal comfort in urban spaces: A cross-cultural study in the hot arid climate. *Int. J. Biometeorol.* **2018**, *62*, 1901–1909. [CrossRef]
20. Salata, F.; Golasi, I.; Vollaro, R.D.L.; Vollaro, A.D.L. Outdoor thermal comfort in the Mediterranean area. A transversal study in Rome, Italy. *Build. Environ.* **2016**, *96*, 46–61. [CrossRef]
21. Huang, J.; Zhou, C.; Zhuo, Y.; Xu, L.; Jiang, Y. Outdoor thermal environments and activities in open space: An experiment study in humid subtropical climates. *Build. Environ.* **2016**, *103*, 238–249. [CrossRef]
22. Xu, M.; Hong, B.; Mi, J.; Yan, S. Outdoor thermal comfort in an urban park during winter in cold regions of China. *Sustain. Cities Soc.* **2018**, *43*, 208–220. [CrossRef]
23. Hirashima, S.Q.D.S.; de Assis, E.S.; Nikolopoulou, M. Daytime thermal comfort in urban spaces: A field study in Brazil. *Build. Environ.* **2016**, *107*, 245–253. [CrossRef]
24. Middel, A.; Selover, N.; Hagen, B.; Chhetri, N. Impact of shade on outdoor thermal comfort—A seasonal field study in Tempe, Arizona. *Int. J. Biometeorol.* **2016**, *60*, 1849–1861. [CrossRef]

25. Chi, G.Q.; Ho, H.C. Population stress: A spatiotemporal analysis of population change and land development at the county level in the contiguous United States, 2001–2011. *Land Use Policy* **2018**, *70*, 128–137. [CrossRef]
26. Chi, J.; Zhu, G. *Spatial Regression Models for the Social Sciences*; SAGE Publications: Thousand Oaks, CA, USA, 2019.
27. Toparlar, Y.; Blocken, B.; Maiheu, B.; van Heijst, G. A review on the CFD analysis of urban microclimate. *Renew. Sustain. Energy Rev.* **2017**, *80*, 1613–1640. [CrossRef]
28. Crank, P.J.; Sailor, D.J.; Ban-Weiss, G.; Taleghani, M. Evaluating the ENVI-met microscale model for suitability in analysis of targeted urban heat mitigation strategies. *Urban Clim.* **2018**, *26*, 188–197. [CrossRef]
29. Ozkeresteci, I.; Crewe, K.; Brazel, A.J.; Bruse, M. Use and evaluation of the ENVI-met model for environmental design and planning. An Experiment on Lienar Parks. In Proceedings of the 21st International Cartographic Conference (ICC), Durban, South Africa, 10–16 August 2003; pp. 10–16.
30. Rodríguez Algeciras, J.A.; Gómez Consuegra, L.; Matzarakis, A. Spatial-temporal study on the effects of urban street configurations on human thermal comfort in the world heritage city of Camagüey-Cuba. *Build. Environ.* **2016**, *101*, 85–101. [CrossRef]
31. Sinsel, T. *Advancements and Applications of the Microclimate Model ENVI-Met*; Johannes Gutenberg-Universitat Mainz: Mainz, Germany, 2022. [CrossRef]
32. Ayyad, Y.N.; Sharples, S. Envi-MET validation and sensitivity analysis using field measurements in a hot arid climate. *IOP Conf. Series Earth Environ. Sci.* **2019**, *329*, 012040. [CrossRef]
33. Acero, J.A.; Arrizabalaga, J. Evaluating the performance of ENVI-met model in diurnal cycles for different meteorological conditions. *Theor. Appl. Clim.* **2018**, *131*, 455–469. [CrossRef]
34. Acero, J.A.; Herranz-Pascual, K. A comparison of thermal comfort conditions in four urban spaces by means of measurements and modelling techniques. *Build. Environ.* **2015**, *93*, 245–257. [CrossRef]
35. Alsaad, H.; Hartmann, M.; Hilbel, R.; Voelker, C. ENVI-met validation data accompanied with simulation data of the impact of facade greening on the urban microclimate. *Data Brief* **2022**, *42*, 108200. [CrossRef]
36. Morakinyo, T.E.; Lai, A.; Lau, K.K.-L.; Ng, E. Thermal benefits of vertical greening in a high-density city: Case study of Hong Kong. *Urban For. Urban Green.* **2019**, *37*, 42–55. [CrossRef]
37. Forouzandeh, A. Numerical modeling validation for the microclimate thermal condition of semi-closed courtyard spaces between buildings. *Sustain. Cities Soc.* **2018**, *36*, 327–345. [CrossRef]
38. Elraouf, R.A.; Elmokadem, A.; Megahed, N.; Eleinen, O.A.; Eltarabily, S. Evaluating urban outdoor thermal comfort: A validation of ENVI-met simulation through field measurement. *J. Build. Perform. Simul.* **2022**, *15*, 268–286. [CrossRef]
39. US Census Bureau. 2022. Available online: https://www.census.gov/quickfacts/fact/table/US/PST045222 (accessed on 23 January 2023).
40. Mayor's Office of Sustainability and ICF International. Growing Stronger: Toward a Climate-Ready Philadelphia. 2015. Available online: https://www.phila.gov/documents/growing-stronger-toward-a-climate-ready-philadelphia/ (accessed on 1 March 2023).
41. Schade, R.S. *Philadelphia Rowhouse Manual: A Practical Guide for Homeowners*; The City of Philadelphia: Philadelphia, PA, USA, 2008. Available online: https://www.phila.gov/media/20190521124726/Philadelphia_Rowhouse_Manual.pdf (accessed on 18 July 2021).
42. Baechler, P.M.; Michael, C.; Williamson, J.L.; Gilbride, T.L.; Cole, P.C.H.; Marye, G.L. Building America Best Practices Series: Volume 7.1: Guide to Determining Climate Regions by County. 2010. Available online: https://www.osti.gov/biblio/1068658 (accessed on 22 July 2021).
43. Centers for Disease Control and Prevention/Agency for Toxic Substances and Disease Registry/Geospatial Research, Analysis, and Services Program. 2018. Available online: https://www.atsdr.cdc.gov/placeandhealth/svi/data_documentation_download.html (accessed on 15 June 2021).
44. U.S. Census Bureau. 2009. Available online: https://www.census.gov/library/publications/2008/compendia/statab/128ed.html (accessed on 21 August 2021).
45. Bruse, M.; Fleer, H. Simulating surface–plant–air interactions inside urban environments with a three dimensional numerical model. *Environ. Model. Softw.* **1998**, *13*, 373–384. [CrossRef]
46. Shaw, E.W. Thermal Comfort: Analysis and applications in environmental engineering, by P. O. Fanger. 244 pp. DANISH TECHNICAL PRESS. Copenhagen, Denmark, 1970. Danish Kr. 76, 50. *R. Soc. Health J.* **1972**, *92*, 164. [CrossRef]
47. Havenith, G.; Holmér, I.; Parsons, K. Personal factors in thermal comfort assessment: Clothing properties and metabolic heat production. *Energy Build.* **2002**, *34*, 581–591. [CrossRef]
48. Pennsylvania Spatial Data Access PASDA. The Pennsylvania State University. 2016. Available online: http://www.pasda.psu.edu/ (accessed on 15 August 2021).
49. Hashemi, F.; Marmur, B.; Passe, U.; Thompson, J. Developing a workflow to integrate tree inventory data into urban energy models. In *SimAUD 2018 Symposium on Simulation for Architecture and Urban Design*; Simulation Series; TU Delft: Delft, The Netherlands, 2018; Volume 50.
50. Hashemi, F.; Iulo, L.D.; Poerschke, U. A novel approach for investigating canopy heat island effects on building energy performance: A case study of Center City of Philadelphia, PA. In Proceedings of the AIA/ACSA Intersections Research Conference: CARBON, Fall Conference Proceedings, Washington, DC, USA, 30 September – 2 October 2020; pp. 197–203. [CrossRef]

51. Roudsari, M.S.; Pak, M.; Smith, A.; Gill, G. Ladybug: A Parametric Environmental Plugin For Grasshopper To Help Designers Create An Environmentally-conscious Design. In Proceedings of the Building Simulation, Chambéry, France, 26–28 August 2013; pp. 3128–3135. [CrossRef]
52. Stewart, I.D.; Oke, T.R. Local Climate Zones for Urban Temperature Studies. *Bull. Am. Meteorol. Soc.* **2012**, *93*, 1879–1900. [CrossRef]
53. Ali-Toudert, F.; Mayer, H. Numerical study on the effects of aspect ratio and orientation of an urban street canyon on outdoor thermal comfort in hot and dry climate. *Build. Environ.* **2006**, *41*, 94–108. [CrossRef]
54. de Abreu-Harbich, L.V.; Labaki, L.C.; Matzarakis, A. Effect of tree planting design and tree species on human thermal comfort in the tropics. *Landsc. Urban Plan.* **2015**, *138*, 99–109. [CrossRef]
55. Gatto, E.; Buccolieri, R.; Aarrevaara, E.; Ippolito, F.; Emmanuel, R.; Perronace, L.; Santiago, J.L. Impact of Urban Vegetation on Outdoor Thermal Comfort: Comparison between a Mediterranean City (Lecce, Italy) and a Northern European City (Lahti, Finland). *Forests* **2020**, *11*, 228. [CrossRef]
56. Zaki, S.A.; Toh, H.J.; Yakub, F.; Saudi, A.S.M.; Ardila-Rey, J.A.; Muhammad-Sukki, F. Effects of Roadside Trees and Road Orientation on Thermal Environment in a Tropical City. *Sustainability* **2020**, *12*, 1053. [CrossRef]
57. Bernabé, A.; Bernard, J.; Musy, M.; Andrieu, H.; Bocher, E.; Calmet, I.; Kéravec, P.; Rosant, J.-M. Radiative and heat storage properties of the urban fabric derived from analysis of surface forms. *Urban Clim.* **2015**, *12*, 205–218. [CrossRef]
58. Elraouf, R.A.; Elmokadem, A.; Megahed, N.; Eleinen, O.A.; Eltarabily, S. The impact of urban geometry on outdoor thermal comfort in a hot-humid climate. *Build. Environ.* **2022**, *225*, 109632. [CrossRef]
59. Zhang, Y.; Du, X.; Shi, Y. Effects of street canyon design on pedestrian thermal comfort in the hot-humid area of China. *Int. J. Biometeorol.* **2017**, *61*, 1421–1432. [CrossRef] [PubMed]
60. Zaki, S.; Azid, N.; Shahidan, M.; Hassan, M.; Daud, M.; Abu Bakar, N.; Salim, S.A.Z.S.; Yakub, F. Analysis of Urban Morphological Effect on the Microclimate of the Urban Residential Area of Kampung Baru in Kuala Lumpur Using a Geospatial Approach. *Sustainability* **2020**, *12*, 7301. [CrossRef]

buildings

MDPI

Article

A Parametric Approach for Optimizing Design Solutions in Urban Regeneration and Reshaping: An Application to a District Block in Bologna, Italy

Cecilia Mazzoli [1,*], Dimitra Papadaki [2] and Annarita Ferrante [1]

1 DA—Department of Architecture, University of Bologna, 40136 Bologna, Italy; annarita.ferrante@unibo.it
2 Group of Building Environmental Studies, Physics Department, University of Athens, 15784 Athens, Greece; dpapadaki@phys.uoa.gr
* Correspondence: cecilia.mazzoli@unibo.it

Abstract: This paper addresses the deep renovation and enhancement of energy efficiency in existing urban areas, aiming to transform them into zero energy districts, utilizing renewable sources. This transformative process is essential to align with contemporary models for new buildings and to elevate existing structures to meet environmental housing standards mandated by current regulations. The proposed densification strategy is motivated by the necessity to augment the real estate value and enhance the architectural and performance quality of the existing building stock, all while minimizing additional land occupation. Its efficacy has been rigorously tested and demonstrated within the European Horizon 2020 project "ABRACADABRA". The key actions advocated for in this approach revolve around the incorporation of renewable energies, volumetric expansion, and completion volumes (urban infill), within the Positive Energy District. The validation of this process occurs on a district scale in a social housing urban block located in Bologna (Italy), with currently poor environmental performance. A multi-criteria analysis was employed to examine the architectural, climatic, and energy context. The meta-design phase successfully attained the energy and economic targets set by the project and by local and European regulations settings, offering an optimized design solutions for reshaping the built environment.

Keywords: urban densification; volumetric additions; urban infill; positive energy district; Bologna; parametric design tools; multi-criteria analysis

1. Introduction

One of the most obvious phenomena manifested that can be seen in the medium- sized and large cities of our time is urban sprawl. The phenomenon of urban sprawl describes a certain peculiar settlement processes, that takes place mainly in the peripheral areas of consolidated urban cores. These processes are the result of a lack of foresight in spatial planning, which leads to high land consumption and considerable costs for society, both in environmental and economic social terms.

This formless urban growth, in other words periphery, peri-urban, agglomeration, urban fog, exurbia (i.e., diffuse city, in which the size of the city has exceeded the recognizable measure), as Ingersoll defines "urban sprawl", occurs in the peripheral areas of cities, and combines uncontrolled urban expansion in these areas with the phenomenon of low population density [1].

After the Second World War, the uncontrolled expansion of cities in Europe was initially curbed due to a new awareness of the historical building fabric damaged by the war, and consolidated but incisive spatially. At the same time, large housing estates were planned and built in the major cities. The European diffusion processes are of various kinds: economic, demographic, housing policy, and infrastructural. Nevertheless, two aspects can be assumed. Firstly, that agricultural areas on the outskirts of cities are subject to diffusion, due to their low cost, compared to already urbanized areas. Secondly, the decrease in the attractiveness of urban centres in contrast to the increase in the attractiveness of rural areas.

There is a clear need for the development of policies and strategies on land consumption and fragmentation of urban areas, that will lead to altered ecological balances, with less impact on local flora and fauna. Nevertheless, with the New Green Deal [2], the European Union (EU) is carefully addressing issues of decarbonization, energy sustainability, and the use of renewable energy sources [3,4] through processes of urban regeneration, as well as the realization of new buildings. However, diffuse processes must be integrated into a legal framework that regulates them according to sustainable planning methods.

In European policy, the role of the city and the management of its resources are key elements for future sustainable growth. Urban regeneration is a coordinated set of actions not only for urban planning and construction, but also for the development of the social, productive, and cultural fabric of urban areas, with a view to environmental sustainability and strengthening the resilience of cities.

In terms of land use, agricultural use and open spaces should be minimized as much as possible. Nevertheless, there is a need for new housing units, but if land is withdrawn, it should be ensured that this can be compensated for elsewhere. For example, unused land could be recultivated or denaturalized so that it can once again provide the ecosystem services of unsealed soils [5].

This process also introduces the principles of recycling and the circular economy, which are already expressed in the Europe 2020 strategy [6], into urban and spatial planning, with the ultimate aim of decoupling urban development from the consumption of land resources [7]. The concept of circular economy aims to transform current production and consumption patterns that place a significant burden on our planet and its environmental capacity [8]. With the Communication Roadmap to a Resource-Efficient Europe [5,9], the European Commission (EC) has determined that European policies should take into account the direct and indirect impact of land use change by setting the ambitious goal of 'zero net land take' by 2050 [5,10]. Comprehensive interventions in buildings, in line with the objectives of the European directives, avoid further land use and can initiate urban and social regeneration processes towards future sustainable cities [11].

Among the various instruments promoted for the regeneration of urban fabric is urban densification, which is seen as a process of selective intervention in pre-existing urban fabrics. These include different construction strategies, such as smart growth, the formation of communities of a manageable size, and the careful development of social needs. This positively influences the containment of the use of green spaces and the expansion of built-up areas at the expense of agricultural areas. A compact city with a high population density is the best answer to the need to accommodate growing population masses. Although Europe's population growth is declining, other forces may be driving the growth in housing demand. One of the causes is the increase in smaller families, a phenomenon that can be attributed to various reasons, including changing lifestyles and an ageing population, which will keep demand for housing active. In addition, urbanisation is increasing significantly; 75% of Europeans already live in or near cities [12].

The challenge for architects and urban planners is to develop strategies for the regulating the expansion of large urban agglomerations, considering land use and housing density.

2. Strategy and Method of Intervention

2.1. *Urban Densification Strategy*

The harmonization of the existing urban fabric and strategically planned and selective expansion of the built heritage can reduce land consumption and the overall environmental footprint [13]. This helps to initiate a positive process of sustainable urban development, especially in cities that are currently characterized by fragmentation and dispersion settlement.

The opportunities associated with urban densification include: the use of all interior spaces of the built-up city, the minimization of further land consumption, the uncontrolled expansion and fragmentation of landscapes, the enhancement of permeable areas, and the intensity of cultural relationships and developments. Potential constraints to densification, on the other hand, include mobility congestion, the vertical development of buildings, reduction in the size of housing units, and reduction in the provision of standards [7]. Moreover, challenges inherent in urban densification, include substantial financial investments, private property issues, and structural limitations of buildings, may impede their execution in urban areas. This is true even as they become integrated into ongoing urban planning policies. Consequently, it becomes imperative to deepen our comprehension of the environmental benefits stemming from physical interventions, emphasizing the positive outcomes over associated costs [14].

The adoption of a holistic approach that synergistically considers the dimensions of environment, society, and economy is of crucial importance for the process of building densification [15]. If high settlement density is desirable as a contribution to sustainable urban development, the question arises: what density is realistically sustainable? This paper aims to confirm that it is possible to design sustainable urban densification as a participatory tool to achieve a new energy and social balance. The identification of precise indicators will help clarify the potential of the proposed planning instruments. Densification processes in the consolidated and compact city will have a greater positive impact when associated with adequate public transportation infrastructures, or in derelict abandoned land close to the city's existing spaces and services. To this end, it is necessary to verify the possible adoption of multi-scalar well-defined solutions, which are more easily and physically connected to the centre areas of the city. From this perspective, the case study investigated in this paper is indeed part of the most consolidated urban areas of the first expansion of the city of Bologna and, therefore, very well-connected and linked to the existing services of the urban community.

2.2. The ABRACADABRA Project

The EU's challenge on energy efficiency in buildings is mainly about energy-efficient renovation and investment in existing building stock. Yet, there is a clear lack of investment in this sector and, particularly, the housing market. This is since high upfront investments are required and characterized by a high degree of risk, with potentially limited return on investment. The housing stock in the EU comprises around 200 million units, accounting for around 27% of energy consumption in the EU. Three quarters of today's buildings stock, including residential stock, is expected to remain in operation until 2050. To date, only 0.4 to 1.2% of existing buildings in the EU are renovated each year [16–18].

Under the Horizon 2020 European funding program, the project "ABRACADABRA—Assistant Buildings' addition to Retrofit, Adopt, Cure And Develop the Actual Buildings up to zeRo energy, Activating a market for deep renovation", under grant agreement No. 696126 [19], promotes the implementation of measures to increase urban density by adding habitable space to existing buildings. This extra space, through the financial benefits that can be generated, can help finance the renovation of the entire building. This proves to be significant, since the cost–benefit assessments of retrofitting actions show excessive payback times, creating a strong and generalized lack of confidence by investors and final users [20]. Moreover, this type of intervention is a useful tool to reshape the urban landscape or transform former non-residential buildings into new housing units. The ABRACADABRA project is based on the assumption that a substantial increase in the property value of existing buildings can play a key role. Indeed, the project has demonstrated the attractiveness of redevelopment based on volumetric additions and the use of renewable energy sources, called Add-ons and Renewable Energy Sources (AdoRES). Volumetric add-ons can be meant as one or more building units, such as side facades, roof extensions, or even new building constructions, that "assist" existing buildings in achieving the near-zero energy goal. This strategy, if adopted at the urban scale, can lead to the implementation of a punctual building densification policy within still transformable areas of cities. In this context, the goal of the project is to induce legislative and market change, accelerating the revolution toward nZEB targets [20].

2.3. Legislative Framework

The European Directive 2018/2001/EU [4] promotes the use of energy from renewable sources and requires that 32% of the EU's energy consumption will be covered by renewable energy sources by 2030. Furthermore, the EU has set itself the target of reducing net greenhouse gas emissions by at least 55% by 2030 compared to 1990 levels [2]. On December 2019, the EU heads of state and government decided to make the EU climate-neutral by 2050 in line with the Paris Agreement [21,22]. In the field of local spatial planning for the Emilia-Romagna region, where the case study of this research work is located, regional law no. 24/2020 stands out, which introduced the new general urban plan (*PUG—Piano Urbanistico Generale*) [23]. The general planning instrument includes environmental resilience, support for the attractiveness of the city and the supply of skilled labour, and the need to ensure habitability and inclusion.

2.4. Positive Energy District (PED)

In Italy, the Ministerial Decree of 16/09/2020 [24] introduces the concept of energy communities on a trial basis, and regulates the implementation of their incentive system. Its focus is on the transition to a more sustainable and decentralized energy system, based on the use of renewable energy and the active participation of citizens.

With the three-year implementation plan 2022–2024 [25] of the Regional Energy Plan 2030 [26], the Emilia-Romagna region has adopted most of the recent measures taken by the Italian government at EU level. These measures have made the climate and energy targets progressively more ambitious, with the aim of increasing energy efficiency and covering an increasing share of consumption with renewable energy. In this context, the new Regional Law no. 5/2022 [27] also establishes the obligation to support the dissemination of renewable energy communities, which is definitively regulated by Legislative Decree no. 199 of 08.11.2021 [28], implementing the EU Renewable Energy Directive 2018/2001 [4].

The common goal of these new models of energy production and consumption is to make a tangible contribution to complete decarbonization, to further exploit the potential of renewable energy sources. This will promote an energy system based on proximity, decentralization, and interconnection, also through a more active role of end customers [29].

2.5. Urban Infill

Soft densification through "urban infill" design interventions is a strategy that can potentially be replicated extensively. This can be performed through a systemic approach, functional to the physical characteristics of the urban void, local needs, and contextual and case-by-case strategic requirements. Urban infills are an effective mode of densification for achieving a compact city, which is a prerequisite for sustainable city development. Architectural infill exhibits significant figurative variability, particularly in response to the morphology of the intervention area, effectively resolving the spatial dynamics. Within the dense urban structure, marked by intensive or semi-intensive buildings, specific void patterns recur. This categorization allows us to delineate five potential infill techniques: expanding and constructing a new building volume, adding a volume adjacent to an existing structure, inserting between blind facades, elevating structures and, ultimately, demolishing and reclaiming vacant areas [30].

The analysis of several recent projects in Europe based on the application of a densification strategy, also by means of urban infill, has made it possible to identify some common characteristics and design choices aligned with the requirements of the various directives on limiting land use, environmental protection, and the maximization of renewable energy use.

Urban sprawl results from high demand for new homes and limited city land. Densification is a recognized solution, but researchers face challenges determining city-wide densification potential. A 2021 study in Tanta, Egypt, focused on densification as a strategy to combat urban sprawl, proving its effectiveness in addressing issues like agricultural land loss. The research not only controlled urban growth, but also fostered sustainable communities in developing countries. Successful densification relies on seamless integration of city planning, regulations, engineering, and social and economic aspects. Using an analytical hierarchy process, the study prioritizes densification sites based on sustainability parameters. Results showed potential for a 428% population increase in Tanta by 2027, while preserving 53% of agricultural land under current regulations [31]. The rise in the global urban population has generated an increasing need for additional residential space, leading to various adverse effects, including accelerated urbanization, urban sprawl, and elevated carbon footprints. In 2019, a study in Vienna delved into the potential of urban densification to counteract heat island effects and enhance outdoor thermal conditions. The study used simulations during hot summers to analyse the urban climate before and after densification using modelling tools. The results showed that adding vertical extensions to existing buildings had a significant solar shielding effect, lowering temperatures and enhancing thermal conditions in shaded urban areas [14]. Morphological changes, demographic shifts, and the deployment of renewable energy systems significantly influence the carbon performance of a precinct. In a 2020 study by Huang et al., an integrated carbon assessment model was developed to encompass overall precinct carbon emissions and the carbon offset derived from precinct-scale renewable energy harvesting. The results of the scenario analysis revealed that redeveloping buildings with higher energy efficiency ratings and increasing the penetration of renewable energy can have a lasting positive impact on the carbon performance of urban precincts. The analysis emphasizes the importance of embodied carbon in the total carbon footprint and the carbon reduction potential of a precinct during densification [32].

While there has been considerable research on urban densification planning to address environmental impacts, land use, and the challenges of decarbonization amidst population growth, there is a notable gap in research focused on densification strategies for existing urban areas also combining energy and financial aspects. Specifically, there is limited exploration of strategies that simultaneously consider deep renovation and improved energy efficiency through renewable sources, with the goal of transforming these areas into zero-energy districts. This paper aims to address this gap by proposing a densification strategy motivated by the need to increase real estate value and improve the architectural and performance quality of existing buildings, all while minimizing additional land occupation and promoting energy efficiency and renewable energy production.

3. Analysis of the Case Study

3.1. Research Methodology

This paper focuses on a specific building block located in the San Donato district in Bologna. The analysis employs a multi-criteria approach to identify socio-cultural obstacles, energy performance deficiencies, and economic barriers. These factors are crucial considerations for designing a comprehensive urban reshape and energy efficiency intervention at an urban scale. The multi-criteria approach, supported by specific digital tools, allowed us to balance all the above-mentioned factors and to select the most successful solution according to the project's target, linked to local needs and regulations. These targets are social inclusion and accessibility, enhancement of green spaces, energy efficiency measures and RES installations, reduction of payback times and, hence, profit-making.

Based on the design concept of urban reshaping through infills and volumetric additions blocks, several intervention scenarios were proposed. The research method adopted pursued the following phases:

(i) Analysis of the architectural and social context, in order to identify the critical points in terms of accessibility to the block, connection with the road system, availability of green areas, presence of buildings with historical documentary value.
(ii) Creation of different intervention scenarios (Concept #0–COURTYARD, Concept #1, Concept #2, and Concept #3).
(iii) For each intervention scenario, a climatic and environmental analysis was conducted using digital simulations with the *LadyBug–Grasshopper* software. This was aimed

at assessing the climate parameters necessary for determining potential blocks to be demolished, with the goal of enhancing the environmental characteristics of the blocks.
(iv) For each intervention scenario, an energy analysis was developed with *Dragonfly–Grasshopper* software. This was aimed at calculating the energy needs and consumption of different volumetric assumptions for urban infill interventions.
(v) Comparison of the different scenarios and selection of the optimal layout of buildings, balancing all the parameters calculated (for the climatic and environmental analysis: SunPath, Shadow Range, Universal Thermal Climate Index (UTCI), Direct Sun Hour Analysis, Incident Radiation, and Solar Envelope; for the energy analysis: End Use Intensity (EUI)).
(vi) Verification of the project targets in terms of profit from the investment before and after renovation.
(vii) Verification of the available surfaces for renewable energy sources (photovoltaic panels on roof and facade) to accomplish the energy balance of the whole settlement by and through the final reshaping of the urban blocks.

3.2. Multi-Scalar and Multi-Criteria Approach

As a first step, a multi-scalar analysis was carried out for all the aspects of the built environment at the urban scale, the block building scale, and at the level of the individual buildings. Given the multiple aspects to be considered in the energy and urban regeneration, the adopted approach has considered the evaluation of multiple criteria, based on the collection and analysis of data related to the environmental, energy and social characteristics of the site. From this perspective, the analysis aimed at "measuring to optimize" the urban context and the climatic context of the city in relation to the environmental characters of the blocks, through environmental analysis using parametric analysis software, as well as the performance aspects of the buildings, through district-scale energy analyses using parametric software.

Given the complexity of these criteria and the multiple indicators embedded in them, the multi-criteria analysis [33] implemented here has necessarily been divided in different steps (see Section 3.3, Section 3.4, Section 3.5), then analysed and compared to feed the next design steps in an optimized manner. This process has been designed to exploit more detailed parametric digital tools and explore various scenarios for an optimized deep renovation considering both volumes and related real estate values, as well as energy performance. In this framework, the optimized design solutions for the reshaped building blocks are those that provide the maximum increase in terms of real estate value while guaranteeing the energy balance and self-sufficiency.

3.3. Analysis of the Architectural and Social Context

The case study is an urban block in the San Donato district in Bologna, located in the Nort-East part of the city. San Donato is one of the city's most diverse residential neighbourhoods, due to the age variability and the cultural backgrounds of the residents. This district stands as one of the city's strategic points regarding vehicular mobility as it has several major roadways and is served by metropolitan rail service (Figure 1).

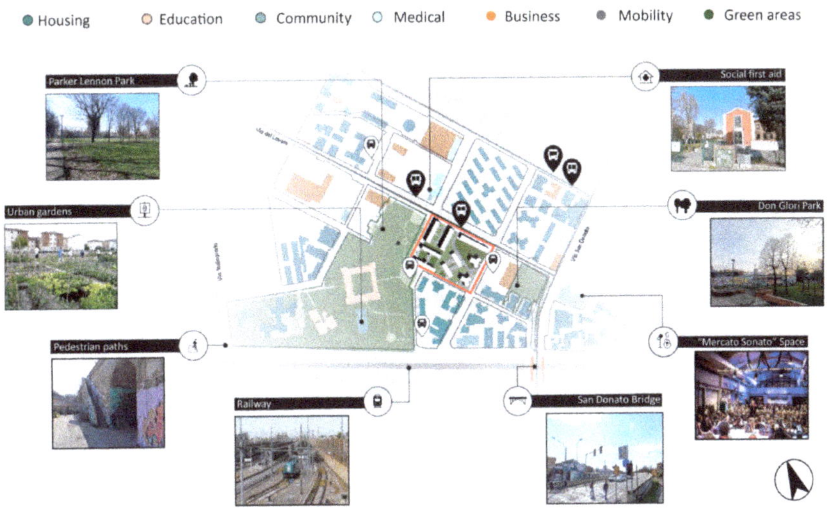

Figure 1. Graphical representation of the different mixed uses in the area, main mobility infrastructures, and green areas. (© 2023, C. Mazzoli after Q. Palazzo and P. Dabbicco).

Nevertheless, one of the most significant concentrations of public housing exists in this district. Nearly 50% of the housing lacks adequate parking spaces. The area also experiences high population density, reaching up to 142 inhabitants per hectare, resulting in increased traffic, heightened demand for parking, elevated noise pollution, and a limited availability of communal facilities.

The green spaces in the San Donato district are primarily dedicated to agriculture. However, there is a recent trend of reclaiming areas for gardens and small parks within the closely woven grid of buildings near the historical centre (see Figure 2). In general, while the amount of green space is considered insufficient, there has been significant redevelopment in recent years.

After a district-scale analysis, the study focused on the specific urban block located in the eastern part of the neighbourhood, closer to the city centre and to the railway.

The social perspective was registered by collecting advice from residents through ethnographic research and interviews. This information was crucial for understanding the residents' needs and, thus, supporting the decision-making process in terms of selection of buildings to be demolished (such as buildings 8 and 9, that were indicated as inadequate for spatial characteristics, functionalities, and indoor comfort levels from residents).

The urban block takes on a rectangular configuration and comprises a total of nine buildings, as illustrated in Figure 3. Seven of these buildings align in a row. The structures include: two "Popolari" blocks (blocks 1 and 2); two "Popolarissime" blocks (blocks 3 and 4); two buildings of lower height than the others, oriented at an incline to the street axes (blocks 8 and 9, proposed for demolition both for the ethnographic analysis and for their negative climatic impact due to the undesired projection of shadows within the courtyard and towards the existing buildings); two tower buildings with a distinctive "Y"-shaped structure (blocks 6 and 7); and a single in-line building (block 5) (Figure 4).

The urban block is bisected by a private driveway, featuring a limited number of condominium parking spaces. This roadway, along with a handful of parking spaces within the inner courtyards, defines the internal pathways, which are degraded and poorly interconnected due to the enclosure of the outdoor spaces of the tower condominiums.

The block is bounded to the west by a one-way street and to the east by major high-traffic streets. Immediately next to the lot is the Parker-Lennon garden which, however, does not appear to have pedestrian connection with the block.

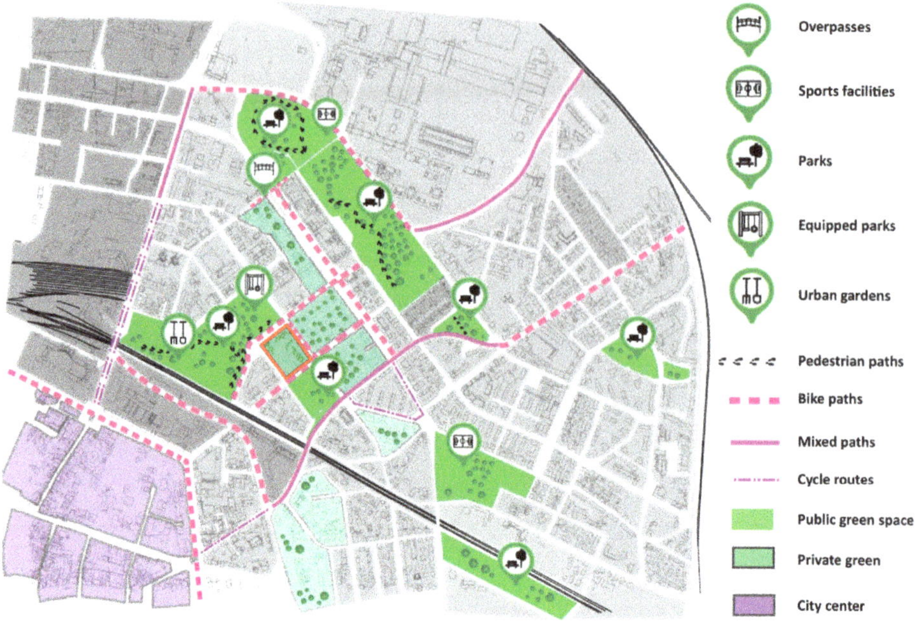

Figure 2. Graphical representation of the pedestrian and cyclable paths, and the most important private and private green areas. (© 2023, C. Mazzoli after Q. Palazzo and P. Dabbicco).

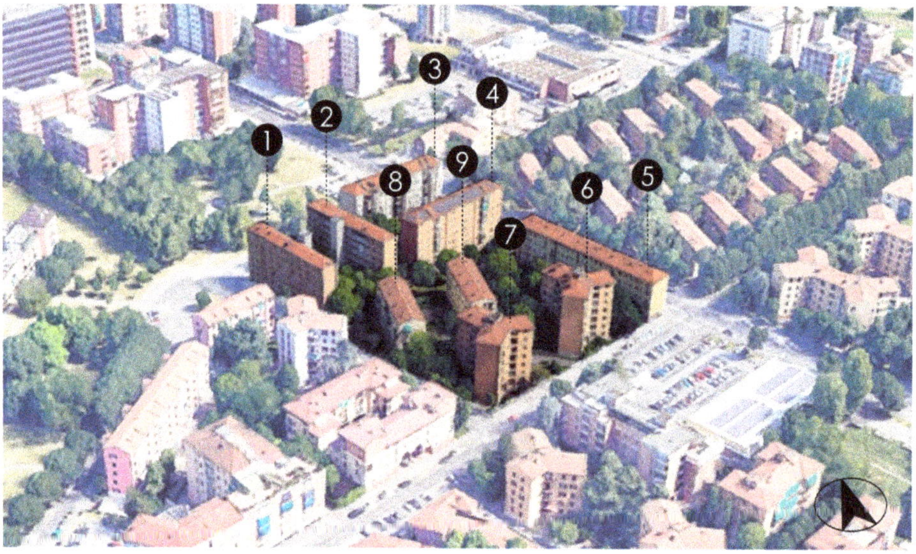

Figure 3. Aerial view of the project block. (© 2023, C. Mazzoli).

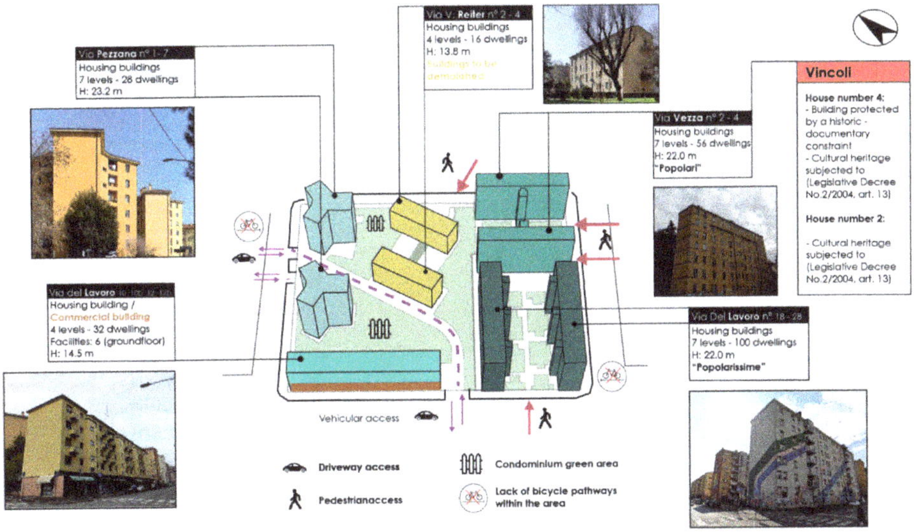

Figure 4. Graphical representation of the nine buildings located in the block. (© 2023, C. Mazzoli after Q. Palazzo and P. Dabbicco).

The construction of the parcel is part of a plan of social housing interventions in the first half of the 20th century, aimed at the weakest classes of the population, with reference to the families of recent immigration in the early post-war period.

The first buildings were constructed in the immediate first post-war period: among them are the "Popolarissime", designed between 1934–1936, and the "Popolari", whose name derived from their social housing function (in Italian, "popolare" alludes to the working class for whom these buildings were built). All these buildings were constructed as an emergency plan by social housing institutions to provide housing for families evicted from the historic centre districts demolished during the fascist era. They are buildings with a reinforced concrete frame structure and brick envelope. The current thermal and structural performance of these buildings is very poor and, therefore, deep renovation interventions to improve performance and to modify some distributional characteristics of the interior spaces are necessary to improve living comfort. The entire built-up area of the block, although it emerged in successive phases over a twenty-year period, shows a clear rationalist imprint regarding its architectural style. Rationalism imparts a design approach characterized by rows of buildings aligned perpendicular to the streets, featuring a uniform orientation. The structures are evenly spaced at a minimum distance to prevent shading between buildings, ensuring that facades receive ample sunlight, even during the winter solstice.

From the naturalistic point of view, the subdivision is characterized by various types of trees, including tall trees (common plane, Atlas cedar, oak, and beech), which provide excellent shading, protection against solar radiation, and low maintenance costs given their performance.

Today, such building stock, built more than fifty years ago, is inadequate with respect to current energy and seismic safety requirements and regulations, and therefore requires consolidation and energy efficiency interventions. Furthermore, considering demographic indicators, there is a necessity for interventions aimed at enhancing accessibility and eliminating architectural barriers within the residential infrastructure.

The energy simulations were developed through the use of digital parametric tools, namely *Dragonfly–Grasshopper*, which uses the simulation platform *URBANopt SDK (Urban Renewable Building And Neighborhood optimization) 0.10.0*, based on *EnergyPlus 23.2.0* and

OpenStudio 3.7.0 SDK, aimed at district-scale thermal and electrical analysis for energy modelling of urban communities and districts. The results from these energy simulations prove the significant deviation of the buildings from the energy requirements standards of a nZEB, mostly due to high insulation deficits in the envelopes and oldness of energy installations accompanied by visible degradation of materials. These characteristics are combined with the complexity of the buildings due to the varying orientations, heights, the almost total absence of adequate interior lighting and a parallel deficient passive shading. In general, the transparent elements of a building play a key role in the heat and mass transfer control, and also in natural light penetration, sound insulation, the thermal and visual comfort of the occupants and their health, and the improvement of aesthetic features [34]. Therefore, the bioclimatic design of this urban block emerges as a particularly challenging project, which the present research endeavours to address. This includes the incorporation of volumetric additions while mitigating potential adverse impacts.

3.4. Climatic and Environmental Analysis

According to the Köppen–Geiger classification, the climate of the city of Bologna falls into the "Warm summer temperate climate" range, typical of the low-altitude areas of northern Italy. In the cool season, temperatures range from 0 °C to 22 °C. The mild seasons range from 8 °C to 25 °C, and during them, about 60% of the daytime hours require no air conditioning. For this reason, passive strategies of ventilation, lighting, and reduction of solar loads in indoor spaces were incorporated in the development of this project. Throughout the summer season, the temperature ranges from 16 °C to 36 °C, with daytime temperatures mostly surpassing 25 °C and experiencing high solar radiation levels. Hence, it is imperative to implement effective external shading in the project, strategically designed to reduce solar exposure on glazed facades and alleviate the workload on air conditioning systems. Examination of climatic data for the Bologna area indicates a robust average of daily sunshine hours, exceeding 12.2; however, this can vary based on the prevalence of cloudy days. The incorporation of site-specific climatic data, combined with a thorough contextual analysis, has proven essential in identifying the optimal environmental and climatic conditions for conducting climate simulations.

All simulations were carried out through the *LadyBug* plug-in of *Grasshopper* software, a parametric workspace within *Rhinoceros 8*, where the geometries then subject to the simulations are modelled. *Ladybug* provides a variety of interactive 2D and 3D climate graphics that support decision making during the early design stages through solar radiation studies, view analysis, and modelling of sunlight hours.

The fundamental concept guiding the design process revolves around the aspiration to establish a unified and expansive urban courtyard that mirrors the dimensions of a private block. This courtyard aims to be easily accessible to residents, incorporating green spaces and communal facilities. As a result, initial design decisions dictated the removal of centrally positioned blocks (blocks 8–9) arranged diagonally in relation to street directions. This choice was driven by the blocks' varying heights and orientations compared to surrounding structures, introducing environmental challenges that make bioclimatic design notably complex and hindering the creation of a usable outdoor area. Additionally, it was deemed necessary to demolish the stairwells of blocks 1–2 (Figure 5) due to their structural inadequacy, as they were constructed later than the buildings themselves. This decision contributes to a proposed demolition area totalling 593 m^2.

Due to the parametric process, it was possible to carry out the simulations on different input geometries by considering not only the existing state of the site, but also assumptions of demolitions and new infill volumes. These were useful to map and compare different morphologies of the block. Specifically, the solar path was initially analysed through the SunPath parameter analysis during the summer season and in two different hours (10 am and 3 pm). Furthermore, the Universal Thermal Climate Index (UTCI) was assessed in terms of Temperature UTCI (measured in perceived temperature in °C) and Comfort UTCI (measured in UTCI thermal stress); this allowed us to obtain insights into the interior microclimate of the urban block and define strategies for open spaces (Figure 6). Other parameters linked to the solar impact on the buildings have been assessed, such as the Shadow Range in summer and winter season and the yearly load (measured in percentage of daytime shadow) (Figure 7). Subsequently, the analysis of the Incident Radiation (measured in kWh/m^2) was carried out on the facades of the volumes in summer and winter season (Figure 8), as well as the Direct Sun Hour analysis during summer in order to calculate the daily sun hours on the envelope of buildings (Figure 9). Through the analysis of all these parameters, combined also with the Solar Envelope study (Figure 9), it was possible to understand the spatial relationships between the buildings and to locate the surfaces with higher and lower illuminance criticality.

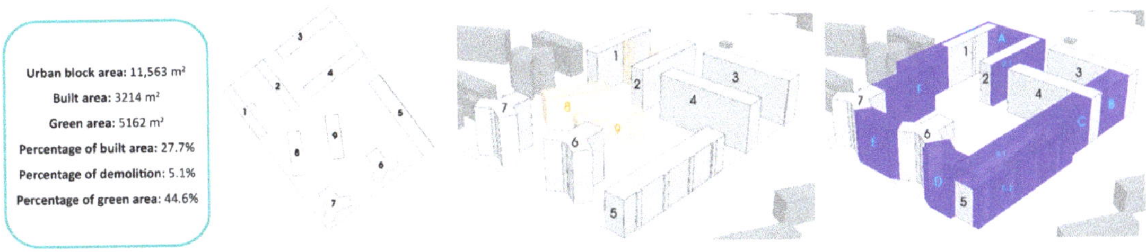

Figure 5. (From left to right) existing state plan of the lot under consideration; perspective axonometry of the existing lot showing demolitions (yellow blocks); perspective view of the urban block under study for scenario Concept #0–COURTYARD with new infill and add-ons blocks (in purple) (© 2023, C. Mazzoli after Q. Palazzo and P. Dabbicco).

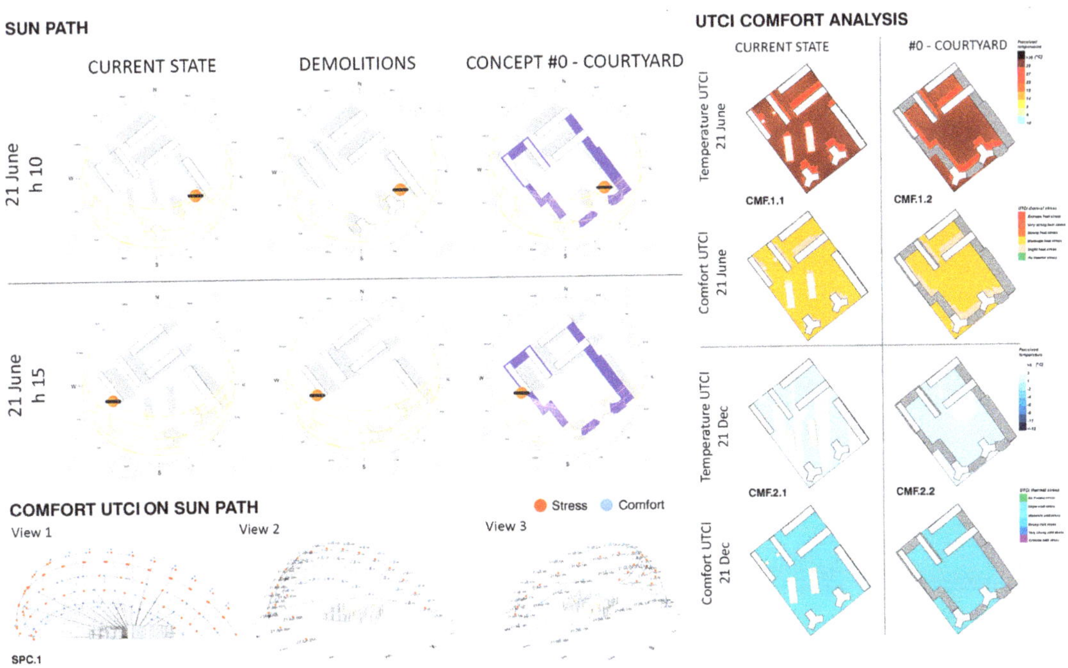

Figure 6. Abacus of the environmental simulations for the current state scenario and the Concept #0–COURTYARD scenario: SunPath and UTCI comfort analysis (© 2023, C. Mazzoli after Q. Palazzo and P. Dabbicco).

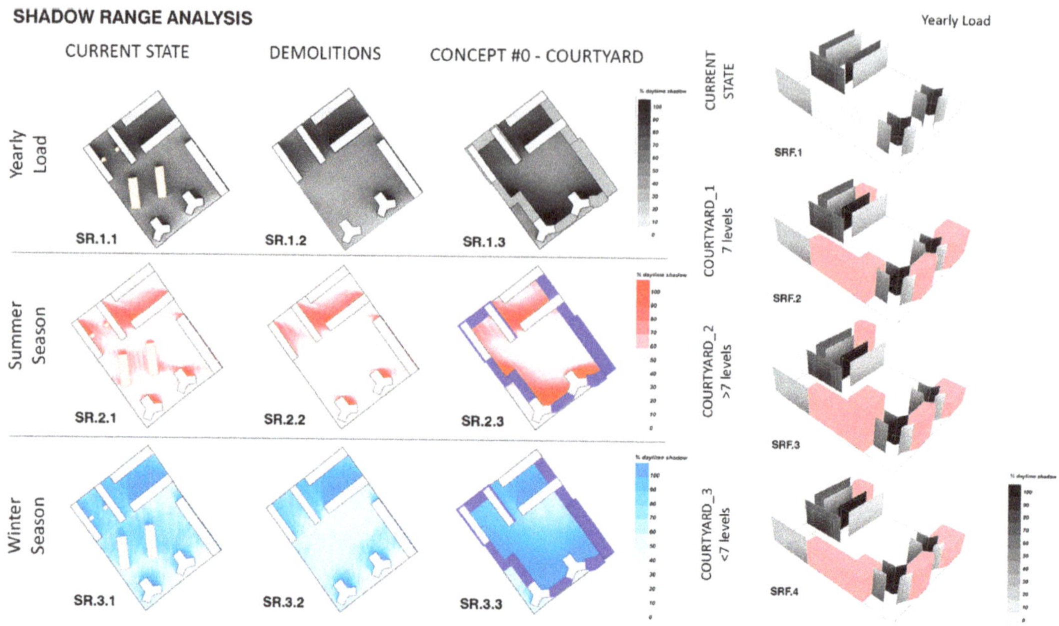

Figure 7. Abacus of the environmental simulations for the current state scenario and the Concept #0–COURTYARD scenario: Shadow Range analysis (© 2023, C. Mazzoli after Q. Palazzo and P. Dabbicco).

All the results obtained in the simulations were then collected in a comparison table and each heatmaps displayed in a general abacus of the environmental simulations (Figures 6–9).

Among the various values mapped, it is of particular interest for considerations on photovoltaic (PV) systems, the share of solar irradiance that invests the facades of the south front of the block: in fact, in the winter season, this value reaches higher peaks than in the roof. In addition, the presence of a full elevation on the south front would allow a significant gain of shading in the inner courtyard (+12%) by defining clearer areas of shade and light and, therefore, be more easily modulated by design (Figure 10).

In the local area, a PV system that is in the optimal tilt (35°) and climatic conditions can produce, for example, around 1200–1300 kWh per year per kWp of power while, when opting for flat PV, it is reduced to 1000–1050 kWh per year per kWp of power.

In particular, the Direct Sun Hour analysis and Incident Radiation analysis climate simulations provided useful data for the positioning and quantification of the PV systems envisioned as the main strategy of ABRACADABRA for self-supply of energy, as well as important indications on the most appropriate buildings to host on top Add-ons.

Roofs turn out to be the surfaces that receive the most solar radiation (in summer) with seasonal peak above 600 kWh/m^2, thus are the most suitable for the installation of PV systems. Among the vertical surfaces, on the other hand, the south-facing facades are the most suitable for PV installation; in particular, the south front of the lot (5-6-7-D-E), which includes two blind facades, exceeds the values in the roofing in winter. In addition, passive shading systems have been provided on the south-east, south, and south-west fronts.

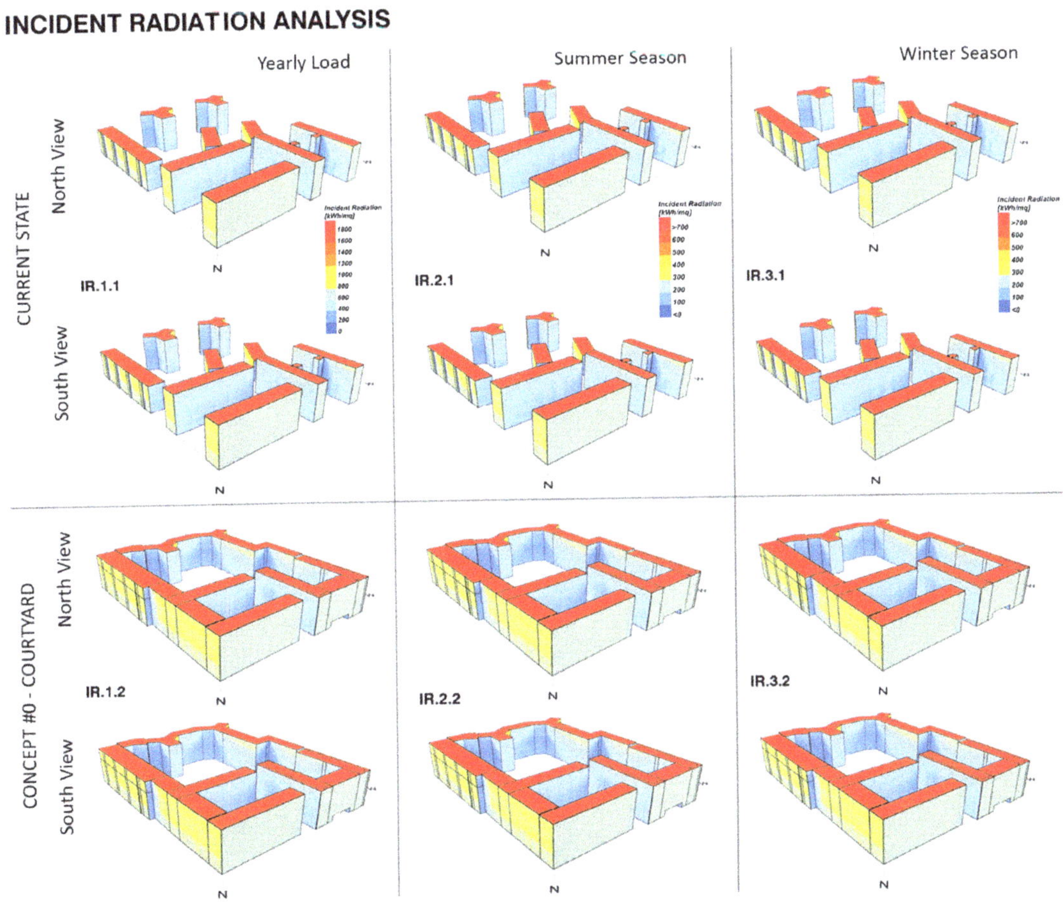

Figure 8. Abacus of the environmental simulations for the current state scenario and the Concept #0–COURTYARD scenario: Incident Radiation analysis (© 2023, C. Mazzoli after Q. Palazzo and P. Dabbicco).

3.5. Energy Analysis at the District Scale

The analysis phase was conducted via district-scale energy simulations carried out by *Dragonfly*, a plug-in of *Grasshopper* software, which, through the simulation platform *URBANopt*, allows the calculation of the energy use of an urban district energy model. These analyses made it possible to parametrically quantify the physical quantities of the geometries processed in the mentioned model, and especially to evaluate and estimate the energy needs and consumption of different volumetric assumptions for urban infill interventions.

In particular, the value of EUI, which is an energy efficiency indicator used in building design and/or construction, mainly used in the United States, was mapped in the different simulations to estimate the energy balance of the geometric models being simulated. The impacts of consumption for heating, cooling and lighting systems were analysed, representing the Monthly EUI and Monthly Energy Balance values (measured in kWh/m^2 for energy intensity).

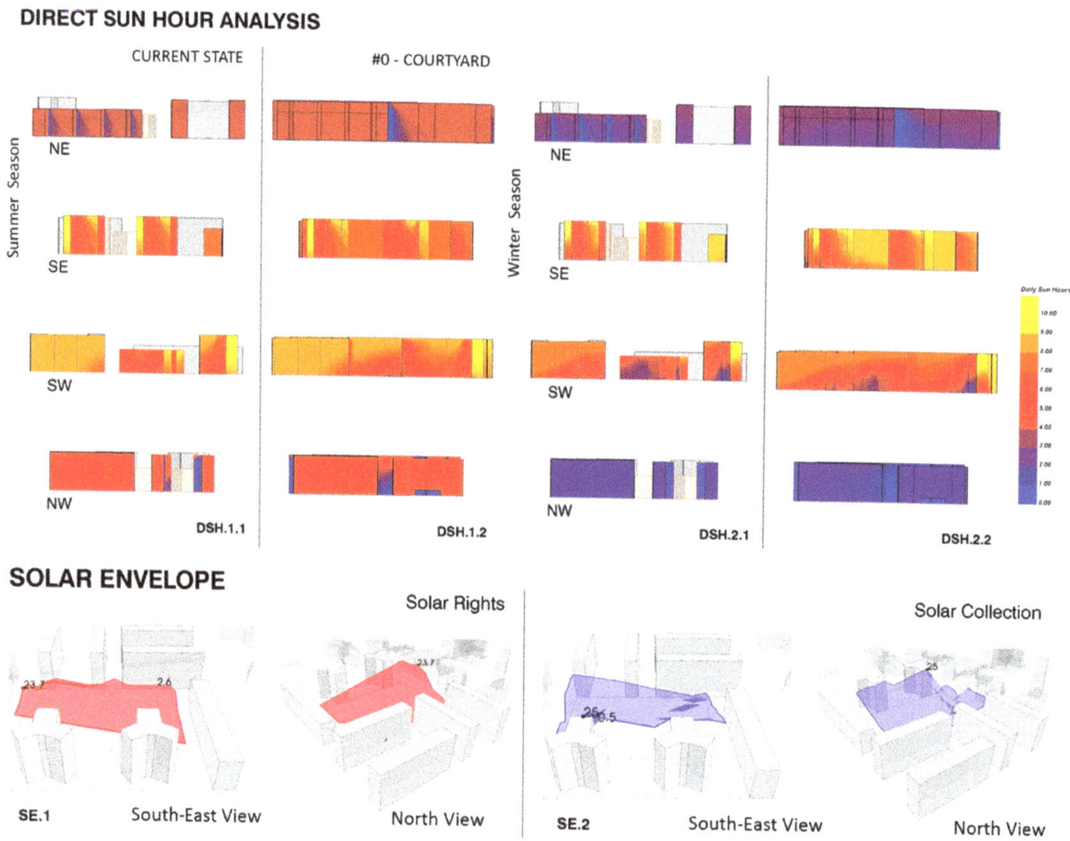

Figure 9. Abacus of the environmental simulations for the current state scenario and the Concept #0–COURTYARD scenario: Direct Sun Hour analysis and Solar Envelope analysis (© 2023, C. Mazzoli after Q. Palazzo and P. Dabbicco).

Calculation of heat demand was carried out based on a heat balance of the environment bounded by the building envelope. Knowing the efficiencies of the plant systems, the primary energy demand can be traced from the EUI value. To evaluate the results of the energy analyses, the values of EUI and more generally of energy consumption provided by the New Building Institute (NBI) [35] were taken into consideration. This provides average consumption targets for different types of building use, dividing them according to their climate class. In addition, NBI sets EUI standards for achieving nZEB. In this case, for the residential typology connoted as "Mid-rise Apartment" by ASHRAE programs (i.e., multi-family residential building higher than five stories), the achievement of the nZEB target occurs for an EUI of less than 75 kWh/m^{2}year. This value was considered as a numerical reference value for the comparison between the simulated design scenarios.

A precise "Construction set" was created within the *Dragonfly* algorithm to describe the architectural elements of the buildings and detail the energy model within the simulation algorithm. Each set contains the stratigraphy of the materials used, each characterized by its respective mechanical and physical properties (Figure 11).

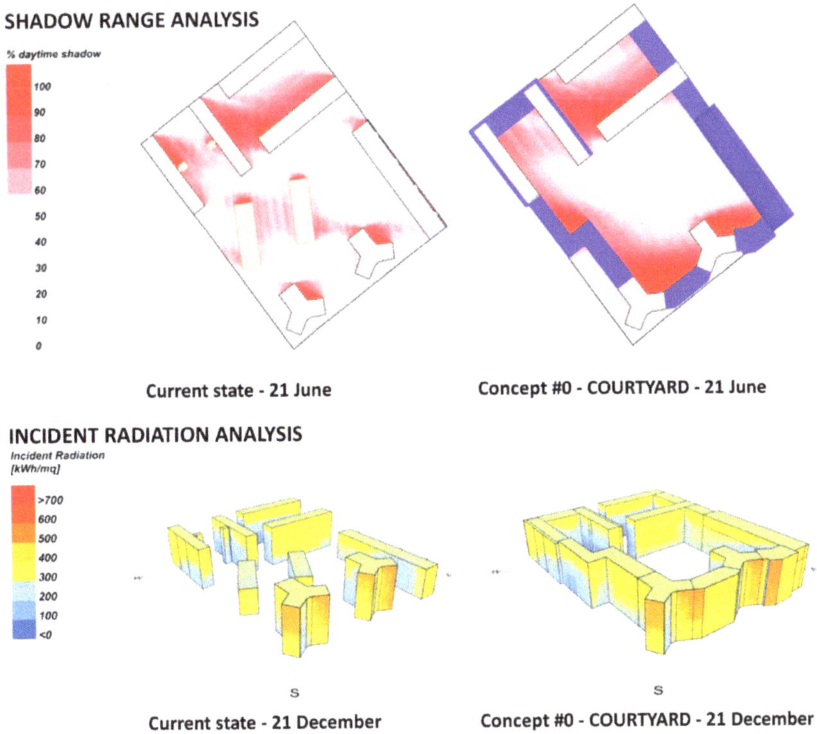

Figure 10. Shadow Range Analysis (**top**) showing a significant gain in shading in the inner court (+12%), and Incident Radiation Analysis (**bottom**) showing that in the winter period the solar incidence is greater on the facades than on the roofs: (**left**) current state and (**right**) Concept #0–COURTYARD scenario. (© 2023, C. Mazzoli after Q. Palazzo and P. Dabbicco).

For the "Construction sets" referring to the current state, the construction stratigraphies of the buildings in the block was considered. In particular, the following traditional stratigraphies of residential construction built in Italy in the 1930s–1970s were evaluated: non-insulated facades in bricks (with a thermal transmittance U of 2.213 W/m^{2}K), hollow-core concrete roofs and ground-floors, and single-glazed windows.

Instead, stratigraphies referring to the deep renovation intervention mainly aimed at the energy refurbishment of the building envelope were associated with the energy simulations carried out on the assumptions of new volumes in the block. In particular, for the deep renovation of existing buildings, the following solutions were considered: the implementation of an External Thermal Insulation Composite System (ETICS) in rock wool (12 cm thick) for external walls, the integration of a thermal insulation layers in recycled textile fibres for roofs, and double/triple-glazed windows. For the new infill blocks, the envelope is considered as a ventilated facade composed of a X-Lam structure, high thermal insulation layer in rock wool and wood fibre (14 cm thick), a light metallic sub-structure for the creation of a micro-ventilated airspace (3 cm thick), and a dry finishing cladding system in ceramic (1 cm thick). The thermal transmittance of the envelope systems considered for these scenarios ranges from 0.250 W/m^{2}K for the renovated facades to 0.120 W/m^{2}K for the more performant new facades.

The "Construction set" uses indicative parameters that refer to average values of high-performance, but still renovation technological solutions; the envelope performance that was considered for new construction has higher performance. All the selected stratigraphies meet the minimum requirements established by the Minimum Environmental Criteria (CAM), introduced in Italy by Law No. 221 of 2015 [36].

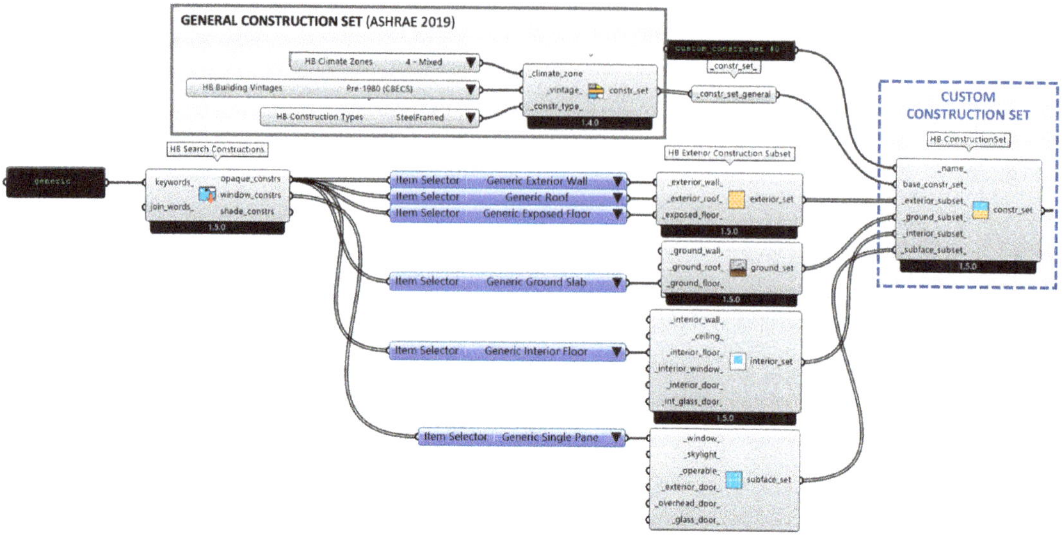

Figure 11. Input parameters considered in the *Dragonfly–Grasshopper* algorithm to define the "Construction set" describing the architectural elements of buildings: each set includes the layers sequence of each internal or external partition, characterized by specific mechanical and physic properties. Starting from the General Construction set based on the ASHRAE 2019, the algorithm has been customized according to standardized values for traditional/innovative construction solutions. (© 2023, C. Mazzoli after Q. Palazzo and P. Dabbicco).

Furthermore, *Dragonfly* needed the setting of precise "Programs", which are data sets describing the intended uses, the times of use of the buildings and the consumption of the main systems (Figure 12). Specifically, for the simulations carried out, two personalized programs were created which refer to the state of the existing building ("Program-Current State") and that of the deeply renovated building ("Program-Deep renovation"). Both programs start from the default program proposed for buildings intended for residential use characterized as Mid-rise Apartment by the ASHRAE programs, which then differ in the values of the parameters regarding electricity and gas consumption, losses due to thermal bridges, and the categories of HVAC systems. Table 1 reports the values considered for the two programs, defined according to standardized values reported in literature for traditional built heritage [37] used to carry out the energy simulations, whose results are shown in Figure 13.

For all the intervention scenarios analysed (Concept #0–COURTYARD, Concept #1, Concept #2, and Concept #3), the same "Program-Deep renovation" and the same "Construction set-Deep renovation" were used; indeed, they differ only in their geometry.

The comparison between the existing urban block and the different intervention scenarios (Figure 13) allowed us to assess the impact of possible energy upgrading strategies on the consumption of the current state. Although the variation in EUI value is minimal, the analyses show that, by using the same "Program" and "Construction set" for all the intervention scenarios, the more compact shape of the volumes leads to an improvement in the overall energy requirements. In conclusion, the Concept #0–COURTYARD scenario is the most efficient, with an EUI value of 123.90 kWh/m^2.

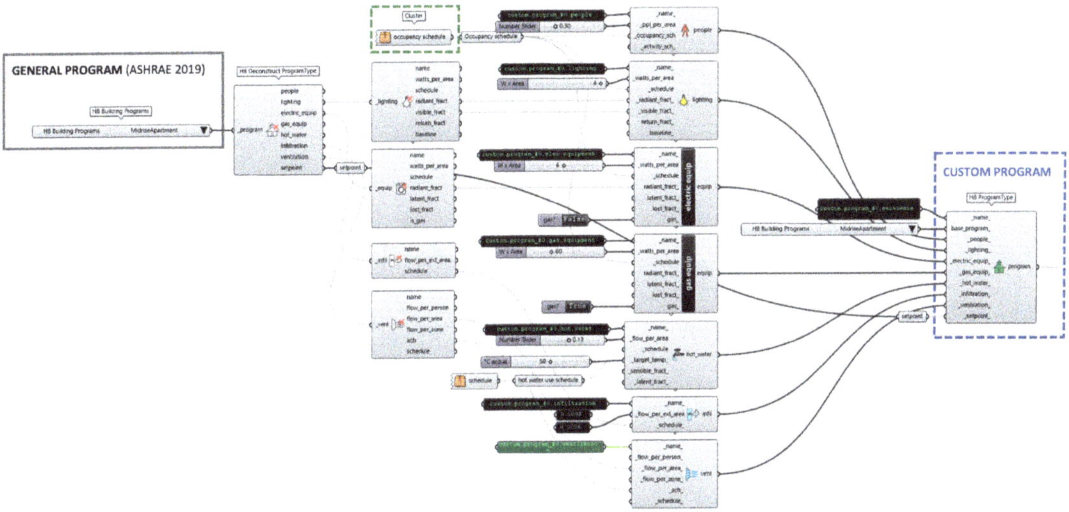

Figure 12. Input parameters considered in the *Dragonfly–Grasshopper* algorithm to define the "Programs" describing the characteristics and the use times of uses of buildings, as well as the consumption of the main system plants. Starting from the General Program based on the ASHRAE 2019, the algorithm has been customized according to ASHRAE standards for Mid-rise Apartment. (© 2023, C. Mazzoli after Q. Palazzo and P. Dabbicco).

Table 1. The values used for the two "Programs" and the simulation sheets summarizing the data, inputs, and results of the simulations.

Parameters	Program-Deep Renovation	Program-Current State
Base program	Mid-rise Apartment	Mid-rise Apartment
People	0.3 per m^2	0.3 per m^2
Lighting	3 W per m^2	4 W per m^2
Electric equipment	6 W per m^2	6 W per m^2
Gas equipment	6 W per m^2	10 W per m^2
Hot water	0.13 L/h per m^2	0.13 L/h per m^2
Ventilation	Generic	Generic
Infiltration	Tight Building 0.0001 m^3/s per m^2	Leaky Building 0.0006 m^3/s per m^2
Setpoint	From base program	From base program
HVAC system	Ideal Air Load (ASHRAE.2021)	Ideal Air Load (Pre.1980)

The comparison of the results shows that a significant reduction in building energy consumption is possible, mainly due to two factors. First, through retrofit of the building envelopes of the existing blocks to solve the lack of insulation in the infills. Second, through the infill and increased volume interventions which, by creating a "compact" morphological arrangement, significantly reduces the overall energy demand (EUI). That overall energy demand is, in its average values, always more than halved compared to the EUI of the current state of 263 kWh/m^2 year.

Figure 13. Analyses of the EUI values for the current state and different intervention scenarios: geometric characteristics, input data (referred to the "Construction set" and "Program" used in Dragonfly–Grasshopper, output data in terms of EUI (with the specific demand for heating, cooling, interior lighting (© 2023, C. Mazzoli after Q. Palazzo and P. Dabbicco).

4. Urban Re-Design Project

4.1. Definition of the Project Targets

After the context analyses were concluded, a meta-design approach was initially adopted on the case study, in which the strategic and theoretical guidelines for sustainable urban densification were resumed. Thanks to the parametric tools used, the architectural and energy design objectives were monitored throughout the concept phase, with the intention of optimizing the design proposals, which were then detailed from a technological and systems point of view.

To manage the meta-design phase, the objectives of economic attractiveness of the intervention and energy self-sufficiency were defined through project targets. These were used as numerical indicators to quantitatively assess and compare the performance of different design scenarios. In fact, through the parameterization of the physical and performance characteristics of the new project volumes, it was possible to calculate in progress the relevant values concerning the set targets.

Specifically, two project targets were identified:

1. The first one is related to the profit from the investment, focusing on the economic feasibility of the intervention and the attractiveness for investors.
2. The second target focuses on the use of renewable energy sources, by sizing the PV area for energy self-sufficiency, monitoring the energy savings and finally quantifying the payback time of the investment.

Regarding the profit of the investment, ABRACADABRA has shown that the implementation of measures to increase urban density, through the volumetric expansion, helps to economically support the renovation of the entire building through the income generated. This contribution has an effect also at the urban scale, proposing a densification of the built environment through urban infill interventions. This can be particularly attractive, both for investors interested in new spaces, but also for the residents who can gain from the energy savings, the space addition and the increase in the commercial value of their property. Regarding the costs per m^2, data was taken from the *Prezzario della Regione Emilia-Romagna* of 2022 [38] and from the ABRACADABRA project database for each type of intervention. In addition, the gains obtained by the resale of nZEB buildings were considered through local market research for the area under study. A target return on investment was set, corresponding to profit over total investment set at a minimum of 15%.

4.2. Design Project Choices

The starting point of the design was the energy performance and bioclimatic potential derived from the comparison of the results between the current state and the Concept #0–COURTYARD hypothesis. The project was dedicated to achieving a condensed configuration for the new subdivision, aiming to facilitate the development of internal courtyards accessible to residents. To fulfil this objective, as mentioned in Section 3.2, the proposal involves the demolition of specific structures within the urban block. The identified elements for demolition include: the external stairwells of buildings 1 and 2, characterized by non-compliance as they were constructed later and did not meet standards; and the removal buildings 8 and 9, which is motivated by their unfavourable orientation and height, posing bioclimatic challenges and acting as an impediment to the establishment of usable outdoor areas (see Figure 5).

With respect to the Concept #0–COURTYARD hypothesis, the idea of implementing infill blocks to connect the existing pairs of buildings 1–2 and 3–4 was excluded. Buildings 1–2 are subject to historic preservation constraints. On the other hand, the intervention in buildings 3–4 would be too invasive for the related interior spaces, also negatively affecting the natural lighting levels.

Nonetheless, the numerous blind facades have strong potentials for infill intervention. Three new volumes (A, B, C) were considered useful to connect buildings 1, 7, and 6, as well as a new building contiguous to block 6 that would architecturally reset the facade of building 5 subject to demolition and allow its lost spaces to be recovered (Figure 14). Other volumetric increases have been planned as facade Add-ons, integrating an exoskeleton on the interior facades of buildings 1, 2, 3, and 4, to recover their vertical connections and to provide new spaces for residents in the form of balconies, verandas, and galleries. In any case, deep renovation is planned for all existing buildings that are not subject to demolition with the intention of reaching the nZEB target and improving their envelopes and systems.

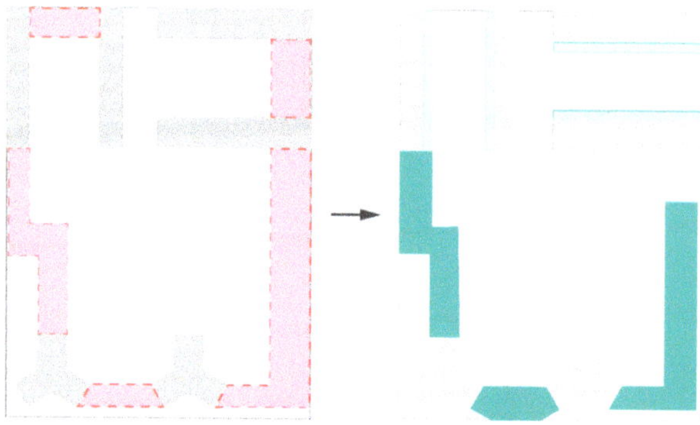

Figure 14. (**Left**) Plan of the hypothesis Concept #0–COURTYARD with the infill blocks in pink and (**right**) plan of the first intervention scenario with the infill and add-ons blocks in light blue. (© 2023, Q. Palazzo and P. Dabbicco).

For this first design scenario, indicators of profit and payback time of the investment were calculated in a preliminary manner. These preliminary results (see Tables 2 and 3) showed that the profit was very low: the considered average selling price per m^2 of about EUR 4100 has been multiplied by the total area increase of 5576 m^2 for a total profit of EUR 1,191,682, corresponding to an investment return rate of 5.49%.

Table 2. The values related to the energy and economic assessment of the final design solutions.

Parameters	Concept #0–COURTYARD	Final Design
Costs for Deep Renovation	260 EUR/m^2	260 EUR/m^2
Costs for new Add-ons on existing buildings (balconies)	400 EUR/m^2	400 EUR/m^2
Costs for new infill of building blocks	1500 EUR/m^2	1900 EUR/m^2
Housing infill building blocks-Income from selling	4100 EUR/m^2	4200 EUR/m^2
Commercial infill building blocks-Income from selling	-	4500 EUR/m^2
Housing infill building blocks–Total area	10,859.00 m^2	10,859.30 m^2
Commercial Infill blocks–Total area	-	3216.10 m^2
Infill blocks–Total area	10,859.00 m^2	14,075.40 m^2
Total increase in net floor area	5576 m^2	8757 m^2
Total Costs for Deep Renovation including demolitions	EUR 3,820,908	EUR 3,820,908
Costs for new Add-ons on existing buildings (balconies)	EUR 1,506,400	EUR 1,506,400
Total Costs for new infill of building blocks	EUR 16,342,200	EUR 26,743,260
Total income from increase in net floor area	EUR 22,861,190	EUR 37,742,130
Profit	EUR **1,191,682**	**5,671,562**
Investment profit percentage	**5.49%**	**17.68%**

Table 3. The values related to the verification of project targets in terms of payback time and PVs quantification.

Parameters	Concept #0–COURTYARD	Final Design
Project End Use Intensity (*pEUI*)	75 kWh/m^{2}y	40 kWh/m^{2}y
Total energy demand–Current state	263 kWh/m^{2}y	263 kWh/m^{2}y

Table 3. *Cont.*

Parameters	Concept #0–COURTYARD	Final Design
Total energy demand–Existing blocks	5,263,866 kWh	5,263,866 kWh
Total energy demand–Deep renovated existing blocks	1,102,185 kWh	587,832 kWh
Total energy demand–After intervention	**2,201,745 kWh**	**1,301,488 kWh**
Peak kW (Bologna, Italy)	1400 kWh	1400 kWh
Peak PV 1 kW	4.15 m^2	4.15 m^2
Renewable Energy needed for deep renovated blocks	787 kWh	420 kWh
Total Renewable Energy needed	1573 kWh	930 kWh
PV area needed for deep renovated blocks	3267 m^2	1743 m^2
PV total area	**6527 m^2**	**3858 m^2**
Costs for PV systems	4.50 EUR/W	4.50 EUR/W
Costs for PV for deep renovation	EUR 3,542,738	EUR 1,889,460
Energy savings	**188 kWh/m^{2}y**	**223 kWh/m^{2}y**
Energy cost	0.50 EUR/kWh	0.50 EUR/kWh
Total costs	EUR 551,093	EUR 293,916
Bill savings	EUR 1,381,405	EUR 1,638,582
Payback time	**2.56 y**	**1.15 y**

Therefore, from the volumes considered, which were totally filled from ground to sky, the general morphology of the blocks was modified, in terms of height and plan shape of blocks A, B, and C (Figure 15). The infill buildings, named A, B, C, were shaped with the aim to increase the residential floor area by adding additional levels, especially at the corner building C. In addition, ground-floor openings were created in blocks A and B to facilitate access to the courtyard. The new volumes allowed for improved energy performance to come closer to the targets set (Table 2).

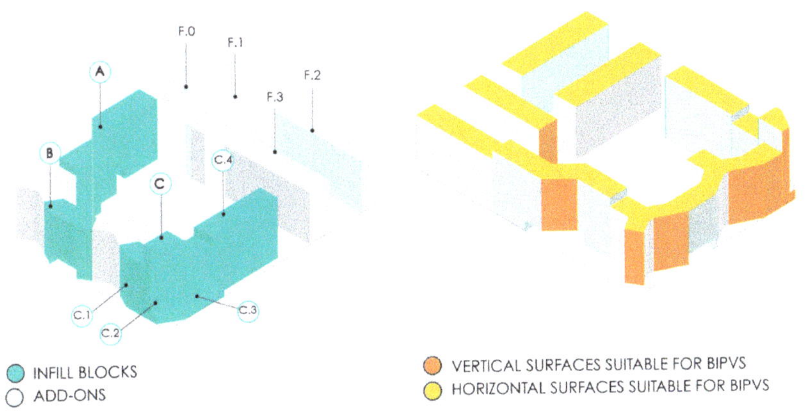

Figure 15. (**Left**) Axonometry of the final design solution selected, characterized by add-ons and infill blocks; (**right**) axonometry with visualization of facades and roofs suitable for PVs installation. (© 2023, C. Mazzoli after Q. Palazzo and P. Dabbicco).

Each of block A, B, and C was analysed individually through the use of *Ladybug*. Climate simulations were conducted to assess shading and incident radiation for the inner courtyard and critical facades in the block. The project targets on the new lot setting were then re-evaluated in order to further lower the benchmark *pEUI* for nZEBs (equal to 75 kWh/m^{2}year, as indicated by NBI). The final design project includes a higher ratio of m^2

of built-up area, which directly affects the profit generated from the resale of space: thus, it is guaranteed to exceed the minimum threshold of 15% set at the beginning of the project.

Regarding the dimensioning of the required PV area, incident radiation analyses have allowed for the more precise definition of the average peak radiation data for Bologna, resulting in a lowering of the required PV area, to be added to the surface area of the most favourable facades in terms of solar exposure, resulting in a total amount of 3858 m^2 of PV surface. The results on payback times are also positive, resulting in 1.15 years, due to the increase of 10% of the commercial area and of the sale profit of the new volumes. The target is thus achieved. Overall, in comparison with the current state of the district block, this proposed retrofit scenario will lead to a reduction of 75% of the total energy demand, now amounting to 1,301,488 kWh, and an energy savings of 223 kWh/m^2y (Table 3).

The analyses carried out, together with current regulations, supported the selection of the specific solutions needed to achieve energy efficiency goals. Among the solutions studied, the following aspects were investigated in depth: structural technology, thermal and acoustic insulation, on-site energy supply, and reuse of energy through facilities that can fully meet the building's overall needs.

Regarding the energy aspects of the concept, it was possible to reassess and lower the estimated *pEUI* (from 75 kWh/m^2y to 40 kWh/m^2y) thanks the minimal energy consumption provided by extremely high performant solutions for the envelope (X-Lam panels and additional thickness of thermal insulating materials) (Figure 16). In particular, the following solutions were identified:

- Use of XLAM panels for the structural system and external envelope, chosen for its significant energy, static and environmental advantages.
- Installation of a PV system on roof and facade, sized to ensure the storage and supply of sufficient renewable energy for the autonomy of the new housing settlement.
- Implementation of a condominium thermal power plant system with heat pumps to supply the cooling, refrigeration and domestic hot water systems.
- Use of extensive green roofs on the roof for water recovery and beneficial effects in terms of reducing the heat island effect of the urban block and evapotranspiration for the large PV plant on the roof.

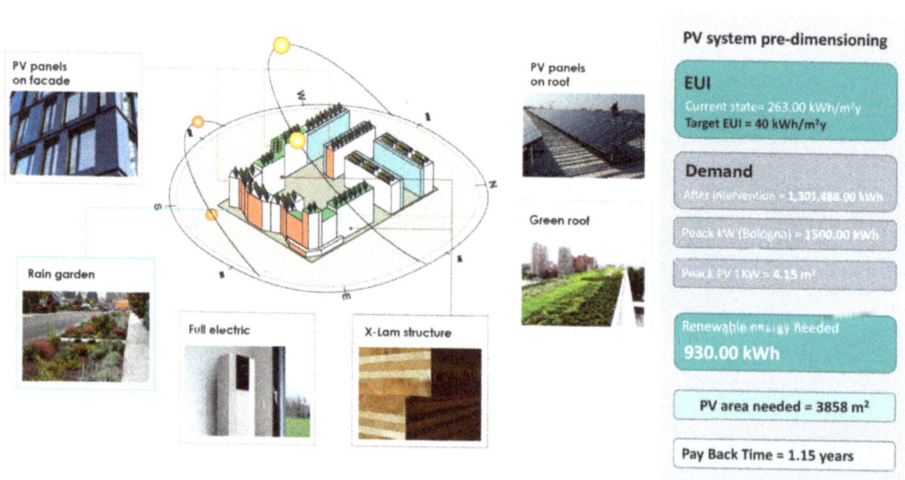

Figure 16. Overview of the different technological solutions implemented to achieve the project targets and data to verify them, in terms of energy efficiency and payback time. (© 2023, C. Mazzoli after Q. Palazzo and P. Dabbicco).

The main objective concerns the elimination of gas-fuelled plants to prefer the use of full-electric, in line with European directives to reduce CO_2 waste.

5. Conclusions

This paper seeks to illustrate the potential strategies for urban densification, employing a sustainable, and economically viable approach. It aligns with the successful model presented by the ABRACADABRA research project, extending its application to a larger scale of intervention beyond individual buildings. The practical efforts undertaken primarily concentrated on achieving energy self-sufficiency for an entire block of buildings located in San Donato district in Bologna. Maximizing economic gains through income generated by new construction, aimed at deep architectural and energy transformation.

The chosen block, identified for densification to create a zero-energy subdivision, possesses several advantageous qualities that align with the pursued strategies for urban infill. This implementation unfolded practically through a meta-design phase, wherein data collected during a multi-fold analysis played a crucial role in calibrating the entity of for the urban infill interventions and energy retrofit actions. This involved establishing numerical project targets that encapsulated the goals of substantial economic gain and a notable enhancement in energy performance of the area (reduction of 75% of the total energy demand, reaching beyond the EU and local regulation standards). Throughout the process, these project targets were continuously monitored and updated. The objective was to craft a design that effectively responded to the urban-scale strategies explored in this contribution, ensuring a harmonious integration of the densification and energy retrofit efforts. The proposed analysis and design method, with the use of parametric software to monitor performance requirements, can be replicated and applied to other similar urban contexts, allowing the test and validation of densification strategies. Although limited to a specific case study, the proposed design solutions demonstrate the technical and economic feasibility of urban regeneration aimed at implementing PEDs that might result in both attractive and profitable interventions for the environment, investors, and residents themselves.

The successful attainment of project target through this strategy underscores the effectiveness of interventions based on the combination of Add-ons and urban infill to enhance urban consolidated settlements, while avoiding further land consumption. Simultaneously, the commitment to deep renovation to align with current energy transformation objectives for existing stock contributes significantly to the broader European agenda. This collective effort aims to achieve climate-neutral cities in Europe by 2050, emphasizing the dual benefits of intensifying urban areas while meeting critical energy efficiency goals at both the local and continental levels.

The limitations of this densification strategy concern the possibility of adopting it only for non-historical building heritage, which is not protected by regulations in terms of reshape and deep renovation of the existing environment. Given the complexity and multiplicity of the criteria to be considered for achieving an optimized design solution, the proposed methodology is based on a multi-scalar and multi-criteria analysis, divided into different steps, supported by parametric digital tools. Further research development could consider the possibility of controlling the design process through the use of holistic digital platform combining and balancing all design criteria and solutions.

Author Contributions: Conceptualization, A.F.; methodology, C.M.; validation, C.M.; formal analysis C.M., D.P. and A.F.; design C.M.; investigation, A.F., C.M. and D.P.; data curation, C.M.; writing—original draft preparation, C.M.; writing—review and editing, C.M., D.P. and A.F.; supervision, A.F.; funding acquisition, A.F. ("ABRACADABRA" G.A. No 696126). All authors have read and agreed to the published version of the manuscript.

Funding: This research was funded by the EU Horizon 2020 Programme ABRACADABRA, G.A. No 696126 (https://cordis.europa.eu/project/id/696126 (accessed on 14 December 2023).

Data Availability Statement: Data are contained within the article.

Acknowledgments: The authors would like to thank Quilvia Palazzo and Pierpaolo Dabbicco for the contribution made in the framework of their master's degree thesis, entitled, respectively, "Strategie di densificazione urbana per comparti a zero energia. Il caso di un nuovo edificio in via Virginia Reiter" and "Strategie di densificazione urbana per comparti a zero energia: Il caso dei nuovi edifici in via del Lavoro" (Alma Mater Studiorum—University of Bologna; tutor: Ferrante, A.; co-tutor: Mazzoli, C.).

Conflicts of Interest: The authors declare no conflict of interest.

References

1. Ingersoll, R. *Sprawltown, Looking for the City on Its Edges Richard*, 1st ed.; Thompson, J.N., Ed.; Princeton Architectural Press: New York, NY, USA, 2006; ISBN 9781568985664.
2. European Commission. *The European Green Deal*; European Commission: Brussels, Belgium, 2019.
3. European Parliament. Directive (EU) 2018/844 of the European Parliament and of the Council of 30 May 2018 Amending Directive 2010/31/EU on the Energy Performance of Buildings and Directive 2012/27/EU on Energy Efficiency. *Off. J. Eur. Union* **2018**, *276*, 75–91. [CrossRef]
4. European Parliament. Directive (EU) 2018/2001 of the European Parliament and of the Council of 11 December 2018 on the Promotion of the Use of Energy from Renewable Sources (Recast). *Off. J. Eur. Union* **2018**, *2018*, 82–209.
5. European Commission. *Future Brief: No Net Land Take by 2050?* European Commission: Brussels, Belgium, 2016; Volume 67.
6. European Commission Europe 2020: Commission Proposes New Economic Strategy in Europe. European Union Website 2010, 1–2. Available online: https://ec.europa.eu/commission/presscorner/detail/en/IP_10_225 (accessed on 14 December 2023).
7. ISPRA. *Il Consumo Di Suolo in Italia*; ISPRA: Ispra, Italy, 2015; ISBN 9788844807030.
8. Papadaki, D.; Nikolaou, D.A.; Assimakopoulos, M.N. Circular Environmental Impact of Recycled Building Materials and Residential Renewable Energy. *Sustainability* **2022**, *14*, 4039. [CrossRef]
9. European Commission. *Roadmap to a Resource Efficient Europe COM(2011) 571 Final*; European Commission: Brussels, Belgium, 2011; Volume COM (2011).
10. European Commission Science for Environment Policy. European Union Website 2016, 2. Available online: https://environment.ec.europa.eu/research-and-innovation/science-environment-policy_en (accessed on 14 December 2023).
11. Mastellone, M.; Ruggiero, S.; Papadaki, D.; Barmparesos, N.; Fotopoulou, A.; Ferrante, A.; Assimakopoulos, M.N. Energy, Environmental Impact and Indoor Environmental Quality of Add-Ons in Buildings. *Sustainability* **2022**, *14*, 7605. [CrossRef]
12. EEA. *Urban Sustainability Issues—What Is a Resource-Efficient City?* EEA: Copenhagen, Denmark, 2015; ISBN 9789292137106.
13. Williams, K. *Does Intensifying Cities Make Them More Sustainable?* Burton, E., Jenks, M., Williams, K., Eds.; E&FN Spon: London, UK, 2000; ISBN 9780203827925.
14. Vuckovic, M.; Loibl, W.; Tötzer, T.; Stollnberger, R. Potential of Urban Densification to Mitigate the Effects of Heat Island in Vienna, Austria. *Environments* **2019**, *6*, 82. [CrossRef]
15. Iacovantuono, A. *Urban Infill L'interpretazione Dei Vuoti Nella Città Compatta, Sapienza Universita di Roma*. 2014. Available online: http://www.arc1.uniroma1.it/dottoratocomposizionearchitettonica/Dissertazioni/29_ArmandoIacovantuonoUrbanInfill.pdf (accessed on 14 December 2023).
16. European Commission. *Financial Support for Energy Efficiency in Buildings COM(2013) 225 Final*; European Commission: Brussels, Belgium, 2013; pp. 1–12.
17. European Parliament. *Directive 2012/27/EU of the European Parliament and of the Council of 25 October 2012 on Energy Efficiency*; European Parliament: Strasbourg, France, 2012; pp. 1–56.
18. Arbulu, M.; Oregi, X.; Etxepare, L.; Hernández-Minguillón, R.J. Barriers and Challenges of the Assessment Framework of the Commission Recommendation (EU) 2019/786 on Building Renovation by European RTD Projects. *Energy Build.* **2022**, *269*, 112267. [CrossRef]
19. European Commission. *Assistant Buildings' Addition to Retrofit, Adopt, Cure and Develop the Actual Buildings up to ZeRo Energy, Activating a Market for Deep Renovation*; European Commission: Brussels, Belgium, 2016.
20. Ferrante, A.; Fotopoulou, A.; Semprini, G.; Cantelli, D.; Ruggiero, S.; Karalis, M.; Efthymiou, C.; Papadaki, D.; Assimakopoulos, M.N. IEQ and Energy Improvement of Existing Buildings by Prefabricated Facade Additions: The Case of a Student House in Athens. *IOP Conf. Ser. Mater. Sci. Eng.* **2019**, *609*, 042047. [CrossRef]
21. European Council. 12–13 December 2019, Main Results. Available online: https://www.consilium.europa.eu/en/meetings/european-council/2019/12/12-13/ (accessed on 2 November 2023).
22. European Council Fit for 55—The EU's Plan for a Green Transition—Consilium. Available online: https://www.consilium.europa.eu/en/policies/green-deal/fit-for-55-the-eu-plan-for-a-green-transition/ (accessed on 14 December 2023).
23. Comune di Bologna Piano Urbanistico Generale Di Bologna. Available online: http://dru.iperbole.bologna.it/piano-urbanistico-generale (accessed on 2 November 2023).

24. Ministero Delle Imprese Decreto Ministeriale—Individuazione Della Tariffa Incentivante per La Remunerazione Degli Impianti a Fonti Rinnovabili Inseriti Nelle Configurazioni Sperimentali Di Autoconsumo Collettivo e Comunità Energetiche Rinnovabili. Available online: https://www.mimit.gov.it/it/normativa/decreti-ministeriali/decreto-ministeriale-16-settembre-2020-individuazione-della-tariffa-incentivante-per-la-remunerazione-degli-impianti-a-fonti-rinnovabili-inseriti-nelle-configurazioni-sperimentali-di-autoconsumo-c (accessed on 1 November 2023).
25. *Regione Emilia-Romagna Piano Triennale Di Attuazione 2022–2024 Piano Energetico Regionale*; 2017; pp. 1–123. Available online: https://energia.regione.emilia-romagna.it/piani-programmi-progetti/programmazione-regionale/piano-energetico-per/piano-energetico-regionale#autotoc-item-autotoc-0 (accessed on 14 December 2023).
26. Emilia-Romagna. *Ervet Piano Energetico Regionale 2030 Emilia-Romagna*; 2017; p. 114. Available online: https://energia.regione.emilia-romagna.it/piani-programmi-progetti/programmazione-regionale/piano-energetico-per/piano-energetico-regionale#autotoc-item-autotoc-0 (accessed on 14 December 2023).
27. Emilia-Romagna, A.L. della R. LEGGE REGIONALE 27 Maggio 2022, n. 5. 2022, n. 5, 1–6. Available online: https://bur.regione.emilia-romagna.it/bur/area-bollettini/bollettini-in-lavorazione/n-162-del-27-05-2022-parte-prima.2022-05-27.2212115799/promozione-e-sostegno-delle-comunita-energetiche-rinnovabili-e-degli-autoconsumatori-di-energia-rinnovabile-che-agiscono-collettivamente/l-r-n-5-del-2022 (accessed on 14 December 2023).
28. Gazzetta Ufficiale Della Repubblica Italiana DECRETO LEGISLATIVO 8 Novembre 2021, n. 199 2021. Available online: https://www.gazzettaufficiale.it/eli/id/2021/11/30/21G00214/sg (accessed on 14 December 2023).
29. ART-ER Emilia-Romagna TRANSIZIONE ENERGETICA FONTI RINNOVABILI COMUNITÀ ENERGETICHE NEUTRALITÀ CARBONICA I Quaderni per La Transizione Energetica: Comunità Energetiche Rinnovabili e Gruppi Di Autoconsumatori. 2022. Available online: https://energia.regione.emilia-romagna.it/notizie/notizie-home/2022/quaderno-comunita-energetiche-1.pdf (accessed on 14 December 2023).
30. Milic, V.; Amiri, S.; Moshfegh, B. A Systematic Approach to Predict the Economic and Environmental Effects of the Cost-Optimal Energy Renovation of a Historic Building District on the District Heating System. *Energies* **2020**, *13*, 276. [CrossRef]
31. Abdrabo, K.I.; Hamed, H.; Fouad, K.A.; Shehata, M.; Kantoush, S.A.; Sumi, T.; Elboshy, B.; Osman, T. A Methodological Approach towards Sustainable Urban Densification for Urban Sprawl Control at the Microscale: Case Study of Tanta, Egypt. *Sustainability* **2021**, *13*, 5360. [CrossRef]
32. Huang, B.; Xing, K.; Pullen, S.; Liao, L. Exploring Carbon Neutral Potential in Urban Densification: A Precinct Perspective and Scenario Analysis. *Sustainability* **2020**, *12*, 4814. [CrossRef]
33. Hoffman, A.; Camagni, R.P.; Latella, F. *Mezzogiorno e Scienze Regionali: L'Analisi e La Programmazione*; Angeli, F., Ed.; Franco Angeli Edizioni: Milano, Italy, 1992.
34. Ruggiero, S.; Assimakopoulos, M.N.; De Masi, R.F.; De Rossi, F.; Fotopoulou, A.; Papadaki, D.; Vanoli, G.P.; Ferrante, A. Multi-Disciplinary Analysis of Light Shelves Application within a Student Dormitory Refurbishment. *Sustainability* **2021**, *13*, 8251. [CrossRef]
35. New Buildings Institute (NBI). Available online: http://newbuildings.org/index.php?q=kpi (accessed on 2 November 2023).
36. Decreto Legislativo. Disposizioni in Materia Ambientale per Promuovere Misure Di Green Economy e per Il Contenimento Dell'uso Eccessivo Di Risorse Naturali. *Gazz. Uff. Della Repub. Ital.* **2015**, *221*, 103.
37. Charan, T.; Mackey, C.; Irani, A.; Polly, B.; Ray, S.; Fleming, K.; El Kontar, R.; Moore, N.; Elgindy, T.; Cutler, D.; et al. Integration of Open-Source Urbanopt and Dragonfly Energy Modeling Capabilities into Practitioner Workflows for District-Scale Planning and Design. *Energies* **2021**, *14*, 5931. [CrossRef]
38. Regione, E.-R. Prezzario Delle Opere Pubbliche Della Regione Emilia-Romagna—Territorio. Available online: https://territorio.regione.emilia-romagna.it/osservatorio/Elenco-regionale-prezzi/prezzari-regionali (accessed on 2 November 2023).

buildings MDPI

Article

Research on the Correlations between Spatial Morphological Indices and Carbon Emission during the Operational Stage of Built Environments for Old Communities in Cold Regions

Fei Zheng *, Yuqing Wang, Zhicheng Shen and Yuetao Wang

College of Architecture and Urban Planning, Shandong Jianzhu University, Jinan 250101, China
* Correspondence: 12914@sdjzu.edu.cn

Abstract: The escalation of the urban population and energy demands has exacerbated the carbon emission intensity at the operational stage of urban old communities. The spatial elements of the built environments comprising building groups, roads and landscape, and the spatial morphology of these elements, are endowed not only with human activities but also impact local microclimates and overall carbon emissions. Nonetheless, little attention has been paid to the correlation mechanism between the spatial morphology of the urban built environments and carbon emissions. In this paper, the aim is to combine carbon emissions simulation and statistical analysis to find the correlation between the spatial morphological indices and carbon emissions and to bridge the gaps. Thus, guided by the principles of urban energy modeling, this research adopts a parametric process of "information model construction–carbon emission simulation–statistical analysis". First, taking 60 typical samples of an old community in Jinan, China, as objects, morphological indices such as density, texture and layout are analyzed through regression analysis to highlight their impacts on carbon emissions. Then, a carbon emission prediction model based on spatial morphological indices is established and verified. The results show that the floor area ratio (FAR), building coverage ratio (BCR), enclosure degree (ED), shape factor (SF) and average road aspect ratio (AS) have significant impacts on carbon emissions during the operational stage. Among these indices, the FAR and the ED are identified as the pivotal influencers. The findings confirm the important role of spatial morphological design of old communities in cold regions in improving urban carbon reduction potential, and they provide theoretical underpinnings and empirical data as references for urban morphology design formulated within the context of low-carbon objectives.

Keywords: urban morphology; carbon emission; old communities; regression analysis; cold region

1. Introduction

Climate change is an issue of common concern to the global society. Extreme climate events caused by excessive carbon emissions have seriously affected the human living environment. The 2015 Paris Climate Conference not only adopted the first fair global agreement to deal with global warming, but it also aimed to limit the increase in temperature to 2 °C before 2100 [1]. In recent years, international standard-writing organizations have published the latest UNI EN ISO 14064:2018 and UNI EN ISO 14064:2019 [2–4] (Parts 1, 2 and 3) to improve the comparability and science of carbon footprint calculations across countries [5].

Human activity in cities is a major source of CO_2 emissions and a contribution to climate change [6]. As an important component of urban residential land, old communities have received much attention due to the influence of their spatial morphology on building energy consumption [7]. Traditional urban planning tends to adopt a large-scale building unit and high-density building layout [8]. This planning mode took less into account the impact of the spatial morphological indices of urban residential land on building energy consumption and carbon emissions [9]. In fact, urban planning and spatial optimization are playing an important role in the mitigation of CO_2 emissions at an urban level [10]. Traditionally, studies assessing the environmental impact of urban morphology had focused on GHG emissions caused by the operational stage of (a) buildings [11] (b), urban transport [12] (c) or both [13,14].

Carbon emission in cities is composed of the carbon source and carbon sink from activities of various land use properties [15]. Carbon source land shows different carbon emission characteristics through different land use properties. Carbon sink land shows different degrees of carbon clearing and carbon absorption capacity according to different land use properties. According to the data of the International Energy Agency, urbanized areas consume 60% of the world's energy and produce more than 70% of the world's carbon emissions [16,17]. The building industry accounts for one-third of global energy consumption, accounting for 28% of CO_2 emissions, of which residential buildings account for the largest proportion of energy consumption [18]. The carbon emission of urban residential land is an important part of urban carbon emission. In 2020, the carbon emissions of urban residential building in China during the operational phase totaled 901 million tons, accounting for 42% of the national building carbon emissions [19]. Studies evaluating the impact of urban spatial morphological indices on carbon emissions mainly focused on the operational stage [20]. Cuéllar-France and Azapagic used the LCA method to evaluate the carbon emissions of three common residential buildings in the UK, which were divided into three stages: the construction stage, operational stage and waste disposal stage. It was found that the carbon emissions in the operational stage contributed the most [21]. Based on the LCA method, Huang et al. calculated the carbon emissions in the whole life cycle of residential buildings in Shanghai and pointed out that the two stages of building materials' production and operation have the largest carbon emissions [22]. In addition, studies have also found that carbon emissions in the operational stage account for 70–90% of the whole life cycle carbon emissions [23–25]. Thus, it can be seen that the key to the peak of building carbon emissions is to control the carbon emissions in the construction operational stage. As the basic unit of energy consumption and carbon emissions in urban residential buildings, old communities produce a large amount of detectable data in their long life cycle, which has important practical conditions for carbon reduction research in the operational stage.

Meanwhile, research shows that there is a close correlation between carbon emissions, energy consumption and energy utilization in old communities, and the correlation is even more pronounced in cold areas [26,27]. A study by Chen Jialiang analyzed the relationship between the energy consumption and morphology of the northern community in the background and found that the distance from the sea, the construction time and the building height are the three most important factors affecting the residential energy consumption [28]. Comparing the energy consumption of Chinese and foreign buildings, Jiang Yi concluded that building energy consumption mainly depends on the difference in energy consumption mode and climate environment [29]. Professor Zhang Jie discovered in his research that in the cold region, where Jinan is located, the highest energy consumption method is electricity, followed by heating energy consumption [30]. Therefore, it is natural for people to believe that achieving energy-saving and emission reduction goals for residential buildings in cold regions will face even greater challenges.

Furthermore, the energy consumption of residential buildings is also affected by the local microclimate [31]. The formation of local microclimate is closely related to the urban spatial form [32]. Therefore, the spatial morphology of the built environment for old communities affects the balance of the internal microclimate, including air temperature, relative humidity, wind speed, wind direction and other microclimate elements, which then affects the energy consumption of heating and cooling and ultimately affects carbon emission during the operational stage. Through the simulation of the physical environment of residential communities with different spatial morphology, Xuan Wei investigated how incorporating a reasonable spatial design can enhance the microclimate environment of residential communities and effectively lower energy consumption and carbon emissions [33].

In addition, the utilization of solar energy has gradually become an important issue for the construction industry to implement the carbon reduction path [34]. On the one hand, the acquisition of sunlight will reduce lighting energy consumption. On the other hand, solar radiant heat reduces heating and cooling energy consumption by increasing solar energy production capacity, which reduces carbon emissions from non-renewable energy use. A large number of studies have verified that the morphological indices of residential buildings have a great influence on the acquisition of solar radiation [35–38]. Tathiane, Adolphe et al. studied the influence of different morphological indices on solar energy utilization potential. The results show that increasing the aspect ratio of the block is more conducive to increasing the acquisition of solar radiation than increasing the floor area ratio [39]. A study by Kruger et al. found that the north–south orientation of the building group can produce shading effects between buildings and reduce carbon emissions from building HVAC systems [40]. It can be seen that the improvement of the morphology of old communities can have a positive effect on energy saving and carbon reduction, whether from the perspective of the influence mechanism of space itself on the comfort of living environment or spatial planning on the implementation of auxiliary technology paths.

As shown above, although the influence of spatial morphology on building performance and energy consumption has been discussed, most of the studies focused on the single aspect of buildings and the urban level. There are few studies on the correlation between the spatial morphological indices of the old communities and carbon emissions, especially in cold regions. Studies are usually targeted at severe cold regions or mild climate zones. If considering only the single building morphology, it is easy to ignore the influence of the layout of the building group on the heat dissipation and heat demand of the building [41,42]. Additionally, the height, shape, shape factor, length of continuous building interface and size of the public area of the internal buildings in old communities may have an impact on the level of building energy consumption [43,44]. Accordingly, in the present study, we focused on the impact of spatial morphological indices of old communities on carbon emissions in cold regions. By quantitatively analyzing its underlying mechanisms, the study reveals the carbon reduction potential of building clusters in spatial planning and renovation within cities. Specifically, this is theoretically significant in guiding the urban building energy conservation and carbon reduction policies from the perspective of urban planning.

Given this background, the aim of this work is to quantitatively analyze the influence mechanism of spatial morphological indices of old communities on carbon emission and determine which factors play the greatest role in carbon reduction in order to provide recommendations to carry out low-carbon-oriented urban planning in the future.

In the cold region city of Jinan, China, 60 typical old communities are selected as research objects. This research starts by emphasizing the influence mechanism of urban morphology on carbon emission and expounding the knowledge gaps in the research and literature on carbon emissions in old communities. The Section 2 selects the study samples and explains the methodological approach. In Section 3, the results of this study are presented through three subsections: simulation of carbon emissions in old communities, calculation of morphological indices and statistical analysis of correlation between variables. The Section 4, further discusses how the spatial morphological indices of the old community affect carbon emission and the perspective of spatial morphological planning strategies towards carbon reduction. The Section 5 summarizes the conclusion of this paper, which is that the design of spatial morphology for old communities in cold regions has a significant impact on reducing carbon emissions during their operational stage.

2. Study Area and Methods

2.1. Study Area

The cold regions of China mainly refer to areas where the average temperature of the coldest month is −10–0 °C, and the daily average temperature of 5 °C is present for 90~145 days. Jinan is one of the representative cities in the cold region. According to the "2023 urban old community renovation plan table" and "Jinan 2023 old community renovation task table" issued by Jinan City, the article selected the old communities in the central urban area of Jinan City as the research object, which can represent the overall basic morphology of old communities and guide planning and renovation design. The distribution points of the sample area of the old community are shown in Figure 1a. The FAR and BCR of the sample area are significantly reduced from the center to the periphery (Figure 1b,c).

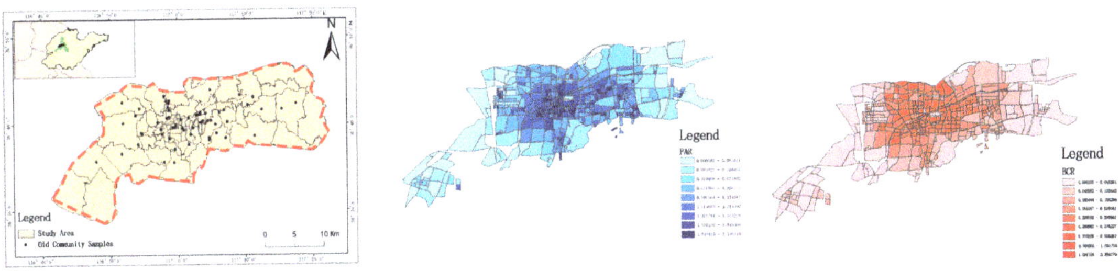

(a) (b) (c)

Figure 1. (**a**) Distribution of old community samples; (**b**) Distribution of FAR in the study area; (**c**) Distribution of BCR in the study area.

2.2. Methods

Currently, the research methods for building carbon emission and energy consumption are relatively mature. With the emphasis on urban energy consumption and carbon emissions, researchers began to expand the concept of building energy conservation and emission reduction design to the field of urban design. Accordingly, the urban energy modeling method (urban building energy modeling, UBEM) emerged. Swan and Ugursal divided the modelling approach into top-down and bottom-up approaches [39–45]. The gray box method as one of the bottom-up methods is used in this study. The gray box method can predict energy demand and energy consumption in urban areas by combining the physical model method and the data analysis method and integrating them into a model [46–51]. Since this model can perform statistical analysis of three types of data, including climate parameters, building geometry and non-geometric information, considering local microclimate and urban morphology, it can be applied to the simulation of energy consumption and carbon emission of old communities in cold regions.

Based on the UBEM method, this research constructs a parametric analysis process of "information model construction–carbon emission simulation–statistical analysis" to study the influence mechanism of spatial morphological indices and carbon emission of old communities.

The process is divided into four modules (Figure 2): (1) Parameter preset module: Use ArchGis10.6.1 software to identify the spatial morphological indices of the old community and obtain the building type, envelope structure, energy load and other parameters of the old community samples combined with the research content; (2) Shape generation module: Based on the identification of morphological indices, UMI-Site [52], Grasshopper and other plug-ins are used to calculate the morphological indices and the construction of the old community model; (3) Performance calculation module: The UWG operator is used to simulate the microclimate of the old residential area. The results are presented in the EPW meteorological data file. Then, the EPW file is imported into UMI for carbon emission simulation; (4) Data analysis module: Based on the statistical analysis method, the calculation value of the spatial morphological indices and the carbon emission simulation value are analyzed accordingly.

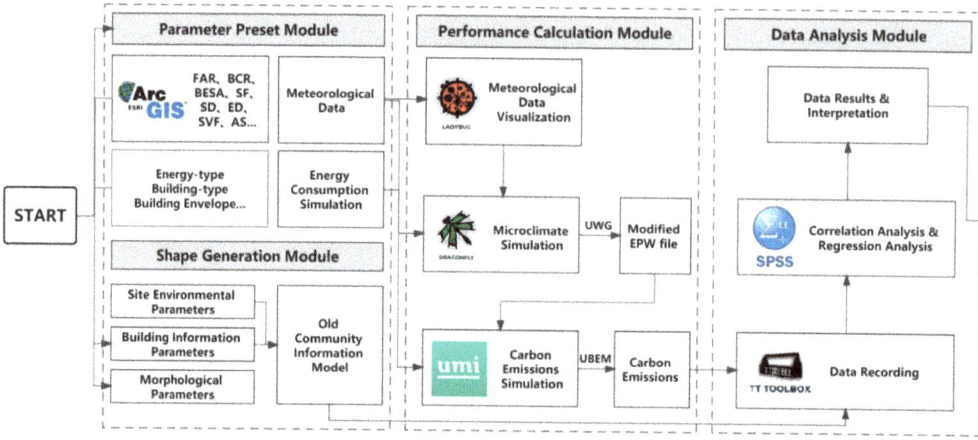

Figure 2. Process of "information model construction–carbon emission simulation–statistical analysis".

2.2.1. The Construction of Building Information Model

The building information model includes the following parts: climate data, building geometric data and non-geometric data of buildings.

- Acquisition of climate parameters

The climate parameters of Jinan City were extracted through the ladybug. The average temperature in Jinan is 14.7 °C, and the coldest month is January, with an average temperature of −0.4 °C. The hottest month is July, with an average temperature of 27.5 °C. The dominant wind direction is southwest and northeast, followed by east, north and south. Northwest wind is the least dominant. The summer and winter climate characteristics in Jinan are different: the summer temperature is hot, with an average temperature of 26.7 °C. This season is not only hot but also rainy, with the characteristics of hot rain in the same season. The average air temperature in the winter is about 1 °C.

- Construction of the geometric model of the old community

At first, geometric data, such as the outline and morphology of old plots, are captured by Google satellite map. Then, we use the Grasshopper1.0.0007 platform to filter and identify 3D data (shp data) of buildings and sites, extracted and classified according to the gis data in the satellite openstreetmap. The 3D information model of the old community was established in Rhino to facilitate editing in the energy consumption simulation platform. Next, we use the @it plug-in to identify and extract the shp file, establish the geometric information of the site and building and use the excel plugin to extract the dbf file containing the building height and layer information, giving the height information to the building block, and finally generate the geometry model of old communities.

- Input of building non-geometric attribute information

Building non-geometric attribute information can be provided by the plug-in Template Library Editor, which is a standalone application in UMI for creating, managing and editing XML Template Library Files (TLF). TLF files can exchange and store information regarding the materials, structures, schedules, heat loads and morphology. By editing the template of building parameters in the TLF file, non-geometric information of old residential areas for carbon emission simulation can be obtained (Figure 3).

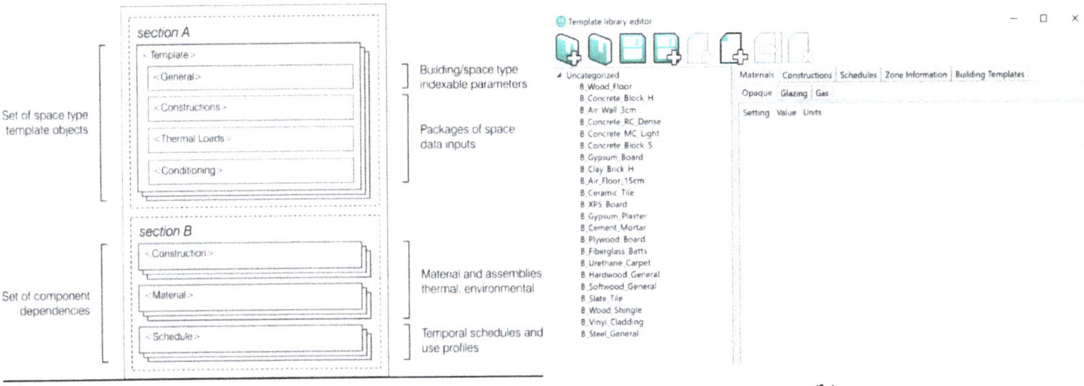

Figure 3. (**a**) Establish TLF template; (**b**) UMI-TLF extractor.

2.2.2. Extraction and Measurement of Morphological Indices

- Extraction of the morphological indices

Prof. Meta Berghauser Pont, University of Delft, The Netherlands, proposed a chart to evaluate the association between morphological density and urban morphology. She combined the four morphological density indices of building floor area ratio, building density, average number of floors and open space rate. She named them as "Spacemate" [53], which can reflect the possibility of climate resources entering inside the complex. Urban morphological texture can reflect the organizational structure between material entities and space in the urban environment, and it has an important impact on the construction of local microclimate within the building complex [54]. Considering the comprehensive description of the space morphology of the old area and the influence of climate resources in the area, this research selected the floor area ratio (FAR) and building coverage ratio (BCR) as the morphological density indices. Additionally, we selected building exterior surface area (BESA), shape factor (SF), enclosure degree (ED), scattered degree (SD), aspect ratio (AS), sky visual factor (SVF) as morphological texture indices. The above eight indices can comprehensively describe and evaluate the overall spatial morphology quality of the old community [55]. In addition, the old community can be divided into four categories—point-type, slab-type, mixed-type and enclosed-type—based on its architectural form.

- Calculation of morphological indices

The advantages of parametric modeling lie in its ability to describe the geometric features of building morphology in the form of data information [56]. FAR and BCR can be directly measured in the site module of the UMI. Gross floor area (m^2) can calculate the total building area. Site ground area can calculate the area of the site. Floor area ratio can calculate the building floor area ratio. Occupants can calculate the bottom area of the building. Thus, BCR can be obtained. BESA, SF, ED, SD, AS need to be obtained from the three-dimensional information of the building and site model. The following table (Table 1) summarizes the connotation and extraction methods of different morphological indices. The extraction results of sample morphology are shown in Table 2.

Table 1. Morphological indices.

Morphological Indices	Formula	Connotation	Extraction Method	Schematic Diagram
Floor Area Ration (FAR)	FAR = Gross Floor Area/Site Area Reflects the spatial distribution characteristics of buildings	FAR = Gross Floor Area/Site Area Reflects the spatial distribution characteristics of buildings	UMI-Site Module	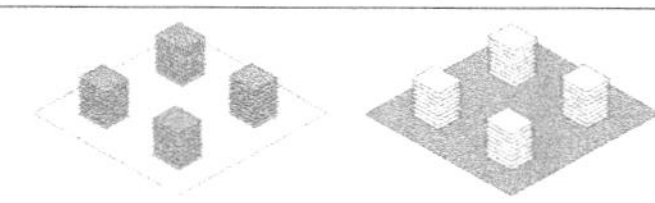
Building Coverage Ratic (BCR)	BCR = Floor Space/Site Area	Reflects the density of urban or regional buildings		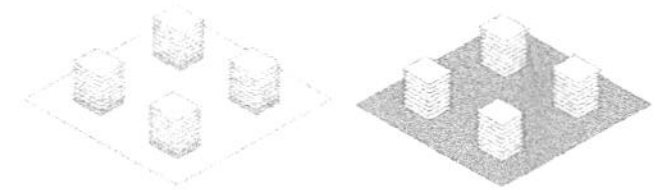
Building Exterior Surface Area (BESA)	BESA = Building Exterior Surface Area/Site Area	Reflects the ability of the outer surface of the building to absorb solar radiation	Grasshopper Module	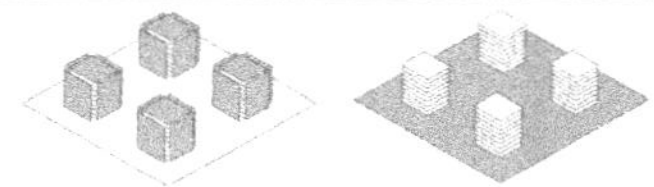
Shape Factor (SF)	SF = Building Exterior Surface Area/Volume	Reflects the complexity of the building form and the surface area of the enclosure structure		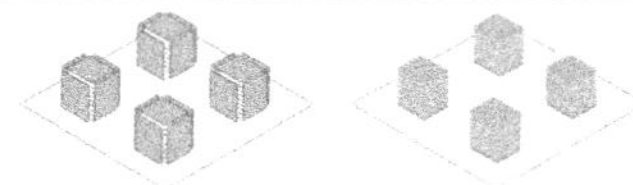
Enclosure Degree (ED)	ED = Building Façade Length/Building Control-Line Length	Reflects the openness and closure of a region or space		-
Scattered Degree (SD)	SD = Height Max-Height Average	Reflects the degree of dispersion of buildings in an area or space		-
Average Road Aspect Rati> (AS)	AS = Height Average /Road Width	Reflects the basic morphological unit formed by buildings and roads on both sides of the road in the plot		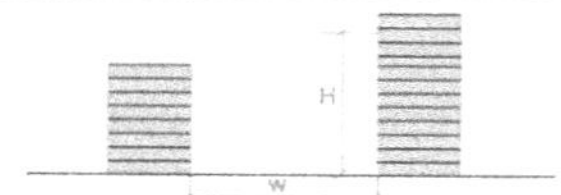
Sky View Factor (SVF)	-	Reflects the shielding degree of building density, height and shape from the surrounding environment	ShadingMask Module	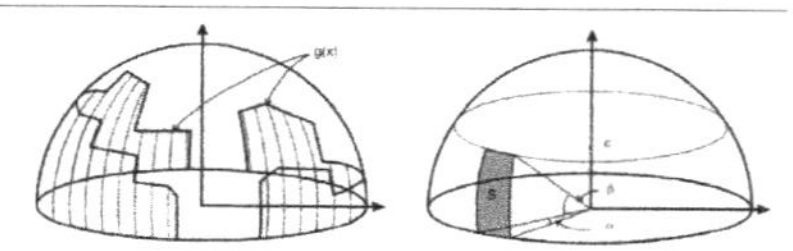

Table 2. Sample morphology.

Type	Samples				
Slab-type			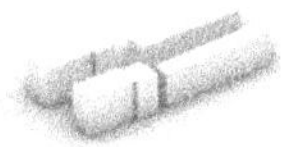		...
Point-type					...
Mixed-type	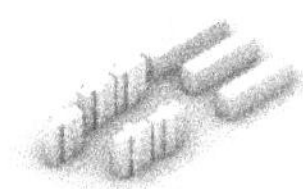				...
Enclosed-type		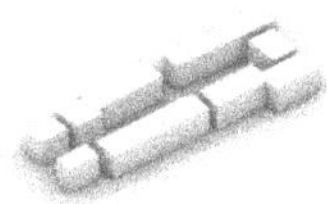			...

2.2.3. Carbon Emission Simulation

- Simulation parameter setting

This research uses the physical model method and the performance calculation module to calculate the carbon emission intensity of the old community.

Based on the sample research and literature review, this research uses the control variable method to characterize the control of building thermal parameters. Parameters related to physical simulation, such as building plane form and window to wall ratio, are excluded to exclude the interference of building monomer from the simulation results. According to the provisions of heat transfer coefficient and thermal inertia in the "Design Standard for Energy Efficiency of Residential Buildings in Severe Cold and Cold Areas (JGJ26-2018)" [57], combined with the investigation results of thermal parameters of old residential areas, this research obtained the main envelope structure parameters of old residential areas for carbon emission simulation (Table 3).

Table 3. Thermal parameters of building envelope.

Enclosure Structure	Exterior Wall	Floor Slab	Window	Roof	Partition Wall
Heat Transfer Coefficient (W/(m^2·K))	1.6	1.5	3.0 (SHGC = 0.6; South/North Window–Wall Ratio = 0.35; East/West Window–Wall Ratio = 0)	1.0	1.0

Secondly, it is necessary to set the energy consumption parameters of the sample area. In the energy consumption simulation, the "schedule" refers to the operational mode of various equipment, lighting, heating, ventilation and other systems in the building in different time periods. The "schedule" can accurately calculate the energy consumption and carbon emissions by simulating the daily use of the equipment. The "schedule" is usually composed of a series of time periods and values in the building simulation software. Each period corresponds to a time range, such as a specific period of the day, and the value of the period represents the running state of the system during that period. This research combines the "Code for Design of Heating Ventilation and Air Conditioning of Civil Buildings (GB50736-2012)" [58], "Green Performance Calculation Standard for Civil Buildings (JGJT449-2018)" [59] and the energy consumption survey results of the sample area to set the energy consumption parameters of the construction equipment during the operational stage (Table 4).

Table 4. Energy consumption parameter setting.

Parameter Type		Parameter Settings	
HVAC System	Type	Domestic Split Air Conditioner	
		Natural Gas Heating	
	Energy Efficiency	Refrigeration Energy Efficiency Ratio	2.3
		Heating Energy Efficiency Ratio	1.9
	Temperature	Refrigeration	7:00 am–8:00 pm, 20 °C 8:00 pm–7:00 am, 18 °C
		Heating	7:00 am–8:00 pm, 26 °C 8:00 pm–7:00 am, 27 °C
Occupants	Density	0.025 per/m^2	
	Indoor Activity	7:00 pm–8:00 am, 1 h 8:00 am–9:00 am and 6:00 pm–7:00 pm, 0.7 h 9:00 am–6:00 pm, 0.3 h	
Domestic Water	Time Period	7:00 am–12:am and 9:00 pm–11:00 pm, 1 h 12:00 am–1:00 pm, 0.6 h 6:00 pm–9:00 pm, 0.2 h	
Lighting System	Illuminance	200 lux	
	Lighting Power Density	7 W/m^2	
	Illuminating Period	7:00 am–8:00 am and 10:00 am–12:00 am and 11:00 pm–12:00 pm, 0.2 h 8:00 am–10:00 am and 7:00 pm–8:00 pm, 0.5 h 8:00 pm–12:00 pm, 0.8 h	
Equipment System	Power	5 W/m^2	
	Time Period	1:00 am–6:00 am, 0.2 h 7:00 am–8:00 am and 6:00 pm and 11:00 pm–12:00 pm, 0.6 h 9:00 pm–11:00 pm, 0.8 h 7:00 pm–9:00 pm, 1 h	

- Simulation result

First, the microclimate conditions need to be the output parameters. The research simulates the microclimate according to the following three steps: building morphology setting, urban morphology setting and simulation operation. Second, the microclimate EPW file of the sample old village will be imported into UMI-Energy to measure the carbon emission. The carbon emission simulation of the UMI plug-in is part of its full life cycle (LC) module. Currently, this module only allows the calculation of the basic implicit environmental impact related to building materials and does not apply to the calculation of the impact of carbon emissions during the operation stage. Hence, carbon emissions in the operational phase must be converted by obtaining energy consumption results in the operational energy module. This module can obtain the annual total energy consumption and energy consumption per unit area of lighting, heating, refrigeration, equipment, domestic water and gas in the sample area (Figure 4). Based on the IPCC method, the simulation results of various energy sources are multiplied by their carbon emission factors (Table 5) to obtain the carbon emissions per unit area of each energy source. Finally, the carbon emissions of various types are summarized to obtain the overall carbon emissions per unit area, namely the carbon emission intensity.

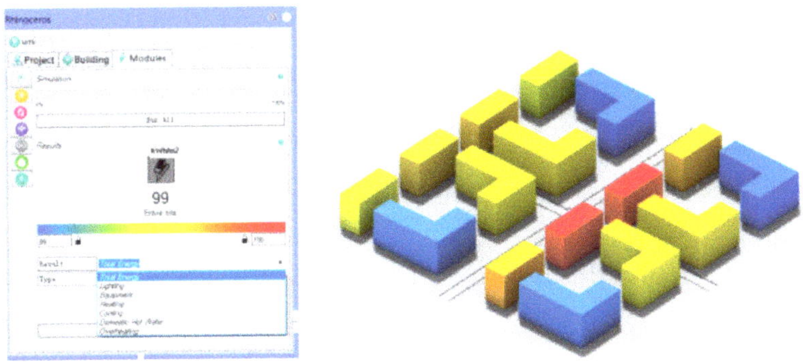

Figure 4. Operating energy computing module.

Table 5. Carbon emission factors.

	Energy Consumption Unit	Factor	Source	Carbon Emission Units
Electrical Energy	kWh/month	0.758	Ministry of Science and Technology of China	$KgCO_2$/kWh
Combustion Gas	m^3/month	0.232	Ministry of Science and Technology of China	$KgCO_2$/kWh
Domestic Water	t/month	2.77	The China Energy Management Network	$KgCO_2/m^3$

2.2.4. Statistical Analyses

This section uses correlation analysis and multiple linear regression analysis of the statistical analysis method to establish a carbon emission prediction analysis model based on the spatial morphological indices of old communities in order to establish the influence mechanism of different morphological indices and carbon emissions.

- Correlation analysis

This research uses the Pearson correlation analysis to test whether there is a correlation between independent variables and dependent variables and to establish a correlation matrix to screen morphological indices, eliminating parameters with strong collinearity. At the same time, correlation analysis can explain the influence of single variables on different types of old communities. Pearson correlation analysis explains the strength and direction of the relationship between the two variables. The Pearson correlation coefficient is usually represented by r, with values ranging from −1 to 1, where 0 indicates no linear correlation, 1 represents a complete positive correlation, and −1 represents a complete negative correlation.

- Multiple linear regression analysis

Using the morphological indices as independent variables, the prediction analysis model of carbon emission was established to predict the overall carbon emission values of samples. Finally, the actual data are selected to test the error of the regression equation to ensure its scientific nature. The regression equation can more specifically show the key morphological indices and their influence laws, which affect the carbon emissions of old communities.

3. Results

3.1. Spatial Morphological Indices

After the establishment of 60 sample models, this section calculates and counts the morphological indices of the old residential areas and draws the box plot of morphological distribution. The statistical morphological parameter information is shown in the following table (Table 6). Categories A, B, C and D refer to slab-type, point-type, mixed-type and enclosed-type, respectively. According to the calculation results and the distribution map of different morphological indices (Figure 5), it can be found that the BESA of the enclosed-type buildings is high, and the SD of point-type buildings is large, with a high body complexity. According to the morphological calculation data, the ED of the enclosed-type building group is the largest; meanwhile, the SD of the mixed-type building group is the largest. In addition, the SVF of the slab-type building group is the largest, indicating that there is a good vision and mutual occlusion level in slab-type old communities. Furthermore, the enclosed-type buildings have the highest FAR and BCR, which indicates that the enclosed-type building group has high land development intensity, while the land development intensity of the other old communities is insufficient. Moreover, the AS of the point-type building group is at a high level, followed by mixed-type buildings.

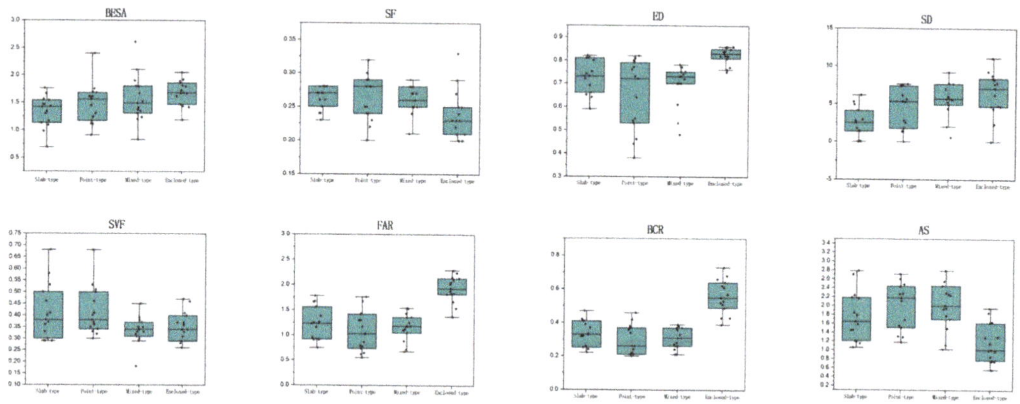

Figure 5. Data distribution of morphological indices for sample old communities.

Table 6. Statistical results of morphological indices.

	BESA	SF	ED	SD	SVF	FAR	BCR	AS
A-1	0.98	0.26	0.74	2.50	0.29	1.25	0.32	1.45
A-2	1.30	0.25	0.81	4.10	0.40	1.36	0.39	1.2
A-3	1.46	0.24	0.8	2.50	0.38	1.67	0.41	1.84
A-4	1.43	0.28	0.82	2.90	0.29	1.07	0.37	1.14
A-5	1.42	0.27	0.72	4.90	0.29	1.42	0.42	1.43
A-6	0.69	0.28	0.7	0	0.58	1.01	0.22	1.88
A-7	1.66	0.27	0.73	0	0.30	1.16	0.26	1.64
A-8	1.47	0.24	0.81	2.50	0.33	1.56	0.41	1.15
A-9	1.34	0.28	0.69	1.4	0.37	0.92	0.25	1.18
A-10	1.54	0.26	0.67	0	0.41	1.21	0.24	1.82
A-11	1.56	0.28	0.81	2.8	0.36	0.91	0.25	1.05
A-12	1.13	0.27	0.73	5.25	0.5	1.23	0.32	1.45
A-13	1.09	0.28	0.79	1.36	0.53	0.92	0.33	1.58
A-14	1.17	0.26	0.75	6.13	0.68	1.26	0.33	1.71
A-15	1.76	0.23	0.82	0	0.46	1.08	0.47	1.62
B-1	1.61	0.29	0.45	1.40	0.37	0.70	0.20	2.18
B-2	1.24	0.29	0.55	0	0.41	0.77	0.22	1.51
B-3	1.11	0.24	0.81	2.80	0.36	1.21	0.42	1.49
B-4	1.75	0.28	0.64	5.25	0.50	0.96	0.24	2.13
B-5	1.12	0.24	0.8	1.36	0.53	1.32	0.37	2.26
B-6	0.91	0.25	0.77	6.13	0.68	1.3	0.34	1.05
B-7	1.69	0.32	0.73	1.75	0.46	1.17	0.26	1.17
B-8	2.40	0.22	0.46	2.53	0.51	0.84	0.21	2.27
B-9	1.68	0.23	0.68	7.63	0.34	1.21	0.46	1.69
B-10	1.56	0.29	0.52	5.32	0.3	0.62	0.21	2.44

Table 6. *Cont.*

	BESA	SF	ED	SD	SVF	FAR	BCR	AS
B-11	1.61	0.25	0.76	7.4	0.35	0.91	0.36	1.31
B-12	1.45	0.20	0.79	5.36	0.4	0.98	0.37	2.43
B-13	1.6	0.28	0.58	7.39	0.38	0.78	0.38	1.55
B-14	1.31	0.28	0.72	5.62	0.32	1.03	0.26	1.14
B-15	1.17	0.30	0.53	7.53	0.33	0.73	0.21	1.24
C-1	1.31	0.24	0.78	5.25	0.35	1.43	0.37	1.28
C-2	0.83	0.28	0.7	4.90	0.38	1.05	0.26	1.36
C-3	1.35	0.26	0.74	0.58	0.37	1.25	0.33	2.26
C-4	2.61	0.28	0.61	7.00	0.18	1.46	0.24	1.02
C-5	1.49	0.27	0.73	4.38	0.35	1.18	0.28	2.79
C-6	1.90	0.21	0.7	19.04	0.32	1.1	0.27	2.45
C-7	2.10	0.28	0.7	2.00	0.33	1.13	0.26	2.55
C-8	1.37	0.25	0.75	5.95	0.34	1.35	0.36	1.11
C-9	1.8	0.29	0.53	5.67	0.31	0.77	0.21	1.93
C-10	1.79	0.27	0.73	9.17	0.39	1.18	0.31	1.95
C-11	1.67	0.25	0.77	5.68	0.45	1.53	0.38	1.49
C-12	1.2	0.24	0.76	5.94	0.29	0.92	0.21	2.79
C-13	1.52	0.25	0.75	7.63	0.33	1.36	0.37	2.24
C-14	1.24	0.27	0.72	5.53	0.35	1.18	0.31	1.7
C-15	1.41	0.25	0.77	7.71	0.3	1.54	0.39	1.78
D-1	1.19	0.24	0.81	2.33	0.37	1.67	0.73	0.75
D-2	1.86	0.27	0.83	4.67	0.29	1.88	0.55	0.99
D-3	1.89	0.22	0.88	0.00	0.28	2.02	0.52	1.85
D-4	2.05	0.33	0.91	4.38	0.31	2.15	0.51	0.85
D-5	1.92	0.20	0.86	6.22	0.33	1.89	0.49	1.64
D-6	1.47	0.23	0.84	8.75	0.29	1.93	0.43	1.95
D-7	1.42	0.21	0.86	4.67	0.46	1.82	0.43	1.3
D-8	1.67	0.29	0.84	11.08	0.26	1.54	0.39	1.62
D-9	1.81	0.25	0.75	8.2	0.47	1.37	0.53	1.01
D-10	1.79	0.25	0.96	7.77	0.4	2.05	0.68	0.62
D-11	1.62	0.23	0.82	9.25	0.37	2.13	0.66	0.76
D-12	1.68	0.23	0.82	7.06	0.41	2.29	0.64	0.77
D-13	1.72	0.21	0.88	8.36	0.34	2.1	0.62	0.54
D-14	1.44	0.20	1	7.63	0.36	1.83	0.61	0.56
D-15	1.67	0.20	0.95	4.74	0.34	2.24	0.57	0.62

3.2. *Results of the Carbon Emission Simulation*

In this section, the annual energy consumption of various types of energy in the sample old communities is simulated through the energy module of the UMI plug-in. Then, the energy consumption is converted into carbon emissions per unit area through the carbon emission factors (Table 7).

Table 7. Simulation results of carbon emissions per unit area during the operational stage.

	Carbon Emissions ($KgCO_2/m^2$)		Carbon Emissions ($KgCO_2/m^2$)		Carbon Emissions ($KgCO_2/m^2$)		Carbon Emissions ($KgCO_2/m^2$)
A-1	74.28	B-1	86.41	C-1	72.77	D-1	63.67
A-2	72.01	B-2	78.83	C-2	75.80	D-2	68.22
A-3	70.49	B-3	70.49	C-3	74.28	D-3	68.98
A-4	72.77	B-4	78.07	C-4	78.07	D-4	69.74
A-5	75.80	B-5	72.77	C-5	75.04	D-5	69.74
A-6	78.83	B-6	73.53	C-6	75.80	D-6	69.74
A-7	75.04	B-7	75.04	C-7	75.80	D-7	70.49
A-8	71.25	B-8	81.11	C-8	73.53	D-8	72.77
A-9	76.55	B-9	70.49	C-9	80.35	D-9	73.53
A-10	78.07	B-10	84.14	C-10	75.04	D-10	65.95
A-11	77.32	B-11	73.53	C-11	72.77	D-11	65.95
A-12	75.04	B-12	72.77	C-12	80.35	D-12	65.95
A-13	76.56	B-13	78.83	C-13	73.53	D-13	65.95
A-14	74.28	B-14	75.80	C-14	75.04	D-14	66.70
A-15	69.74	B-15	79.59	C-15	72.77	D-15	67.46

According to the box plot of carbon emissions per unit area for old communities (Figure 6), it can be observed that the overall distribution of carbon emissions per unit area of the slab-type building group is relatively uniform, with a median value of approximately 74 $KgCO_2/m^2$, which suggests that there is not much variation in carbon emissions between the slab-type buildings in the sample. Additionally, the box plot shows that the median carbon emissions per unit area for point-type buildings are relatively high, at approximately 76 $KgCO_2/m^2$, indicating that there are more individuals with high carbon emissions for the point-type building group. Furthermore, the median carbon emissions per unit area for mixed-type neighborhoods are approximately 75 $KgCO_2/m^2$, similar to slab-type and point-type buildings, and the distribution of carbon emissions is also relatively uniform. Notably, the overall carbon emissions per unit area for enclosed-type old neighborhoods are relatively low, with a relatively concentrated distribution and a median value of approximately 67 $KgCO_2/m^2$.

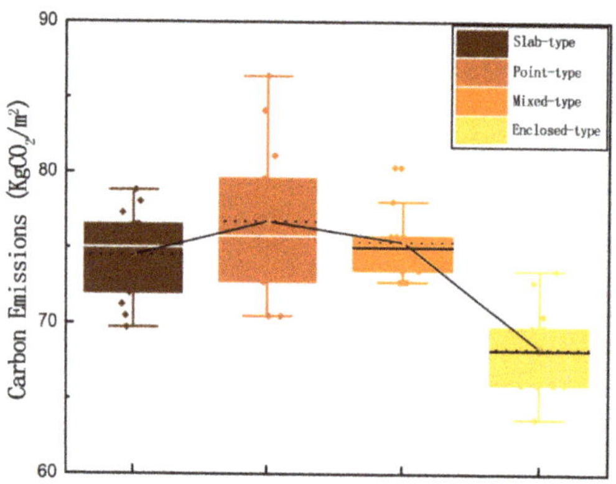

Figure 6. Box plots of carbon emission distribution of four types of old communities.

3.3. Statistical Analysis

3.3.1. Correlations between Morphological Indices and Carbon Emission

The calculated results of the eight spatial morphological indices obtained with correlation analysis are shown in Table 8.

- FAR

According to the research sample data, the overall FAR is in the range of 0.70–2.24, which is slightly lower than the standard level. The scatter plot shows that there is a significant correlation between carbon emissions and FAR. The carbon emissions per unit area for samples with FAR between 1.5 and 2.5 exhibit a relatively large degree of dispersion. The R2 and partial regression coefficient (B) values of the fitted equation are 0.758 and −7.07, respectively. This suggests that there is a certain negative linear correlation between the carbon emission intensity per unit area and the FAR for samples within the range of 0.35 to 2.77 in terms of FAR. On the whole, with the increase in FAR, the overall building aggregation degree in the old community increases, resulting in the reduction in heat island effect and heat dissipation, which reduces the heating energy consumption in the community, thus affecting the overall carbon emission level of the old community.

Furthermore, from the quadratic regression results, there is a strong negative linear correlation between the point-type and the mixed-type building group and the carbon emissions per unit area. On the whole, the influence degree of floor area ratio on the four building types is ranked as follows: point-type > mixed-type > slab-type > enclosed-type.

Table 8. Correlation analysis of morphological indices and carbon emission.

Impact on Carbon Emissions	The Degree of Influence on Four Types of Old Communities	Impact on Carbon Emissions	The Degree of Influence on Four Types of Old Communities
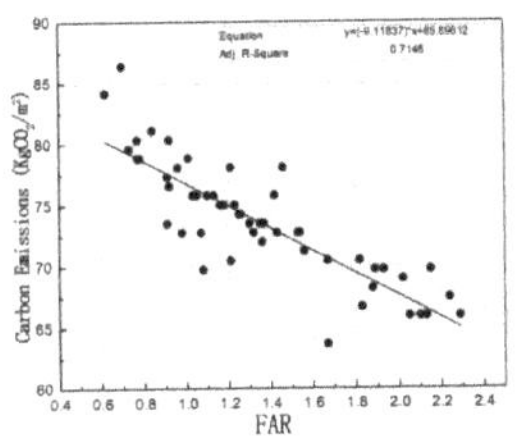	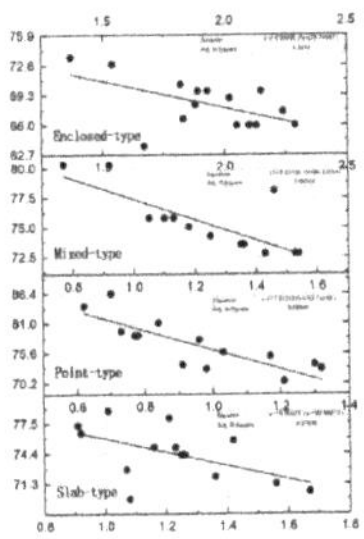	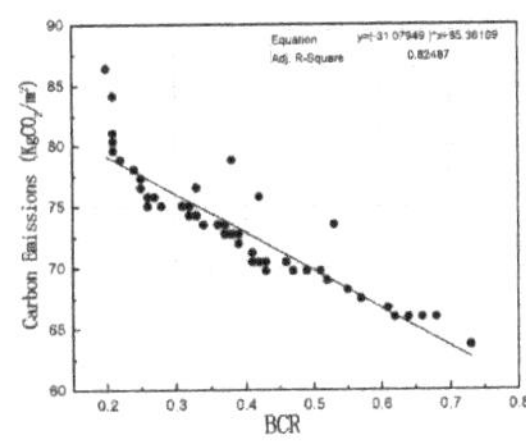	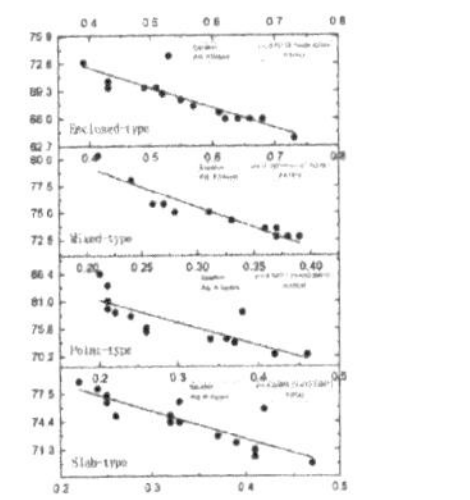
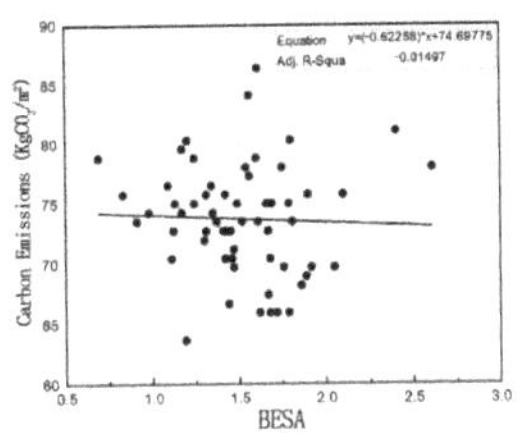	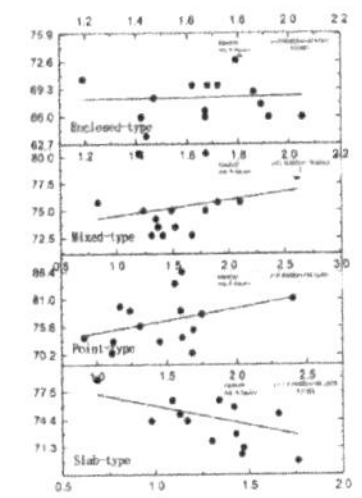	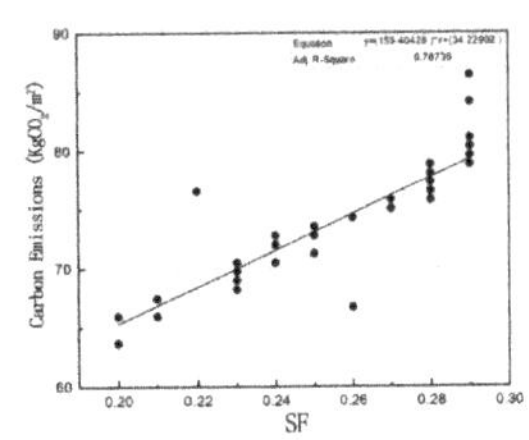	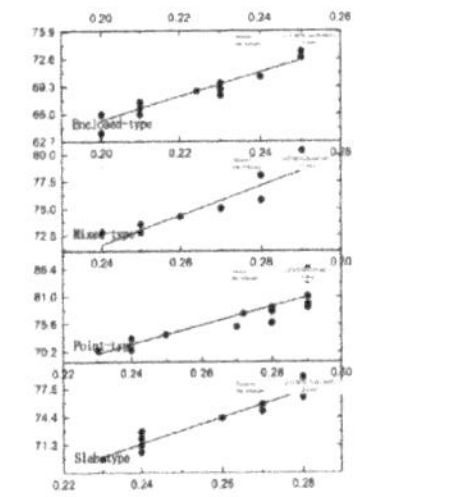
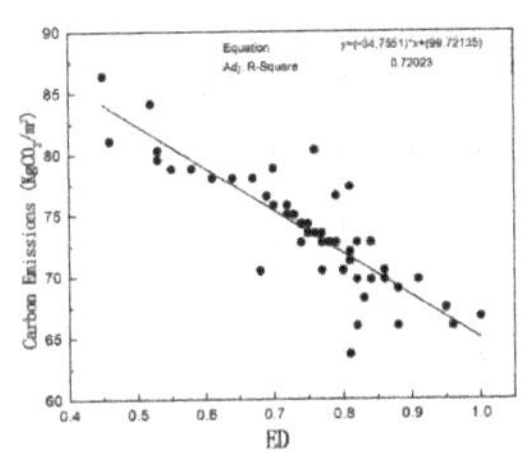	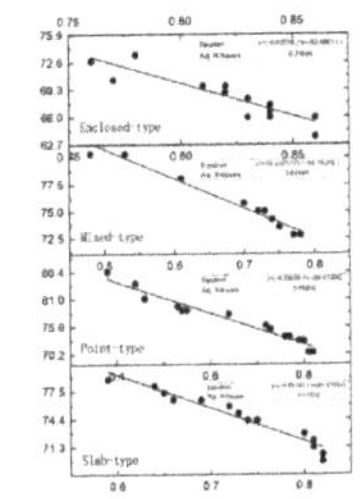	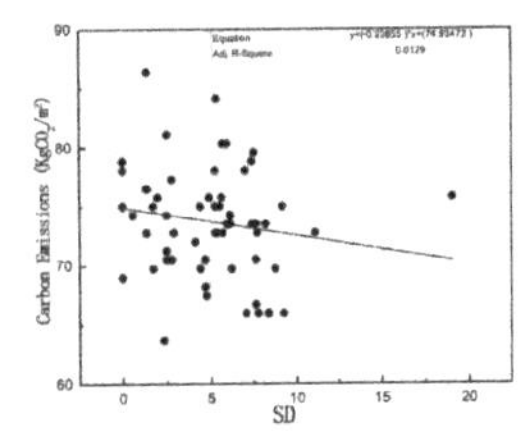	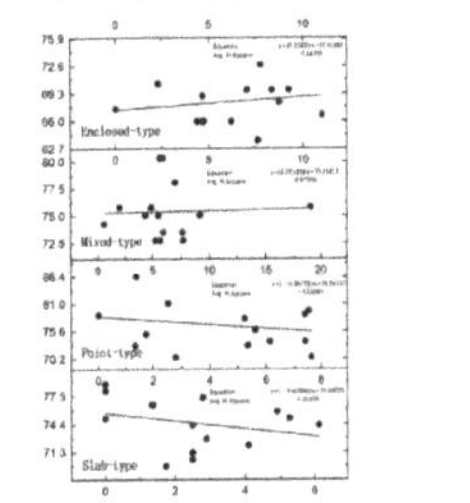
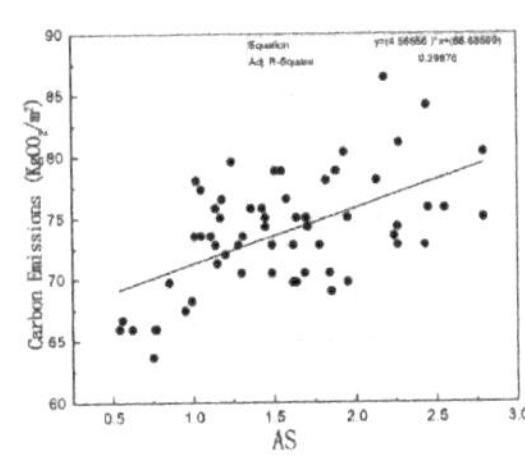	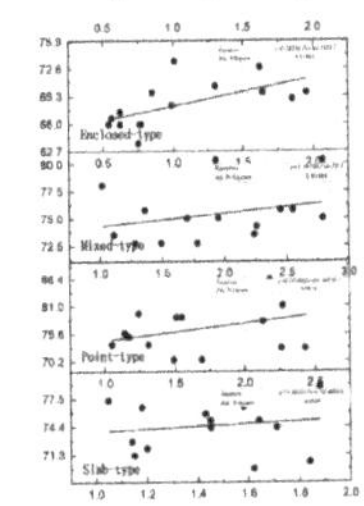	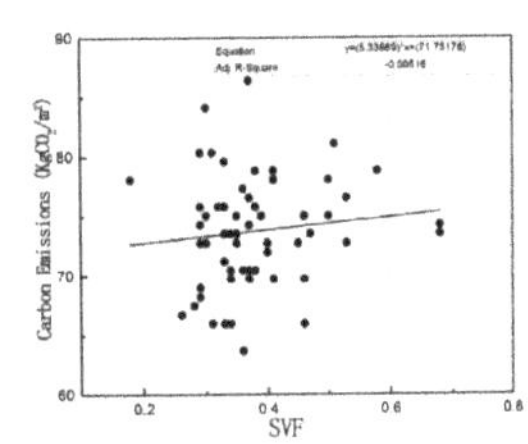	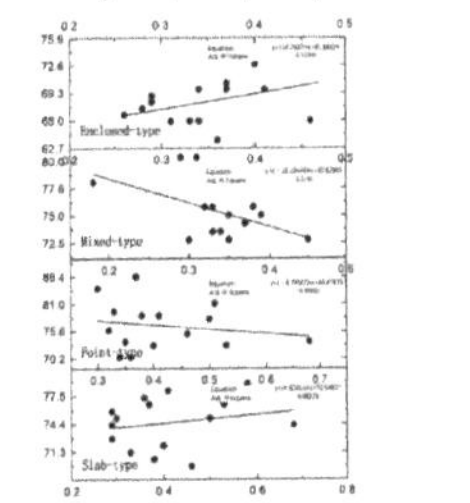

- BCR

There is a clear linear fitting relationship between BCR and carbon emissions. The R2 and partial regression coefficient (B) values of the fitted equation are 0.824 and -31.107949, showing a high linear negative correlation. Based on the quadratic regression results of BCR and carbon emissions for the four building types, the influence degree of BCR on carbon emissions is ranked as follows: point-type > mixed-type > slab-type > enclosed-type.

- BESA

The results show that there is a linear trend between carbon emission intensity and BESA, but the R2 of the two fitting equations is only 0.015, indicating that the linear relationship between the two is not significant enough, and the data are more scattered. From the quadratic regression analysis results, there are differences in the relationship between the BESA and the carbon emission intensity. The BESA fits the carbon emission intensity in different cell types. This means that the BESA may not be a major factor affecting the carbon intensity.

- SF

The SF refers to the relationship between the shape of the building and the aerodynamic performance of the building, and it is also an important index used to measure the heat transfer ability of the buildings. The R2 and partial regression coefficient (B) values of the fitted equation are 0.78736 and 155.40428, which indicates that there is a significant positive correlation between the SF and the carbon emission intensity. From the quadratic regression results, it can be found that the influence degree of SF on the carbon emissions of four building types is ranked as follows: point-type > slab-type > mixed-type > enclosed-type.

- ED

The results show that the R2 and partial regression coefficient (B) values of the fitted equation are 0.72 and −34.755, indicating that there is a significant negative correlation between them. The larger the ED, the lower the heat loss of the building. This means that increasing the ED to a certain degree can reduce the heating energy consumption of buildings, thereby lowering the overall carbon emissions of the old communities. From the quadratic regression results, it can be found that the influence degree of SF on the carbon emissions of four building types is ranked as follows: point-type > slab-type > mixed-type > enclosed-type.

- SD

The R2 of the fitting equation of carbon emission per unit area and SD in old communities is 0.0129, and the partial regression coefficient (B) is −0.23955, indicating that there is no obvious linear correlation between the two. From the perspective of the quadratic regression result, the patchwork of the four types of buildings does not show an obvious linear trend of carbon emission per unit area and a large dispersion degree, which means that the SD is not the main factor affecting the overall carbon emission per unit area of the old communities.

- AS

The scatter plot between the AS and the carbon emission intensity shows a certain linear trend. However, the R2 and partial regression coefficient (B) values of the fitted equation are 0.299 and 4.56656, indicating that there is a certain degree of positive correlation between the AS and carbon emissions. In general, with the increase in the AS, the carbon emission intensity of old communities increases.

- SVF

The results show that the R2 of the linear fitting equation between the SVF and the carbon emission intensity is 0.005, while the partial regression coefficient (B) is 5.33669, which means that there is a certain positive correlation between the carbon emission intensity and the SVF.

3.3.2. Regression Analysis of the Morphological Indices and Carbon Emission

- Establishing the correlation indices

After identifying the linear relationship between morphological indices and unit carbon emissions per unit area during the operational stage, the research establishes a Pearson correlation coefficient matrix based on morphological indices and unit carbon emissions per unit area. Finally, highly correlated non-collinear morphological indices are selected for multiple linear regression analysis (Table 9).

Table 9. Correlation matrix between morphological indices and carbon emissions.

		BESA	SF	ED	SD	SVF	FAR	BCR	AS	Carbon Emission
BESA	Correlation Coefficient	1	−0.075	−0.052	0.232	−0.371 **	0.231	0.13	0.01	−0.047
	Significance		0.571	0.69	0.074	0.004	0.076	0.322	0.94	0.72
	Numbers	60	60	60	60	60	60	60	60	60
SF	Correlation Coefficient	−0.075	1	−0.453 **	−0.254	−0.042	−0.461 **	−0.517 **	0.044	0.547 **
	Significance	0.571		0	0.05	0.751	0	0	0.74	0
	Numbers	60	60	60	60	60	60	60	60	60
ED	Correlation Coefficient	−0.052	−0.453 **	1	0.128	−0.057	0.772 **	0.736 **	−0.477 **	−0.851 **
	Significance	0.69	0		0.329	0.664	0	0	0	0
	Numbers	60	60	60	60	60	60	60	60	60
SD	Correlation Coefficient	0.232	−0.254	0.128	1	−0.197	0.201	0.212	−0.068	−0.172
	Significance	0.074	0.05	0.329		0.132	0.123	0.104	0.608	0.188
	Numbers	60	60	60	60	60	60	60	60	60
SVF	Correlation Coefficient	−0.371 **	−0.042	−0.057	−0.197	1	−0.16	−0.076	0.025	0.064
	Significance	0.004	0.751	0.664	0.132		0.221	0.565	0.852	0.629
	Numbers	60	60	60	60	60	60	60	60	60
FAR	Correlation Coefficient	0.231	−0.461 **	0.772 **	0.201	−0.16	1	0.830 **	−0.525 **	−0.848 **
	Significance	0.076	0	0	0.123	0.221		0	0	0
	Numbers	60	60	60	60	60	60	60	60	60
BCR	Correlation Coefficient	0.13	−0.517 **	0.736 **	0.212	−0.076	0.830 **	1	−0.600 **	−0.910 **
	Significance	0.322	0	0	0.104	0.565	0		0	0
	Numbers	60	60	60	60	60	60	60	60	60
AS	Correlation Coefficient	0.01	0.044	−0.477 **	−0.068	0.025	−0.525 **	−0.600 **	1	0.563 **
	Significance	0.94	0.74	0	0.608	0.852	0	0		0
	Numbers	60	60	60	60	60	60	60	60	60
Carbon Emission	Correlation Coefficient	−0.047	0.547 **	−0.851 **	−0.172	0.064	−0.848 **	−0.910 **	0.563 **	1
	Significance	0.72	0	0	0.188	0.629	0	0	0	
	Numbers	60	60	60	60	60	60	60	60	60

** At the 0.01 level, the correlation was significant.

As shown in the figure below (Figure 7), there is no significant correlation observed between SD, SVF and the unit area carbon emissions of old communities. The remaining indices have some correlation with carbon emissions, which can be used as independent variables for regression analysis. To avoid the collinearity problem of the morphological indices, the BCR with strong collinearity is removed. Finally, multiple linear regression analysis is conducted on the four independent variables, namely AS, FAR, ED and SF.

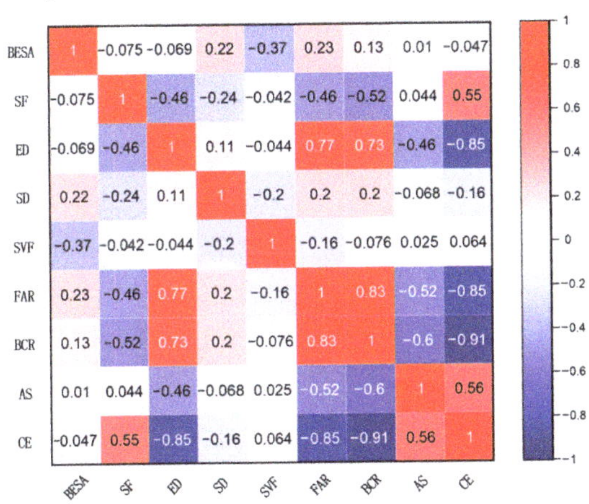

Figure 7. Heat map of correlation between morphological indices and carbon emission.

- Multiple linear regression analysis

In this section, a Pearson multiple linear regression equation including AS, FAR, ED and SF is established to predict the carbon emissions per unit area of old communities. According to Table 10, the R2 of the model is 0.855, which indicates that the change in the independent variable could explain 85.5% of the variance of the dependent variable. The adjusted R2 value of 0.923 reflects the model's fitting effect, which takes into account the number of independent variables used in the model.

Table 10. The fitting degree of model.

Model	R	R^2	Adjusted R^2	Standard Error of Estimate	Durbin–Watson
1	0.925	0.855	0.845	1.813167	1.971

The regression coefficient table provides an indication of the degree to which each independent variable affects the dependent variable (Table 11). The absolute values of the standardized coefficients for FAR and ED are relatively larger, at 0.241 and 0.628, respectively, indicating that they have a greater impact on unit area carbon emissions. By contrast, the absolute values of standardized coefficients for AS and SF are relatively smaller, at 0.149 and 0.111, respectively, indicating that they have a smaller impact on unit area carbon emissions (Figure 8).

Table 11. The regression result of morphological indices.

Model		Partial Regression Coefficient		Standardized Regression Coefficient	t	Statistical Significance	Collinearity	
		B	Standard Error	Beta			Tolerance	VIF
1	(constant)	81.927	4.21		19.458	0		
	FAR	− 3.731	0.936	− 0.347	− 3.987	0	0.348	2.877
	SF	28.76	9.527	0.186	3.019	0.004	0.696	1.437
	ED	− 17.078	3.358	− 0.42	− 5.086	0	0.386	2.59
	AS	1.44	0.513	0.178	2.808	0.007	0.655	1.527

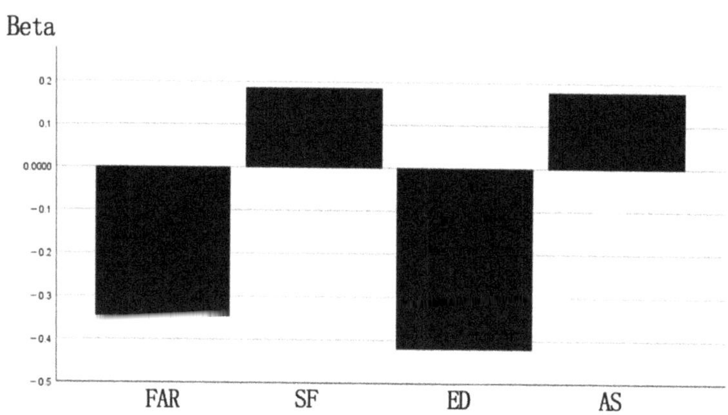

Figure 8. Comparison of standardization coefficients.

Furthermore, the regression coefficients of FAR, SF, ED and AS in this model are −3.731, 28.76, −17.078 and 1.44, respectively. Thus, the regression equation is

$$y = -3.731*FAR + 28.76*SF - 17.078*ED + 1.44*AS + 81.927 \quad (1)$$

After the regression equation was determined, the sample data of 15 old communities in Tianjin [60], which belong to cold regions, were selected to verify the validity of the regression equation, including 3 point-type, 7 slab-type and 5 enclosed-type building groups (Table 12). The results indicate that the average error between the measured value and predicted value is 1.09%, suggesting that the prediction model is reasonably accurate (Figure 9). In other words, the regression results can provide a useful reference for studying the relationship between the spatial morphological indices and carbon emission at the operational stage of old communities in cold regions.

Table 12. The comparison of measured carbon emissions and predicted carbon emissions.

No.	Layout Type		FAR	SF	ED	AS	Measured Value of Carbon Emission ($KgCO_2/m^2$)	Prediction of Carbon Emission ($KgCO_2/m^2$)
A01		Slab-type	1.91	0.26	0.74	1.45	78.58	80.95
A02		Enclosed-type	1.33	0.25	0.81	1.20	75.56	77.33
A03		Enclosed-type	1.51	0.24	0.82	1.84	74.49	78.05
A04		Slab-type	1.38	0.28	0.80	1.14	71.77	78.15
A05		Slab-type	1.29	0.27	0.72	2.18	79.80	80.96
A06		Slab-type	1.30	0.25	0.77	2.71	76.43	79.96
A07		Slab-type	1.87	0.27	0.73	1.64	80.04	81.50
A08		Enclosed-type	1.36	0.24	0.81	2.70	71.25	78.91
A09		Point-type	1.75	0.28	0.69	1.18	76.64	81.88
A10		Enclosed-type	1.58	0.20	0.86	1.54	73.14	76.18
A11		Slab-type	1.67	0.24	0.81	2.11	79.39	78.99

Table 12. *Cont.*

No.	Layout Type		FAR	SF	ED	AS	Measured Value of Carbon Emission ($KgCO_2/m^2$)	Prediction of Carbon Emission ($KgCO_2/m^2$)
A12		Point-type	1.23	0.27	0.73	1.45	75.04	79.74
A13		Point-type	1.09	0.28	0.66	1.58	76.56	81.50
A14		Slab-type	1.65	0.26	0.75	2.78	78.48	81.56
A15		Enclosed-type	1.78	0.23	0.82	2.23	77.74	78.97

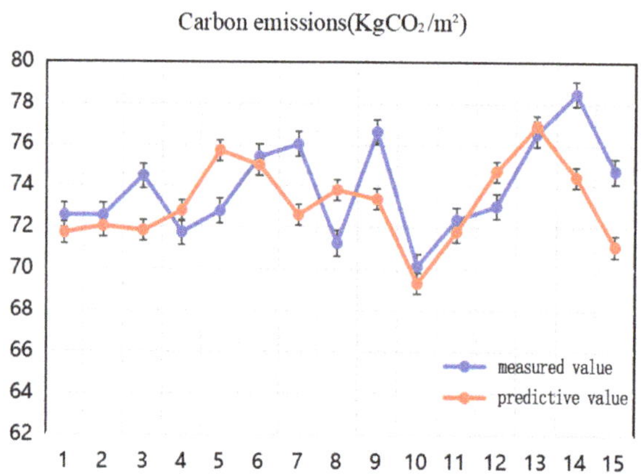

Figure 9. Comparison of measured carbon emissions and predicted carbon emissions.

4. Discussion

4.1. Result Analysis

4.1.1. Impact of Morphological Density Indices on Carbon Emission

Morphological density indices are important baeses to measure the capacity of the residential area. Observing the sample distribution of old communities, it can be found that different FAR values will bring diversified spatial morphology characteristics. An increase in FAR will lead to an increase in building volume, which has an impact on ventilation and lighting. For the old communities with low development intensity, appropriately increasing the FAR of the old communities can contribute to the natural ventilation effect inside the organization, which is conducive to the reduction in energy consumption and carbon emission. The results also show that appropriately increasing the FAR of the old communities can reduce the carbon emission per unit area of the whole community, and the impact degree increases sharply after rising to 1.25. Among them, the FAR has the greatest impact on the carbon emission of the mixed-type building group and the least impact on the enclosed-type building group. Therefore, it is suggested to set the FAR above 1.25 fot old communities. In the face of the current situation of low density in the old communities, in the low-carbon renewal of the old communities, we can consider extending the façade function of the old communities and appropriately increasing the basic function facilities and other operations to increase the FAR. BCR also has an important impact on ventilation and lighting. High building density will lead to a smaller distance between the buildings, and the crowded space will have a negative impact on ventilation and natural lighting, aggravating the energy consumption and carbon emissions of the buildings themselves.

4.1.2. Impact of Morphological Texture Indices on Carbon Emissions

The morphological texture indices are the reflection of the three-dimensional spatial state of the old community. From the perspective of energy flux, the spatial morphological indices of the community play a great role in the simulation of carbon emission for the community.

The statistical results confirm that the AS is closely related to the carbon emissions of the old community. High buildings will affect the surrounding wind speed and ventilation, resulting in heat accumulation and poor circulation, so as to increase the energy consumption of air conditioning and carbon emission intensity of old communities. If the AS of the built environment is large, the height of the building is relatively low, which usually means that the air circulation is not smooth. It means that buildings require more energy consumption to maintain a comfortable indoor temperature. to maintain a comfortable indoor temperature. Based on the results of this research, the AS is positively correlated with the carbon emission of old communities.

According to the results, the SF has an obvious negative correlation with the carbon emission per unit area of the old communities. The smaller the SF of the building, the smaller the surface area of the building, the lower the heat transfer efficiency, and the lower the energy consumption. On the contrary, the larger the SF of the building, the greater the surface area of the building, the higher the heat transmission efficiency, and the higher the energy consumption, which affects the overall carbon emission of the old community. Therefore, in the process of low-carbon renewal of the old community, it is appropriate to reduce the complexity of residential buildings as far as possible and to select a more regular volume to reconstruct the space of the old community. The design of the building should minimize the SF to reduce energy consumption and improve energy efficiency.

In addition, the research shows that there is a correlation between ED and carbon emissions of the old community. On the one hand, the suitable open space layout can adjust the microclimate environment of old communities and reduce the carbon emission intensity of buildings. On the other hand, it can open the spacing between the buildings and avoid aggravating the shielding effect between them.

4.1.3. Impact of Morphological Layout Type on Carbon Emissions

Morphological layout types have a significant impact on carbon emissions from old residential areas in cold regions. From the perspective of the building types proposed in this paper, the carbon emission of enclosed-type buildings is lower than that of other types of old communities. The fundamental reason is that the high heating demand in cold regions is large, and the high enclosure degree can avoid the heat loss in residential areas in the winter. Therefore, it is suggested that the degree of buildings should be increased as much as possible in the design of old communities in cold regions.

4.2. Perspectives

Urban morphology has gradually become an important factor, which architects pay attention to in studying urban low-carbon renewal [61]. Nowadays, scholars are not only limited to research of the low-carbon renewal strategy of the building monomer, but they also pay much more attention to the influence of different regions or types of urban morphology on building energy consumption and carbon emission. In this research, we verified the close connection between the spatial morphological indices of the built environment of old communities and carbon emission in the operational stage, and we also emphasized that urban morphology can be used as an effective way of saving building energy and reducing carbon emission in cities in cold regions. Thus, urban planners and policy makers can help better understand the carbon emissions of old communities, formulate more effective emission reduction measures and ultimately achieve the goal of sustainable urban development.

On the other hand, with the development of big data and artificial intelligence, digital technology has had an enormous impact on the field of architectural design and urban planning. The enhanced computing power enables UBEM technology to integrate more performance simulation technologies and apply them in the urban design field [62]. For example, Leng Hong et al. proposed the application of "auxiliary energy-saving-oriented urban design" as the application of urban morphology research results [63], that is, to assist the formulation of building energy-saving-oriented urban design scheme by studying the influence of urban morphology on building energy consumption and carbon emission. In this regard, this research established a statistical prediction model of carbon emission based on spatial morphological indices to evaluate the influence of spatial morphology on carbon emission at different mesoscales and finally obtained morphological indices and influence laws with significant effects. Based on this, the study provides guidance for low-carbon-oriented urban morphology renewal strategies by quantitatively analyzing the relationship between eight specific spatial morphological indices and carbon emissions: in cold regions, residential area renewal can increase the compactness of neighborhood buildings while ensuring basic functional requirements are met, increasing block enclosure and reconfiguring building façades to reduce the shape factor. Additionally, this research model can incorporate the construction status and morphological characteristics of different old cities to expand the types of spatial morphological indices, thus providing more targeted guidance for the formulation of control indices related to spatial morphology in low-carbon-oriented new urban planning.

Furthermore, there are still some limitations to be solved and discussed. Due to the difficulty in obtaining the data of carbon emissions in a large range of old communities, the representative small sample size of this paper is selected, and the regression model is verified by converting the energy consumption data into carbon emissions. In addition, based on the influence mechanism of urban morphology on carbon emission from the perspective of spatial design, this research selects eight morphological indices to form the spatial morphology parameter system of old communities. While these indices possess a degree of typicity, they nonetheless remain insufficiently comprehensive. Moreover, with respect to environmental factors, such as the greening rate of the site, underlying surface materials, as well as human factors, such as users' energy consumption habits, this research adopted relative standard values in the model instead of conducting an in-depth exploration at this level. Nevertheless, the method of this research still provides a new possibility to study the relationship between spatial morphological indices and carbon emission of old communities with limited sample statistics and data acquisition conditions.

5. Conclusions

This research quantitatively examines the correlation between spatial morphological indices and carbon emission during the operational stage of built environments for old communities in cold regions. Specifically, 60 samples of old residential areas located in Jinan, a cold area, were selected for analysis. By combining numerical simulation and statistical analysis, this research conducted a comprehensive examination of the eight spatial morphological indices comprising three types of indices characterizing the morphological density, morphological texture and morphological layout of old communities. The final results show that

1. Of the four types of building groups outlined in this research, namely point-type, slab-type, mixed-type and enclosed-type, it was found that the total carbon emissions of the point-type and mixed-type building groups were relatively high, while the carbon emissions of the slab-type building group fell in the middle. Conversely, the carbon emissions per unit area of the enclosed-type building group were relatively low;
2. Among the morphological density and texture indices, the FAR, BCR and ED have obvious negative correlation with the carbon emission of old communities, while the

SF and AS are positively correlated with the carbon emission of old communities and have different degrees of influence on different layout types of old communities. Specifically, FAR and ED play a pivotal role in shaping point-type communities, although their influence on enclosed-type communities is minimal. BCR and SF have the greatest influence on point-type communities and the least influence on enclosed-type communities. In addition, AS only has an obvious impact on old communities with an enclosed layout. The other morphological indices have a weak impact on the carbon emissions of the old communities;

3. A multiple linear regression analysis was performed to develop a statistical predictive model of carbon emissions in old communities. The fitted equation yielded an R2 value of 0.855, signifying that the variations in independent variables, such as FAR, SF, ED and AS, can account for up to 85.5% of the variance in carbon emissions per unit area of the dependent variable;
4. FAR and ED are the primary indices, which significantly impact carbon emissions in old communities, with ED demonstrating the most pronounced effect. Keeping all other indices constant, each unit increase in FAR results in a 3.7% reduction in the carbon emission per unit area, while a unit increase in ED leads to an even greater reduction of 17.1%;
5. In essence, enhancing the spatial morphology of built environments in cold regions is a crucial factor in bolstering the carbon reduction potential of older communities in these areas.

Author Contributions: Conceptualization, F.Z. and Y.W. (Yuqing Wang); methodology, Y.W. (Yuqing Wang); software, Z.S.; validation, F.Z. and Y.W. (Yuetao Wang); formal analysis, Y.W. (Yuetao Wang); investigation, Z.S.; resources, Z.S.; data curation, F.Z. and Y.W. (Yuqing Wang); writing—original draft preparation, Y.W. (Yuqing Wang) and Z.S.; writing—review and editing, Y.W. (Yuqing Wang); visualization, F.Z.; supervision, Y.W. (Yuetao Wang); project administration, F.Z.; funding acquisition, F.Z. All authors have read and agreed to the published version of the manuscript.

Funding: This research received no external funding.

Data Availability Statement: Data available on request due to restrictions, e.g., privacy or ethics.

Acknowledgments: The authors are grateful for the support from the College of Architecture and Urban Planning at Shandong Jianzhu University. In addition, the authors would like to thank all the students of the Forward Architecture Study Society for their support in this research.

Conflicts of Interest: The authors declare no conflict of interest.

References

1. Laurent Fabius, Opening Speech by Laurent Fabius—Paris Climate Conference. 2015. Available online: http://www.diplomatie.gouv.fr/en/french-foreign-policy/climate/events/ (accessed on 2 May 2023).
2. *ISO 14064-1:2018*; Greenhouse Gases-Part 1: Specification with Guidance at the Organization Level for Quantification and Reporting of Greenhouse Gas Emissions and Removals. ES-UNE: Madrid, Spain, 2018.
3. *ISO 14064-2:2019*; Greenhouse Gases—Part 2: Specification with Guidance at the Project Level for Quantification, Monitoring and Reporting of Greenhouse Gas Emission Reductions or Removal Enhancements. ES-UNE: Madrid, Spain, 2019.
4. *ISO 14064-2:2019*; Greenhouse Gases—Part 3: Specification with Guidance for the Verification and Validation of Greenhouse Gas Statements. ES-UNE: Madrid, Spain, 2019.
5. Cano, N.; Berrio, L.; Carvajal, E.; Arango, S. Assessing the carbon footprint of a Colombian University Campus using the UNE-ISO 14064–1 and WRI/WBCSD GHG Protocol Corporate Standard. *Environ. Sci. Pollut. Res.* **2023**, *30*, 3980–3996. [CrossRef]
6. United States Environmental Protection Agency. *Overview of Greenhouse Gases*; United States Environmental Protection Agency: Washington, DC, USA, 2014.
7. Hui, E.C.; Liang, C.; Yip, T.L. Impact of semi-obnoxious facilities and urban renewal strategy on subdivided units. *Appl. Geogr.* **2018**, *91*, 144–155. [CrossRef]
8. Cubi, E.; Doluweera, G.; Bergerson, J. Incorporation of electricity GHG emissions intensity variability into building environmental assessment. *Appl. Energy* **2015**, *159*, 62–69. [CrossRef]
9. Fang, C.L.; Wang, S.J.; Li, G.D. Changing urban forms and carbon dioxide emissions in China: A case study of 30 provincial capital cities. *Appl. Energy* **2015**, *158*, 519–531. [CrossRef]

10. Wang, S.; Fang, C.; Wang, Y.; Huang, Y.; Ma, H. Quantifying the relationship between urban development intensity and carbon dioxide emissions using a panel data analysi. *Ecol. Indic.* **2015**, *49*, 121–131. [CrossRef]
11. Ishii, S.; Tabushi, S.; Aramaki, T.; Hanaki, K. Impact of future urban form on the potential to reduce greenhouse gas emissions from residential, commercial and public buildings in Utsunomiya, Japan. *Energy Policy* **2010**, *38*, 4888–4896. [CrossRef]
12. Kim, J.; Brownstone, D. The impact of residential density on vehicle usage and fuel consumption: Evidence from national samples. *Energy Econ.* **2013**, *40*, 196–206. [CrossRef]
13. Burgalassi, D.; Luzzati, T. Urban spatial structure and environmental emissions: A survey of the literature and some empirical evidence for Italian NUTS 3 regions. *Cities* **2015**, *49*, 134–148. [CrossRef]
14. Lee, S.; Lee, B. The influence of urban form on GHG emissions in the U.S. household sector. *Energy Policy* **2014**, *68*, 534–549. [CrossRef]
15. Hachem, C. Impact of neighborhood design on energy performance and GHG emissions. *Appl. Energy* **2016**, *177*, 422–434. [CrossRef]
16. Grubler, A.; Bai, X.; Buettner, T.; Dhakal, S.; Fisk, D.; Ichinose, T.; Keirstead, J.; Sammer, G.; Satterthwaite, D.; Schulz, N.; et al. Urban energy systems. In *Glob Energy Assess—Toward a Sustainable Future*; Cambridge University Press: Cambridge, UK, 2012.
17. UNEP-DTIE Sustainable Consumption and Production Branch. In *Cities and Buildings*; UNEP: Nairobi, Kenya, 2016.
18. IEA. Buildings-Tracking Clean Energy Progress. 2019. Available online: https://www.iea.org/tcep/buildings/ (accessed on 25 March 2019).
19. Energy Consumption Statistics Committee of China Building Energy Conservation Association. *Research Report on Building Energy Consumption in China*; Energy Consumption Statistics Committee of China Building Energy Conservation Association: Xiamen, China, 2022.
20. Roldán-Fontana, J.; Pacheco-Torres, R.; Jadraque-Gago, E.; Ordónez, J. Optimization of CO_2 emissions in the design phases of urban planning, based on geometric characteristics: A case study of a low-density urban area in Spain. *Sustain. Sci.* **2015**, *12*, 65–85. [CrossRef]
21. Cuéllar-Franca, R.M.; Azapagic, A. Environmental impacts of the UK residential sector: Life cycle assessment of houses. *Build. Environ.* **2012**, *54*, 86–99. [CrossRef]
22. Huang, P.J.; Cui, H.; Song, J.L.; Yu, H.Y. Research on carbon emission accounting of buildings in Shanghai. *J. Shanghai Univ. Sci. Technol.* **2022**, *44*, 343–350.
23. Chen, S.; Cui, D.G.; Zhang, H.J. Building carbon emission calculation method and case study. *J. Beijing Polytech.* **2016**, *42*, 594–600.
24. Li, J.; Liu, Y. Carbon emission calculation model of construction project based on the whole life cycle. *J. Eng. Manag.* **2015**, *29*, 12–16.
25. She, J.Q.; Zhang, Y.B.; Qi, S.J. Life-cycle carbon emission characteristics and emission reduction strategies of public buildings in hot summer and warm winter zone—A case study of Xiamen City. *Build. Sci.* **2014**, *30*, 13–18.
26. Zhou, N.; McNeil, M.A.; Levine, M. *Energy for 500 Million Homes: Drivers and Outlook for Residential Energy Consumption in China*; Lawrence Berkeley National Laboratory: Berkeley, CA, USA, 2009.
27. Madlener, R.; Sunak, Y. Impacts of urbanization on urban structures and energy demand: What can we learn for urban energy planning and urbanization management? *Sustain. Cities Soc.* **2011**, *1*, 45–53. [CrossRef]
28. Chen, J.L.; Li, Z.W.; Wang, X.; Ruan, Y.J. Research on the optimization strategy of northern residential area form based on energy consumption survey. *Build. Energy Effic.* **2018**, *46*, 8–14.
29. Yang, X.; Jiang, Y. Comparison of building energy consumption between China and foreign countries. *Energy China* **2007**, *29*, 21–26.
30. Zhang, J.; Yang, Y.; Chen, X.; Mao, Q.Z. Study on the impact of residential built environment on household travel energy consumption in Jinan City. *Urban Dev. Stud.* **2013**, *20*, 83–89.
31. Liu, Z.; Chen, X.F.; Li, S.X. Ambient Microclimate: Collaborative Evaluation Process for Outdoor Thermal Comfort in the Early Design Stage. *Chin. Overseas Archit.* **2023**, 42–48.
32. Jin, H.; Cui, P. Research on the correlation between morphological elements and temperature of Harbin Central Street commercial district. *Build. Sci.* **2019**, *35*, 11–17.
33. Wei, X. Research on Reducing Carbon Consumption in Residential Community Spaces as Influenced by Microclimate Environments. *J. Urban Plan. Dev.* **2021**, *147*, 04021037. [CrossRef]
34. Yang, P.R.; Quan, J.G. Eco-volume ratio (EAR): Energy consumption assessment and carbon reduction method for urban redevelopment in high-density environment. *Urban Plan. Forum* **2014**, *3*, 61–70.
35. Zhu, R.; Wong, M.S.; You, L.; Santi, P.; Nichol, J.; Ho, H.C.; Lu, L.; Ratti, C. The effect of urban morphology on the solar capacity of threedimensional cities. *Renew. Energy* **2020**, *153*, 1111–1126. [CrossRef]
36. Zhu, R.; You, L.; Santi, P.; Wong, M.S.; Ratti, C. Solar accessibility in developing cities: A case study in Kowloon East, Hong Kong. *Sustain. Cities Soc.* **2019**, *51*, 101738. [CrossRef]
37. Vermeulen, T.; Merino, L.; Knopf-Lenoir, C.; Villon, P.; Beckers, B. Periodic urban models for optimization of passive solar irradiation. *Sol. Energy* **2018**, *162*, 67–77. [CrossRef]
38. Hachem, C.; Fazio, P.; Athienitis, A. Solar optimized residential neighborhoods: Evaluation and design methodology. *Sol. Energy* **2013**, *95*, 42–64. [CrossRef]

39. De Lemos Martinsa, T.A.; Adolphe, L.; Bastos, L.E.G.; de Lemos Martins, M.A. Sensitivity analysis of urban morphology factors regarding solar energypotential of buildings in a Brazilian tropical context. *Sol. Energy* **2016**, *137*, 11–24. [CrossRef]
40. Krüger, E.; Pearlmutter, D.; Rasia, F. Evaluating the impact of canyon geometry and orientation on cooling loads in a high-mass building in a hot dry environment. *Appl. Energy* **2010**, *87*, 2068–2078. [CrossRef]
41. Tsirigoti, D.; Bikas, D. A cross scale analysis of the relationship between energy efficiency and urban morphology in the Greek city context. *Procedia Environ. Sci.* **2017**, *38*, 682–687. [CrossRef]
42. Li, C.; Song, Y.; Kaza, N. Urban form and household electricity consumption: A multilevel study. *Energy Build.* **2018**, *158*, 181–193. [CrossRef]
43. Ahn, Y.J.; Sohn, D.W. The effect of neighbourhood-level urban form on residential building energy use: A GIS-based model using building energy benchmarking data in Seattle. *Energy Build.* **2019**, *196*, 124–133. [CrossRef]
44. Mouzourides, P.; Kyprianou, A.; Neophytou, M.K.A.; Ching, J.; Choudhary, R. Linking the urban-scale building energy demands with city breathability and urban form characteristics. *Sustain. Cities Soc.* **2019**, *49*, 101460. [CrossRef]
45. Swan, L.G.; Ugursal, V.I. Modeling of end-use energy consumption in the residential sector: A review of modeling techniques. *Renew. Sustain. Energy Rev.* **2009**, *13*, 1819–1835. [CrossRef]
46. Howard, B.; Parshall, L.; Thompson, J.; Hammer, S.; Dickinson, J.; Modi, V. Spatial distribution of urban building energy consumption by end use. *Energy Build.* **2012**, *45*, 141–151. [CrossRef]
47. Das, A.; Paul, S.K. CO_2 emissions from household consumption in India between 1993–94 and 2006–07: A decomposition analysis. *Energy Econ.* **2014**, *41*, 90–105. [CrossRef]
48. Robinson, D.; Campbell, N.; Gaiser, W.; Kabel, K.; Le-Mouel, A.; Morel, N.; Page, J.; Stankovic, S.; Stone, A. SUNtool–A new modelling paradigm for simulating and optimising urban sustainability. *Sol. Energy* **2007**, *81*, 1196–1211. [CrossRef]
49. Robinson, D.; Haldi, F.; Leroux, P.; Perez, D.; Rasheed, A.; Wilke, U. CitySim: Comprehensive micro-simulation of resource flows for sustainable urban planning. In Proceedings of the Eleventh International IBPSA Conference, Glasgow, UK, 27–30 July 2009; pp. 1083–1090.
50. Abbasabadi, N.; Ashayeri, M. Urban energy use modeling methods and tools: A review and an outlook. *Build. Environ.* **2019**, *161*, 106270. [CrossRef]
51. Xu, X.; Taylor, J.E.; Pisello, A.L.; Culligan, P.J. The impact of place-based affiliation networks on energy conservation: An holistic model that integrates the influence of buildings, residents and the neighborhood context. *Energy Build.* **2012**, *55*, 637–646. [CrossRef]
52. Letellier-Duchesne, S.; Nagpal, S.; Kummert, M.; Reinhart, C. Balancing demand and supply: Linking neighborhood-level building load calculations with detailed district energy network analysis models. *Energy* **2018**, *150*, 913–925. [CrossRef]
53. Pont, M.B.; Haupt, P. The relation between urban form and density M. Berghauser Pont and P. Haupt. *Urban Morphol.* **2007**, *11*, 62–65. [CrossRef]
54. Ratti, C.; Baker, N.; Steemers, K. Energy consumption and urban texture. *Energy Build.* **2005**, *37*, 762–776. [CrossRef]
55. Liu, K. *Energy Efficiency-Driven Urban form Optimization Strategy in Hot Summer and Cold Winter Areas: A Case Study of Residential Blocks in Jianhu County*; Southeast University: Nanjing, China, 2022.
56. Leng, H.; Xiao, Y.T. The influence of residential area form on residential energy consumption in cold city. *J. Harbin Inst. Technol.* **2020**, *52*, 147–156+163.
57. *JGJ26-2018*; Design Standard for Energy Efficiency of Residential Buildings in Severe Cold and Cold Zones. China Architecture & Building Press: Beijing, China, 2018.
58. *GB50736-2012*; Design Code for Heating Ventilation and Air Conditioning of Civil Buildings. China Architecture & Building Press: Beijing, China, 2012.
59. *JGJ/T449-2018*; Standard for Green Performance Calculation of Civil Buildings. China Architecture & Building Press: Beijing, China, 2018.
60. Dong, N. *Research on Low-Carbon Renewal Strategy of Old Residential Areas in Tianjin Based on Carbon Emission Accounting*; Tianjin University: Tianjin, China, 2021.
61. Privitera, R.; Palermo, V.; Martinico, F.; Fichera, A.; La Rosa, D. Towards lower carbon cities: Urban morphology contribution in climate change adaptation strategies. *Eur. Plan. Stud.* **2018**, *26*, 812–837. [CrossRef]
62. Yang, F.; Jiang, Z.D. Urban Building Cluster Energy Simulation (UBEM) and Environmentally Sustainable Oriented Urban Planning and Design: Methods, Tools and Paths. *Build. Sci.* **2019**, *37*, 17–24.
63. Leng, H.; Chen, X.; Ma, Y.H. Research progress and enlightenment of the influence of urban form on building energy consumption. *Archit. J.* **2020**, *617*, 20–126.

buildings

MDPI

Article

A Sustainable Evaluation Study Based on Emergy–Geographic Information System (GIS) Methodology in Hangzhou City from 2010 to 2035

Lan Liu [1], Runhui Cai [2] and Junxue Zhang [3,*]

[1] School of Architecture, Sanjiang University, Nanjing 210012, China; liulan0930@163.com
[2] Yangzhou Survey Design Research Institute Co., Ltd., Yangzhou 225000, China; cairunhui1019@163.com
[3] School of Civil Engineering and Architecture, Jiangsu University of Science and Technology, Zhenjiang 212100, China
* Correspondence: zjx2021@just.edu.cn

Abstract: Sustainability studies are vital for the long-term development of ecological cities. For cities, single qualitative or quantitative evaluation studies cannot effectively illustrate the ecological and sustainable status of a city. This study employs the emergy–geographic information system (GIS) method to conduct a sustainability evaluation of a city, so as to effectively verify the evaluation results. The emergy method and GIS are both commonly used approaches to address urban issues, but their synergistic effect has rarely been considered, explored, and utilized in urban planning. This study aims to investigate this effect on Hangzhou city through comparative analysis. The results show that rain (geopotential energy) and rain (chemical potential energy) have the highest emergy values from 2000 to 2035, followed by solar emergy, wind emergy, and geothermal heat emergy. These findings are also supported by the GIS map that shows a similar pattern with renewable emergy. Using the five plots (cropland, woodland, grassland, water area, and built-up land) on the GIS map as examples, the accuracy of emergy calculation results for Hangzhou city can be verified, demonstrating the effectiveness of the emergy–GIS methodology. This research provides practical recommendations for city designers and professionals worldwide on urban sustainability. By incorporating both emergy and GIS methods, cities can make informed decisions toward achieving environmentally sustainable development.

Keywords: sustainability; emergy-GIS method; urban system; predictive analysis

1. Introduction

At present, the vast majority of people reside in urban areas, leading to numerous challenges such as urban heat islands, solid and water pollution, air pollution, traffic congestion, and more. Cities in China are facing a variety of issues, including outdated infrastructure, urban congestion, water and air pollution, public health crises, waste gas and solid pollution, urban heat island effects, and others [1]. To address these challenges, it is urgent to apply and practice the concept of sustainable development. Before initiating sustainable urban renewal and governance, it is crucial to evaluate the state of urban development. Therefore, urban sustainability assessment has become a critical research topic.

Urban sustainability analysis needs to assess the environmental, social, and economic impacts of urban development and operations. It aims to find balanced development strategies that meet current needs without compromising the ability of future generations to meet their own needs. Urban sustainability analysis typically involves the following aspects: (1) Environmental sustainability: evaluating resource efficiency, waste management, air quality, water resource management, and other factors to ensure that urban development does not exert excessive pressure on ecosystems. (2) Social sustainability: considering the social welfare, equity, and inclusivity of urban residents. This includes assessing education, healthcare services, community engagement, cultural preservation, and other factors to ensure that everyone can share in the benefits of urban development. (3) Economic sustainability: analyzing economic growth, employment opportunities, and innovation capacity in cities. This includes evaluating industrial diversity, entrepreneurial environments, and sustainable economic models to ensure long-term prosperity and development. Through urban sustainability analysis, governments, planners, and decision-makers can understand the overall impact of urban development and formulate corresponding policies and measures to promote sustainable development. This comprehensive analytical approach helps achieve a balance between the environment, society, and economy of cities, creating more sustainable and livable urban environments.

However, due to the complexity of urban sustainability and the wide range of aspects it encompasses, researchers have conducted their analysis and exploration from various different perspectives. These include ecological security [2–4], ecological footprint [5,6], urban symbiosis [7], ecological civilization construction [8], ecological network [9], environmental quality [10], ecological smart city [11,12], ecological livability assessment [13], emergy analysis [14], ecological literacy [15], ecological pattern [16], ecological risks [17], and more. These analytical perspectives provide theoretical support for the design and development of sustainable cities, offering various insights for urban managers and designers. They contribute to the improvement and enhancement of urban sustainability.

In the field of urban sustainability, GIS methods and emergy methods are two mainstream research approaches that assess the sustainable state of cities from quantitative and qualitative perspectives. The GIS approach has been widely used in urban research, encompassing areas such as bibliometric analysis [18], dynamic evaluation [19], disaster susceptibility [20], smart perspective [21,22], ecological risk [23], urban greenspace accessibility [24], ecological security [25], land-use change assessment [26], and more. The emergy method is also frequently utilized in urban research [27–40]. Its advantage lies in the ability to quantitatively calculate various inputs and outputs of urban systems, which facilitates a clear identification of the sustainability status of the city system.

Therefore, some researchers have combined emergy and GIS methods to analyze and explore the sustainability of urban systems. This methodological approach effectively leverages the advantages of both approaches and enables precise positioning of urban sustainability. Emergy–GIS is a methodology based on ecology, economics, and GIS. This method can quantitatively and qualitatively evaluate the flow of materials and energy in a specific area, as well as the environmental and social impacts of these flows. Using this method, resource utilization efficiency and costs can be determined, and the consequences of various policy and technological choices can be predicted. Through the emergy–GIS method, the city can be divided into different subsystems, such as energy, water resources, waste treatment, and transportation, and each subsystem can be assessed. In this way, different aspects od a city's sustainability, such as resource utilization efficiency, environmental impact, and social justice, can be determined. Considering the two aspects of the emergy–GIS evaluation method, the quantitative evaluation more readily enables comparisons and is more accurate, while the qualitative evaluation is more suitable for determining long-term goals and planning. Therefore, both are necessary for the assessment of urban sustainability.

Until now, the emergy–GIS method mainly focused on family farm analysis [41], resource monitoring applications [42], evaluation of renewable resources [43], sustainability studies of rivers [44], land use analysis [45], landscape classification [46], product productivity analysis [47], sponge city design [48], and ecosystem structure [49]. The current research lacks studies and analysis on urban sustainability, and this article precisely fills this gap. The purpose of this article is to use the emergy–GIS method to quantify and qualify the sustainability of the entire urban system, to compare and analyze the similarities and differences between the two, to identify key influencing factors, and to provide improvement measures and efficacy analysis, in order to provide city managers and designers with new choices and references.

2. Materials and Methodology

2.1. Emergy–GIS Framework Design

In this study, an emergy–GIS framework has been built by integrating the emergy method and GIS approach for evaluating eco-cities. The framework is presented in Figure 1. The emergy method provides quantitative calculation of all resource and service inputs in the city, while GIS simulation offers clear images to validate the accuracy of emergy calculations.

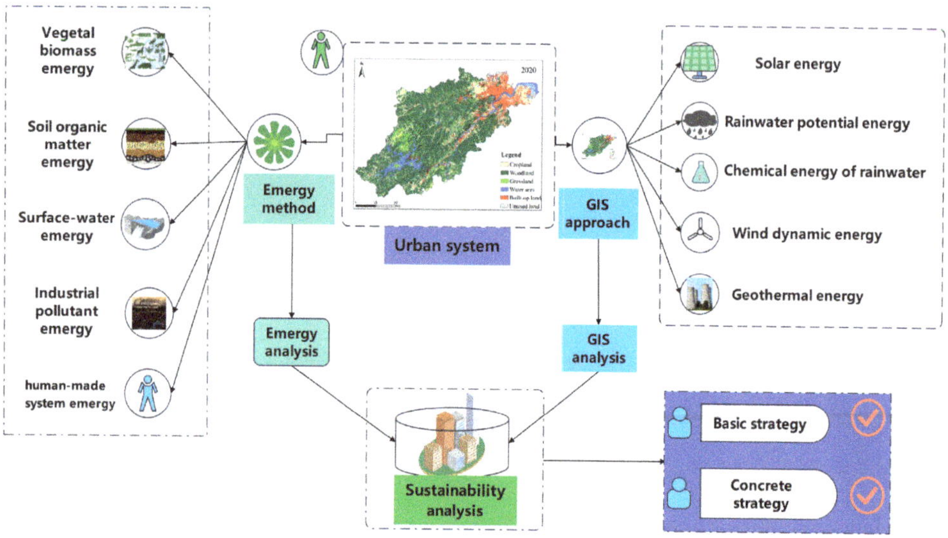

Figure 1. Research framework design.

In the overall research framework, urban studies serve as the core focus. Through the combined assessment and computation using the energy valuation method and GIS method, the sustainability changes of the entire urban system are analyzed, and specific improvement measures are proposed. The energy valuation method involves input factors such as biodiversity energy value, soil organic matter energy value, water body energy value, industrial waste energy value, and artificial energy value. The GIS method incorporates five elements: solar energy, potential energy of rainwater, chemical energy of rainwater, wind energy, and geothermal energy.

The GIS software (10.3) used in this study is ArcGIS version 10.3. The core steps of utilizing the emergy–GIS methodology include confirmation of area boundary conditions, simulation of various types of pictures (such as water, buildings, forest, grass, etc.), and the emergy assessment of each type of land in the city. In order to predict changes in urban form, the Cellular automata–Markov methodology was adopted. The cellular automata Markov methodology is a method used for simulating and predicting spatial changes. This approach is based on the cellular automaton model, where the study area is divided into small units called cells. Each cell has a specific state and transition rules. By using the probability transition matrix of Markov chains, it models the state changes between cells and predicts future spatial patterns. This methodology is widely applied in fields such as land use, urban development, and environmental management, providing strategies and decision support for understanding future spatial changes.

2.2. Emergy Introduction and Model

2.2.1. Emergy Introduction

The concept of emergy was first proposed by H.T. Odum, and has the unique advantage of evaluating the resource and service value based on available energy from solar energy. The use of transformity allows for the comparison of different types of energy, expressed as sej/g or sej/$ for one unit of each product or service, respectively [50]. Using the emergy method, several systems have achieved efficient comparative calculations on a unified platform, providing valuable management strategies for resources [51–53]. In this paper, the emergy baseline is set at 12.0×10^{24} sej/year [54].

2.2.2. Emergy Calculation

The specific emergy calculation equation can be found in (1)

$$E_i = \sum (M_i \times UEV_i) \tag{1}$$

where E_i is the total emergy; M_i is the inputs (mass, energy, and service); and UEV_i is the transformity for the inputs.

Calculations of biodiversity require special instructions, as they involve long-term estimates of deoxyribonucleic acid (DNA) diversity spanning approximately one billion years. Previous research suggests that over two billion years, there have been roughly 1.5×10^9 species. Based on this estimate, the unit emergy value (UEV) of species can be calculated using the latest emergy baseline (12.0×10^{24} sej/yr), as follows:

$$UEV = \frac{(12 \times 10^{24}\ \text{sej/yr}) \times (2 \times 10^9\ \text{yr})}{1.5 \times 10^9\ \text{species}} = 1.6 \times 10^{25}\ \text{sej/species} \tag{2}$$

2.2.3. Emergy Diagram

Figure 2 illustrates the emergy flow in Hangzhou city and is divided into four parts. On the left is renewable energy, which has five types: sunlight, rain-chemical, rain-geopotential, wind, and geothermal heat. The upper side shows the inputs of the city system elements, including four natural aspects (cropland, woodland, grassland, and water area) and two human elements (labor and service in built-up land). In the middle are two subsystems, the natural ecological system and the manual design system, respectively. Finally, both subsystems are summarized into the environment on the right.

2.2.4. Emergy Indicators

All the metrics considered in the analysis are listed in Table 1.

Table 1. Sustainable indices of emergy.

Item	Index	Meaning
Renewability rate	Re	Renewable proportion
Non-renewability rate	Nr	Non-renewable proportion
Non-renewability rate of purchased resource	Np	Purchased resource rate
Environmental loading ratio	ELR	Environmental pressure
Emergy yield ratio	EYR	Ability to obtain emergy
Emergy sustainability indicator	ESI	Sustainability degree

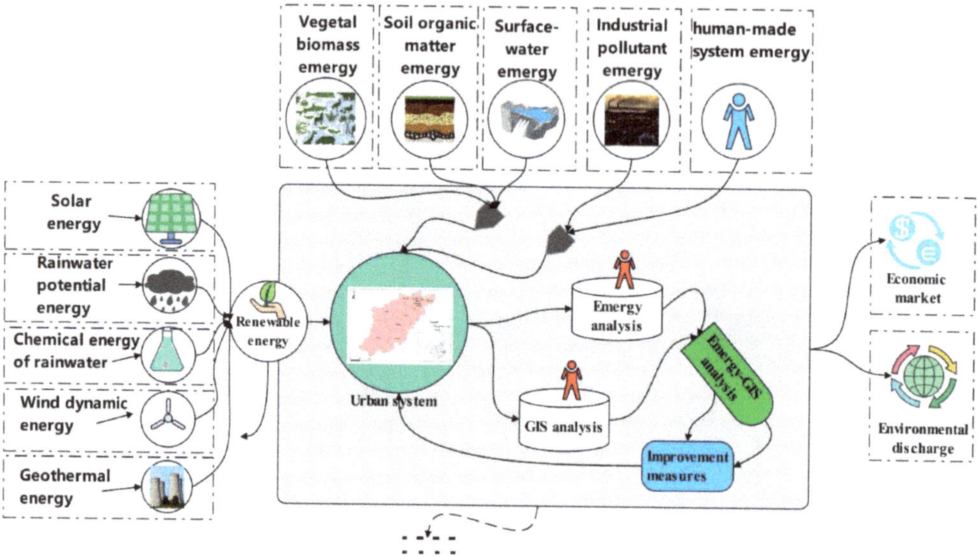

Figure 2. Emergy diagram of Hangzhou city.

2.3. *Natural Input Evaluation*

In the evaluation of Hangzhou city using the emergy–GIS framework, only three parts were considered for analysis: vegetal biomass, soil organic matter, and surface water. Mineral and fossil energy sources in Hangzhou were not included in the calculation.

2.3.1. Vegetal Biomass

The functional distribution of vegetal biomass has been analyzed on the basis of different land uses [55]. The use of each land has been calculated by GIS software. The related raw data can be found in Table 1.

2.3.2. Soil Organic Matter

As a part of an urban system, soil organic matter should be considered in this paper. The equation used specifically to calculate organic matter was as follows:

$$E_{\text{om}} = O_{M}g \times 3.5\ \frac{\text{kcal}}{\text{g}} \times 4186\ \frac{\text{J}}{\text{kcal}} \times UEV_{OM} \quad (3)$$

where E_{om} is the emergy of organic matter; O_M is the amount of organic matter; and UEV_{OM} is the unit emergy value of the organic matter, which is 1.18×10^4 sej/j.

2.3.3. Surface-Water Emergy Calculation

In this paper, there is a hypothesis that the annual runoff volume is equal to the accumulated surface water for the city. The surface water emergy is represented by Equations (4) and (5)

$$F_i = V \times \rho_W \times G \quad (4)$$

where F_i is the energy of the surface water (J); V is the volume of water (m^3); ρ_w is water density (1000 kg/m^3); and G represents the Gibbs free energy of surface water, 4940 J/kg; and

$$E_f = F_i \times U_i \quad (5)$$

where E_f is the emergy of surface water in the city (sej); and U_i(sej/J) is the unit emergy value (2.05×10^4 sej/J) of surface water [56].

2.3.4. Industrial Pollutant Emergy Calculation

Industrial pollutants will hurt human health, so their related damage degree needs to be assessed. In general, the following equations can be used:

Firstly, the human health effect level can be found from Equation (6),

$$L = W_i \times D_i \times \alpha \tag{6}$$

where L, i, W_i, D_i, and α show the human health effect in terms of emergy loss, amount of exhaust gas, disability-adjusted life year, and emergy loss to humans per year (1.68×10^{16} sej/(a·person)), respectively.

The exhaust gas amount is obtained from Formula (7):

$$M_i = c \times \left(\frac{U_i \times 10^6}{S_i}\right) \tag{7}$$

where M_i is the dilution air amount (kg/a); i is the gas type; C is the air density (1.23 kg/m^3); U_i is annual air pollutants mass, kg/a; and s_i is the acceptable concentration, in mg/m^3.

The ecological service emergy can be represented as (8)

$$R_{air,i} = 0.5 M_i v^2 T_w \tag{8}$$

where $R_{air,i}$ is the environmental emergy; v is the wind speed (3 m/s); and T_w is the wind transformity (1.86×10^3 sej/j).

Mud emergy can be found from Equation (9)

$$SLM = Z_{mud} P_L \beta_L \tag{9}$$

where SLM, Z_{mud}, P_L, and β_L are the mud emergy, mud value (2.32 ha), land demand, and transformity of land, respectively.

2.3.5. Emergy of Human-Made System

In this paper, the human-made system is mainly the buildings, which include the transportation facilities. The specific emergy of land use can be calculated from Equation (10):

$$E_{building} = \frac{A_1 \times N_f \times h_f}{A_2} \times UEV_{building} \tag{10}$$

where $E_{building}$ is the emergy of building in Hangzhou city; A_1 is the area of urban land use; N_f is the number of the average floor in the region; h_f is the average height; and $UEV_{building}$ is the transformity, of value 1.41×10^{15} sej/m^3 [57].

2.4. GIS Method (CA–Markov)

GIS methodshave been widely used in the analysis and prediction of land use changes in urban areas, most notably the cellular automata–Markov chain (CA–Markov) model. This approach combines the spatial analysis capabilities of GIS with the predictive power of the CA–Markov model to assess urban sustainability.

The CA–Markov model utilizes historical data on land use patterns and transition probabilities to simulate future scenarios of land use change. By integrating various spatial and non-spatial factors such as population growth, economic development, environmental constraints, and policy regulations, the model can provide valuable insights into the potential impacts of different land use planning strategies on urban sustainability.

Through the application of the CA–Markov model within a GIS framework, researchers can evaluate the efficiency and effectiveness of current and proposed urban development plans. The model enables precise identification of areas at high risk of unsustainable land use practices and aids in the formulation of targeted interventions for sustainable urban growth.

Overall, the integration of GIS methods, particularly the CA–Markov model, provides a powerful tool for analyzing and predicting land use changes in urban systems. It supports decision-making processes towards achieving more sustainable and resilient cities.

2.5. Study Area

Hangzhou is located in the southern part of the Yangtze River Delta, at the western end of Hangzhou Bay, downstream of the Qiantang River, and at the southern end of the Beijing–Hangzhou Grand Canal. It is an important central city in the Yangtze River Delta and a transportation hub in southeastern China. Its geographical coordinates are approximately 29°11′ to 30°34′ north latitude and 118°20′ to 120°37′ east longitude. The city center is situated at 120°12′ east longitude and 30°16′ north latitude, with a total area of 16,850 square kilometers. Hangzhou is located in the southern part of the Yangtze River Delta and the Qiantang River basin, characterized by diverse terrain. The western part of Hangzhou belongs to the Zhejiang–Western Hills region, with major mountain ranges such as Tianmu Mountain. The eastern part belongs to the Zhejiang-North Plain, characterized by low-lying terrain, dense river networks, abundant lakes, and rich natural resources. Hangzhou lies in the subtropical monsoon zone, experiencing a subtropical monsoon climate with distinct seasons and abundant rainfall. The average annual temperature is 17.8 °C, with an average relative humidity of 70.3%. The annual precipitation is 1454 mm, and the annual sunshine hours are 1765. Hangzhou boasts a natural environment of rivers, lakes, mountains, and hills. The city's hilly and mountainous areas account for 65.6% of the total area, plains occupy 26.4%, and rivers, lakes, and reservoirs cover 8% (Figure 3).

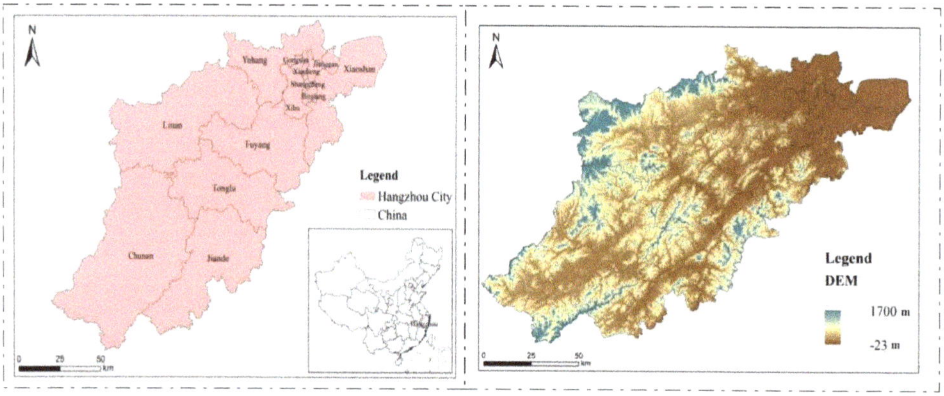

Figure 3. The map of Hangzhou City.

In 2020, Hangzhou achieved a regional gross domestic product (GDP) of 1610.6 billion yuan, representing a growth of 3.9% compared to the previous year. Breaking it down by sectors, the value added by the primary industry was 32.6 billion yuan, a decrease of 1.1%; the value added by the secondary industry was 4821 billion yuan, an increase of 2.3%; and the value added by the tertiary industry was 10,959 billion yuan, showing a growth of 5.0%. The structure of the three industries was 2.0:29.9:68.1.

Table 2 presents data on six types of terrain for the years 2000, 2010, and 2020 in Hangzhou. However, it is uncertain what will happen to each plot after 2020. Therefore, this paper will simulate plot change maps using GIS software for the years 2025, 2030, and 2035. Unfortunately, due to the lack of data, only a general assessment can be made regarding these future years.

Table 2. Data on the type and size of each plot.

	Year	Cropland	Woodland	Grassland	Water Area	Built-Up Land	Unused Land
Area (m^2)	2000	4,341,224,700	10,372,600,800	677,226,600	896,639,400	600,055,200	117,000
	2010	4,116,312,900	10,335,283,200	685,984,500	846,707,400	903,460,500	115,200
	2020	3,329,491,500	10,343,995,200	683,514,900	1,051,451,100	1,479,392,100	18,900
	2025	3,400,155,000	9,032,335,200	1,359,180,900	1,263,180,600	1,833,006,600	5400
	2030	3,199,225,500	9,020,878,200	1,241,289,900	1,360,538,100	2,065,931,100	900
	2035	3,012,943,500	8,994,467,700	1,192,640,400	1,434,662,100	2,253,144,600	5400

Figure 2 shows the area of each land type. In order to improve the efficiency of the simulation, the years selected were 2000, 2010, 2020, 2025, 2030, and 2035, with six land types: cropland, woodland, grassland, water area, built-up land, and unused land. Due to the high level of development in the local area, land has a relatively small proportion and will not be the focus of discussion in this section. By analyzing the data set, we can determine a prediction model to forecast the future trends in land use changes for the different land types.

3. Results and Discussion

3.1. Emergy Part Result

3.1.1. Basic Emergy Calculation

Table 3 presents the raw data inputs for Hangzhou city in 2020. The whole table has two parts: one is natural ecological input, which involves biodiversity, surface water and groundwater; the other is artificial input, comprising roads, rails and buildings.

Table 3. The input of Hangzhou city in 2020.

Item	Raw Amount	Data Ref.	UEV	UEV Ref.	Emergy (sej)
		Artificial systems storage			
Roads	16,919 km	c	3.04×10^9 sej/g	[46]	1.64×10^{21}
Rails	539 km	c	3.04×10^9 sej/g	[46]	1.13×10^{21}
Buildings	822,000 m^2	c	1.49×10^9 sej/g	[50]	2.46×10^{28}
		Natural capital storage			
Biodiversity	1237 species	a	1.6×10^{25} sej/species	[50]	1.98×10^{28}
Groundwater	38.75×10^9 m^3	b	2.26×10^6 sej/j	[58]	4.33×10^{23}
Surface water	216.69×10^9 m^3	b	2.05×10^4 sej/j	[59]	2.19×10^{22}
		Total			4.44×10^{28}

Note: [a] refers to the Plant List of Hangzhou in 2020. For specific calculations, refer to Section 2.3. [b] refers to the Hangzhou Water Resources Announcement in 2020. The specific calculation can be found in Section 2.3. [c] refers to the Hangzhou Yearbook in 2020.

3.1.2. Renewable Emergy Calculation for Six Types of Plots

Table 3 represents the input-type calculation for Hangzhou city in 2020, involving the natural spatial storage capacity and the artificial system storage capacity. This provides data support for calculating the renewable energy value and non-renewable energy value of the entire city. Additionally, it further calculates key indicators such as energy load factor, energy generation rate, and environmental sustainability parameters.

In Table 4, the renewable emergy data has been calculated and the specific calculations can be found in the Appendix A.

Table 4. Renewable emergy calculation from 2000 to 2035.

Item	Cropland	Woodland	Grassland	Water Area	Built-Up Land	Year
Solar energy emergy (unit: sej)	1.65×10^{19}	3.94×10^{19}	2.57×10^{18}	3.41×10^{18}	2.28×10^{18}	2000
	1.56×10^{19}	3.93×10^{19}	2.61×10^{18}	3.22×10^{18}	3.43×10^{18}	2010
	1.27×10^{19}	3.93×10^{19}	2.59×10^{18}	3.99×10^{18}	5.68×10^{18}	2020
	1.29×10^{19}	3.43×10^{19}	5.17×10^{18}	4.8×10^{18}	6.97×10^{18}	2025
	1.22×10^{19}	3.43×10^{19}	4.72×10^{18}	5.17×10^{18}	7.85×10^{18}	2030
	1.15×10^{19}	3.42×10^{19}	4.53×10^{18}	5.45×10^{18}	8.56×10^{18}	2035
Rain (geopotential energy) emergy (unit: sej)	1.07×10^{20}	2.54×10^{20}	1.66×10^{19}	2.20×10^{19}	1.47×10^{19}	2000
	1.01×10^{20}	2.54×10^{20}	1.68×10^{19}	2.08×10^{19}	2.22×10^{19}	2010
	8.17×10^{19}	2.54×10^{20}	1.68×10^{19}	2.58×10^{19}	3.63×10^{19}	2020
	8.34×10^{19}	2.22×10^{20}	3.32×10^{19}	3.10×10^{19}	4.49×10^{19}	2025
	7.84×10^{19}	2.21×10^{20}	3.04×10^{19}	3.32×10^{19}	5.08×10^{19}	2030
	7.39×10^{19}	2.21×10^{20}	2.93×10^{19}	3.52×10^{19}	5.52×10^{19}	2035
Rain (chemical potential energy) emergy (unit: sej)	1.66×10^{20}	3.97×10^{20}	2.59×10^{19}	3.43×10^{19}	2.29×10^{19}	2000
	1.58×10^{20}	3.96×10^{20}	2.63×10^{19}	3.24×10^{19}	3.46×10^{19}	2010
	1.28×10^{20}	3.96×10^{20}	2.62×10^{19}	4.03×10^{19}	5.67×10^{19}	2020
	1.30×10^{20}	3.46×10^{20}	5.21×10^{19}	4.84×10^{19}	7.02×10^{19}	2025
	1.23×10^{20}	3.45×10^{20}	4.75×10^{19}	5.21×10^{19}	7.91×10^{19}	2030
	1.15×10^{20}	3.44×10^{20}	4.57×10^{19}	5.49×10^{19}	8.63×10^{19}	2035
Wind energy emergy (unit: sej)	6.28×10^{10}	1.48×10^{11}	9.69×10^{9}	1.28×10^{10}	8.59×10^{9}	2000
	5.89×10^{10}	1.48×10^{11}	9.82×10^{9}	1.21×10^{10}	1.29×10^{10}	2010
	4.76×10^{10}	1.48×10^{11}	9.78×10^{9}	1.50×10^{10}	2.12×10^{10}	2020
	1.28×10^{10}	3.40×10^{10}	5.11×10^{9}	4.75×10^{9}	6.90×10^{9}	2025
	1.20×10^{10}	3.39×10^{10}	4.67×10^{9}	5.12×10^{9}	7.77×10^{9}	2030
	1.13×10^{10}	3.38×10^{10}	4.49×10^{9}	5.40×10^{9}	8.48×10^{9}	2035
Geothermal heat emergy (unit: sej)	5.22×10^{12}	1.24×10^{13}	8.15×10^{11}	1.08×10^{12}	7.22×10^{11}	2000
	4.95×10^{12}	1.24×10^{13}	8.25×10^{11}	1.02×10^{12}	1.09×10^{12}	2010
	4.01×10^{12}	1.24×10^{13}	8.22×10^{11}	1.27×10^{12}	1.78×10^{12}	2020
	4.09×10^{12}	1.09×10^{13}	1.64×10^{12}	1.52×10^{12}	2.21×10^{12}	2025
	3.85×10^{12}	1.09×10^{13}	1.49×10^{12}	1.64×10^{12}	2.49×10^{12}	2030
	3.63×10^{12}	1.08×10^{13}	1.43×10^{12}	1.73×10^{12}	2.71×10^{12}	2035

Table 4 and Figure 4 present the renewable emergy data and change for Hangzhou city. Rain (both geopotential energy and chemical potential energy) have the largest emergy values, followed by solar emergy. Wind emergy and geothermal heat play a minor role in the total renewable emergy in 2000. In Figure 4, the size of each emergy component is clearly shown in the cloud image. Compared to Figure 4a–c, the trends in variation are similar. The entire emergy tends to decrease because part of the ecological land has been occupied by building land from 2000 to 2020. These changes are consistent with the trend observed in the variation of natural ecological input and artificial input over time.

In addition to the calculations based on actual data in 2000, 2010, and 2020, emergy calculations for predicted renewable parts in 2025, 2030, and 2035 were also performed. Figure 4d–f reveals the concentration of emergy degrees across the five aspects considered. These sub-figures demonstrate that the emergy is concentrated in solar energy and rain (both geopotential energy and chemical potential energy), which account for more than 95% of the total renewable emergy for Hangzhou city. The GIS maps of 2025, 2030, and 2035 are predictive maps generated using the CA–Markov prediction model. Although these maps are based on predictions, they exhibit similar change trends to the actual data from 2000, 2010, and 2020. This similarity indicates the reliability of the predictive model.

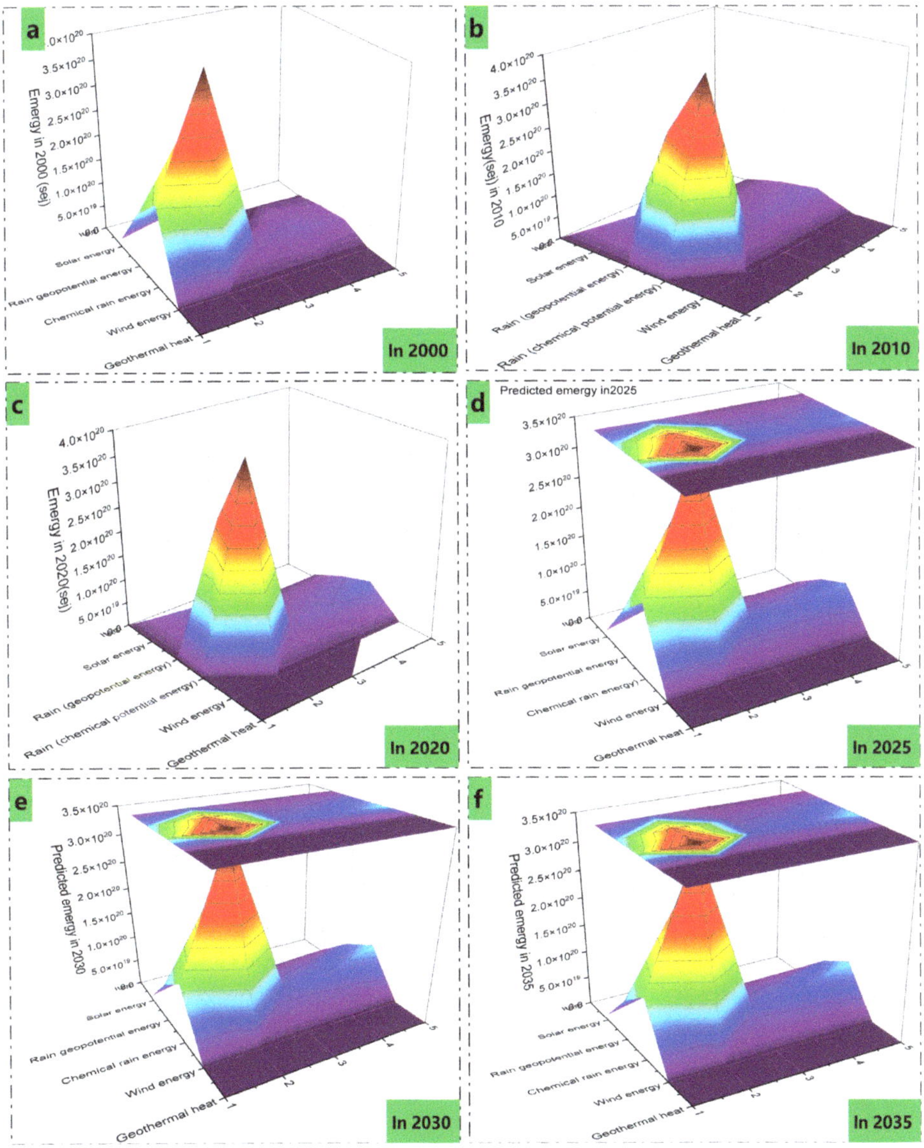

Figure 4. Concentration degree of renewable emergy from 2010 to 2035. (**a**)—Emergy in 2000; (**b**)—Emergy in 2010; (**c**)—Emergy in 2020; (**d**)—Predicted emergy in 2025; (**e**)—Predicted emergy in 2030; (**f**)—Predicted emergy in 2035; 1—Crop land; 2—Wood land; 3—Glass land; 4—Water area; 5—Built-up land.

3.2. GIS Part Discussion

3.2.1. Validation of the GIS Method

To ensure the accuracy of the research results, the effectiveness of the GIS simulation was analyzed before its application in this study. The year 2020, with the most available data, was selected for GIS simulation validation. In this comparison, the error standard between the actual situation and the predicted effect should be within 3% to indicate that the GIS simulation calculation is valid.

Figure 5 shows the six types of feature plates simulated according to the real data of 2020, which are cropland, woodland, grassland, water area, built-up land, and unused land. Based on the total calculation data, the area error of the two drawings is within 3%, indicating that the GIS simulation is accurate. Taking the water area as an example, a comprehensive comparative analysis was conducted based on position, size, and shape aspects. It was found that there was little change in the position of the water area, mainly in the northeast and southwest regions. Regarding the size of the water area, there was also no significant difference between the actual and simulated data. The overall shape of the water area did not change much, except for slight differences, and the main variation was concentrated in the small rivers.

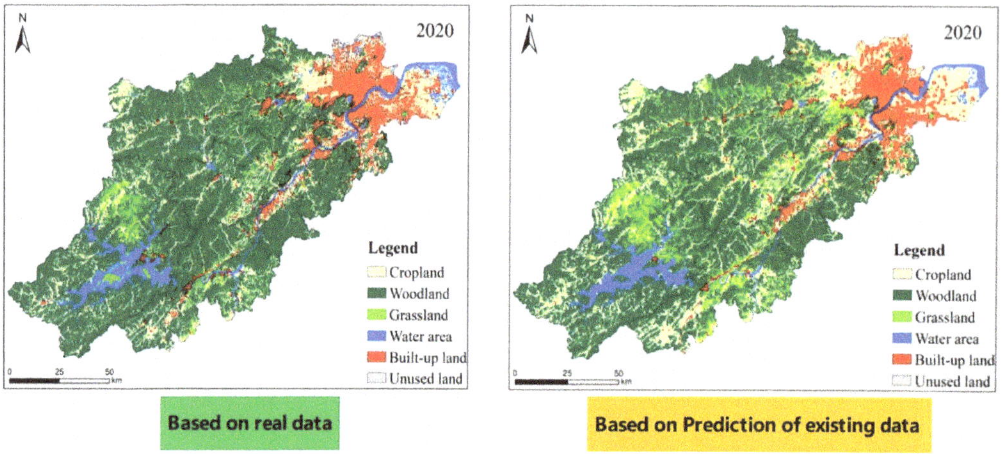

Figure 5. Preliminary validation verification of GIS method.

To sum up, the errors of the five sub-items were calculated, which were 4.2% for cropland, 3.9% for woodland, 4.1% for grassland, 1.3% for water area, and 5.1% for built-up land. Overall, the GIS software simulation is effective and can be used for further simulation analysis. In conclusion, the effectiveness of the GIS simulation was analyzed to ensure the accuracy of the research results. The results demonstrate that the GIS simulation is reliable and can be used to inform sustainable development strategies for Hangzhou city and other urban areas.

3.2.2. The Comparison Analysis Based on Emergy and GIS from 2000 to 2010

Figure 6(1) presents the analysis of Hangzhou city from 2000 to 2020 based on real data. The area of farmland is seen to be decreasing, while the floor space is increasing rapidly from 2000 to 2020, especially in the northeast region of Hangzhou. The cropland area decreased by 5.2% (23.3%) from 2000 to 2010 (2020), while built-up space in 2010 was 50% higher than in 2000 and had increased nearly 1.5 times by 2020. However, the area of woodland appears to be almost unchanged visually across the GIS maps, which is supported by specific plate data. For example, the woodland was 100 square meters in 2000, while in 2010 (respectively 2020), 99.6% (respectively 99.7%) of the area was retained, highlighting the effectiveness of Hangzhou's forest protection policies. In contrast, the grassland areas demonstrate an opposite trend, with grassland planting greatly encouraged to improve the ecological environment. Therefore, the area of grassland increased by 1.3 percent in 2010 and by 0.9 percent in 2020, positively contributing to the restoration of the ecological environment. From 2000 to 2010, the water area decreased by 5.57 percent, highlighting the need for sustainable water resource management policies. To ensure sustainable water resources, the government departments in Hangzhou city have implemented relevant policies and regulations to restrict the development and utilization of water resources. In 2020, the water area had increased by 17.3 percent compared to the data recorded in 2010, indicating that these policies and regulations are effective for maintaining sustainable water resources.

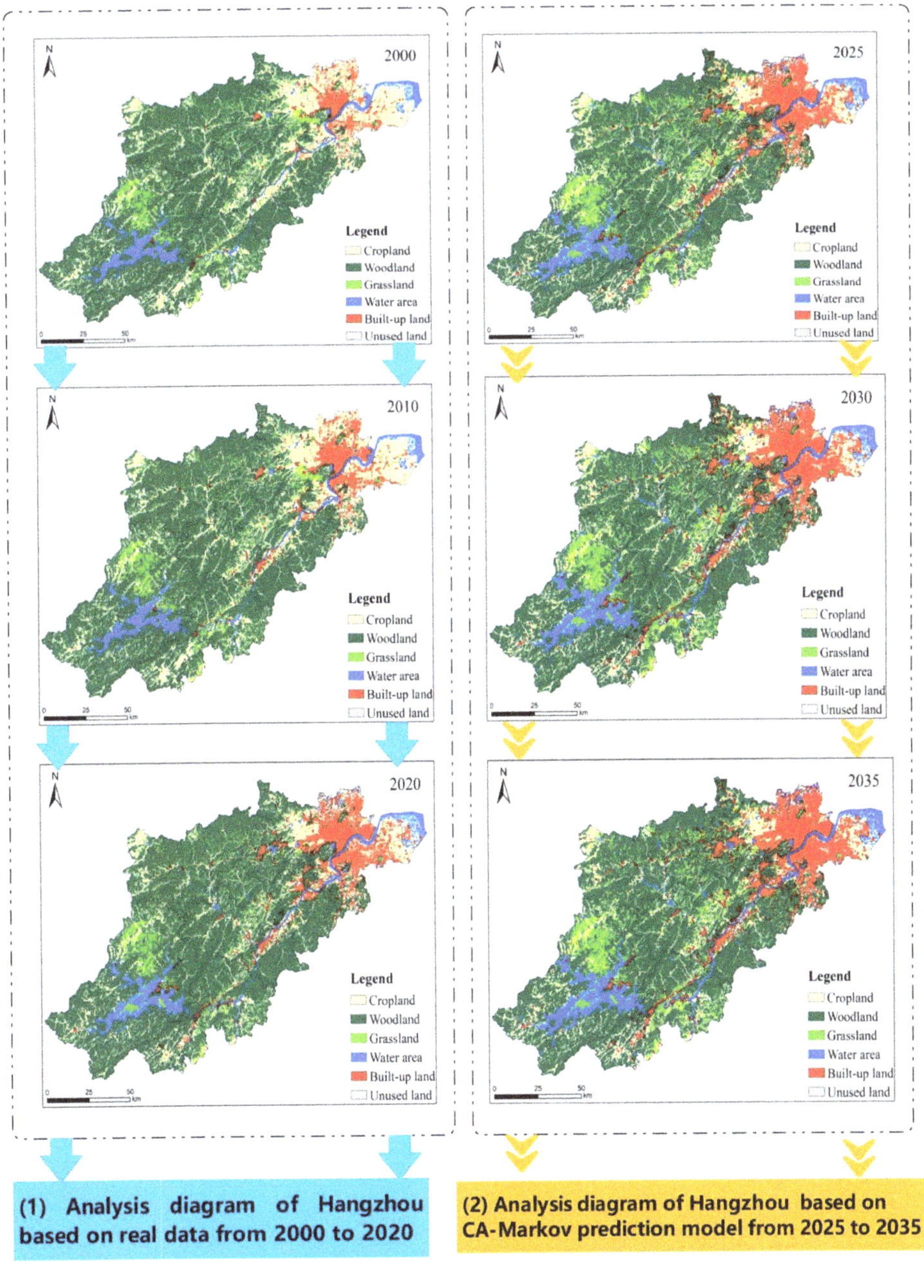

Figure 6. Analysis diagram of Hangzhou.

Using the CA–Markov prediction model, change trends have been displayed for Hangzhou city in 2025, 2030, and 2035 (in Figure 6(2)). The building area continues to expand, while the cropland area continues to decrease. By 2035, the built-up area has expanded to the southwest of Hangzhou, particularly along the Qiantang River. However, there is little change in the GIS maps of woodland, grassland, and water area, indicating that although the green area of Hangzhou has not changed significantly, the expansion of the city has had a negative impact on the sustainability of the city as a whole.

3.2.3. Single Factor Analysis

Figures 7–9 show the predicted changes in cropland, woodland, and built-up land for Hangzhou city from 2020 to 2035. Overall, the most significant changes are anticipated in built-up land, followed by woodland and cropland. As a developed city in China, Hangzhou primarily relies on industry and service industries for urban development, with agriculture being secondary. For example, in 2020, the GDP ratio of agriculture, industry, and services was 1:16.5:36.9, accounting for only about 0.18% of the entire GDP in Hangzhou city. Therefore, the modelled cropland changes from 2020 to 2035 have hardly changed on the GIS maps.

With the rapid growth of the population, ecological space is being squeezed, resulting in the reduction of woodland, which is evident on the GIS map from 2020 to 2035. These findings emphasize the need for sustainable development strategies that protect ecological resources while promoting urbanization and economic growth. Furthermore, these results can inform policymakers and stakeholders in Hangzhou city and other urban areas to design effective sustainable development strategies that balance urban development with environmental protection.

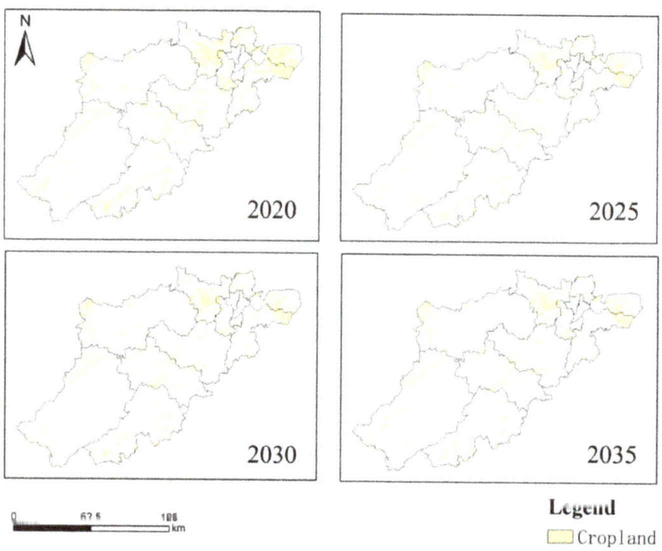

Figure 7. Cropland changes on GIS map in 2020, 2025, 2030 and 2035.

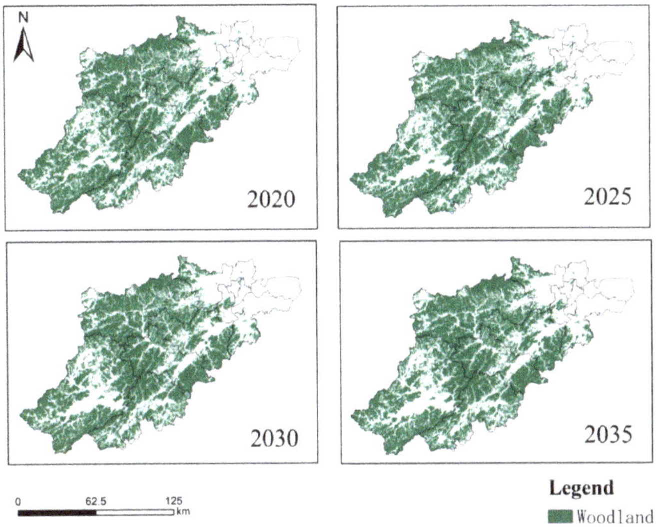

Figure 8. Woodland changes on GIS map in 2020, 2025, 2030 and 2035.

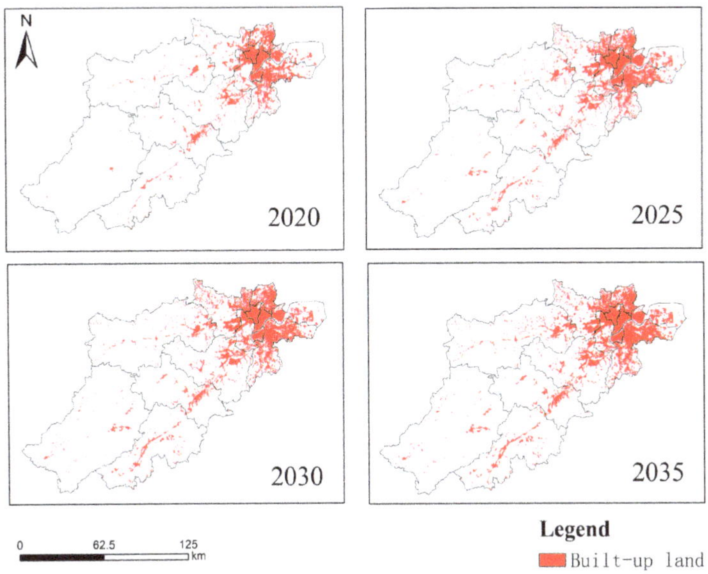

Figure 9. Built-up land changes on GIS map in 2020, 2025, 2030 and 2035.

4. Discussion

4.1. Main Contributor Analysis

Table 4 presents the emergy calculation for Hangzhou city in 2020. The table shows that biodiversity and buildings display the largest emergy quantities, with building emergy having the largest value of 1.98×10^{28} sej, followed by biodiversity emergy with a value of 2.46×10^{28} sej. This suggests that sustainable change for 2020 will be determined by these two factors. Other terms have relatively small emergy values, playing a minor role in the sustainability of the Hangzhou system in 2020.

The emergy values for 2000 and 2010 were also calculated using the same calculation model, and similar patterns can be observed in 2020. These results highlight the importance of biodiversity and buildings in determining the sustainability of Hangzhou city over time, emphasizing the need for sustainable development strategies that protect biodiversity while promoting urbanization.

4.2. Ecological Indicator Analysis

Figure 10 presents the sustainable degree of Hangzhou city from 2000 to 2020. From the perspective of a single indicator (Re, Nr, Np), the renewable rate of Hangzhou is decreasing, while the non-renewable rate is increasing. This trend is due to the reduction of ecological areas and the expansion of urban building areas as Hangzhou continues to develop.

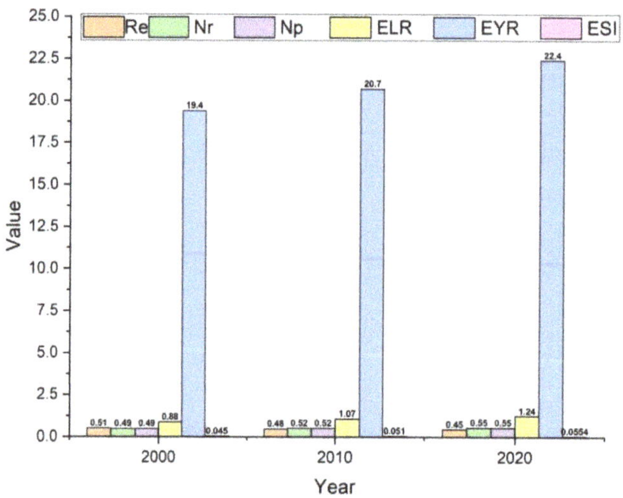

Figure 10. Comparison between and change from 2000 to 2020.

Based on a comprehensive analysis of ELR, EYR, and ESI, ELR and EYR tend to increase from 2000 to 2020, which reflects continued urban construction during these decades. As a general rule, when these two values (ELR and EYR) continue to increase, the final sustainability index (ESI) will become smaller. However, in this study, the most critical sustainability indicators are tending to increase from 2000 to 2020. This phenomenon demonstrates that despite environmental pressures, the urban development of Hangzhou is still in a healthy development trend in the long run.

4.3. Sensitivity Analysis

Due to the involvement of a large amount of data, the sensitivity of the data needs to be designed and analyzed to ensure the accuracy of the results. Scholars in the field have offered in-depth discussions and already conducted a range sensitivity analyses [60,61]. In turn, this study provides sensitivity analysis on five types of land use, including arable land, forest land, grassland, water area, and construction land, as follows:

The sensitivity of different land types, including farmland, forests, grasslands, water bodies, and construction land, was analyzed based on solar data from the five types of land. As shown in Figure 11A–E, the data sensitivity of water bodies and construction land is better, while that of farmland, forests, and grasslands is relatively poor. Figure 11F shows the magnitude of changes in data for these five land types, and the trend analysis on the right side indicates that the sensitivity of farmland has changed the most, followed by grasslands and forests. This result is consistent with the corresponding Figure 11A–E.

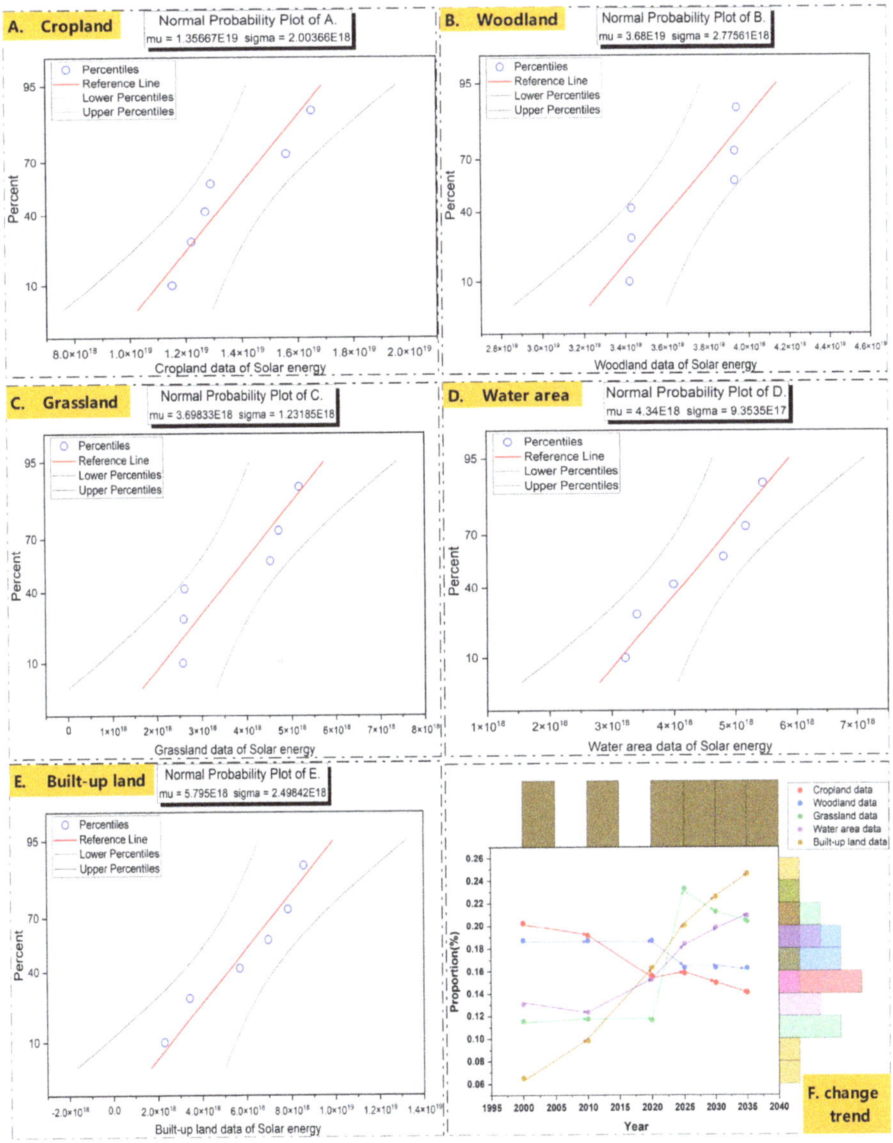

Figure 11. Sensitivity analysis of solar energy data. (**A**) Cropland data; (**B**) Woodland data; (**C**) Grassland data; (**D**) Water area data; (**E**) Built-up data; (**F**) Comprehensive analysis.

In addition to analyzing the sensitivity of solar energy data for the five types of land, the study also explored the sensitivity of four other types of energy: rain potential energy, rain chemical energy, wind energy, and geothermal energy. Figure 12 includes four violin plots that show the trend of sensitivity changes indicated by the red lines, which connect the mean data of the five types of land. The four plots reveal that the sensitivity of rain potential energy, rain chemical energy, and wind energy data fluctuate the least. On the other hand, the sensitivity of geothermal energy fluctuates more, which may be attributed to difficulties in extracting geothermal energy data and large errors in actual data.

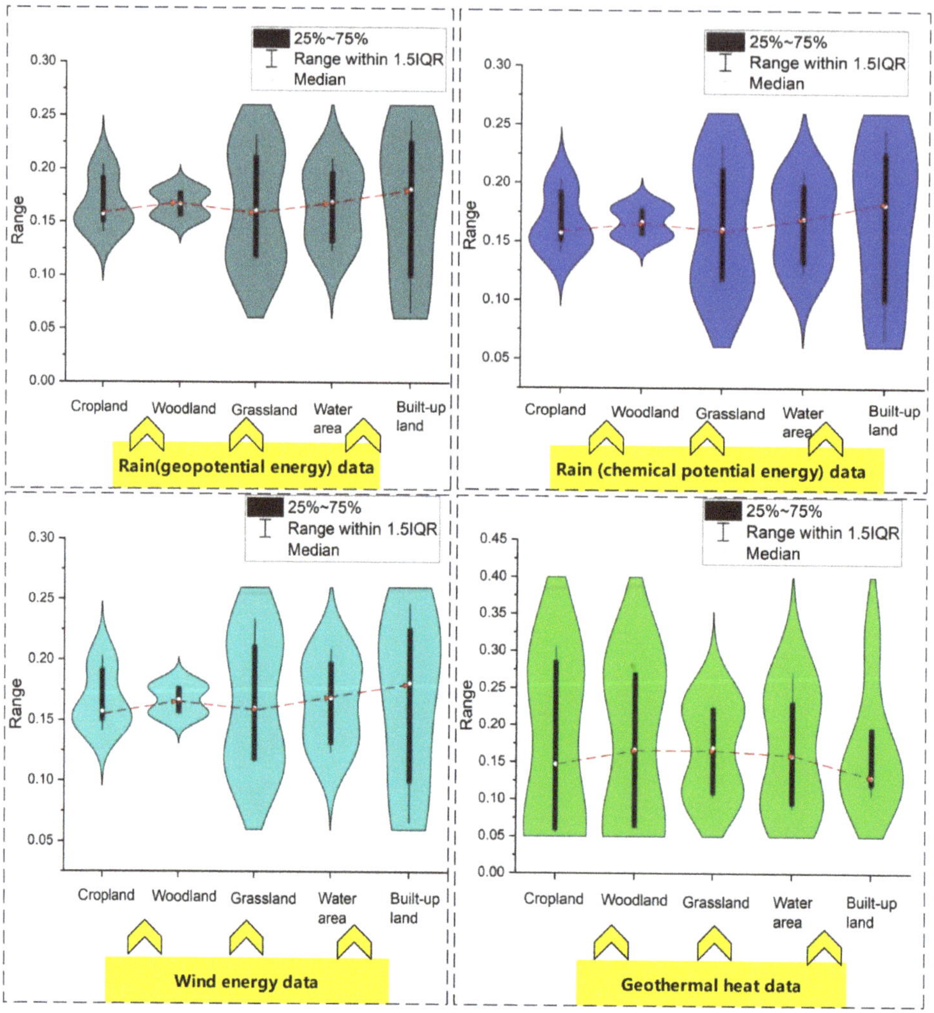

Figure 12. Sensitivity analysis of the remaining four types of data.

4.4. Comprehensive Sustainability Analysis

The emergy method and the GIS method are two distinct approaches for studying urban problems. The emergy method is an ecological economic approach, while the GIS method is a geographic information system method that examines morphological changes from a spatial perspective. Additionally, the emergy method is a quantitative evaluation model, whereas the GIS method is a qualitative research model.

This paper aims to investigate ecological sustainability in Hangzhou city by coupling these two methods. Using the emergy method, the ecological sustainability of the city was calculated and assessed, while the GIS map was used to display the ecological changes of the city (i.e., the five types of plots) to verify the accuracy of the emergy calculation results.

Using data from 2000, 2010, and 2020 based on the GIS method, it is evident that the area of ecological elements is decreasing, while the building area is increasing rapidly, indicating an increase in ecological pressure throughout the city. This trend is also supported by the data presented in Figure 7, which indicate that the environmental loading ratio increased from 0.88 in 2000 to 1.07 in 2010 and to 1.24 in 2020, showing that the entire environmental burden of the city is gradually increasing with its sustainable development.

Taken together, the emergy method and the GIS method can be used to comprehensively analyze urban ecological changes and provide valuable insights into how to promote sustainable development strategies that balance urbanization and ecological protection.

5. Improvement Measures

5.1. Increasing Green Area Measures

Green areas play a significant role in promoting urban ecology, as they not only reduce carbon emissions but also alleviate the urban heat island effect, thereby enhancing the sustainable development of cities. In this study, based on the green area of Hangzhou city in 2020, if the green area is increased by 20%, 30%, and 40%, the corresponding indicators can be obtained according to emergy calculation. Figure 13 shows the changes in various indicators after improvement, clearly indicating the positive effects of green areas on urban ecology. Taking the ELR index as an example, with a 20% increase in urban green areas, the negative value load ratio will decrease by 1.62%; with a 30% increase in green areas, the index will decrease by 4.01%; with a 40% increase in green areas, the index will decrease by 6.45%. These results fully demonstrate that increasing green areas can reduce the ecological burden of cities, and the larger the green area, the more significant the reduction in the ecological burden. To enhance the ecological sustainability of cities, green areas should be increased based on the specific circumstances of each city.

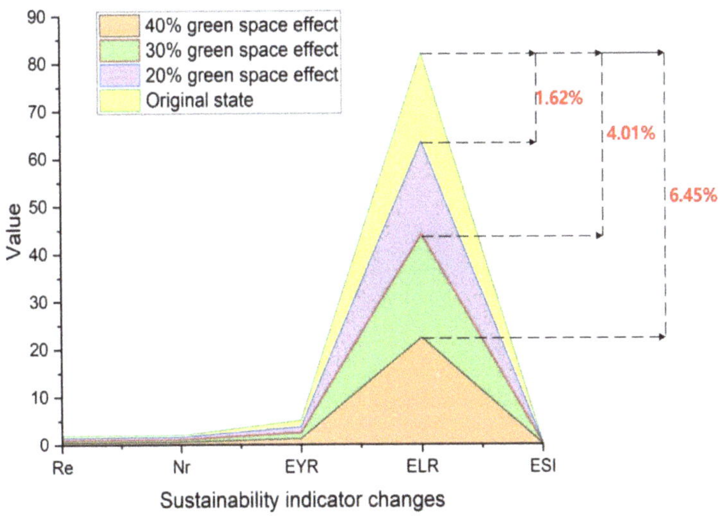

Figure 13. Sustainability indicator changes between original state and scenarios of increasing green space.

5.2. Improving Urban Water Areas

Water bodies are also one of the core elements for enhancing urban ecology. To improve the overall sustainable level of Hangzhou city, measures to increase water area have been proposed. As there are not many large river gathering areas in Hangzhou city and they are mostly concentrated in the southwest, in order to verify the effect of increasing water area on the sustainability of the city, the study used the water area of Hangzhou city in 2020 as the basic data and quantitatively analyzed its changes by assuming incremental increases in water area based on emergy evaluation indicators.

In Figure 14, the results of four hypothetical scenarios were considered. As the water area increases by 20%, 25%, 30% and 40%, the environmental loading ratio of Hangzhou city decreases significantly (as shown in Figure 14A–D), with reductions of 5.4%, 8.1%, 19.2%, and 23.7%, respectively. The increasing difference in area indicates that water bodies promote urban sustainability and reduce environmental pressure.

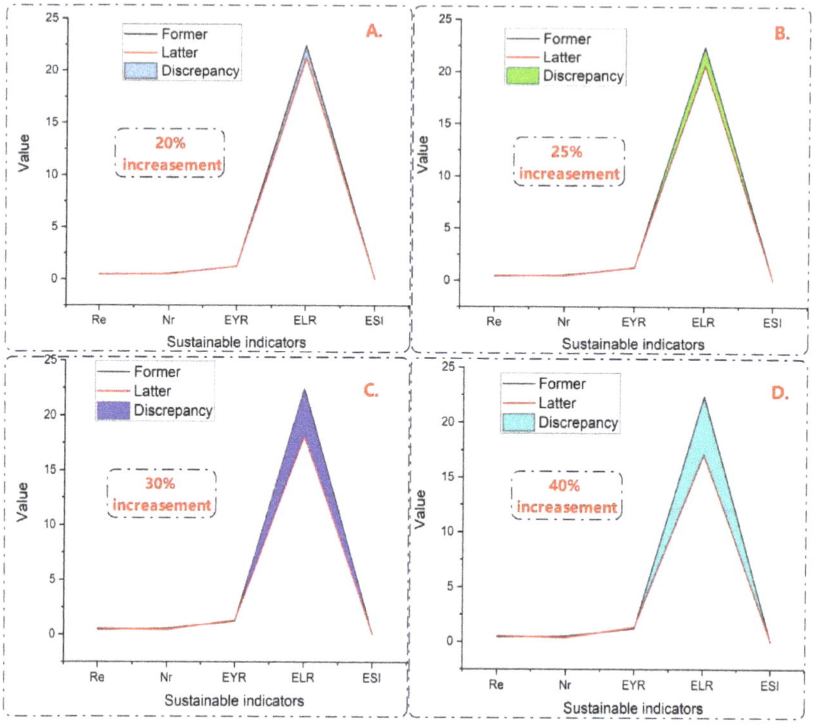

Figure 14. Sustainability indicator changes between original state and scenarios of increasing water space. (**A**) 20% increasement design; (**B**) 25% increasement design; (**C**) 30% increasement design; (**D**) 40% increasement design.

6. Conclusions

In summary, this study utilized the emergy–GIS method to assess urban sustainability in Hangzhou city from 2000 to 2035, which demonstrates that combining the emergy with the GIS approach can effectively promote each other to validate the sustainability status of a city. Therefore, this study provides a comprehensive methodology for researchers and designers to assess urban sustainability from a new perspective in the future.

However, these methods also have their limitations. Although the emergy method and GIS method can be integrated for urban sustainability research, they belong to two different fields, and this can present obstacles to comparative studies of urban sustainability. Firstly, the GIS method is a large-scale urban spatial analysis approach that involves multiple elements on the GIS map. This presents challenges for all emergy calculations for the city because not all urban elements currently have a unit emergy value. For instance, the emergy of microbes is difficult to calculate accurately due to an inaccurate unit emergy value, which results in uncertainty in the calculation results. Secondly, the lack of urban data leads to inconsistent results between the two methods, particularly in larger cities. The more detailed the data results, the more consistent the calculation results become based on the emergy method and GIS method. To address this issue and further promote the development of the emergy–GIS method, comprehensive urban data and existing unit emergy values are essential.

In addition, multi-criteria decision analysis (MCDA) and life cycle assessment (LCA) contribute to enhancing the integration of emergy analysis with GIS, enabling comprehensive qualitative and quantitative results. MCDA techniques provide a systematic framework for evaluating multiple criteria and making informed decisions. By combining emergy analysis with MCDA, decision-makers can incorporate environmental, economic, and social factors into a unified evaluation process. This integration allows for prioritizing different alternatives or interventions based on their emergy performance and spatial distribution. Life cycle assessment is a widely used methodology for assessing the environmental impacts of products, processes, or systems throughout their entire life cycle. Integrating emergy analysis with LCA enables a more comprehensive evaluation of resource use and environmental impacts. Leveraging the strengths of both approaches provides a holistic perspective on sustainability and facilitates decision-making towards more sustainable practices.

Future research should strive to overcome the obstacles previously noted by utilizing more reliable and accurate data sources to improve the precision and accuracy of the emergy–GIS method. Additionally, researchers could explore new ways to integrate the emergy method and GIS method more effectively to achieve a more comprehensive understanding of urban sustainability and its associated challenges.

Author Contributions: Conceptualization, L.L. and J.Z.; investigation, R.C.; formal analysis, J.Z.; methodology, J.Z.; resources, J.Z.; writing—review and editing, J.Z. All authors have read and agreed to the published version of the manuscript.

Funding: The work described in this paper was supported by the XJTLU Urban and Environmental Studies University Research Centre (UES) (UES-RSF-23030601).

Data Availability Statement: The data presented in this study are available within this article.

Conflicts of Interest: The authors declare no conflict of interest.

Appendix A

Taking the cropland (in 2000) as the calculation example, calculations are as follows:

1. Solar energy calculation:

- ➢ Area of plot = 4,341,224,700 m^2;
- ➢ Insolation = 5.43×10^9 J/m^2/yr;
- ➢ Albedo = 0.30;
- ➢ Energy = (insolation) × (1-albedo) × (area) = (5.43×10^9 J/m^2/yr) × (1−0.30) × (4,341,224,700 m^2) = 1.65×10^{19} J/yr;
- ➢ UEV = 1.00 sej/J by definition;
- ➢ Emergy = 1.65×10^{19} J/yr × 1.00 sej/J = 1.65×10^{19} sej.

2. Rain (geopotential energy) calculation:

- ➢ Area of plot = 4,341,224,700 m^2;
- ➢ Rainfall (annual average, n = 5) = 0.71 m/year;
- ➢ Average elevation = 316 m; water density = 1.00×10^3 kg/m^3;
- ➢ Runoff rate = 40.00%;
- ➢ Energy = (area) × (rainfall) × (runoff rate) × (water density) × (average elevation) × (gravity) = (4,341,224,700 m^2) × (0.71 m/yr) × (40.00%) × (1.00×10^3 kg/m^3) × (316 m) × (9.8 kg/m^2) = 3.82×10^{15} J/yr;
- ➢ UEV = 2.79×10^4 sej/J;
- ➢ Emergy = 3.82×10^{15} J/yr × 2.79×10^4 sej/J = 1.07×10^{20} sej.

3. Rain (chemical potential energy) calculation:

- ➢ Area of plot = 4,341,224,700 m^2;
- ➢ Rainfall (annual average, n = 5) = 0.71 m/year;
- ➢ Water density = 1000 kg/m^3;

- ➢ Evapotranspiration rate = 60%;
- ➢ Gibbs free energy of water = 4.94×10^3 J/kg;
- ➢ Energy = (area) × (rainfall) × (evapotranspiration rate) × (water density) × (Gibbs free energy of water) = (4,341,224,700 m^2) × (0.71 m/yr) × (1.00×10^3 kg/m^3) × (60%) × (4.94×10^3 J/kg) = 9.14×10^{15} J/year;
- ➢ UEV = 18199 sej/J;
- ➢ Emergy = 9.14×10^{15} J/yr × 18,199 sej/J =1.66×10^{20} sej.

4. Wind energy calculation:

- ➢ Area of plot = 4,341,224,700 m^2;
- ➢ Air density = 1.29 kg/m^3;
- ➢ Wind velocity (annual average, n = 2) = 1.95 m/s;
- ➢ Velocity of geostrophic wind = 3.25 m/s (surface winds are considered as 0.6 of geostrophic wind);
- ➢ Drag coefficient = 1.00×10^{-3};
- ➢ Energy = (area) × (air density) × (drag coefficient) × (velocity of geostrophic wind)^3 = (4,341,224,700 m^2) × (1.29 kg/m^3) × (1.00×10^{-3}) × (1.95 m/s)^3 = 4.2×10^7 J/yr;
- ➢ UEV = 1496 sej/J;
- ➢ Emergy = 4.2×10^7 J/yr × 1496 sej/J = 6.28×10^{10} sej.

5. Geothermal heat calculation:

- ➢ Area of plot = 4,341,224,700 m^2;
- ➢ Heat flow (average) = 3.50×10^{-2} J/m^2/s. Energy = (area) × (heat flow) = (4,341,224,700 m^2) × (3.50×10^{-2} J/m^2/s) = 1.52×10^8 J/yr;
- ➢ UEV = 34,377 sej/J;
- ➢ Emergy = 1.52×10^8 J/yr × 34,377 sej/J = 5.22×10^{12} sej.

References

1. China Urban Development Report in 2021. Available online: http://bj.ieaschina.org/organization/introduce.html (accessed on 10 September 2023).
2. Wang, D.; Huang, Z.; Wang, Y.; Mao, J. Ecological security of mineral resource-based cities in China: Multidimensional measurements, spatiotemporal evolution, and comparisons of classifications. *Ecol. Indic.* **2021**, *132*, 108269. [CrossRef]
3. Chai, J.; Wang, Z.; Yu, C. Analysis for the Interaction Relationship between Urbanization and Ecological Security: A Case Study in Wuhan City Circle of China. *Int. J. Environ. Res. Public Health* **2021**, *18*, 13187. [CrossRef]
4. Zhang, H.; Li, J.; Tian, P.; Pu, R.; Cao, L. Construction of ecological security patterns and ecological restoration zones in the city of Ningbo, China. *J. Geogr. Sci.* **2022**, *32*, 663–681. [CrossRef]
5. Wu, J.; Bai, Z. Spatial and temporal changes of the ecological footprint of China's resource-based cities in the process of urbanization. *Resour. Policy* **2022**, *75*, 102491. [CrossRef]
6. Zhang, S.; Li, F.; Zhou, Y.; Hu, Z.; Zhang, R.; Xiang, X.; Zhang, Y. Using Net Primary Productivity to Characterize the Spatio-Temporal Dynamics of Ecological Footprint for a Resource-Based City, Panzhihua in China. *Sustainability* **2022**, *14*, 3067. [CrossRef]
7. Lu, C.; Wang, S.; Wang, K.; Gao, Y.; Zhang, R. Uncovering the benefits of integrating industrial symbiosis and urban symbiosis targeting a resource-dependent city: A case study of Yongcheng, China. *J. Clean. Prod.* **2020**, *255*, 120210. [CrossRef]
8. Zhang, L.; Wang, H.; Zhang, W.; Wang, C.; Bao, M.; Liang, T.; Liu, K. Study on the development patterns of ecological civilization construction in China: An empirical analysis of 324 prefectural cities. *J. Clean. Prod.* **2022**, *367*, 132975. [CrossRef]
9. Yang, Y.; Feng, Z.; Wu, K.; Lin, Q. How to construct a coordinated ecological network at different levels: A case from Ningbo city, China. *Ecol. Inform.* **2022**, *70*, 101742. [CrossRef]
10. Sun, F.; Ye, C.; Zheng, W.; Miao, X. Model of Urban Marketing Strategy Based on Ecological Environment Quality. *J. Environ. Public Health* **2022**, *2022*, 8096122. [CrossRef]
11. Xu, S.; Hou, Y.; Mao, L. Application Analysis of the Ecological Economics Model of Parallel Accumulation Sorting and Dynamic Internet of Things in the Construction of Ecological Smart City. *Wirel. Commun. Mob. Comput.* **2022**, *2022*, 8770859. [CrossRef]
12. Ma, Q.; Yang, Y. Analysis of Ecological Environment Evaluation and Coupled and Coordinated Development of Smart Cities Based on Multisource Data. *J. Sens.* **2022**, *2022*, 5959495. [CrossRef]
13. Fan, Z.; Wang, Y.; Feng, Y. Ecological Livability Assessment of Urban Agglomerations in Guangdong-Hong Kong-Macao Greater Bay Area. *Int. J. Environ. Res. Public Health* **2021**, *18*, 13349. [CrossRef]
14. Zhang, J.; Ma, L. Urban ecological security dynamic analysis based on an innovative emergy ecological footprint method. Environment. *Environ. Dev. Sustain.* **2021**, *23*, 16163–16191. [CrossRef]

15. Ha, C.; Huang, G.; Zhang, J.; Dong, S. Assessing ecological literacy and its application based on linguistic ecology: A case study of Guiyang City, China. *Environ. Sci. Pollut. Res.* **2022**, *29*, 18741–18754. [CrossRef]
16. Huang, L.; Wang, D.; He, C. Ecological security assessment and ecological pattern optimization for Lhasa city (Tibet) based on the minimum cumulative resistance model. *Environ. Sci. Pollut. Res.* **2022**, *29*, 83437–83451. [CrossRef] [PubMed]
17. Wang, D.; Ji, X.; Li, C.; Gong, Y. Spatiotemporal Variations of Landscape Ecological Risks in a Resource-Based City under Transformation. *Sustainability* **2020**, *13*, 5297. [CrossRef]
18. Xia, H.; Liu, Z.; Efremochkin, M.; Liu, X.; Lin, C. Study on city digital twin technologies for sustainable smart city design: A review and bibliometric analysis of geographic information system and building information modeling integration. *Sustain. Cities Soc.* **2022**, *84*, 104009. [CrossRef]
19. Xu, W.; Yi, J.; Shuai, J.; Yu, Z.; Cheng, J. Dynamic evaluation of the ecological civilization of Jiangxi Province: GIS and AHP approaches. *PLoS ONE* **2022**, *17*, e0271768. [CrossRef] [PubMed]
20. Wang, X.; Zhang, C.; Wang, C.; Liu, G.; Wang, H. GIS-based for prediction and prevention of environmental geological disaster susceptibility: From a perspective of sustainable development. *Ecotoxicol. Environ. Saf.* **2021**, *226*, 112881. [CrossRef]
21. Di Pinto, V.; Rinaldi, A.M.; Rossini, F. Learning from the Informality. Using GIS Tools to Analyze the Structure of Autopoietic Urban Systems in the "Smart Perspective". *ISPRS Int. J. Geo-Inf.* **2021**, *10*, 202. [CrossRef]
22. Zhu, J.; Wu, P. Towards Effective BIM/GIS Data Integration for Smart City by Integrating Computer Graphics Technique. *Remote Sens.* **2021**, *13*, 1889. [CrossRef]
23. Xie, H.; Wen, J.; Chen, Q.; Wu, Q. Evaluating the landscape ecological risk based on GIS: A case-study in the Poyang Lake region of China. *Land Degrad. Dev.* **2021**, *32*, 2762–2774. [CrossRef]
24. Pallathadka, A.; Pallathadka, L.; Rao, S.; Chang, H.; Van Dommelen, D. Using GIS-based spatial analysis to determine urban greenspace accessibility for different racial groups in the backdrop of COVID-19: A case study of four US cities. *GeoJournal* **2021**, *87*, 4879–4899. [CrossRef] [PubMed]
25. He, D.; Hou, K.; Wen, J.F.; Wu, S.Q.; Wu, Z.P. A coupled study of ecological security and land use change based on GIS and entropy method—A typical region in Northwest China, Lanzhou. *Environ. Sci. Pollut. Res.* **2022**, *29*, 6347–6359. [CrossRef] [PubMed]
26. Parveen, M.T.; Ilahi, R.A. Assessment of land-use change and its impact on the environment using GIS techniques: A case of Kolkata Municipal Corporation, West Bengal, India. *GeoJournal* **2022**, *87*, 551–566. [CrossRef]
27. Fan, Y.; Fang, C. Evolution process analysis of urban metabolic patterns and sustainability assessment in western China, a case study of Xining city. *Ecol. Indic.* **2020**, *109*, 105784. [CrossRef]
28. Chen, Y.; Peng, L.; Cao, W. Health evaluation and coordinated development characteristics of urban agglomeration: Case study of Fujian Delta in China. *Ecol. Indic.* **2021**, *121*, 107149. [CrossRef]
29. Wang, Q.; Liu, M.; Tian, S.; Yuan, X.; Ma, Q.; Hao, H. Evaluation and improvement path of ecosystem health for resource-based city: A case study in China. *Ecol. Indic.* **2021**, *128*, 107852. [CrossRef]
30. Liu, Y.; Qu, Y.; Cang, Y.; Ding, X. Ecological security assessment for megacities in the Yangtze River basin: Applying improved emergy-ecological footprint and DEA-SBM model. *Ecol. Indic.* **2022**, *134*, 108481. [CrossRef]
31. Voukkali, I.; Zorpas, A.A. Evaluation of urban metabolism assessment methods through SWOT analysis and analytical hierocracy process. *Sci. Total Environ.* **2022**, *807*, 150700. [CrossRef]
32. Tang, M.; Hong, J.; Wang, X.; He, R. Sustainability accounting of neighborhood metabolism and its applications for urban renewal based on emergy analysis and SBM-DEA. *J. Environ. Manag.* **2020**, *275*, 111177. [CrossRef] [PubMed]
33. Li, Q.; Wu, J.; Su, Y.; Zhang, C.; Wu, X.; Wen, X.; Huang, G.; Deng, Y.; Lafortezza, R.; Chen, X. Estimating ecological sustainability in the Guangdong-Hong Kong-Macao Greater Bay Area, China: Retrospective analysis and prospective trajectories. *J. Environ. Manag.* **2022**, *303*, 114167. [CrossRef]
34. Lee, Y.-C.; Liao, P.-T. The effect of tourism on teleconnected ecosystem services and urban sustainability: An emergy approach. *Ecol. Model.* **2021**, *439*, 109343. [CrossRef]
35. Alizadeh, S.; Zafari-koloukhi, H.; Rostami, F.; Rouhbakhsh, M.; Avami, A. The eco-effificiency assessment of wastewater treatment plants in the city of Mashhad using emergy and life cycle analyses. *J. Clean. Prod.* **2020**, *249*, 119327. [CrossRef]
36. Zhou, H.; Li, H.; Zhao, X.; Ding, Y. Emergy ecological model for sponge cities: A case study of China. *J. Clean. Prod.* **2021**, *296*, 126530. [CrossRef]
37. Li, J.; Sun, W.; Song, H.; Li, R.; Hao, J. Toward the construction of a circular economy eco-city: An emergy-based sustainability evaluation of Rizhao city in China. *Sustain. Cities Soc.* **2021**, *71*, 102956. [CrossRef]
38. Zhang, J.; Asutosh, A.T.; Zhang, H.; Yan, Y.; Zhang, Y.; Wei, G.; Ma, C.; Shi, Y.; Gao, Y.; Yan, X.; et al. Environmental sustainability in the city of Shanghai municipal solid waste treatment system: An integrated framework of artifcial neural network (ANN) and LCA-emergy methodology. *Arab. J. Geosci.* **2022**, *15*, 1271. [CrossRef]
39. Farzaneh, H.; Dashti, M.; Zusman, E.; Lee, S.-Y.; Dagvadorj, D.; Nie, Z. Assessing the Environmental-Health-Economic Co-Benefifits from Solar Electricity and Thermal Heating in Ulaanbaatar, Mongolia. *Int. J. Environ. Res. Public Health* **2022**, *19*, 6931. [CrossRef]
40. Liu, W.; Zhan, J.; Li, Z.; Jia, S.; Zhang, F.; Li, Y. Eco-Efficiency Evaluation of Regional Circular Economy: A Case Study in Zengcheng, Guangzhou. *Sustainability* **2018**, *10*, 453. [CrossRef]

41. Agostinho, F.; Diniz, G.; Siche, R.; Ortega, E. The use of emergy assessment and the Geographical Information System in the diagnosis of small family farms in Brazil. *Ecol. Model.* **2008**, *210*, 37–57. [CrossRef]
42. Pulselli, R.M. Integrating emergy evaluation and geographic information systems for monitoring resource use in the Abruzzo region (Italy). *J. Environ. Manag.* **2010**, *91*, 2349–2357. [CrossRef] [PubMed]
43. Mellino, S.; Ripa, M.; Zucaro, A.; Ulgiati, S. An emergy–GIS approach to the evaluation of renewable resource flows: A case study of Campania Region, Italy. *Ecol. Model.* **2014**, *271*, 103–112. [CrossRef]
44. Arbault, D.; Rugani, B.; Tiruta-Barna, L.; Benetto, E. A first global and spatially explicit emergy database of rivers and streams based on high-resolution GIS-maps. *Ecol. Model.* **2014**, *281*, 52–64. [CrossRef]
45. Mellino, S.; Buonocore, E.; Ulgiati, S. The worth of land use: A GIS–emergy evaluation of natural and human-made capital. *Sci. Total Environ.* **2015**, *506–507*, 137–148. [CrossRef]
46. Mellino, S.; Ulgiati, S. Mapping the evolution of impervious surfaces to investigate landscape metabolism: An Emergy–GIS monitoring application. *Ecol. Inform.* **2015**, *26*, 50–59. [CrossRef]
47. Wang, C.; Zhang, S.; Yan, W.; Wang, R.; Liu, J.; Wang, Y. Evaluating renewable natural resources flow and net primary productivity with a GIS-Emergy approach: A case study of Hokkaido, Japan. *Sci. Rep.* **2016**, *6*, 37552. [CrossRef]
48. Zhao, Y.-B.; Yang, M.-Z.; Ni, H.-G. An emergy-GIS method of selecting areas for sponge-like urban reconstruction. *J. Hydrol.* **2018**, *564*, 640–650. [CrossRef]
49. Lee, D.J.; Brown, M.T. Estimating the Value of Global Ecosystem Structure and Productivity: A Geographic Information System and Emergy Based Approach. *Ecol. Model.* **2021**, *439*, 109307. [CrossRef]
50. Odum, H.T. Environmental accounting. In *Emergy and Environmental Decision Making*; John Wiley & Sons: New York, NY, USA, 1996.
51. Agostinho, F.; Ambrósio, L.A.; Ortega, E. Assessment of a large watershed in Brazil using Emergy Evaluation and Geographical Information System. *Ecol Model.* **2010**, *221*, 1209–1220. [CrossRef]
52. Brown, M.T.; Martinez, A.; Uche, J. Emergy analysis applied to the estimation of the recovery of costs for water services under the European Water Frame-work Directive. *Ecol. Model.* **2010**, *221*, 2123–2132. [CrossRef]
53. Zhang, L.X.; Ulgiati, S.; Yang, Z.F.; Chen, B. Emergy evaluation and economic analysis of three wetland fifish farming systems in Nansi Lake area, China. *J. Environ. Manag.* **2011**, *92*, 683–694. [CrossRef]
54. Brown, M.T.; Ulgiati, S. Updated evaluation of exergy and emergy driving the geobiosphere: A review and refinement of the emergy baseline. *Ecol. Model.* **2016**, *221*, 2501–2508. [CrossRef]
55. Hangzhou Yearbook. 2021. Available online: http://tjj.hangzhou.gov.cn/col/col1229453450/index.html (accessed on 1 September 2023).
56. Campbell, E.T.; Brown, M.T. Environmental accounting of natural capital and ecosystem services for the US National Forest System. *Environ. Dev. Sustain.* **2012**, *14*, 691–724. [CrossRef]
57. Pulselli, R.M.; Simoncini, E.; Marchettini, N. Energy and emergy based cost-benefifit evaluation of building envelopes relative to geographical location and climate. *Build. Environ.* **2009**, *44*, 920–928. [CrossRef]
58. Dong, X.B.; Yu, B.H.; Brown, M.; Zhang, Y.S.; Kang, M.Y.; Jin, Y.; Zhang, X.S.; Ulgiati, S. Environmental and economic consequences of the overexploitation of natural capital and ecosystem services in Xilinguole League, China. *Energy Policy* **2014**, *67*, 767–780. [CrossRef]
59. Brown, M.T.; Ulgiati, S. A geobiosphere baseline for LCA-emergy evaluations. In *Emergy Synthesis 7: Theory and Applications of the Emergy Methodology, Proceedings of the 7th Biennial Emergy Conference Center for Environmental Policy*; Brown, M.T., Sweeney, S., Campbell, D.E., Huang, S., Kang, D., Rydberg, T., Tilley, D., Ulgiati, S., Eds.; University of Florida Gainesville: Gainesville, FL, USA, 2013.
60. Zhang, J.; Asutosh, A.T. A Sustainability Analysis Based on the LCA–Emergy–Carbon Emission Approach in the Building System. *Appl. Sci.* **2023**, *13*, 9707. [CrossRef]
61. Cao, J.; Zhu, Y.; Zhang, J.; Wang, H.; Zhu, H. The Sustainability Study and Exploration in the Building Commercial Complex System Based on Life Cycle Assessment (LCA)–Emergy–Carbon Emission Analysis. *Processes* **2023**, *11*, 1989. [CrossRef]

buildings

MDPI

Article

Research on the Industrial Heritage Community Retrofitting Design Based on Space Network Model of Carbon

Yuetao Wang [1], Ruidong Zhu [1], Jinming Liu [2,*], Fei Zheng [1] and Chengbin Wu [1]

[1] School of Architecture and Urban Planning, Shandong Jianzhu University, Jinan 250101, China; wyeto@163.com (Y.W.); zrd17806168220@163.com (R.Z.); zf1667@163.com (F.Z.); wu1311817055@163.com (C.W.)

[2] China Shanghai Architectural Design & Research Institute Co., Ltd., Shanghai 200333, China

* Correspondence: 13816583595@163.com; Tel.: +86-13816583595

Abstract: The low-carbon retrofitting of industrial heritage communities is an important issue for reducing urban carbon emissions. Previous research on the judgment of heritage elements and carbon emission factors of industrial heritage communities lacked the construction of elements within the space, and the value judgment of conservation and the determination of low-carbon factors lacked a systematic network analysis. Carbon spatial networks as a systematic approach can systematically harmonize the contradictions between "conservation–transformation–low carbon" while considering the spatial and temporal carbon emissions of nodes. This research uses hierarchical analysis to analyze the value of heritage elements and locate them in space, then combine the elements that affect carbon activities in space to form carbon space nodes integrated with heritage conservation elements, and analyze the links between nodes to form a carbon space network. Then, this research uses a carbon spatial network to dissect the structure of carbon emissions, calculate the carbon activity at nodes, and reflect it into a parameterized platform to guide designers. After that, this research selected 16 industrial heritage communities in the severe cold region for the carbon activity measurement of carbon spatial network node elements in the operation stage and conducted a correlation analysis to obtain a correlation matrix model of node elements and carbon activities. Within the constraints of heritage element protection, the results show that the volume ratio, green area ratio, and building density of the carbon spatial network model have the greatest influence on its carbon performance, followed by porosity and road density. Through case simulation verification, the floor area ratio of industrial heritage communities should be controlled between 2.1–2.5, the number of residential building floors should be 7–14, the road network density should be 4.16–4.50%, the green space ratio should be 20–35%, and the porosity should be 35–45%. Taking the three major power road communities in Harbin as an example, retrofit measures were decided by reference to relevant parameter control intervals, resulting in a 21.1% reduction in energy consumption, an approximate 32.7% reduction in carbon emissions, and a 7.3% reduction in the annual percentage of hours in extreme hot and cold environments.

Keywords: industrial heritage communities; carbon space networks; historic community preservation; performance simulation; microclimate

1. Introduction

Today, the main cause of global warming is the emission of GHGs (e.g., CO_2, CH_4, etc.), and its reduction has become an urgent issue. The heat island effect and deteriorating air quality in urban areas are having a serious impact on human habitat, and the public health of people in urban areas is being seriously affected. With urbanization, about 55% of the world's population currently lives in urban areas [1], and the number of urban dwellers is expected to grow to 68% by 2050 [2]. Industrial heritage communities with high population densities and building density will continue to be a significant contributor to global GHG emissions, accounting for approximately 50–60% of total GHG emissions and approximately 71–76% of total energy use [3,4] (Figure 1). Currently, urban development is shifting to stock development, and industrial heritage communities, as the basic units that make up the city, contain people's productive lives and show all aspects of carbon activities. Industrial heritage communities are both residential living areas for urban residents and symbols of the city's industrial civilization. In order to achieve industrial culture preservation and enhance the living environment, it is necessary to improve their energy efficiency in order to achieve the carbon neutral construction goals set by the government.

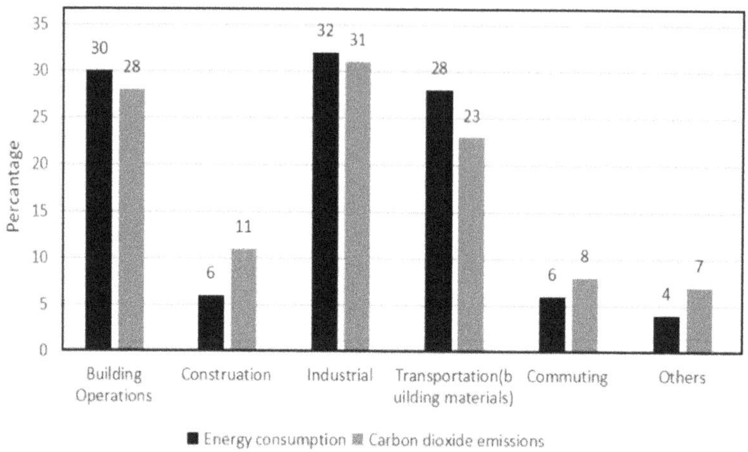

Figure 1. Energy consumption and carbon emission share of industrial heritage communities.

According to preliminary statistics, nearly 160,000 old communities exist in China [3], and 39,000 old communities will be retrofitted by 2025, with plans to achieve the full retrofitting of old communities by 2025 [3,4], with industrial heritage communities accounting for 1/4 of the target. The energy consumption and carbon emissions of industrial heritage communities are about 23% higher than those of new communities, so it is important to clarify the carbon activities of industrial heritage communities to reduce energy consumption and carbon emissions. Industrial heritage communities often consume large amounts of energy and produce large carbon emissions due to their inefficient and high-emitting heating, ventilation, and air-conditioning (HVAC) systems, making it more difficult to properly retrofit old buildings than new ones [5–7]. Today's low-carbon retrofit upgrades are merely superficial piles of new technologies and ideas, such as retrofitting landscapes, adding rainwater harvesting systems, and other green retrofit technologies [8,9]. However, these approaches do not dig deeper into the carbon activities of the community and do not systematically consider the elements on the nodes of the carbon space network. At the same time, the specificity of industrial heritage communities represents a myriad of retrofitting constraints, with fixed energy use, the nature of the site, and the difficulty of changing the building form. In the retrofitting process, it has to adapt to the needs of living and production, meet the requirements of heritage conservation, and, at the same time, adapt to the direction of low-carbon development, which makes the low-carbon improvement process challenging [10,11].

Therefore, the main purpose of this paper is to combine heritage conservation and low-carbon retrofitting, integrate these two elements into a carbon spatial network of industrial heritage communities, and then measure the basic values of the carbon spatial network elements, consider the optimization of human settlements, and reconcile the contradictions between heritage conservation and low-carbon renovation, so as to guide the conservation and low-carbon retrofitting efforts. In a carbon spatial network, nodes of elements are interconnected, and these elements act as information points in the network system, driving conservation and low-carbon retrofitting actions along the network connectivity, and are control points for retrofitting in industrial heritage communities. Conservation and low carbon are the broad framework of the whole system, and these control points connect the decision points of designers and parameterized platforms—the whole carbon space network. Low carbon and conservation decisions flow repeatedly, multiple times, and in multiple cascades along the network node linkages, and, when this cascade of flows and interconnected information feedback loops are combined, a complete carbon spatial network is formed [12,13]. The carbon space network will, therefore, revolutionize the preservation and transformation of industrial heritage communities.

In the harsh cold regions of China (Figure 2), the long cold winters and high heating demand have led to particularly high energy consumption and high carbon emissions, and the existing low-carbon retrofit studies have ignored the situational and networked characteristics of carbon activities. In the carbon space network, the overall carbon activities of the community can be considered in a contextualized and systematic perspective, defining carbon activities into space, optimizing the nodal elements of the carbon space network, and considering the interrelationship between nodes to achieve carbon reduction in industrial heritage communities. At the same time, the analytic hierarchy process is used to analyze the conservation and utilization elements of industrial heritage communities, and, on this basis, the carbon space network is combined to determine reasonable transformation measures to achieve the goal of sustainable development while preserving the historical heritage and ecological environment.

Figure 2. Thermal design zoning of buildings in China.

The UN IPCC and international institutions such as the World Bank have pointed out in their studies that CO_2 emissions come from seven main areas: energy, transportation, buildings, industry, agriculture, forestry, and waste disposal. Among them, the five main elements of the carbon cycle system affecting the community are energy, transportation, buildings, production, and consumers [14]. Industrial heritage communities are an important type of industrial heritage with high historical, artistic, scientific, social, and cultural values, as well as being an important element of urban cultural identity and a part of contemporary urban life. The keyword search of the relevant literature review in recent years is shown with Figure 3.

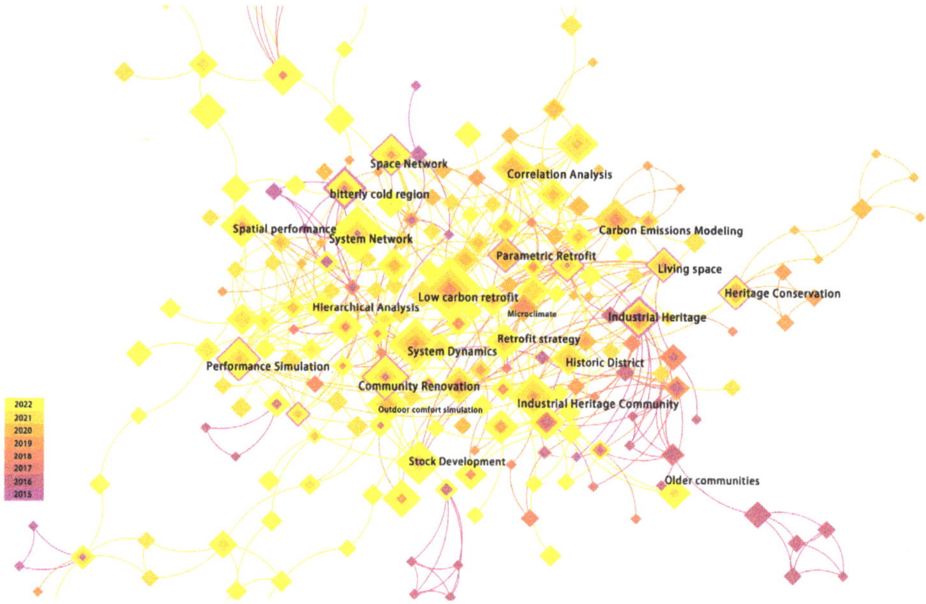

Figure 3. Schematic diagram of keyword search for related literature review.

In terms of industrial heritage community conservation, Liu Lihua et al. [15] selected the workers' village in Tiexi District, Shenyang, as a research object to analyze the mutual influence of urban industrial heritage and community memory sites, to explore the means of revitalizing urban industrial heritage buildings, and to strongly protect the community memory sites so that they can become community public activity sites and realize urban space regeneration. In the FAW industrial heritage community regeneration project in Changchun, Liu et al., (2019) [16] proposed a design method based on context-aware weighted multi-criteria decision making. The study focused on the analysis and application of quantitative indicators while considering factors such as historical and cultural heritage preservation, environmental protection, and social development to achieve sustainable industrial heritage community regeneration. However, these approaches are vaguer and have not been analyzed quantitatively.

Zhao et al. [17–21] explored the strategies and methods of the green transformation of industrial heritage communities, and proposed a sustainable green transformation model including, for example, building energy conservation, environmental protection, and landscape greening. These approaches are more mature and actual cases exist, but they do not think deeply about the community's microclimate environment and carbon network, and they think less about the nature of carbon activities. The common practice of certifications and standards for energy efficiency and emission reduction in buildings (such as the European Union's Near Zero Energy Building Standard) is to optimize energy use in isolation, without considering the urban environment or interaction with the surrounding environment. The carbon networks of communities are often complex and poorly understood, so it is particularly important to understand the carbon fabric in communities. Jin, Shi, Sun, and Tan et al. [22,23] used an integrated approach including a combination of data collection, statistical analysis, and modeling techniques to study multiple medium variables including carbon emissions, land use intensity, urbanization level, population density, industrial structure, and economic development. These studies used various statistical methods such as correlation analysis, principal component analysis, and regression analysis, and found that carbon emissions were positively correlated with land use intensity. This indicates that a higher land use intensity is associated with higher carbon emissions, and economic development and industrial structure are also found to have significant effects on carbon emissions. RD. et al. [24] in 2019 sorted out and summarized the research results about the sustainability of industrial heritage regeneration in recent years, and analyzed and reviewed them from multiple perspectives. The study focused on sustainability, urban planning, cultural heritage preservation, and socio-economic aspects, and summarized the main results and problems of the research. However, previous studies only summarized and analyzed the research on the sustainability of industrial heritage regeneration in recent years, but did not discuss the experience and application results of specific practices in depth. From the review of the literature, previous studies have placed the regeneration element indices in a relatively single way and have not formed a systematic theory and method.

Network science and system dynamics have become powerful frameworks for understanding and analyzing complex systems in a variety of fields. Havlin S. et al. explore the challenges faced in diverse fields including infrastructure, climate, social systems, applied network science, and economics. These challenges include modeling dynamic interactions in multilayer networks, predicting cascading failures and system resilience, and incorporating real-time data to improve decision making [25]. This is also a challenge for the low-carbon retrofitting of industrial heritage communities. Industrial heritage communities are complex systems that require a network approach to thinking about conservation and retrofitting. Labanca N. et al. delve into the connections between complex systems theory and social practices, exploring how the combination of the two can provide innovative and effective decarbonization strategies. Research has shown that complex systems theory, with its emphasis on interconnections and non-linear interactions within systems, can explain low-carbon strategies under the influence of social norms and cultural factors and can understand the role of user behaviors and habits in technological change, as well as develop a range of practical pathways [26]. The complex dynamics of communities in networks have been studied in depth by Porter M A. et al. By exploring the underlying structure of networks, they have revealed the emergence and characterization of communities, providing valuable insights into the complex interactions between the components of networks. In the process of the low-carbon transformation of industrial heritage communities, the conservation and low-carbon elements can be better guided by dissecting them through the idea of a system network [27]. However, most of today's research is directed at theory and not put into practice. In other words, the existing practice and research only consider the information points in the network, but not the control points and decision points. Therefore, one of the aims of this paper is to construct a complete carbon spatial network of industrial heritage communities, which can be used to guide subsequent research and practice.

The green regeneration of industrial heritage communities is the embodiment of sustainable development, which can promote the cultural and economic development of the city while preserving the historical and cultural heritage, as well as maintaining the ecological balance. Constructing a heritage conservation value judgment system, carbon space network model, renewal theoretical framework, and method for industrial heritage communities will greatly contribute to the resilient and low-carbon development of cities. The innovative aspects of this study are as follows:

1. The conservation and low-carbon retrofitting of industrial heritage communities are considered from a network perspective. The elements of conservation and low-carbon retrofit are connected to form a carbon space network to systematically guide the conservation and renewal of existing industrial heritage communities, and to meet the basic needs of heritage conservation, low-carbon, and human settlements.

2. We combine theory and practice: we integrate the information points, control points, and decision-making points in the network to better promote the research and practice of heritage community retrofitting.

3. We build a carbon space network from a spatial perspective, combining the living space, green space, and road space in the community, which is conducive to clarifying the carbon activity texture of the community, and then targeted retrofitting.

2. Materials and Methods

2.1. Research Subjects

In the Nizhny Tagil Charter for Industrial Heritage, it is clearly stated that the value of industrial heritage communities, which include places for social activities related to industry such as housing, should be recognized and preserved through appropriate means, and that residential amenities are usually expressed as industrial heritage communities, which generally refer to those containing industrial heritage surrounded by amenities that meet the needs of employees and surrounding residents for housing, living, interaction, and leisure. These community spaces contain artistic values, cultural values, and strong use functions, and are tools for improving the image of the city and generating a stronger identity, as well as being a catalyst for urban regeneration and contributing to the local community and economy [28]. In China, it refers to the communities built and managed by large industrial enterprises after the founding of the country, during the First Five-Year Plan and Second Five-Year Plan (1949–1962), and subsequent developments, to meet the needs of employees and their families in terms of housing, living, interaction, and leisure. Within this category, communities with rich historical remains, intact life forms, and complete spatial structures, and belonging to historic areas are collectively referred to as industrial heritage communities [29,30]. Currently, the land use of industrial heritage communities is no longer in line with the needs of the times; many hidden carbon sinks are not utilized, and there are problems such as poor habitat, low energy efficiency, high carbon emissions, and poor climate resilience [31]. The distribution of industrial heritage communities in the harsh regions of China selected for this paper is shown in Figure 4, with data from provincial government plans, Google satellite maps, GIS data, and field research.

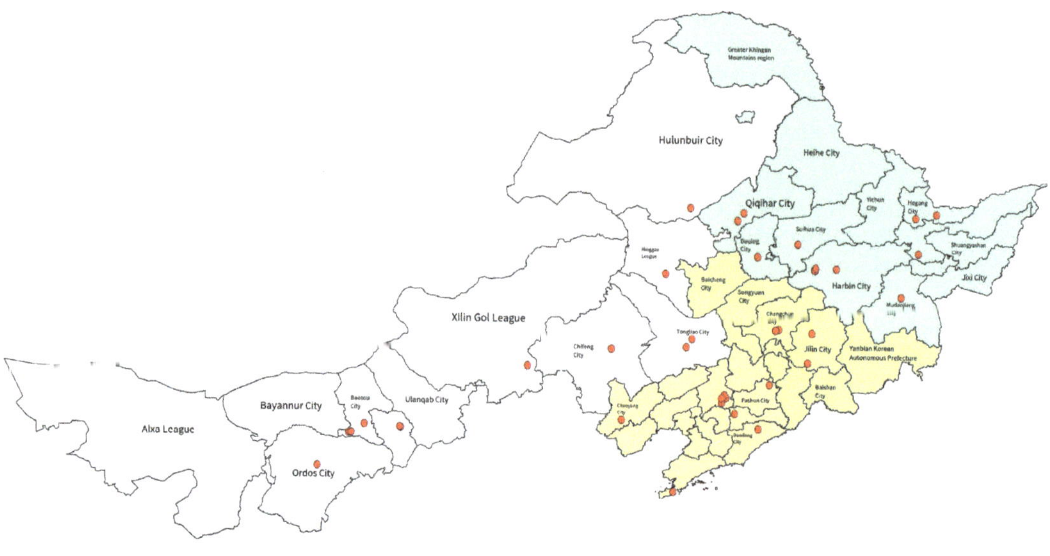

Figure 4. Distribution of industrial heritage communities.

It is important to note that, in the vast majority of cases, clearly defined boundaries of a community are generally the administrative boundaries that are accepted by residents and often used by public agencies for planning, zoning, and data collection [32]. Within the industrial heritage community, there are two kinds of scopes that need to be protected and those that do not need to be protected. Based on China's cultural relics protection law and related regulations, this study divides the industrial heritage community into the scope of ontological protection, style co-ordination zone, and modern living zone, using the boundary of morphological zoning, roads, and landmark building nodes as the dividing line, which is used to further clarify the boundary conditions for protection and renewal, so as to facilitate further retrofitting and renewal.

2.2. Research Methodology

In a study in Famagusta, Cyprus, Ozarisoy B. et al. identified relationships between different relevant variables based on residential patterns and household habit adaptation behaviors on household energy performance, which will deepen our understanding of these relative impacts and predict the interactions between these variables, which can ultimately be calibrated for certain groups of people's energy consumption [33]. These variables include households with different social and demographic characteristics, energy efficiency methods based on household awareness, heating and cooling systems, different modes of energy use, and household energy use, which reflect the low-carbon elements of the system network perspective and are a necessary starting point for realizing low-carbon retrofits. In other studies on the impact of settlement layout on energy consumption, Deng Q. et al. considered the impact of building layout on energy consumption and explored the impact of enclosed layout, staggered layout, and orientation on energy consumption [34], but another important piece of data that affects energy consumption and carbon emission in settlements is land use, which should be considered systematically. It is worth emphasizing that the above studies consider the real energy performance and the interconnections between some of the elements, which can be a guide for related low-carbon retrofit studies. Therefore, in this study, based on the measurement of real energy performance, carbon activities in time and space as well as the main elements affecting carbon activities in industrial heritage communities were considered to further guide the low-carbon retrofit of industrial heritage communities. It is also important to note that the carbon spatial network is a scenario-based projection of carbon-saving potential for the future based on the current situation. The root principle is to improve the overall carbon-saving potential through the transformation of elements with high relevance to carbon emission elements and the improvement of the linkage of information points, decision points, etc., in order to achieve the purpose of carbon reduction.

This study aims to provide theoretical support and tools for the implementation of industrial heritage community renewal in practice based on the carbon space network. First, establish a value assessment and protection system to determine the boundary conditions for retrofitting, with heritage value judgment and community livability as the preconditions; second, on the basis of climate change, as well as community microclimates, construct a carbon space network model, considering situational and networked community carbon emissions, and construct a community carbon space network to address low-carbon community issues at the system level; thirdly, based on the element data of 16 industrial heritage communities, a correlation analysis was conducted to determine the relationship between the elements of community carbon space network nodes and carbon emissions on the basis of carbon space network, and we apply it to the retrofitting method (Figure 5). Therefore, this study applies not only to the cold regions of China, but also to parts of the European Union and other countries and regions with similar continental climates that are also in cold regions.

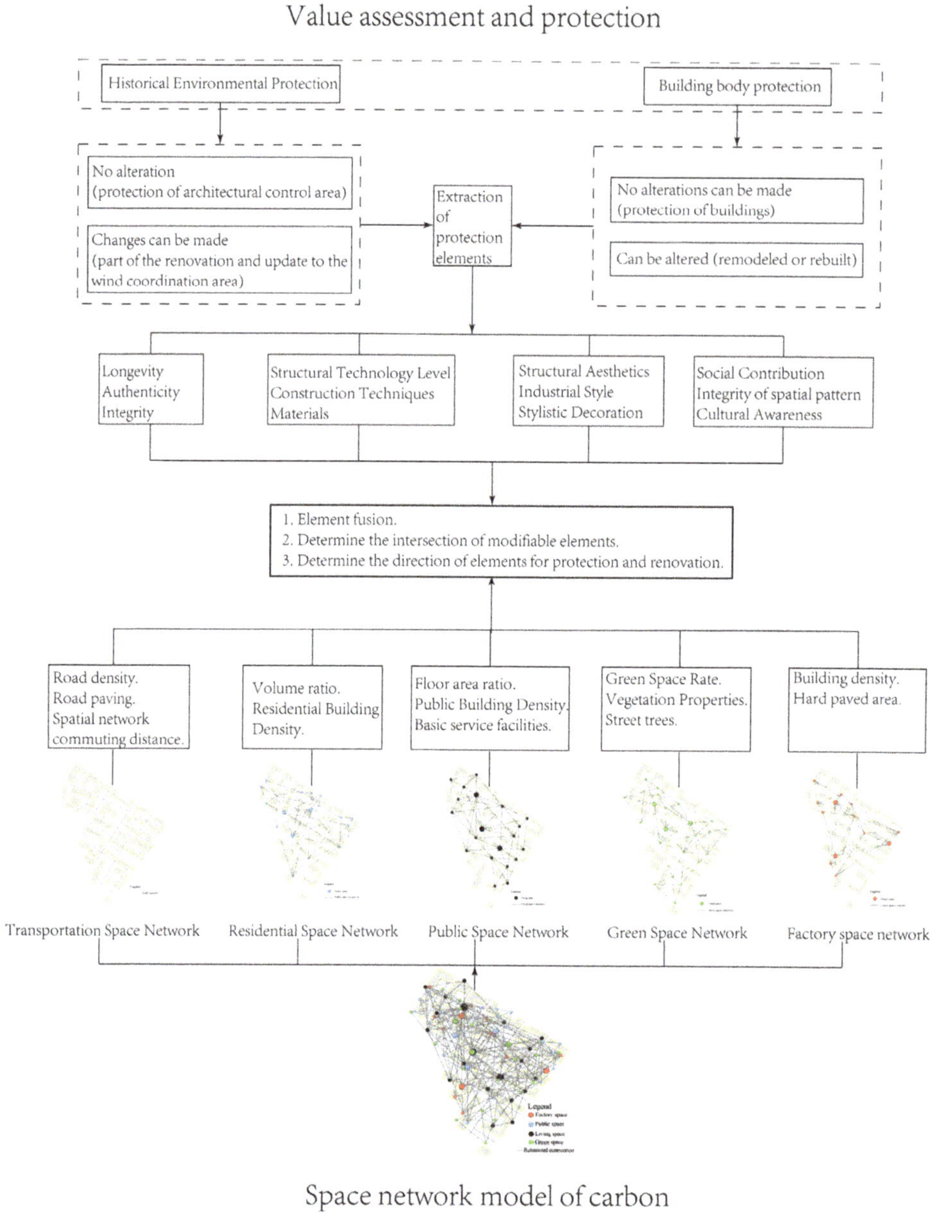

Figure 5. A framework for renovating and renewing industrial heritage communities based on carbon space networks.

2.2.1. AHP-Based Value Assessment

Usually, listed industrial heritage communities are only listed to some extent to meet heritage regulations—the aim is for them to be properly maintained and protected, which usually means keeping the façade of the building and changing other parts. The problem with this approach, however, is not only the loss of the valuable design of the building itself, but also the loss of the specific sense of place that the former industrial area must provide [35,36]. In previous conceptions, the retrofitting of historically significant communities was seen as a cultural risk, and, therefore, local legislation placed restrictions on the retrofitting that might apply; i.e., the higher the relevance or uniqueness of the heritage asset of the building or environment, the more restrictive the conditions for regenerative design. Analytic hierarchy process (AHP) is a quantitative analysis method that decomposes the elements related to decision making into various levels such as objectives, criteria, and options, and calculates a comprehensive index by determining the weights of each element [37]. The specific process is shown in Figure 6.

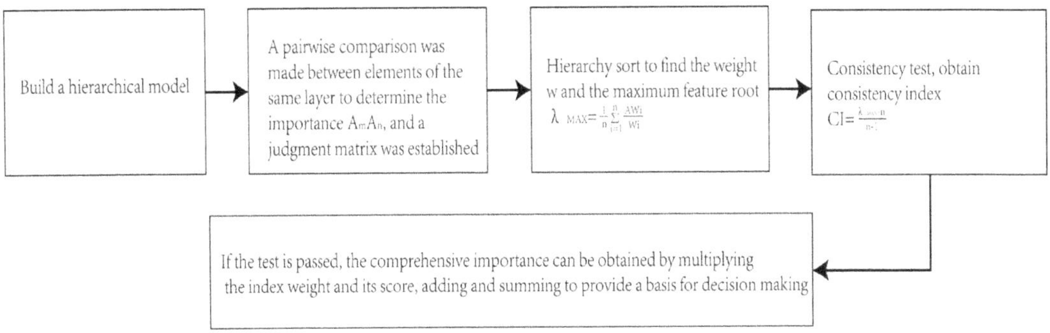

Figure 6. The specific process of analytic hierarchy process.

In the context of stock planning, value determination of industrial heritage communities is a crucial part of conservation and development [38]. When assessing the value of industrial heritage, people tend to focus on the structural form, material, and space of the building itself, while underestimating the living environment and urban value of the heritage. Based on the existing research, we applied the method of spatial morphological zoning to carry out an overall morphological zoning of the community. Moreover, this study uses the existing community roads and other obvious boundaries as the boundary of morphological zoning, and takes into account the similarity of functions and different stages of historical development to further form "morphological areas with obvious spatial morphological characteristics". Finally, AHP was used to make value judgments on different morphological partitions of the community.

In order to make a relatively objective value judgment of morphological subdivisions, a sub-objective evaluation system for morphological subdivisions is first established, which mainly includes the content of the index system, weights, and scoring system for the evaluation of material elements of the community [38]. In this study, morphological zoning is considered as an overall unit, and the heritage value of industrial heritage communities is divided into six A-level indicators based on the evaluation criteria of heritage guidelines, research, and expert opinions, combined with the typical characteristics of communities, and 15 B-level indicators are established on this basis as primary evaluation indicators (Table 1). The evaluation index system is not only applicable to the evaluation of the whole industrial heritage plant area, but also applicable to the evaluation of the industrial heritage individual units and even structures, facilities, and equipment. After that, the judgment matrix was established by two-by-two comparison, and each evaluation index at the same level was compared two-by-two, and the evaluation index m was assigned with n for importance comparison (Table 2), after which the results were obtained by two-by-two comparison of the industrial heritage community A-level indices after integrating the questionnaires from experts.

Table 1. Industrial heritage community value indicator system.

A-Level Indicators	B-Level Indicators
Historical value A1	Longevity B1 Historical events B2 Relationship with the development of industrial culture B3
Value of science and technology A2	Industry pioneering B4 Engineering technology B5
Architectural artistic value A3	Architectural engineering aesthetics B6 Industry style characteristics B7 Landscape integrity B8
Social and cultural value A4	Social emotion B9 Industrial culture B10
Economic utilization value A5	Structure utilization B11 Space utilization B12 Reasonableness of current use B13
Environmental location value A6	Distance from other historic buildings B14 Degree of integration with the surrounding environment B15

Table 2. Schematic of the 1–9 level scale method.

Scale	Meaning
1	m is as important as n
3	m is slightly more important than n
5	m is significantly more important than n
7	m is strongly more important than n
9	m is definitely more important than n
2,4,6,8	Intermediate values in the judgment

2.2.2. Carbon Space Network Construction

The relationship between humans, communities, buildings, urban microclimate, and global climate is complex. Therefore, in this study, microclimate is the basis for sorting out the texture of carbon activity. Climate—especially microclimate—affects a community's carbon emissions, energy efficiency, and outdoor comfort. Oke identified three main causes of changes in urban microclimate: (1) interception of short- and long-wave radiation by buildings, (2) reduction of long-wave thermal radiation due to reduced sky visibility, and (3) increased sensible heat storage in buildings [39]. These elements influence the nodal elements in the carbon space network and have a large impact on various properties of the communities.

In the carbon space network of industrial heritage communities, we divided them into residential space network, transportation space network, factory space network, public space network, and green space network based on human behavior, spatial morphological zoning, and functional elements (Table 3, Figure 7). These spatial network node elements partially overlap with the heritage value assessment elements and complement each other for conservation and transformation. The advantage of considering carbon emissions in a scenario is that the fixed, mobile, and hidden carbon activities can be considered using the closest realistic approach, and these elements are integrated in the carbon spatial network of the industrial heritage community, with the nodal elements shown in Figure 8. The carbon activities influenced by these elements mainly include direct and hidden carbon sources, with direct carbon sources being carbon generated from activities such as fossil fuel consumption, transportation fuel consumption, and waste disposal. The hidden carbon sources include mainly the microclimate impact on green vegetation and, thus, the difference in the activity of carbon sources and sinks, but also the carbon-saving potential of solar photovoltaic panels, etc. and the impact of heat exchange between the external environment and the building. Hidden carbon sources also include indirect carbon sinks from the influence of the overall community form, such as the impact of community building porosity and volume ratio on light, ventilation, and heat production. These elements, as carbon space network nodes, are central to the conservation and renewal of the community, and, together, they have an impact on community carbon activities.

In this paper, the sample selection of carbon space network node elements is carried out through 16 representative industrial heritage communities. The 16 samples were chosen for this reason [40–43]:

1. They were built in the same era.
2. The overall size of the community is similar and the land use intensity of the residential buildings is similar.
3. The aging of the building envelope, public facilities, etc. is similar.
4. The master plans are similar and they are in the same climatic zone, so their microclimates are also relatively similar.
5. Their existing methods of renovation are also relatively close, especially under the general rules of heritage conservation.

And, because the other samples were not columnarly additive to the 16 samples selected for this study, we ultimately chose these 16 industrial heritage communities for our sample pool.

Table 3. Composition of industrial heritage community network.

Composition of Industrial Heritage Community Network		
Residential space network	 Dwelling	 Hotel
Traffic space network	 Roadway	 Sidewalk
Public space network	 Public building	 Square
Greening space network	 Park	 Green space beside street
Factory space network	 Workshop	 Factory Square

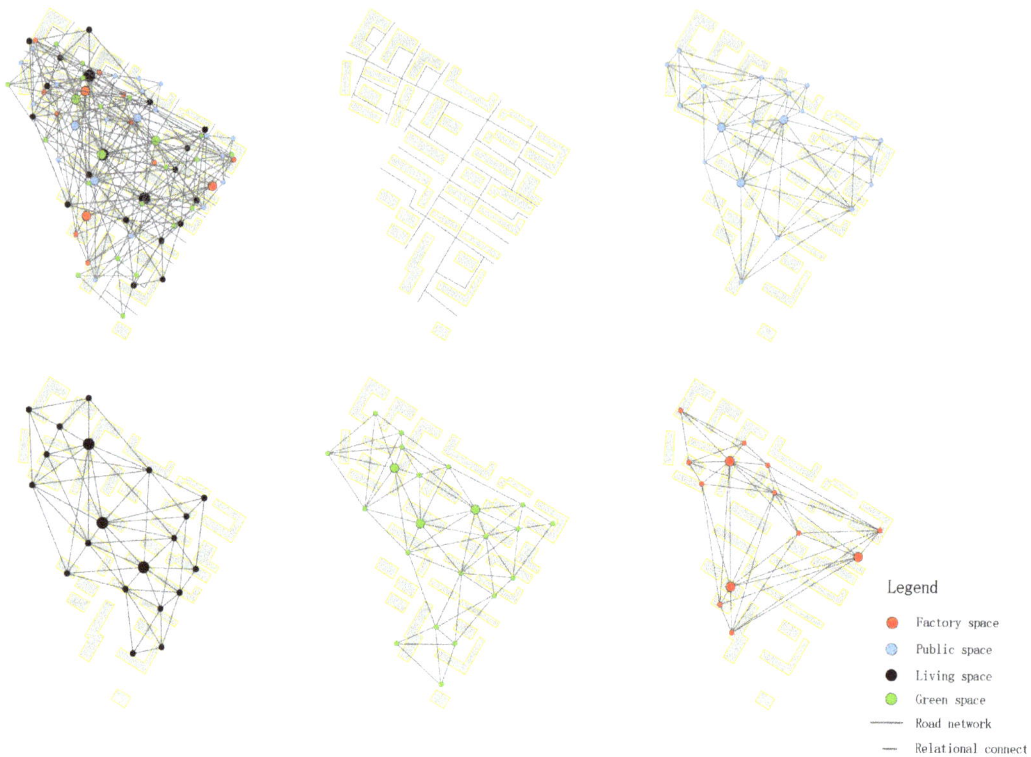

Figure 7. Composition of industrial heritage community network.

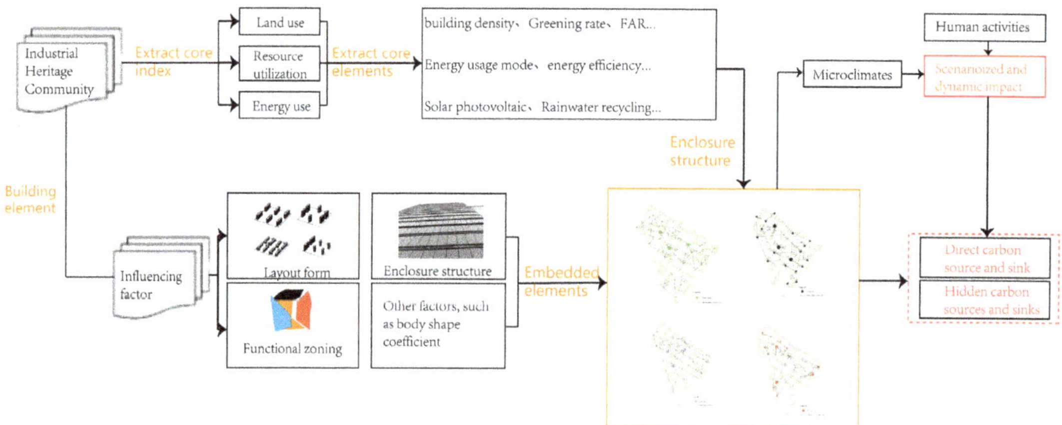

Figure 8. Carbon space network model of industrial heritage communities.

2.2.3. Application of Network Model

The process of regeneration of industrial heritage communities aims to optimize the existing retrofitting framework, address climate change, optimize the habitat, and balance heritage preservation with carbon emission reduction. At the community scale, the layout form, land use, and land cover are key to the thermal environment, overall energy consumption, and carbon emissions [18]. This study is based on the idea of system network, dissecting the key node elements of the network, integrating these elements into each step of the retrofit design process, and improving the efficiency and accuracy of the retrofit design (Figure 9). In the model application, we simulated close to realistic scenarios and combined them with real scenario measurements to achieve the closest-scenario-based understanding of the carbon activity texture.

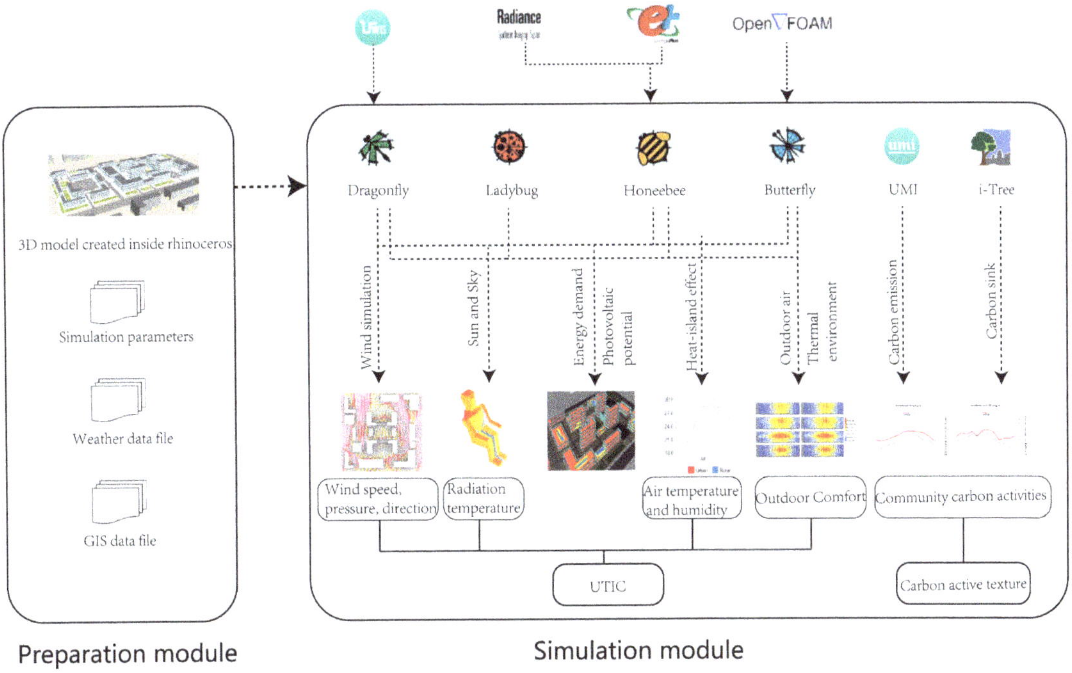

Figure 9. Parameterization-based transformation process.

Microclimate is the basis for the conservation and renewal of industrial heritage communities, and, in order to obtain accurate ground simulation data, the Indra urban future weather algorithm based on the Indra urban future weather algorithm was used in this paper to generate future weather files, which combines historical weather data with the output of climate change models (global and regional climate models) to create files with near-realistic future weather [44]. However, due to the problems inherent in the algorithm, the method can only generate future weather files for typical cities, which may partially deviate from reality to some extent. Therefore, this study uses an average distribution of historical weather data samples to generate future weather files and uses the algorithm to generate more stable future EnergyPlus/ESP-r weather (EPW) files.

For microclimate, the grasshopper-based dragonfly urban weather generator (UWG) was used to generate accurate community microclimate data. UWG can be used as data pre-processing alone or in combination with existing programs to build joint simulation methods. It uses several key geometric and physical variables (building height, roads, roofs, maintenance structures, etc.) to construct the data files [45]. They provide the basic model and weather files for the subsequent performance simulation and optimization of the retrofit process, and are the original database for the entire retrofit process. On the other hand, due to the dynamic nature of the urban environment, the outdoor thermal environment depends on many relevant factors: the characteristics of the conditions inside and outside the building, the relationship between materials and energy use, global climate change, and local microclimate. Also, relevant factors include temperature, solar radiation, wind distribution and velocity, absolute and relative humidity, building density, height, etc. The above complex elements require us to use the universal thermal climate index (UTCI) to model the outdoor thermal environment simulation. In the study, the grasshopper-based ladybug plug-in was used for outdoor wind, light, and thermal environment simulations.

Carbon emissions and energy consumption are a key part of retrofitting industrial heritage communities. For carbon emissions and energy consumption from building operations, this study uses the UMI plug-in developed by MIT Labs to assess the environmental quality and performance of the community. In the measurement process, we also consider the impact of building form, envelope materials, window-to-wall ratio, and other factors on carbon emissions and energy consumption. Compared to other carbon and energy simulation plug-ins, UMI runs faster and produces more accurate results.

The carbon sink in the community mainly includes greenfield carbon sink and soil carbon sink, and the carbon sink is measured by i-Tree method. i-Tree is an urban forestry analysis and eco-efficiency evaluation model developed by the U.S. Forest Service in 2006. For community green space carbon sink estimation, this study uses i-Tree Streets and i-Tree Eco for measurement. i-Tree Streets is a street tree-specific analysis tool for urban forest managers that quantifies the structure, function, and value of annual tree benefits using tree single data, and i-Tree Eco quantifies annual tree benefits based on field data, local hourly air pollution, and meteorological data on urban forest structure, environmental impacts, and community values; both models measure the storage and sequestration of CO_2 by the study population, and also include the impact of the study population on building energy consumption and indirect reductions in CO_2 emissions and air pollution removal [46].

3. Simulation Process and Data Analysis

3.1. Parameter Settings for Community Simulation

In order to facilitate the control of variables in the numerical simulation process and reduce the interference of other factors to the data, three benchmark data of land use intensity, resource use, and energy use are selected in the simulation of the relationship between core community indicators and carbon activities.

The intensity of land use in a community is the extent to which land is used for various purposes within that community. It can also refer to the extent to which human activities and development occur on land in a given area. It is a measure of the intensity with which land is developed or used for different activities, such as residential, commercial, industrial, agricultural, or open-space purposes [47]. The intensity of land use can be assessed by factors such as floor area, number of buildings or structures, population density, floor area ratio (FAR), and the proportion of land designated for different uses [30,47]. In order to strike a balance between land use intensity, community needs, environmental sustainability, and other relevant factors, the following elements were considered along with the setting of the above parameters: (1) Land Use Types: Define the different land use types in the community, such as residential, commercial, industrial, agricultural, recreational, etc. Assign a specific intensity value to each land use type based on the level of activity or development associated with it. (2) Density: Set the population density of a residential area. This parameter determines the number of people living in a given land area and can influence the intensity of land use. (3) Environmental factors: Consider environmental restrictions and conservation goals. For example, if there are sensitive ecological areas or protected targets, further adjust the intensity of land use in these areas. (4) Porosity represents the ability to ventilate and collect light in an area, and refers specifically to the rate of scenic passage over a barrier formed by the enclosure of buildings, structures, trees, etc. This factor is a reflection of both land use intensity and community form, and has a greater impact on community carbon activity.

When setting the parameters of resource utilization and energy use settings at the same time, we consider the following elements: (1) Population: Determine the size and growth rate of the simulated population. This will affect the demand for resources, such as housing, transportation, water, and energy. (2) Land Use: Define the allocation of land for different uses, such as residential, commercial, industrial, and open space. Consider the intensity of land use required and the proportion dedicated to each type of land. (3) Resource Demand: Estimate the consumption patterns and resource demands of the simulated population. This includes energy use, water consumption, waste generation, and transportation needs. (4) Building Energy Efficiency: Define the energy efficiency standards for buildings in the community. This includes the envelope, HVAC system efficiency, lighting standards, and electrical energy ratings. (5) Environmental Factors: Consider environmental factors such as the renewable energy potential, availability of natural resources, and vulnerability to the effects of climate change. These factors can guide decisions on how resources are allocated and utilized. (6) Climate and Weather: Consider the climate and weather patterns of the community. Different climatic conditions can affect energy consumption for heating, cooling, and other uses. Considering the impact of seasonal variations and extreme weather events on energy use, we primarily consider community energy consumption and carbon emissions in winter due to climate conditions in harsh regions.

For the purpose of a correlation analysis, a data set was created in this study. This data set includes the values of carbon emissions and land use intensity of different communities in four provinces in the severe cold areas of China (Appendix A); also included are different forms of building layouts, which include homes, as well as other public buildings within the neighborhood. Among them, in this paper, energy consumption and carbon emissions are simulated for a period of 50 years and the basic units are kWh/m^2 and kgCO$_2$/m^2. These elements play an important role in optimizing the performance of communities as nodal elements of the carbon spatial network. The correlation elements among these variables are shown in Figure 10.

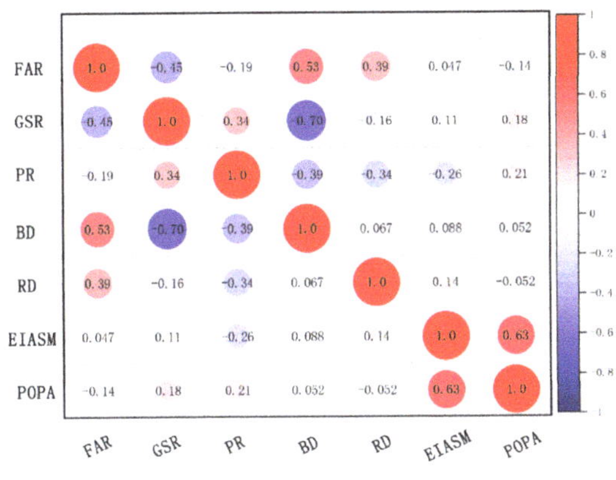

Figure 10. Correlation analysis of each element.

The necessary data for the study came from a series of government announcements, research surveys and GIS, and simulation data that were consistent with the spatial and temporal scales of the analysis. In this study, the building performance was simulated for energy intensity and carbon emission according to local climatic conditions, green building design standards, and retrofitting constraints, and the residential geometric model was set to a 20% window-to-wall ratio in the east–west direction, 50% and 40% window-to-wall ratio in the north–south direction, and building floor height was set to 3 m [45,46]. The input parameters for the energy simulation are based on Chinese regulations on carbon emissions and energy use, assuming an annual radiant lighting rate of 600 kWh/m^2 for exterior walls and 800 kWh/m^2 for roofs, and heating from October to December and January to March. The heat transfer coefficient is U = 0.95 W/m^2-K for exterior walls, U = 0.5 W/m^2-K for roofs, and U = 5.5 W/m^2-K for windows. Other public buildings are set according to the existing state, and the default window opening rate for plants is 10%.

To ensure accuracy and consistency, missing values, outliers, and inconsistencies in the data were handled centrally. Pearson's correlation coefficient in SPSS was used to calculate the correlation coefficient between land use intensity and carbon emissions, ranging from −1 (strong negative correlation) to +1 (strong positive correlation), with 0 indicating no correlation. A positive correlation indicates that carbon emissions tend to increase as land use intensity increases, while a negative correlation indicates the opposite. The strength of the correlation is indicated by the absolute value of the correlation coefficient, with values closer to 1 indicating a stronger relationship.

3.2. Correlation Analysis of Carbon Emissions and Land Use Intensity

Land use intensity provides insights into the spatial organization of communities, guiding the creation of sustainable, efficient, and livable community environments, and is an important part of the carbon space network [47]. Determine the floor area ratio and land cover to represent the land use intensity, and land cover using the green space ratio, building density, and road density. Use energy consumption and industrial activities as carbon emission indicators to calculate carbon emissions for each community in the data set.

Because carbon space networks contain multiple elements that influence community carbon emissions, the mechanisms of influence among networks are complex. For example, the species of trees and the species of soil micro-organisms vary in each community. On the basis of meeting basic usage needs, livability, and inter-network complexity, this study carefully considered this complex mechanism and came up with more accurate correlation analysis results. As shown by the correlation analysis (Figure 11), there is a positive relationship between carbon emissions and land use intensity. Areas with a higher land use intensity tend to have higher carbon emissions due to increased industrial activity and energy consumption. Among them, communities with a higher floor area ratio and road density have high carbon emissions and show a significant positive correlation. It should be noted that vegetation cover, bare land area, and carbon emissions showed a negative correlation, indicating the potential role of green plants and soil micro-organisms in reducing carbon emissions. Porosity, as an important indicator of the community landscape's thermal environment, also correlates with the community morphology at the same time. The greater the porosity, the higher the ventilation and light capacity of the community, which can affect the scenic environment of the community, so porosity is also one of the important factors affecting potential carbon sources.

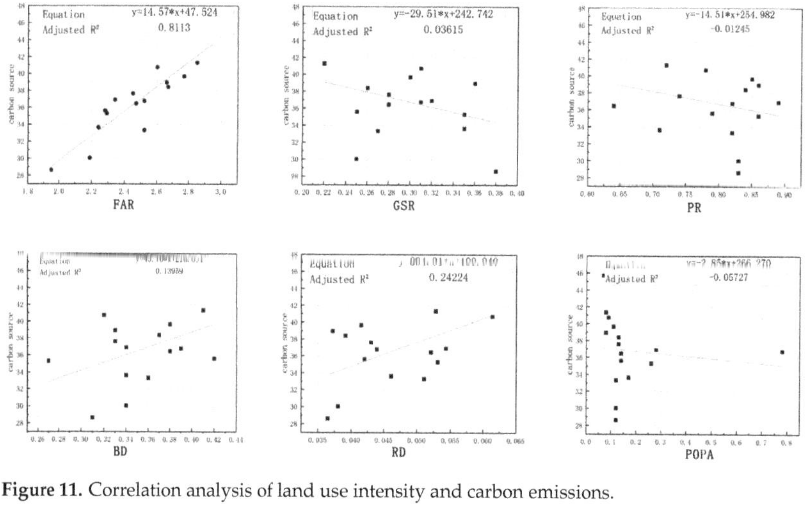

Figure 11. Correlation analysis of land use intensity and carbon emissions.

Based on the measured simulation data and analysis results, and based on the statistics in Appendix A, we believe that the floor area ratio of the industrial heritage community should be controlled between 2.1 and 2.5; building density between 35–40% except for industrial buildings; the number of floors of residential buildings between 7–14, and this limitation of the number of floors is conducive to the preservation of the style of heritage buildings; and the road network density between 4.16–4.50%, and the roads should be mostly permeable paving. This approach could lead to an increase in the area of bare land and, thus, the development of potential carbon sinks. Porosity should be 35–45%; the green space ratio should be 20–35%; and street trees are not counted in this part.

3.3. Correlation Analysis of Communities' Form and Carbon Emissions under the Influence of Land Use Intensity

The building layout form of a community not only directly affects aspects such as the ventilation of residential interiors and the comfort of outdoor public spaces, but also determines the energy consumption of the building [48,49]. To further understand the impact of community form on carbon emissions, the building frontal spacing and side spacing were clarified with reference to the <<Urban Residential Area Planning and Design Standards>> (GB50180-2018) and the <<Technical Regulations on Urban and Rural Planning Management of various regions>>, taking into account the differences of multi-story, mid-rise, and high-rise buildings, and mainly considering the impact of the three layout forms on community carbon emissions, as well as livability.

We measured the impact on carbon activity caused by three types of layout patterns—the enclosed layout, row layout, and staggered layout—within each community. The enclosed layout and staggered layout have less impact on carbon emissions, which is because this layout can improve the internal scenery and heat environment, reducing the heat exchange between indoor and outdoor, so that the hidden carbon sources and sinks can be developed to achieve the purpose of reducing carbon emissions. We further considered the potential carbon sinks of the three layouts, such as the potential carbon reduction from solar photovoltaic panels, and this study concluded that buildings with staggered layouts could receive more daylight and, thus, have greater carbon reduction potential (Figure 12).

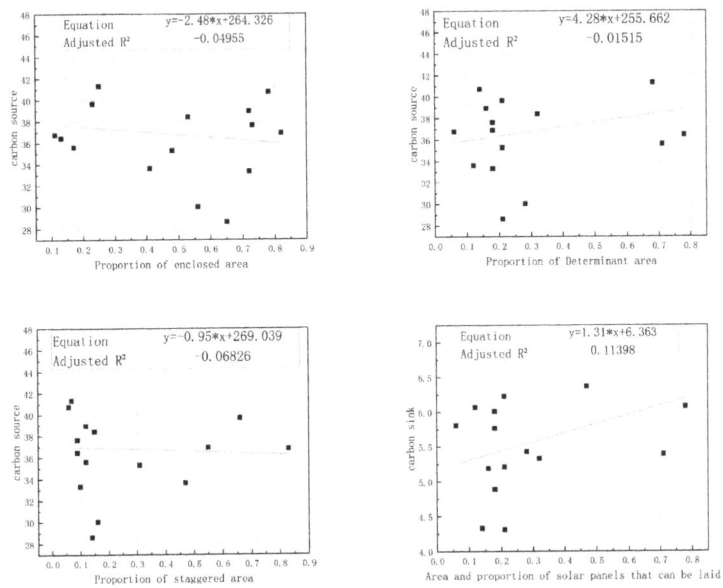

Figure 12. Correlation analysis of community form and carbon emissions.

4. Design Case Validation for the Renewal of an Industrial Heritage Community in a Severe Cold Region

4.1. Conservation Component of Industrial Heritage Communities

This research verifies the impact of each element in the carbon space network on carbon activities, and focuses on the land use intensity and layout pattern of the community. The aim is that designers can retrofit industrial heritage communities with a network perspective. In the renewal design of the industrial heritage community, the morphological zoning and value assessment of the architectural heritage in the community were carried out first. Using the existing heritage protection area and obvious boundaries such as community roads as the boundaries of morphological zoning, the entire community was divided into seven morphological zoning areas for protection according to the similarity of functions and the different stages of historical development (Figure 13). The other areas are the landscape protection areas and modern development areas. For the morphological partitioning results, AHP is used to split the overall objective of comprehensive value judgment into primary and secondary index systems, after which we perform a comprehensive evaluation score and weighted decision analysis according to the different levels.

Figure 13. Case scope and morphological zoning.

According to the methodology identified in Section 3.1, we compared m with n in a two-by-two comparison, and, after combining the questionnaires from experts, we tallied the results of the two-by-two comparison of A-level indicators for heritage buildings in industrial heritage communities, as shown in Table 4.

Table 4. Industrial heritage community A-level indicator judgment matrix expert statistics.

Assign	A1	A2	A3	A4	A5	A6
A1	1.000	2.083	2.100	2.700	3.933	3.725
A2	0.480	1.000	1.008	1.296	1.888	1.788
A3	0.476	0.992	1.000	1.286	1.873	1.774
A4	0.370	0.771	0.778	1.000	1.457	1.380
A5	0.254	0.529	0.534	0.686	1.000	0.947
A6	0.268	0.559	0.564	0.725	1.056	1.000

After that, the judgment matrix is normalized to obtain the weight vector of each index (this research adopts the root method). The process of calculating the weight of each indicator of industrial heritage is as follows:

In the first step, the elements of each row of the judgment matrix are multiplied together, as $M_1 = 1 \times 2.083 \times 2.1 \times 2.7 \times 3.933 \times 3.725 = 173.030$. Similarly, $M_2 = 2.116$, $M_3 = 2.071$, $M_4 = 0.446$, $M_5 = 0.047$, and $M_6 = 0.064$.

In the second step, the product of these products is divided five times, such as $w_1 = 2.803$. Similarly, $w_2 = 1.162$, $w_3 = 1.151$, $w_4 = 0.851$, $w_5 = 0.542$, and $w_6 = 0.578$.

In the third step, vector normalization is performed to derive the A-level indicator weights $w_i = \frac{w1}{\sum_{j=1}^{n} wj}$, $w_1 = 0.164$, $w_2 = 0.164$, $w_3 = 0.162$, $w_4 = 0.120$, $w_5 = 0.076$, and $w_6 = 0.082$.

The maximum characteristic roots of the industrial heritage value index system were obtained and tested for consistency. Maximum characteristic root $\lambda_{MAX} = \frac{1}{n}\sum_{i=1}^{n}\frac{Awi}{wi} = 6.050$. Afterwards, the consistency index is calculated for $CI = \frac{\lambda MAX-1}{n-1} = 0.01 < 0.1$. Since this matrix is a sixth-order matrix, it needs to be substituted into the consistency index RI = 1.24, and we obtain $\frac{CI}{RI} = 0.008 < 0.1$. The judgment matrix meets the consistency requirement.

For the B-level indicators, weights were calculated in the same way as for the A-level indicators, and the results were [0.272, 0.358, 0.370], [0.692, 0.308], [0.752, 0.248], [0.476, 0.223, 0.301], [0.551, 0.266, 0.183], and [0.732, 0.268]. By calculation, the weights of the indicators of the three major power road industrial heritage communities at each level are summarized in Table 5.

Table 5. Industrial heritage community value evaluation index weights.

A-level Indicators	Weights	B-Level Indicators	Weights
A1	0.396	B1	0.108
		B2	0.142
		B3	0.146
A2	0.164	B4	0.113
		B5	0.051
A3	0.162	B6	0.122
		B7	0.040
A4	0.120	B8	0.057
		B9	0.027
		B10	0.036
A5	0.076	B11	0.042
		B12	0.020
		B13	0.014
A6	0.082	B14	0.060
		B15	0.022

Each indicator of the assessment object is assigned a score, and the indicator assignment is multiplied by its weight to obtain a comprehensive evaluation index by aggregating the scores of each indicator. We judge the industrial heritage grade according to the comprehensive evaluation index and determine the renewal method. Industrial heritage with comprehensive evaluation indices above 80 points has the greatest industrial heritage value and belongs to the first-class industrial architectural heritage (immovable cultural relics level), which should be protected in a comprehensive way. Industrial heritage with a comprehensive evaluation index between 60 and 80 points has a greater value and belongs to the second level of industrial heritage (historical building level), and it is appropriate to adopt the means of "protection as the main focus and renovation as a supplement" for renewal. Industrial heritage with a comprehensive evaluation index between 40 and 60 points has a fair value and belongs to the third level of industrial architectural heritage (higher value industrial heritage level), whose reuse should adhere to the principle of "transformation as the main focus and protection as a supplement". Industrial heritage with a comprehensive evaluation index between 0–40 points has a very low industrial heritage value and belongs to the fourth level of industrial architectural heritage (general industrial heritage level), which can be dismantled and rebuilt as needed, while other parts of modern forms can be dismantled and rebuilt according to the situation.

4.2. Retrofitting and Renewal of Industrial Heritage Communities

In order to better explore the livability and various performances within the community, we selected morphological zoning V-2 and its surrounding buildings for performance simulation (Figure 14). In order to improve the performance of the community's environment and, thus, improve the comfort of the community, based on the principles of heritage preservation and morphological zoning, the community has a total area ratio of 2.10, a building density of 42%, a porosity ratio of 40%, and a green ratio of 35%. The density of permeable pavement road is 4.47%, and the area of impervious hard pavement road accounts for 15% of the total road area, which is mostly the hard pavement of urban roads and factories. According to the plan, the area where solar photovoltaic panels can be laid is 30% (Figure 15). The building spacing and orientation are implemented according to the Urban Residential Planning and Design Standards and the General Urban Design Rules. Other parameters are shown in the Appendix B.

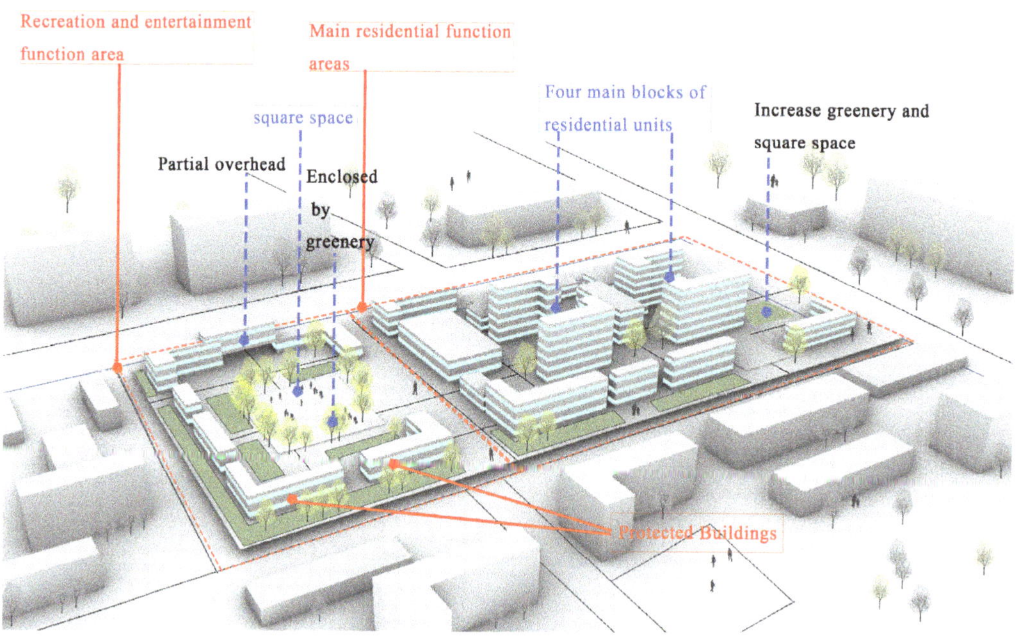

Figure 14. Morphological subzone V-2 and its surrounding buildings.

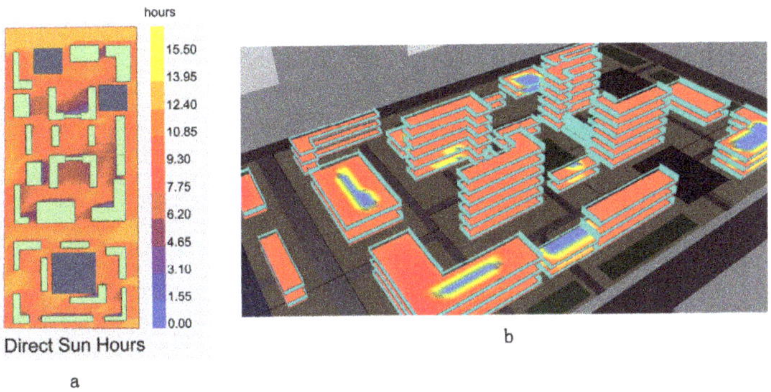

Figure 15. Direct sun hours (**a**) and indoor light environment (**b**).

The performance of a community's energy and carbon activities involves many aspects, including the form of the building layout, the floor area ratio, and the building envelope. The study considered energy performance and carbon emissions at the urban design level, and the results are shown in Figure 16. We also simulated the community elements before the retrofit, which resulted in a 16.1% reduction in energy consumption and an approximately 29.9% reduction in carbon emissions compared to the pre-retrofit period. The reason for this phenomenon is the change in the form of the community, i.e., the change in the form of the layout of the building from an incomplete enclosed first floor to a double enclosed form, which allows the current building layout to significantly reduce the heat loss caused by unstable airflow. The other side of the reason is the change in land use intensity and land cover form. For example, the increase in green space area and the reduction in the hard impervious road pavement area also significantly improve the carbon space network nodes, thus reducing energy consumption and carbon emissions. In order to adapt to climate change, we also projected the future carbon emissions and energy consumption of the community, which increased with climate change, increasing population, and aging of the envelope, but with smaller changes.

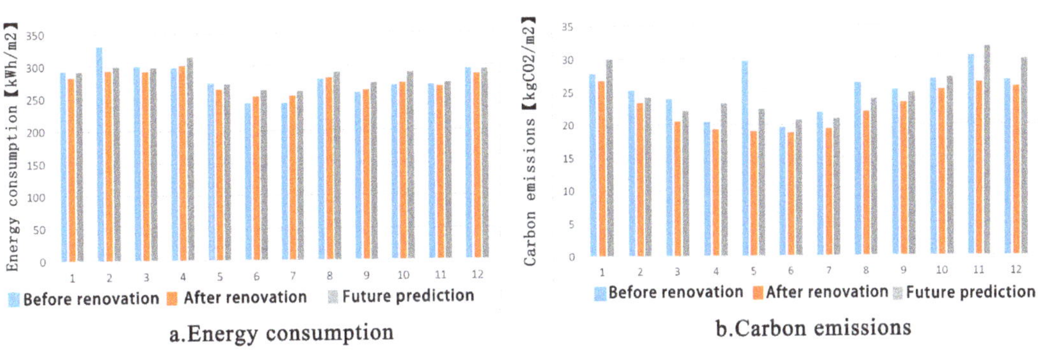

Figure 16. Energy and carbon emissions of research subjects.

The community has a green space ratio of 35%, and, including street trees, 11.3% of the total community area is bare land. According to the field research and simulation data, we counted the condition of greenery and the area of bare land for different specifications, and the specific parameters are shown in Table 6.

Table 6. Greening conditions of the study area.

Vegetation Elements	Total Shrub and Tree Area	Grassland Area	Total
i-Tree: Ground trees: Vegetation > 1 m and 30 mm or above DBH [1]	7400	--	8500
Online Carbon Calculator: Height of shrubs/trees > 1 m, or <30 mm DBH	--	410	410
Soil carbon [2]	24,000	62,000	87,000
Subsurface carbon in the root mass of trees < 20 mm	110	--	110
Total			96,000

Note 1: Some shrubs and trees contain ground cover; Note 2: The figures in columns 2 and 3 calculate soil carbon using Thomason's (2011) [50] conversion factors of 6.93 kgC/m^2 and 5.43 kgC/m^2, respectively. All figures are rounded.

Based on the calculations of the i-Tree model, we calculated the carbon sink capacity of the relevant carbon activities, as shown in Table 7, the air pollution removal potential of the project site vegetation >1 m and >30 mm diameter is estimated to be about 16,000 g, which can reduce the average temperature inside the community by about 1.22 degrees Celsius.

Table 7. Estimated carbon sink capacity according to the i-Tree model.

Number	Category	Existing and Potential Carbon Stocks (kg)
1	Existing carbon content of shrubs and trees, and all soil carbon including root biomass.	102,680
2	Above-ground carbon potential of 8728 square meters.	2200
3	Additional soil carbon from shrubs/trees in original grassland.	14,290
4	New root biomass carbon in new tree areas.	2380
	Realizable potential (2–4).	38,670

In conducting the livability assessment, we measured using seven criteria within the scope of the retrofit, as shown in Figure 17, which represents a schematic representation of outdoor comfort in four extreme environments: with wind and daylight, daylight and no wind, wind and no daylight, and no wind and daylight. It can be seen that the extreme hot and cold environments of the four retrofitted environments were significantly reduced, with a significant increase in comfort and being better suited for improving the community environment. According to the comparative analysis of chromatographic extraction, the extreme hot and cold environments were all reduced by 5.3% compared to the pre-retrofit situation. One of the reasons for this phenomenon is that the building layout and building height have affected the wind environment within the site, accelerating the exchange of hot and cold air and allowing the site to receive more reasonable sunlight. At the same time, the reduction of hard paving and the increase in the amount of greenery in the site have also led to changes in the outdoor thermal environment.

The layout pattern of the community can have a significant impact on the carbon activity as well as the scenic thermal environment. As shown in Figure 18, the wind environment is explored for the pedestrian height of the community (1.5 m) and the southwest wind direction. Due to the change in the form of the building layout, the wind distribution within the site has also changed significantly. The interior of the renovated site is free of uncomfortable areas of intense air movement and narrow ventilation corridors. It can be seen that the long façade of the building accelerates the horizontal vortex airflow at the edges, thus increasing the efficiency of the convective exchange of hot air at low altitude and cold air at high altitude, reducing unnecessary heat exchange and obtaining a pleasant wind environment and thermal comfort at the pedestrian level.

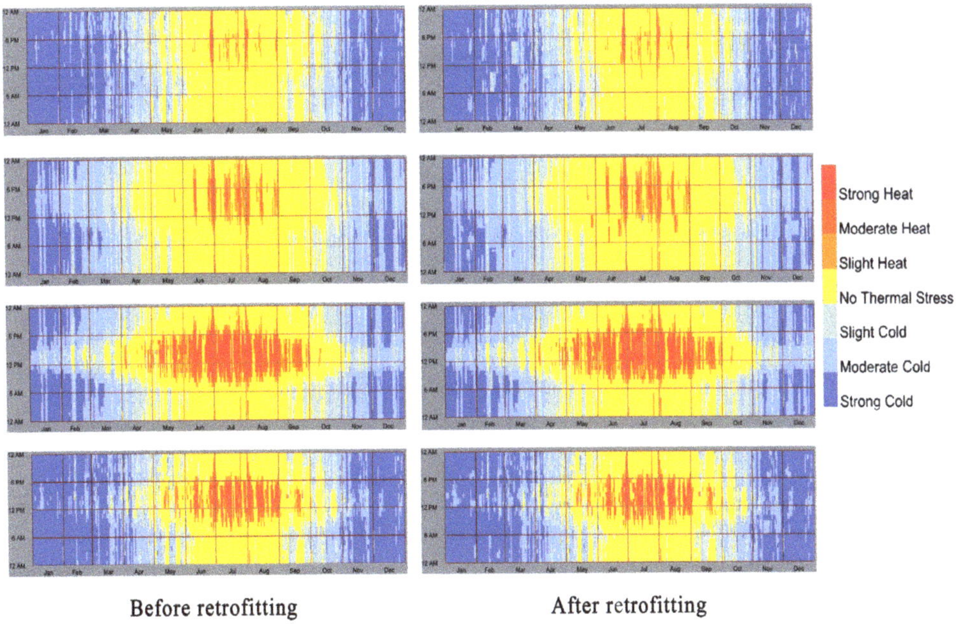

Figure 17. Outdoor thermal comfort labeling map of the study area.

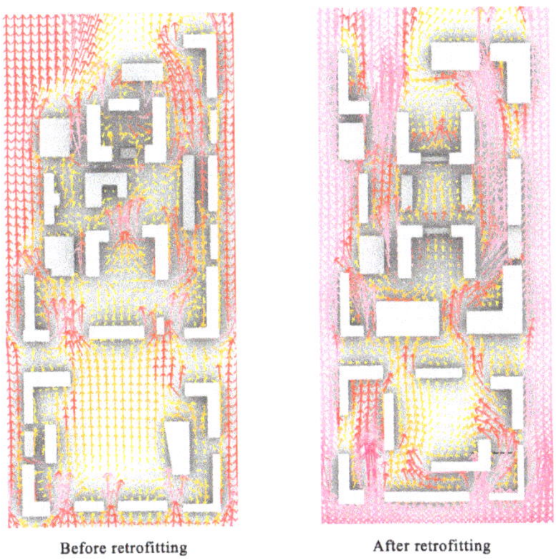

Figure 18. Comparison of the wind environment before and after the retrofitting.

5. Discussion

In the process of retrofitting industrial heritage communities, designers should pay attention to improving the environmental form of the community, providing residents with a good living environment, and addressing the issue of balancing heritage preservation and reducing carbon emissions on the basis of addressing climate change. In order to balance heritage conservation and the low-carbon retrofit, this paper first uses AHP to analyze the value of heritage elements and locate them in space. Then, the elements affecting carbon activities in space are integrated to form carbon space nodes combined with heritage conservation elements, and the links between the nodes are analyzed to form a carbon space network. At the same time, this paper uses a combination of qualitative and quantitative methods to analyze and research the elements affecting community carbon emissions in these aspects: appropriate land use intensity, resource and energy use, and spatial morphology on the basis of heritage conservation.

5.1. Recommendations for Conservation in Industrial Heritage Communities

In terms of protection based on morphological subdivisions, we suggest that, on the basis of the above scope, sufficient space should be allowed for morphological subdivisions that require protection to facilitate further morphological protection. Based on the hierarchical analysis method, this study adopts a step-by-step approach to protect the scope of architectural ontology protection and the scope of geomorphology protection within the industrial heritage community in the following ways [51,52]:

1. Classify and score according to the first- and second-degree indices to confirm the degree of importance of the different morphological subdivisions.

2. Calculate the weighting values of the indices, and divide the community as a whole into four grades of morphological subdivisions.

3. Adopt the corresponding protection and renovation measures: a. Being immovable, for comprehensive protection; b. Having protection as the focus, supplemented by transformation; c. Having transformation as the focus, supplemented by protection; d. Dismantling and rebuilding as needed, but paying attention to the overall morphology zoning style.

5.2. Proposals for Decarbonizing Industrial Heritage Communities

Through case simulation verification, for the renewal of industrial heritage communities, this research recommends that the floor area ratio of industrial heritage communities should be controlled between 2.1–2.5, the number of residential building floors should be 7–14, the road network density should be 4.16–4.50%, building density of 35–40%, and the green space ratio should be 30–35%. In terms of the layout form of the community, this paper recommends the adoption of more enclosed and staggered layouts on the basis of ensuring livability [53,54]. This will make the space within the community have better scenic and thermal conditions, which will enable the community to develop more potential carbon sinks.

6. Conclusions

Existing industrial heritage community retrofitting lacks systematic network thinking, aiming only to increase the intensity of land use and the profitability of real estate development, and low-carbonization retrofitting is also limited to the surface stacking of technology. This study integrates various elements that affect carbon emissions into a network, considers the interconnections between the elements, quantifies the carbon activities of the community, and elucidates the relationship between the network elements and carbon activities. For industrial heritage communities, the most important point is to combine conservation with low carbon to build a network of "heritage conservation—low carbon renovation—habitat enhancement—climate change response", which is academically important for the conservation of industrial heritage, control of community heating energy consumption and carbon emissions, and mitigation of global climate change. This study utilizes a carbon spatial network to retrofit industrial heritage communities based on a range of data measured in cold regions of China and other similar climatic zones, which can result in a 21.1% reduction in energy consumption, an approximate 32.7% reduction in carbon emissions, and a 7.3% reduction in the proportion of time spent in extreme heat and cold environments per year for similarly sized communities. This avoids the unnecessary energy consumption and carbon emissions associated with some other retrofitting approaches and results in a significant improvement in the quality of the habitat. The final part of the paper builds on this research to propose strategies for the conservation and low-carbon retrofitting of communities that have practical implications for the conservation and retrofitting of such communities.

But the correlation analysis in this research may not capture all of the complex factors that influence carbon emissions and land use intensity. Factors such as transportation infrastructure, energy sources, policy interventions, and socioeconomic factors can also play an important role. Follow-up studies consider incorporating additional variables into the analysis or conducting regression analyses to account for the effects of complex factors.

Author Contributions: Conceptualization, F.Z. and R.Z.; funding acquisition, J.L.; methodology, Y.W. and R.Z.; writing—original draft, Y.W., R.Z. and C.W.; writing—review and editing, J.L. and C.W. All authors have read and agreed to the published version of the manuscript.

Funding: This research was funded by the "Study on hierarchical protection of rural settlement units based on GIS spatial analysis technology", grant number ZR2020ME213. This research was funded by the general program of Shandong Natural Science Foundation, grant number ZR2020ME213.

Data Availability Statement: Not applicable.

Conflicts of Interest: The authors declare no conflict of interest.

Appendix A. Industrial Heritage Community Data Sheet

Test Points	FAR	GSR	PR	BD	RD	POPA	Carbon Source	EA	DA	SA	AAPOSP
Baotou First Thermal Power Plant Family Area	2.19	25%	83%	34%	3.80%	12%	30.03	56%	28%	16%	29%
Baotou Textile General Factory Textile District	2.85	22%	72%	41%	5.28%	8%	41.32	25%	68%	7%	31%
Dandong District 204 Road North Community	2.60	31%	78%	32%	6.14%	9%	40.73	78%	14%	6%	23%
Dandong District 204 Road South Community	2.45	28%	74%	33%	4.30%	13%	37.63	73%	18%	9%	24%
Harbin Electric Factory Community	2.77	30%	85%	38%	4.15%	11%	39.66	23%	21%	66%	22%
Harbin Boiler Plant Community	2.28	25%	79%	42%	4.20%	14%	35.61	17%	71%	12%	30%
Inner Mongolia Liberty Road No.1 and No.2 Neighborhood Community	2.52	27%	82%	36%	5.11%	12%	33.33	72%	18%	10%	31%
Haqi Community	2.34	32%	89%	34%	5.43%	28%	36.90	82%	18%	55%	27%
Starlight Family Community	2.67	26%	84%	37%	3.91%	13%	38.41	53%	32%	15%	30%
Sandaizi Workers Community	2.29	35%	86%	27%	5.31%	26%	35.29	48%	21%	31%	19%
Dalian Shipyard Workers Community	2.95	27%	73%	41%	6.01%	7%	45.71	43%	47%	10%	38%
Changchun Diesel Engine Plant Community 1	2.47	28%	64%	38%	5.21%	14%	36.46	13%	78%	9%	31%
Changchun Diesel Engine Plant Community 2	2.52	31%	82%	39%	4.39%	78%	36.77	11%	6%	83%	29%
Changchun FAW 3 Block	2.24	35%	71%	34%	4.61%	17%	33.63	41%	12%	47%	25%
Changchun FAW 6 Block	2.66	36%	86%	33%	3.72%	8%	38.95	72%	16%	12%	26%
Changchun FAW 8 Block	1.95	38%	83%	31%	3.65%	12%	28.63	65%	21%	14%	16%

FAR: Floor area ratio; GSR: Green space ratio; PR: Porosity rate; BD: Building density; RD: Road density; POPA: Percentage of plant area; EA: Proportion of enclosed area; DA: Proportion of determinant area; SA: Proportion of staggered area; AAPOSP: Area and proportion of solar panels that can be laid.

Appendix B. Envelope Parameters

a. Non-transparent envelope parameters

Setting	Value	Units
Conductivity	0.9	W/mK
Cost	0	
Density	2500	kg/m^3
Embodied carbon	5.06	$kgCO_2/kg$
Embodied energy	96.1	MJ/kg
Substitution rate pattern	0.2	
Substitution timestep	50	
Transportation carbon	0.067	$kgCO_2/kg/km$
Transportation distance	500	km
Transportation energy	0.94	MJ/kg/km
Dirt factor	1	
Back-side IR emissivity	0.84	
Front-side IR emissivity	0.84	
IR transmittance	0.01	
Back-side Solar reflectance	0.43	
Front-side Solar reflectance	0.27	
Solar transmittance	0.11	
Back-side visible reflectance	0.35	
Front-side visible reflectance	0.31	
Visible transmittance	0.14	

b. Transparent envelope parameters

Setting	Value	Units
Conductivity	1.75	W/mK
Cost	0	
Density	2400	kg/m^3
Embodied carbon	0.24	$kgCO_2/kg$
Embodied energy	2.12	MJ/kg
Substitution rate pattern	1	
Substitution timestep	100	
Transportation carbon	0.067	$kgCO_2/kg/km$
Transportation distance	500	km
Transportation energy	0.94	MJ/kg/km
Moisture diffusion resistance	50	
Roughness	Rough	
Solar absorptance	0.7	
Specific heat	840	
Thermal emittance	0.9	
Visible absorptance	0.7	

References

1. UN-Habitat Economy. Available online: https://unhabitat.org/topic/urban-economy-and-finance (accessed on 28 June 2023).
2. United Nations. World Urbanization Prospects. Available online: https://www.un.org/development/desa/pd/ (accessed on 28 June 2023).
3. UN-Habitat. Energy. Available online: https://unhabitat.org/topic/energy (accessed on 28 June 2023).
4. UN-Habitat. Climate Change. Available online: https://unhabitat.org/topic/climate-change (accessed on 28 June 2023).
5. Säynäjoki, A.; Heinonen, J.; Junnila, S. A scenario analysis of the life cycle greenhouse gas emissions of a new residential area. *Environ. Res. Lett.* **2012**, *7*, 34–37. [CrossRef]
6. Hui, E.C.; Liang, C.; Yip, T.L. Impact of semi-obnoxious facilities and urban renewal strategy on subdivided units. *Appl. Geogr.* **2018**, *91*, 144–155. [CrossRef]
7. Zou, Q.; Xue, B.; Li, Z.; Ye, X. Correlation analysis of carbon emissions and land use intensity in China's provinces. *Ecol. Indic.* **2016**, *67*, 278–287.
8. Deng, X.; Jiang, Y.; Zhao, X.; Li, X. Correlation analysis of carbon emissions and land use intensity: A case study of 32 Chinese cities. *J. Clean. Prod.* **2017**, *148*, 581–588.

9. Hui, E.C.-M.; Chen, T.; Lang, W.; Ou, Y. Urban community regeneration and community vitality revitalization through participatory planning in China. *Cities* **2021**, *110*, 103072. [CrossRef]
10. Luo, X.; Ren, M.; Zhao, J.; Wang, Z.; Ge, J.; Gao, W. Life cycle assessment for carbon emission impact analysis for the renovation of old residential areas. *J. Clean. Prod.* **2022**, *367*, 1329. [CrossRef]
11. Peng, L.; Li, X.; Xu, Y.; Li, Z. Correlation analysis of carbon emissions and land use intensity in the Pearl River Delta region, China. *J. Clean. Prod.* **2017**, *141*, 42–53.
12. Forrester, J.W. Policies, decisions and information sources for modeling. *Eur. J. Oper. Res.* **1992**, *59*, 42–63. [CrossRef]
13. Forrester, J.W. System dynamics—The next fifty years. *Syst. Dyn. Rev.* **2007**, *23*, 359–370. [CrossRef]
14. Sun, S.; Xu, S.; Aoki, X. Research on Heritage Community Identity in Industrial Land Transformation. *Chin. Cult. Herit.* **2022**, 41–48.
15. Liu, L. Research on the activation of the memory places of urban industrial heritage communities—A case study of Workers' Village in Tiexi District, Shenyang. *J. Shenyang Norm. Univ.* **2022**, *46*, 99–106. [CrossRef]
16. Liu, L. Research on Path dependence and Path Innovation of Urban Industrial Heritage Community Protection. *Mod. City Res.* **2015**, *11*, 41–46. [CrossRef]
17. Girvan, M.; Newman, M.E.J. Community structure in social and biological networks. *Proc. Natl. Acad. Sci. USA* **2002**, *99*, 7821–7826. [CrossRef] [PubMed]
18. Huang, L.; Krigsvoll, G.; Johansen, F.; Liu, Y.; Zhang, X. Carbon emission of global construction sector. *Renew. Sustain. Energy Rev.* **2018**, *81*, 1906–1916. [CrossRef]
19. Cigola, M. Digital tools for Urban and Architectural heritage. In *Geospatial Research: Concepts, Methodologies, Tools, and Applications*; IGI Global: Hershey, PA, USA, 2016; pp. 743–764.
20. Landorf, C. A framework for sustainable heritage management: A study of UK industrial heritage sites. *Int. J. Heritage Stud.* **2009**, *15*, 494–510. [CrossRef]
21. Zhao, X.; Qiu, S.; Sun, J. Research on the Green Transformation Strategy of Industrial Heritage Communitie—Taking Mianshan Dormitory in Tianjin as an Example. *Build. Energy Conserv.* **2019**, *47*, 77–80.
22. Jin, H.; Shi, J.; Sun, G. Correlation analysis of carbon emissions and land use intensity in the Yangtze River Delta region, China. *J. Environ. Plan. Manag.* **2020**, *63*, 1196–1215.
23. Tan, S.; Chen, S.; Liu, Z.; Wang, Y.; Zhao, D. Correlation analysis of carbon emissions and land use intensity in different urban areas: A case study of Beijing. *Sustainability* **2020**, *12*, 4016.
24. Rodwell, D. Urban heritage, communities, and environmental sustainability. In *Securing Urban Heritage*; Routledge: Abingdon, UK, 2019; pp. 62–79.
25. Havlin, S.; Kenett, D.Y.; Ben-Jacob, E.; Bunde, A.; Cohen, R.; Hermann, H.; Kantelhardt, J.W.; Kertész, J.; Kirkpatrick, S.; Kurths, J.; et al. Challenges in network science: Applications to infrastructures, climate, social systems and economics. *Eur. Phys. J. Spéc. Top.* **2012**, *214*, 273–293. [CrossRef]
26. Labanca, N.; Pereira, G.; Watson, M.; Krieger, K.; Padovan, D.; Watts, L.; Moezzi, M.; Wallenborn, G.; Wright, R.; Laes, E.; et al. Transforming innovation for decarbonisation? Insights from combining complex systems and social practice perspectives. *Energy Res. Soc. Sci.* **2020**, *65*, 101452. [CrossRef]
27. Porter, M.A.; Onnela, J.P.; Mucha, P.J. Communities in Networks. *arXiv* **2009**, arXiv:0902.3788.
28. Douet, J. *Industrial Heritage Re-Tooled: The TICCIH Guide to Industrial Heritage Conservation*; Routledge: Abingdon, UK, 2016.
29. Wang, X.; Zhang, J.; Cenci, J.; Becue, V. Spatial distribution characteristics and influencing factors of the world architectural heritage. *Heritage* **2021**, *4*, 2942–2959. [CrossRef]
30. Wang, J.; Lu, C. Research on the Development and Practice of Digital Technology in Architectural Heritage. In Proceedings of the 2022 International Conference on Culture-Oriented Science and Technology (CoST), Lanzhou, China, 18–21 August 2022; pp. 146–150.
31. Lu, N.; Liu, M.; Wang, R. Reproducing the discourse on industrial heritage in China: Reflections on the evolution of values, policies and practices. *Int. J. Herit. Stud.* **2020**, *26*, 498–518. [CrossRef]
32. Guerra, P.; Meira, W., Jr.; Cardie, C.; Kleinberg, R. A measure of polarization on social media networks based on community boundaries. In Proceedings of the International AAAI Conference on Web and Social Media, Cambridge, MA, USA, 8–11 July 2013; Volume 7, pp. 215–224.
33. Ozarisoy, B.; Altan, H. Significance of occupancy patterns and habitual household adaptive behaviour on home-energy performance of post-war social-housing estate in the South-eastern Mediterranean climate: Energy policy design. *Energy* **2022**, *244*, 122904. [CrossRef]
34. Deng, Q.; Wang, G.; Wang, Y.; Zhou, H.; Ma, L. A quantitative analysis of the impact of residential cluster layout on building heating energy consumption in cold IIB regions of China. *Energy Build.* **2021**, *253*, 111515. [CrossRef]
35. Zhang, J.; Cenci, J.; Becue, V.; Koutra, S.; Liao, C. Stewardship of Industrial Heritage Protection in Typical Western European and Chinese Regions: Values and Dilemmas. *Land* **2022**, *11*, 772. [CrossRef]
36. Alfrey, J.; Putnam, T. *The Industrial Heritage: Managing Resources and Uses*; Routledge: Abingdon, UK, 2003.
37. Russo, R.d.F.S.M.; Camanho, R. Criteria in AHP: A systematic review of literature. *Procedia Comput. Sci.* **2015**, *55*, 1123–1132. [CrossRef]

38. Liu, F.; Zhao, Q.; Yang, Y. An approach to assess the value of industrial heritage based on Dempster—Shafer theory. *J. Cult. Herit.* **2018**, *32*, 210–220. [CrossRef]
39. Oke, T.R. *Boundary Layer Climates*; Routledge: Abingdon, UK, 2002.
40. Dong, L.; Wang, Y.; Lin, J.; Zhu, E. The Community Renewal of Shantytown Transformation in Old Industrial Cities: Evidence from Tiexi Worker Village in Shenyang, China. *Chin. Geogr. Sci.* **2020**, *30*, 1022–1038. [CrossRef]
41. Wang, S.-H.; Huang, S.-L.; Huang, P.-J. Can spatial planning really mitigate carbon dioxide emissions in urban areas? A case study in Taipei, Taiwan. *Landsc. Urban Plan.* **2018**, *169*, 22–36. [CrossRef]
42. Mazmanian, D.A.; Kraft, M.E. (Eds.) *Toward Sustainable Communities: Transition and Transformations in Environmental Policy*; MIT Press: Cambridge, MA, USA, 2009.
43. Gu, T.; Hao, E.; Ma, L.; Liu, X.; Wang, L. Exploring the determinants of residents' behavior towards participating in the Sponge-Style old community renewal of china: Extending the theory of planned behavior. *Land* **2022**, *11*, 1160. [CrossRef]
44. Rastogi, P.; Andersen, M. Embedding Stochasticity in Building Simulation through Synthetic Weather Files. In Proceedings of the BS2015: 14th Conference of International Building Performance Simulation Association, Hyderabad, India, 7–9 December 2015.
45. Kamal, A.; Abidi, S.M.H.; Mahfouz, A.; Kadam, S.; Rahman, A.; Hassan, I.G.; Wang, L.L. Impact of urban morphology on urban microclimate and building energy loads. *Energy Build.* **2021**, *253*, 111499. [CrossRef]
46. Nowak, D.J.; Maco, S.; Binkley, M. i-Tree: Global tools to assess tree benefits and risks to improve forest management. *Arboric. Consult.* **2018**, *51*, 10–13.
47. Erb, K.-H.; Haberl, H.; Jepsen, M.R.; Kuemmerle, T.; Lindner, M.; Müller, D.; Verburg, P.H.; Reenberg, A. A conceptual framework for analysing and measuring land-use intensity. *Curr. Opin. Environ. Sustain.* **2013**, *5*, 464–470. [CrossRef]
48. Energy Efficiency Design Standards for Residential Buildings in Severe Cold and Cold Regions. JGJ26-2010. 2010-03-18. Available online: https://kns-cnki-net-443.webvpn.sdjzu.edu.cn/kcms/detail/detail.aspxFileName=SCSD000005978883&DbName=SCSD (accessed on 28 June 2023).
49. Sonta, T.R.A.; Dougherty, R.K.; Jain, R.K. Data-driven optimization of building layouts for energy efficiency. *Energy Build.* **2021**, *238*, 110815. [CrossRef]
50. Lin, C.; Thomson, G.; Popescu, S.C. An IPCC-compliant technique for forest carbon stock assessment using airborne LiDAR-derived tree metrics and competition index. *Remote Sens.* **2016**, *8*, 528. [CrossRef]
51. Fragner, B. Adaptive re-use. In *Industrial Heritage Re-Tooled*; Routledge: Abingdon, UK, 2016; pp. 110–117.
52. Mo, C.; Wang, L.; Rao, F. Typology, preservation, and regeneration of the post-1949 industrial heritage in China: A case study of Shanghai. *Land* **2022**, *11*, 1527. [CrossRef]
53. Baiocchi, G.; Creutzig, F.; Minx, J.; Pichler, P.P. A spatial typology of human settlements and their CO_2 emissions in England. *Glob. Environ. Change* **2015**, *34*, 13–21. [CrossRef]
54. Makido, Y.; Dhakal, S.; Yamagata, Y. Relationship between urban form and CO_2 emissions: Evidence from fifty Japanese cities. *Urban Clim.* **2012**, *2*, 55–67. [CrossRef]

buildings

MDPI

Article

Comparative Urban Performance Evaluation of Sustainable Urbanization in Four Typical Megalopolises in China

Xue Fang [1,2,*], Xinyu Shi [3], Tyson Keen Phillips [4] and Weijun Gao [1,5,*]

1 iSMART (innovation institute for Sustainable Maritime Architecture Research & Technology), Qingdao University of Technology, Qingdao 266061, China
2 Research Institute for Science and Technology, Tokyo University of Science, Tokyo 125-8585, Japan
3 College of Architecture and Urban Planning, Qingdao University of Technology, Qingdao 266061, China
4 Piaggio Fast Forward, Boston, MA 02129, USA
5 Faculty of Environmental Engineering, The University of Kitakyushu, Kitakyushu 808-0135, Japan
* Correspondence: fangxue@rs.tus.ac.jp (X.F.); gaoweijun@me.com (W.G.)

Abstract: The essence of sustainable urbanization is to take a holistic approach to the harmonious development of economic, social, cultural and environmental protection. This paper applies the urban sustainability assessment system to analyze the characteristics of indicators related to the quality of the built environment and environmental pressure of 91 cities in four major megalopolises in China from 2010 to 2018. It also combines statistical methods to summarize the general features of urban development through a comprehensive urban performance evaluation by comparative and classification analysis for the purpose of scientific guidance on sustainable urbanization. The comparative results showed that in terms of urban sustainability, the Yangtze River Delta performed best, followed by JingJinJi, Pearl River Delta and Shandong Peninsula. Of which, the quality of built environment in JingJinJi and the environment pressure in the Shandong Peninsula require particular attention to improve and decrease, respectively. Moreover, cities can be grouped into six development types through performance clustering including three positive and three negative types. The characteristics of all types are summarized, and the performance of the specific indicators are detailed compared to serve as a guiding basis for making generic recommendations of sustainable urbanization.

Keywords: sustainable urbanization; urban sustainability assessment system; urban performance; comparative analysis; the quality of built environment; environmental pressure

1. Introduction

The urbanization trend is a consensus that is a precondition for development in the developing world [1]. By 2015, 16 of the world's 24 megacities (cities with more than 10 million people) had located in Asia [2]. Twelve of the world's twenty-one megacities (with more than 10 million people) are in the Asian region and are still expanding [3]. These highly urbanized areas have not only attracted significant investment but have also developed into economic centers. They have even become attractive cultural complexes with a good infrastructure and resource, making for a dynamic and diverse urban environment with potential for development [4]. However, in recent decades, the pace of population decline is faster in the rest of Asia, and growth led by a booming population is coming to an end [5–7]. In China, urban shrinkage has gradually become an emerging phenomenon in recent years. The research on the Chinese context is not yet comprehensive, but the existing reports show that cities in some regions are experiencing population decline, particularly in the less developed cities and border areas, and even in the most developed coastal regions [8–12]. Moreover, Japan, a developed country, is also experiencing significant population decline in its cities and generating many urban problems, especially in economic downturns and increased employment pressures. The population of less urbanized

Southeast Asian cities, such as Thailand and Vietnam, are also likely to start decreasing earlier than expected [13]. So, many experts [8–11,13] argue that developing countries in Asia are probably going to face a longer period of lower population growth than Japan in the coming period, unless sustainable levels of urban development can be improved through effective approaches.

Since the 21st century, China has been experiencing the fastest urbanization rate in the world. The urbanization rate increased from 10.64% in 1949 to 59.58% in 2018, with an average annual growth rate of over 0.7% [14,15]. Although researchers hold different opinions on the beginning and specific stages of China's urbanization process, most can agree that the adoption of the reform and opening-up policy and China's urban population exceeding half of the country's population nearly a decade ago were two important turning points in the process [16–18]. The period from 2010 to the present can be considered a period of accelerated urbanization due to the urbanization rate has not exceeded 70% with the possibility of further development [18]. Some urbanization responses have misled developing countries to think that they should shift most of their traditional rural population to cities, or they have over-emphasized massive investment in urban infrastructure and the development of land to build housing to accommodate more people, etc. These actions seem to have transformed the appearance of cities in a short period of time [19,20].

The superposition of various urban environmental problems is complicated and the distribution differences between regions are significant. The layout and environmental governance of megalopolises to promote urbanization, however, has become the focus of the government and researchers [21,22]. Therefore, making a coordinated plan for the larger agglomeration faces huge challenges. Effective quantitative assessment is critical for finding coping strategies in areas with complex urban environments such as China. Sustainable urbanization does not only emphasize the dynamic balance of the quality of the built environment and its environmental pressure during urbanization [23–25]. It goes beyond the narrow scope of urban areas into individual sustainable cities—and beyond that into the development of each megalopolis. It aims to bring about the harmonious development of different regions in the whole country. China's large population base, its high speed of economic growth, its energy shortage and its fragile environment are key factors that must be taken into account when establishing a sustainable urbanization model with Chinese characteristics [26]. It is clear that sustainable urbanization must adhere to the national policy based of coordinated development between all parties in China.

Therefore, based on a detailed understanding of the sustainable development of the four major megalopolises, it is possible to compare and discuss the universal development characteristics of cities throughout the country so as to achieve the ultimate goal of providing a basis for decision makers to plan a national sustainable urbanization path. For this, we need a detailed understanding of the current urban performance of cities in the four megalopolises through the application of urban sustainability assessment system. Comparative and categorical analyses were conducted to extract general performance patterns for different types of cities and to make targeted recommendations. We expect that the research process and results of this study will not only provide reference for the future development of Chinese cities but can also be applied to the study of sustainable urbanization in other developing countries at the same stage.

2. Materials and Methods

2.1. Study Area

The Yangtze River Delta, Pearl River Delta, JingJinJi and Shandong Peninsula are four of 11 megalopolises that have approved Chinese government plans since 2015 (Figure 1) [27]. Among them, the first three megalopolises have gained consensus from many parties and are considered to be pioneering regions in China to promote economic, social and regional integration with strong development momentum [28–30]. The Shandong Peninsula has also attracted domestic and overseas attention both in terms of its international location relations and its own comprehensive strength. It is the fourth largest megalopolis in China with the most development potential after the above three megalopolises [21].

Figure 1. The location of study areas in China.

These four comparable megalopolises have been chosen as representative for several reasons. Firstly, they have good economic foundations which is an important base for the feasibility of implementation of strategic recommendations, advanced concepts and technologies related to sustainable urbanization. This allows us to apply the results of our evaluation in a practical way. In 2018, the per capita gross domestic product (GDP) of those four regions all exceeded $10,000/person (Figure 2), and their economic strength ranked among the top five in China with the largest increase in 9 years [31,32]. The increase in Shandong Peninsula is much higher than other megalopolises. As a result of this growth, the Chinese National Development and Reform Commission explicitly calls for the promotion of a multi-level, permanent consultation and coordination mechanism for megalopolises from now on [33]. These representative megalopolises, which have entered a phase of rapid urbanization earlier, will provide rich experiences for other megalopolises that are going through or have not yet developed to the same phase. Meanwhile, the future development of these urban agglomerations deserves sufficient attention. Lastly, the locational characteristics of cities in the megalopolises were taken into consideration, leading to the identification of two northern and two southern megalopolises, each of which includes coastal and inland cities, allowing our study to be more extensive and generalized. Along with economic development and social progress, their environmental problems have become increasingly serious. Some cities have long-term pursuits of high GDP growth at the expense of ecological resource protection which has led to significant impact on the environment [34,35]. So, the study of the sustainable urbanization of these typical megalopolises is particularly necessary and valuable for replication in other regions.

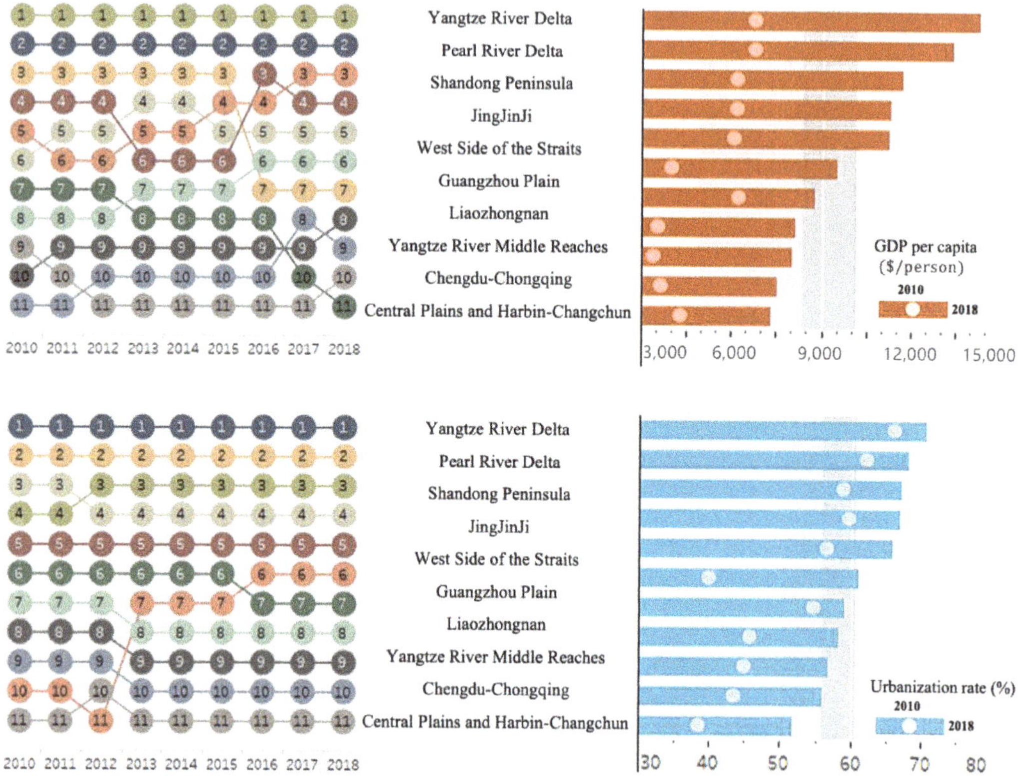

Figure 2. Comparative analysis of GDP and Urbanization rate among 11 megalopolises in China from 2010 to 2018. (Data source: China Statistical Yearbook 2011–2019 [36]).

All applied databases for cities within these four megalopolises are derived from authoritative government statistics from 2010 to 2018 [36]. This data is available on public platforms for easy access and to allow all parties to replicate and build on their results by referencing our research.

2.2. Urban Sustainability Indicators for Performance Evaluation

To accurately compare the performance of sustainable urbanization in four typical megalopolises in China, our proven assessment system of urban sustainability has been used in this research. In it, urban sustainability is developed in the context of China's rapid urbanization, based on sustainable development goals (SDGs) and a detailed critical literature review. It involves four dimensions of the sustainability: economic, social, cultural and environmental. This is then summarized into two main focuses, namely quality of built environment (Qu) and environmental pressures (Pu), which are two urgent concerns for achieving sustainable urbanization in China at the current time. The urban sustainability assessment system is positioned as an easy-to-understand evaluation criteria, not only to highlight the importance of these two key dimensions and to facilitate reach consensus. It is expected to be easily applied by all parties and to gain consensus for joint promotion of sustainable urban policies. For this reason, the ratio of these two categories is defined as the urban sustainability (Su), which is inspired by a classic input-output efficiency model of economics [37,38]. This means the criteria by which the degree of sustainability of the city can be measured as follows:

$$S_U = \frac{Q_U}{P_U} = \frac{\sum_{i=1}^{n} Q_{Ui}}{\sum_{i=1}^{m} P_{Ui}} \tag{1}$$

The selection of evaluation indicators is based on 14 existing reports on sustainable urban assessment tools formulated by the Chinese government and official scientific research institutions from 2000 to the present shown in Table 1. Moreover, several well-known international evaluation systems, such as Leadership in Energy and Environmental Design (LEED), Comprehensive Assessment System for Built Environment Efficiency (CASBEE), the City Development Index (CDI) etc., were used as reference materials for the above 14 assessment tools. Their focus, scoring methods and number of indicators of each evaluation system are different with no uniform standard. The indicator categories mostly focus on economic growth, human settlements, resource utilization, energy conservation, and environmental quality. Based on the above tools, the indicators have been simplified and sorted out as follows: (1) ranking those that appear more frequently and excluding those that appear less than 20% of the time; (2) excluding subjective indicators that cannot be objectively quantified; (3) excluding indicators for which data cannot be obtained from publicly available government platforms; and (4) merging indicators that are repetitive or have co-linearity with others. Through this process, 17 indicators were used in the assessment system of urban sustainability, grouped by subordinate categories and listed in Table 2. The quality of built environment (Qu) consists of urbanization economies (four sub-indicators), infrastructure development (four sub-indicators) and urban attraction (three sub-indicators). Environmental pressures (Pu) consist of resource consumption (two sub-indicators) and environmental pollution (four sub-indicators).

Table 1. The general list of 14 Chinese urban sustainability assessment tools.

No.	Assessment Tools of City Scale by Chinese Authority Agencies	Version Year	Scoring Methods	Indicator Categories	Number of Indicators
1	Ecological Counties, Ecological Cities Construction Evaluation System [39]	2003 (Trial version)	Summation of scores achieved	Economic development; Environmental protection; Social progress	219
2	China's Livable Cities Scientific Evaluation Standards [40]	2007	Weighted standardized values summation	Social civilization; Economic prosperity; Beautiful environment; Resource carrying; Convenience of life; Public safety	83
3	Low-carbon City Evaluation System of China [41]	2011	Standardized values summation	Economy; Energy; Infrastructure; Environment; Society	15
4	Green and Low-carbon Cities Evaluation Indicators [42]	2011 (Trial version)	Summation of scores achieved	Social economy; Planning management; Construction land; Resources and environment; Energy conservation and emission reduction; Infrastructure and greening; Public services; Urban characteristics	62
5	National Environmental Protection Model City Assessment Index and Implementation Rules [43]	2011	Summation of scores achieved	Economic society; Environmental quality; Environmental construction; Environmental management	24

Table 1. *Cont.*

No.	Assessment Tools of City Scale by Chinese Authority Agencies	Version Year	Scoring Methods	Indicator Categories	Number of Indicators
6	National Ecological Civilization Pilot Policy Evaluation System [44]	2013 (Trial version)	Summation of scores achieved	Economy; Natural Environment; Built environment; Regulations; Culture	30
7	Technical Criterion for Ecosystem Status Evaluation [45]	2015 (Draft)	Weighted standardized values summation	Environmental quality; Pollution load; Ecological construction	18
8	Evaluation System for the Construction of National Ecological Civilization Model City [46]	2016 (Trial version)	Summation of scores achieved	Space; Economy; Environment; Life; Regulations; Culture	35
9	China Green Development Index System [47]	2016	Weighted standardized values summation	Resource utilization; Environmental governance; Environmental quality; Ecological protection; Growth quality; Green life; Public satisfaction	56
10	Evaluation System for Ecological Civilization Construction [48]	2016	Summation of scores achieved	Resource utilization; Environmental protection; Annual evaluation results; Public satisfaction; Environmental incidents	23
11	Evaluation indicators for green cities [49]	2017 (Trial version)	Weighted standardized values summation	Green production; Green life; environmental quality	67
12	Evaluation System of the Construction Target of the National Ecological Civilization Pilot Zone [50]	2018	Summation of scores achieved	Economic quality; Resource and energy conservation and utilization; Ecological construction and environmental protection; Ecological culture cultivation; Mechanism construction	49
13	Indicators for City Services and Quality of Life [51]	2019	Weighted standardized values summation	Economy; Education; Energy; Environment; Finance; Fire and emergency response; Governance; Health; Leisure; Safety; Shelter; Solid waste; Communication and innovation; Transportation; Urban planning; Wastewater; water and sanitation	100
14	New-type Urbanization-Evaluation Index System of New Urbanization Quality City [52]	2020 (Trial version)	Weighted standardized values summation	Economic development; Social culture; Ecological environment; Public services; Residents' lives; Reward indicators	76

Table 2. List of indicators and their weights.

Basic Dimensions	Label	Description of Indicators	Weight (%)
Qu Quality of built environment	Qu_1	**Urbanization economies**	
	q1	Urban population percentage (% of total population)	8.56
	q2	GDP per capita ($/person)	8.52
	q3	Services value added (% of GDP)	8.51
	q4	Income per capita ($/person)	7.74
	Qu_2	**Infrastructure development**	
	q5	Construction land per capita (m^2/person)	9.30
	q6	Residential area per capita (m^2/person)	9.00
	q7	Green area per capita (m^2/person)	9.01
	q8	Road area per capita (m^2/person)	6.02
	Qu_3	**Urban attraction**	
	q9	Number of tourists (1000 person)	10.99
	q10	Foreign direct investment ($$10^6$)	10.09
	q11	Export volume of trade ($$10^6$)	12.25
Pu Environmental pressure	Pu_1	**Resource consumption**	
	p1	Energy consumption per capita (GJ/person)	25.00
	p2	Water consumption per capita (ton/person)	25.00
	Pu_2	**Environmental pollution**	
	p3	Wastewater discharged per capita (ton/person)	10.47
	p4	SO_2 emission per capita (kg/person)	15.03
	p5	NOx emission per capita (kg/person)	9.56
	p6	Soot and dust discharged per capita (kg/person)	14.94

Notes: This table is derived from our proven assessment system [25] and have been edited.

2.3. Weighting Method of Principal Component Analysis (PCA)

Determining weighting criteria is important for scientific evaluation. Most scholars prefer quantitative methods that assign weights after standardizing the statistical values of relevant indicators, which can avoid excessive influence or emphasize individual subjectivity. Principal component analysis (PCA) has proven to be very effective and reasonable in some classic evaluation systems, such as the ecological footprint published by Wackernagel and Rees in 1992 [53], City Development Index published by UN-habitat in 1997 [54], etc. The weights are generally determined in two ways: one is the weight of the indicator that is equal to the weight of the variance contribution of the principal components, and the other is a normalization of the weighted average of the coefficients of the indicator in the linear combination of the principal components [55–57]. The application of principal component analysis in this study makes use of the principle of information condensation of data, using the former variance interpretation ratio for the weight calculation. The specific steps are: (1) preparing the standardized data using the min-max scaling method [58]; (2) calculating the principal component scores of the indicator in each category separately; (3) taking the percentage of the scores as the weight of corresponding indicator to represent its contribution to the principal evaluation category. All indicators with weighted scores are shown in Table 2 and are calculated as follows:

$$W_{(p/q)i} = \frac{|\lambda_{(\frac{p}{q})i}|}{\sum_{i=1}^{\frac{n}{m}} |\lambda_{(\frac{p}{q})i}|} \tag{2}$$

$$Q_{Ui} = W_{(q)i} \times u_{(q)i}{}' ; \; P_{Ui} = W_{(p)i} \times u_{(p)i}{}' \tag{3}$$

where $\lambda_{(\frac{p}{q})i}$ is the component score coefficient matrix; $W_{(p/q)i}$ is the weight of all representative indicators in each category of Qu and Pu; $u'_{(q)i}$ *or* $u_{(p)i}{}'$ is the standardized statistical value of sub-indicators. Further, Q_{Ui} *or* P_{Ui} is the weighted score of each indicator.

2.4. Hierarchical Clustering for Comparative Analysis of Classifications

Hierarchical clustering, also known as hierarchical cluster analysis, was used to classify 91 cities in the study area for comparative analysis. It is an algorithm that groups similar objects into what are called clusters [59,60]. In theory, hierarchical clustering begins by treating each observation as a separate cluster. Each cluster is different from the others, and the objects in each cluster are roughly similar to each other. The clustering process executes the following two steps in an iterative manner. (1) the identification of the two nearest clusters, and (2) the joining of the two most similar clusters. This process of iteration lasts until all clusters have been grouped into one type. Data for cluster analysis using SPSS should be prepared as follows: (1) the rows are the observations (individuals) and the columns are the variables; (2) all missing values, if any, in the data must be deleted; (3) the data must be standardized to make the variables comparable [61–63].

$$\Delta(A,B) = \sum_{i \in A \cup B} ||\underset{x_i}{\rightarrow} - \underset{m\,A \cup B}{\rightarrow}||^2 - \sum_{i \in A} ||\underset{x_i}{\rightarrow} - \underset{m\,A}{\rightarrow}||^2 - \sum_{i \in B} ||\underset{x_i}{\rightarrow} - \underset{m\,B}{\rightarrow}||^2 = \frac{n_A n_B}{n_A + n_B} ||\underset{m\,A}{\rightarrow} - \underset{m\,B}{\rightarrow}||^2 \tag{4}$$

where $\underset{m_i}{\rightarrow}$ is the center of cluster i, and n_i is the number of points. Δ is the combining cost of joining the clusters of A and B.

3. Descriptive Statistics Analysis of Urban Performance Indicators

To understand the development of each of the urban environment indicators before assessment, the most important megalopolises (Yangtze River Delta, Pearl River Delta, JingJinJi and Shandong Peninsula) and the highest administrative level cities in China (Beijing and Shanghai) were extracted and analyzed together with China's average data (except autonomous regions). All data are the national statistical data from 2010 to 2018, originally collected from the China Statistical Yearbook (2011–2019) and publicly available on government websites.

3.1. Urbanization Economies

Since the establishment of megalopolises in the 2000s, the Pearl River Delta, the Yangtze River Delta and the JingJinJi have grown rapidly in a stable urbanization process since before 2010 (Figure 3). This is the result of guidance from the national authorities and the most developed first-tier cities (Beijing, Shanghai, Guangzhou and Shenzhen), which are densely populated agglomerations with enormous economic and political influence. Although the start time of the Shandong Peninsula's development is far behind other megalopolises, the growth of the urban population is quite significant after the opening up, especially with the implementation of the outline of urbanization development starting in 2012 [64], and the level of urbanization in the city surged within a year. All four metropolises have reached the upper middle ranking of the world economy since 2014 with a GDP per capita of over $10,000/person. The economic status of Yangtze River Delta and Pearl River Delta cities are relatively better and the GDP per capita, and have been in excess of the average national level for a long time. The economic divide between China's lagging north and its booming south is rapidly widening, due to rapid gains in export trade and high-tech industries in southern cities during accelerated urbanization [65]. The economic development speed of Shandong Peninsula and JingJinJi are about the same. The slowdown in the growth of the secondary industry and the decline in investment demand caused the overall slowdown of economic growth in China around 2015, then the economy returned to a state of rapid growth. Although the proportion of China's tertiary industry is increasing year by year and approaching 50%, it still lags behind the world average of 63.0% (statistical value in 2017). The unsatisfactory development of the service sector leads to a lag in the transformation of the employment structure and a low urbanization level [66].

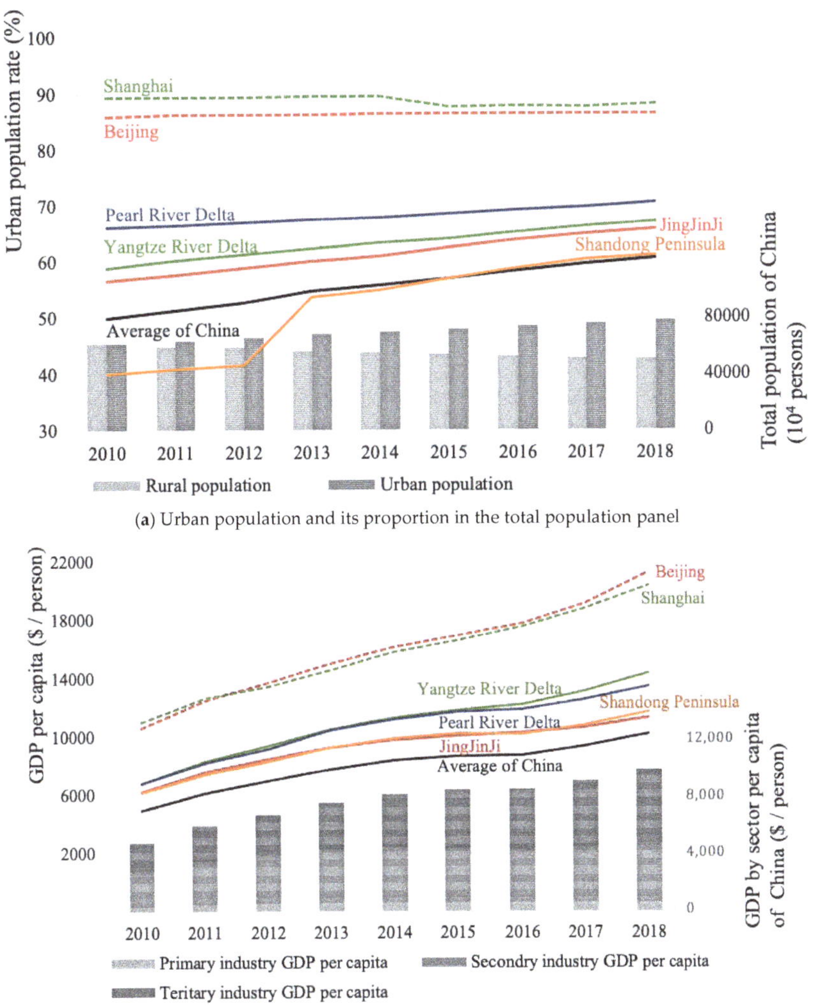

(**a**) Urban population and its proportion in the total population panel

(**b**) Total GDP and GDP by sector per capita

Figure 3. The trend of urbanization economies in China from 2010 to 2018. (Data source: China Statistical Yearbook 2011–2019 [36]).

3.2. *Infrastructure Development*

Over the past 30 years of reform, China's infrastructure stock has ranked first in the world through moderately advanced large-scale infrastructure construction. In particular, the level of infrastructure development in the four megalopolises is generally higher than China's average, but in JingJinJi it is relatively insufficient (Figure 4). Despite almost 30% of infrastructure investment being spent on building railroads and highways, urban transit infrastructure is still underdeveloped in many cities of these megalopolises [67]. The area of construction land per capita in China just exceeded the world average in 2014, but there is still a gap of nearly twice as much residential area per capita as in developed countries [68]. With the influx of urban population in a short period of time, a series of problems have emerged, such as insufficient public activity space, inconvenient transport networks, inadequate housing supply, etc. There is a significant decline in GDP shown in the figure at around 2014, especially in the first-tier cities. Moreover, the residential land area has not increased significantly in the past 9 years. According to the 2010 census, the per capita housing area of 55% of Chinese households is less than 30 m^2. The per capita housing area in the more developed areas of the southeast coast is relatively low, and the demand for housing improvement is growing stronger. The good phenomenon is that there has been a growing awareness of the ecological and social value of urban green spaces in recent years [69,70]. Despite the fact that cities are still under great pressure from construction, they are attempting to increase green space such as the "plant where possible" policy implemented in Beijing [71]. However, the energy consumption during infrastructure development cannot be ignored for the changing lifestyle expectations and demographic trends, especially the demand for household energy demand.

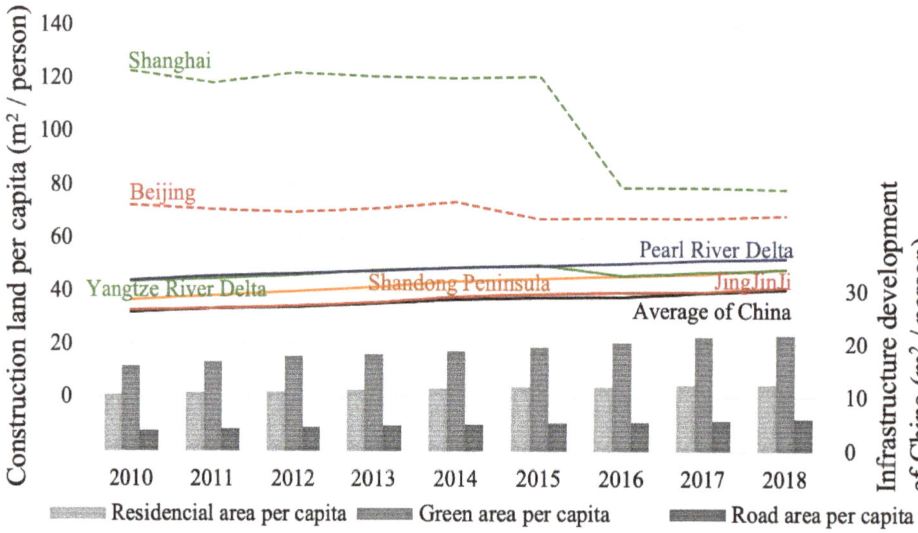

Figure 4. The trend of infrastructure development in China from 2010 to 2018. (Data source: China Statistical Yearbook 2011–2019 [36]).

3.3. *Urban Attraction*

Foreign direct investment (FDI) in a country, attracting high-quality human resources and advanced technologies, is a key indicator that could represent the environmental improvement of the innovation and technology industry to some extent [72,73]. It is pointed out that an attractive investment environment is one of the most essential conditions for the promotion of FDI. With the support of a series of policies to stabilize foreign investment, the four megalopolises have increased their exposure to the world and enhanced their attractiveness to foreign investment and are now all exceeding the Chinese average shown in Figure 5. In particular, the FDI in the Yangtze River Delta is two to four times higher than in other regions in 2018, thanks to the pioneering planning for the international financial center, making a free and convenient environment for investment, trade and transportation [74]. The measures include a further reduction in investment restrictions and continued optimization of the investment conditions. Foreign trade growth has declined or grown slowly, mainly due to two shifts: (1) consumption has transformed from physical to service consumption; (2) attractive products have changed from resource-consuming and sloppy products to technology-intensive and high-performance products since 2015. In 2016, driven by policy measures such as the reform of the commercial system and the reduction in foreign investment restrictions, foreign-invested companies showed a rapid recovery. However, they immediately fell back in JingJinJi and Shandong Peninsula regions mainly due to low income and technological innovation industry trade protections. In addition, investment in tourism is seen as a strategic industry. The extent of its development can directly reflect a city's potential to shape its cultural identity and the attractiveness of international cooperation.

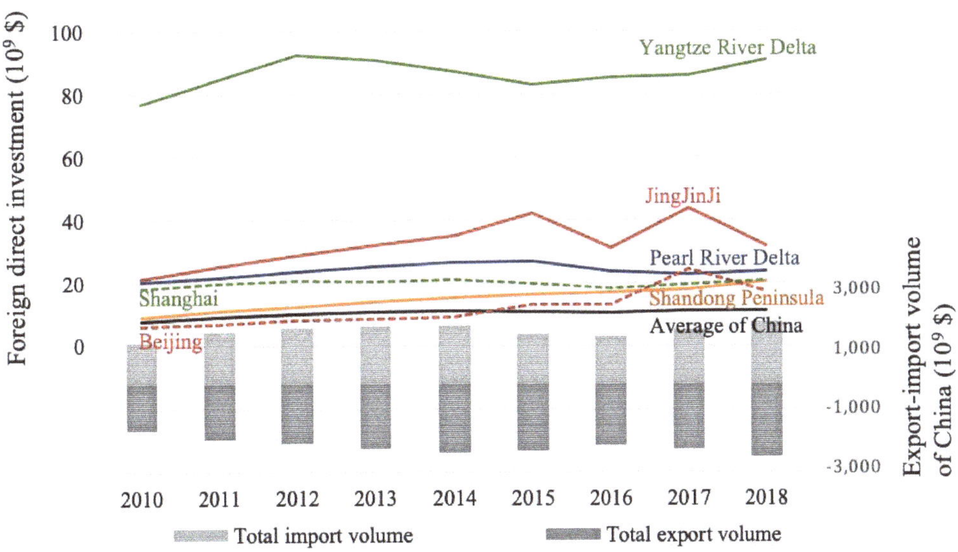

Figure 5. The trend of urban attraction in China from 2010 to 2018. (Data source: China Statistical Yearbook 2011–2019 [36]).

3.4. Resource Consumption

Figure 6 shows that there is a significant linear relationship between energy consumption and GDP. The Shandong Peninsula is the largest resource consumption megalopolis in China. The intensity of its energy extraction far exceeds that of other regions and the output is already exceeding storage. This kind of massive consumption of resources is a detrimental factor to sustainable urbanization. It is followed by JingJinJi, whose energy intensity is also greater than the average level of China, but it is already being controlled and beginning to decrease. From the national average curve, the value of energy consumption per capita began to decrease when per capita GDP was USD 10,000, similar to the environmental Kuznets curve [75]. The results of the third economic census in 2010 showed that the energy consumption intensity of the transportation, storage and postal industries was 1.35 tons of standard coal per CNY 10,000, which has surpassed other industries to become the sector with the highest energy consumption intensity. Therefore, in the Yangtze River Delta and Pearl River Delta regions where these industries are prominent, their energy intensity continues to rise and even grow faster. In contrast, the pressure on water consumption improves when the GDP reaches USD 7000–10,000 per capita in both China and its four megalopolises. However, the Yangtze River Delta, where high-tech industries are concentrated, is over-intensive in water consumption. Although the efficiency increased, much more water is consumed for economic growth than is saved. Its water stress is extremely higher than other megalopolises. In comparison, the intensity of water resources in the other three megalopolises is lower than or close to the national average due to different industrial structures, lack of water resources or water conservation concerns, etc.

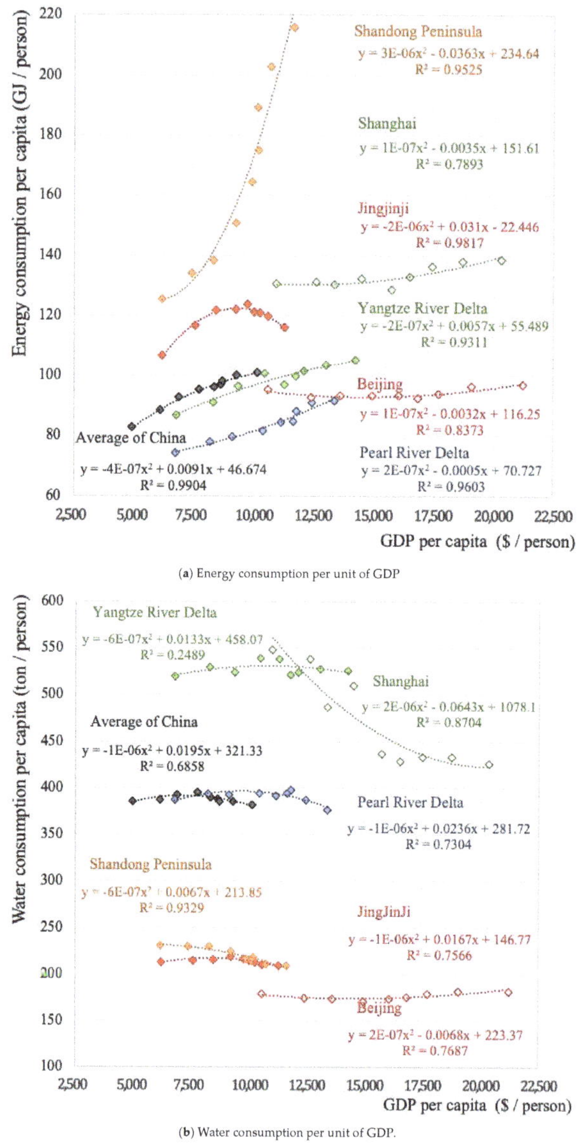

(a) Energy consumption per unit of GDP

(b) Water consumption per unit of GDP.

Figure 6. The trend of resource consumption in China (each point represents an average value of the variable for the cities in that group for each year in the data set from 2010 to 2018). (Data source: China Statistical Yearbook 2011–2019 [36]).

3.5. Environmental Pollution

The enjoyment of a healthy environment has been recognized as a human right by the UN General Assembly in the face of environmental pollution and climate change [76]. The Chinese government's policy to deal with water quality issues mainly relies on strengthening monitoring capabilities and enforcement mechanisms. As early as 1986, the "Regulations on Technical Policies for the Prevention and Control of Water Pollution" were promulgated. In addition, the "Water Pollution Law," distributed in 2008, tried to increase the effect of early legislation by increasing penalties [77,78]. When the GDP reaches about USD 10,000/person, wastewater discharge intensity begins to decline in most regions of China (Figure 7). Although the national average wastewater discharge intensity is low, the wastewater discharge intensity remains high in the Yangtze River Delta and the Pearl River Delta, especially in their large coastal cities. The JingJinJi and Shandong Peninsula regions with more inland cities have better sewage control than the national average. Since the Chinese government promulgated "Ten Articles on Air Quality" in 2013, the overall national air quality has been improved [79], especially the air pollution far lower than the domestic average in Yangtze River Delta and Pearl River Delta. Most cities have made active efforts to control emissions of harmful pollutants such as SO_2, NOx, soot and dust by upgrading detection and filtration technologies and using clean energy. In Yangtze River Delta and Pearl River Delta, the air pollution is far lower than the domestic average. As the curve shows in Figure 7, SO_2 emissions are in a rapid decline across urban agglomerations, with their peak emissions all occurring when the GDP reached about USD 7000/person. From the data released by NASA in 2018 [80], China remains the third largest emitter of SO_2 in the world, and environmental pollution still cannot be ignored.

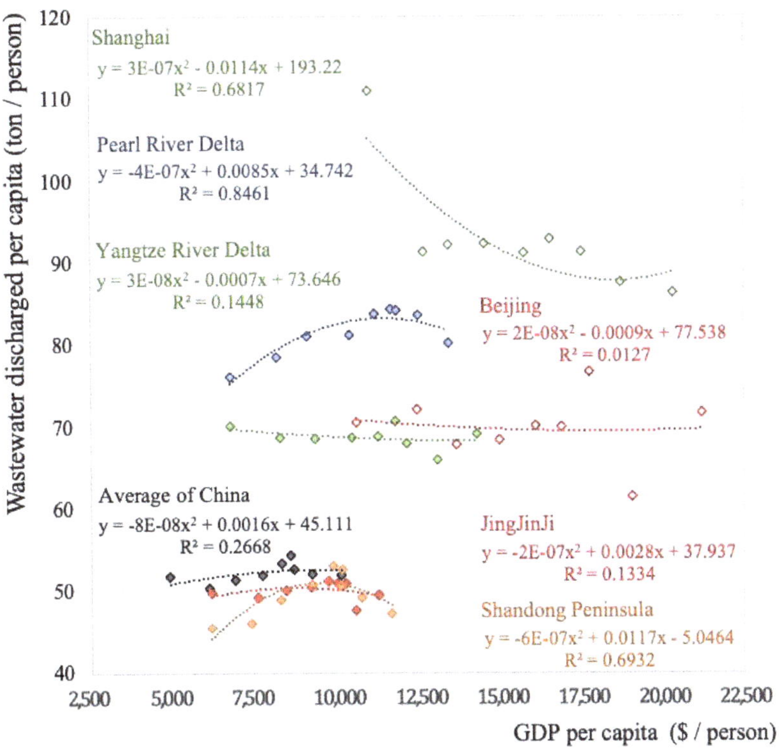

(**a**) Wastewater discharged per unit of GDP

Figure 7. *Cont.*

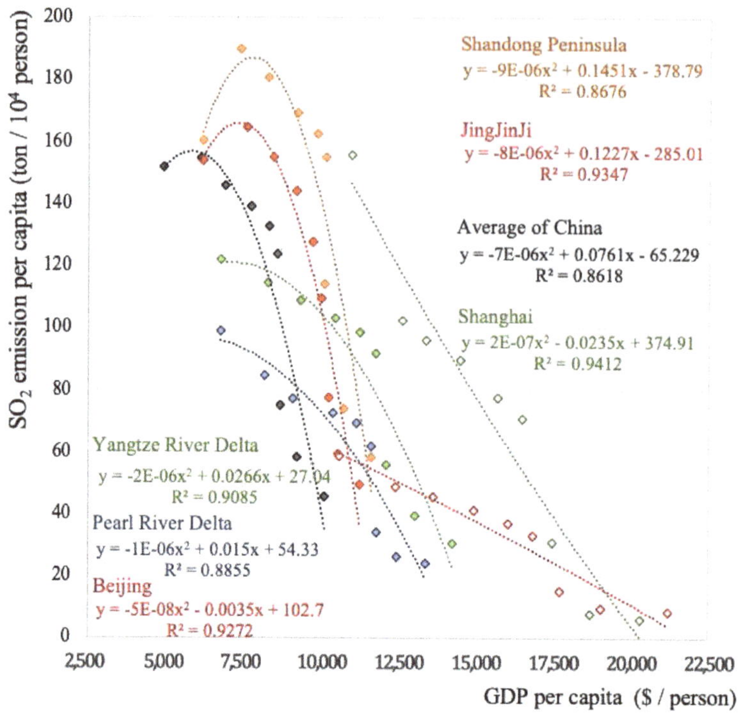

(b) SO_2 emission per unit of GDP

Figure 7. The trend of environmental pollution in China (each point represents an average value of the variable for the cities in that group for each year in the data set). (Data source: China Statistical Yearbook 2011–2019 [36]).

4. Comparative Assessment Results with Classification Analysis

4.1. Assessment Results Analysis among Four Megalopolises

Rapid urbanization makes rapid economic growth in emerging economies and cities play crucial roles in driving megalopolises' development. In terms of the quality of the built environment (Figure 8 and Table 3), all four megalopolises show continuously positive trends from 2010 to 2018 with a steady growth rate. The Qu score in Pearl River Delta was 0.182 in 2010 and 0.262 in 2018, always staying in the leading position. It is followed by the Yangtze River Delta, Shandong Peninsula, and JingJinJi. Although the ranking of the cities has not changed in 9 years, the growth rates of the Shandong Peninsula and the Yangtze River Delta are relatively fast, as seen from the slopes of their regression lines. Their compound annual growth rate (CAGR) reached 7.31% and 6.04%, respectively. The gap between the top three megalopolises is getting smaller, but the gap between them and JingJinJi (in fourth) is increasing. Geographically, the southern regions of China have generally prospered whereas the northern regions have lagged, and the gap between them will likely continue to increase in the coming years. This is due to the dynamic economic activity in the south, which has created a boost to urban development. However, despite their abundance of resources, the northern regions have been adversely affected by slower fixed investment, transformation, and upgradation. Therefore, it is still necessary to deal with horizontal disparities in domestic regions, as they have huge implications for sustainable growth and the corresponding policymaking.

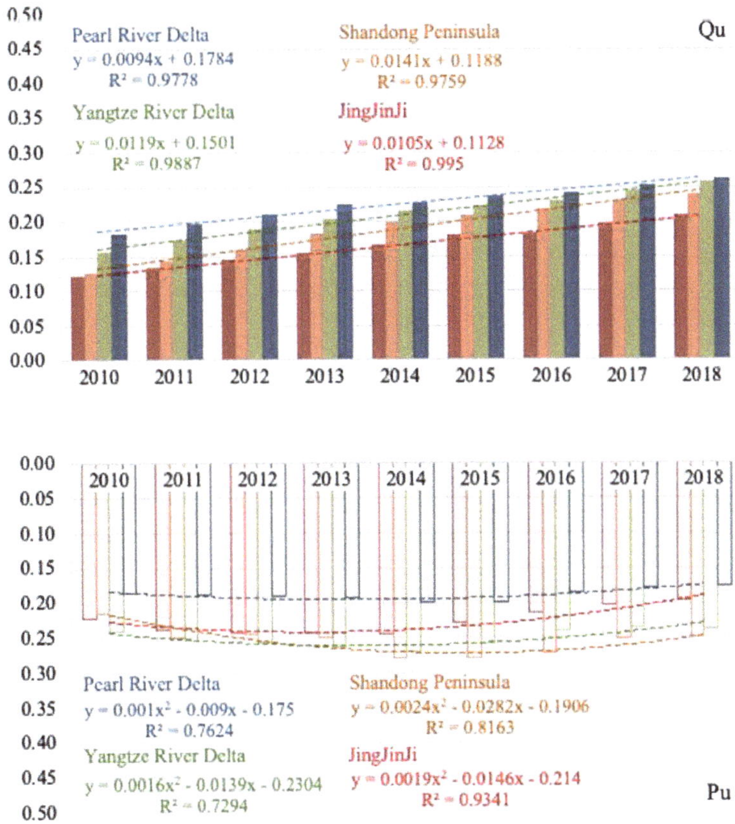

Figure 8. Data comparative analysis of Qu and Pu among four megalopolises.

Table 3. Data comparative analysis among four megalopolises.

	Yangtze River Delta	Pearl River Delta	Shandong Peninsula	JingJinJi
Qu (2010)	0.157	0.182	0.127	0.122
Ranking	2	1	3	4
Qu (2018)	0.257	0.262	0.239	0.208
Ranking	2	1	3	4
Mean	0.21	0.23	0.19	0.17
CAGR (%)	5.62	4.12	7.31	6.04
Pu (2010)	0.240	0.185	0.214	0.222
Ranking	1	4	3	2
Pu (2018)	0.238	0.176	0.250	0.195
Ranking	2	4	1	3
Mean	0.25	0.19	0.25	0.23
CAGR (%)	−0.08	−0.57	1.74	−1.40
Su (2010)	0.741	0.953	0.643	0.607
Ranking	2	1	3	4
Su (2018)	1.283	1.495	1.067	1.345
Ranking	3	1	4	2
Mean	0.97	1.19	0.82	0.89
CAGR (%)	6.29	5.12	5.79	9.24

Note: CAGR = Compound annual growth rate.

From the comparison of Pu (Figure 8 and Table 3), one can see that the current environmental pressure of the Pearl River Delta megalopolis is the smallest and is significantly better than other megalopolises, followed by JingJinJi, Yangtze River Delta and the Shandong Peninsula in 2018. The U-shaped regression characteristic indicates that the environmental pressures of the four megalopolises increase with time at the initial of the study period, but later these pressures diminish, starting just before 2016. Among them, the upward shift of the JingJinJi and Pearl River Delta regions occurred earlier, probably around 2012. The Yangtze River Delta and the Shandong Peninsula occur relatively late, resulting in a more significant gap between the two sets of regions. From the perspective of environmental pressure alleviation trends, JingJinJi has seen the fastest environmental improvement, with an average annual compound decline rate of 1.4%. Other regions are less than 1%. The regional cooperation in JingJinJi proved to be an efficient governance model for the protection and restoration of ecosystems and the prevention and control of pollution. However, the environmental load of the Shandong Peninsula, which has relied on heavy industry, has increased during this time period. The ranking of the Shandong Peninsula has changed from third to first in 9 years, indicating that its urban environmental problems have become increasingly prominent.

4.2. The Sustainability Performance Clustering Classification

According to sustainability performance analysis, the 91 sample cities are formed into 6 clusters by using the hierarchical clustering of Ward Linkage Method, which was originally presented in 1963 [81]. From the dendrogram, different characteristics of these six clusters of cities emerge, which we could summarize into six different cluster types (Table 4 and Figure 9). Furthermore, when considering the sustainability performance combined with the results in Figure 10, it can be seen that the six types are:

(1) Advanced positive type: there are seven cities in this type, including Shanghai, Beijing, Shenzhen, Guangzhou, Dongguan, Qingdao and Hangzhou. These cities have advanced economies with high urbanization and high-level administrative division. This cluster has the highest Qu points among all these six types, and the Pu points are relatively low. Therefore, the difference between Qu and Pu is significant and the sustainability performance is remarkable. The reason is that most of China's mega cities are included in this type, such as Beijing, Shanghai and Shenzhen. Their urban population, GDP and income are incredibly high compared to other cities. The control of SO_2 discharge, soot and dust discharge, water and energy consumption is done well, which makes those cities grow efficiently and sustainably. Such cities are on the right track and would likely continue to have high performance in the near future. Cities such as Beijing, Shanghai and Shenzhen are already recognized as international mega cities with world-leading infrastructure and outstanding urban attractiveness, and their aggregation attracts young talent which further improves the GDP and incomes, etc. These cities should be aware, however, of the additional challenges they are about to face, which need to be carefully handled. These challenges include the inadequacy of residential areas, green spaces and infrastructure. For cities such as Hangzhou, Qingdao and Dongguan, industrial upgrading will be a priority, as well as how to increase their GDP and income while effectively controlling their NOx emission and wastewater discharges.

(2) Coordinated positive type: there are 17 cites in this type, which can be represented by Wenzhou, Zhoushan, Shantou, Liu'an, Maoming, Heyuan and Weifang. These cities are in a well-coordinated but relatively low position in terms of urban quality and environment. It has the lowest Pu value among all six types, even though its Qu value is low, but the combined values are positive and the sustainability performance is remarkable. Most cities in this type are small, or are sub-cities to mega cities with quite a clear function. For this reason, there is not much pressure from political leaders, and the GDP value is low. However, the SO_2 discharge, soot and dust discharge are highly controlled, making their sustainability performance outstanding. Most of these

cities are non-heavy industry based, for example the modern agriculture industry in Weifang or the world-famous small commodity trading market for Wenzhou, etc. Their reasonable industrial structures make them even more sustainable than most of those larger cities. Despite their low population, GDP and income, they will maintain a high performance in sustainable development if they remain in control of environmental pressures and manage their industrial structure.

(3) Smart positive type: there are 31 cities in this type (the largest number of cities of any type). Some of the representative cities are Haozhou, Jinhua, Shaoxing, Ningbo, Weihai, Tianjin, Jiaxing, Tai'an, etc. Both Qu and Pu values for this type are at a moderate level, but there are subtle differences between the two dimensions in this type of city, meaning that the sustainable performance of these cities is still in a positive state of development. These cities are steadily improving, with joint implementation of urban construction and environmental control strategies. Most of them are the middle cities of China. Although their statistics values vary, the urban population, GDP, services, and incomes are generally high enough, whereas the SO_2 discharge, soot and dust, energy consumption, etc., seem to be well controlled. Their status, however, is still sensitive. Compared to cities in Type I, their Qu values are much lower, and Pu are relatively high. Therefore, cities such as Tai'an and Huangshan should make full use of their tourism resources to develop non-heavy industries and reduce environmental pressures. Some other cities need to focus on controlling environmental pressures, such as Xuzhou, Foshan, Ningbo and Linyi. Other heavy industries, such as Linyi's building materials and panels industry, need to be upgraded to more green industries such as Ningbo's petroleum processing and coking industries in order to maintain a more positive sustainability performance.

(4) Transitional negative type: there are 10 cities in this type, include Nanjing, Suzhou, Changzhou, Hefei, Jinan, Wuxi, Zibo, and Dongying, etc. These cities are in the process of industrial transformation and upgrading but are relatively developed. They have very high Qu values, but at the same time have a significantly high value of Pu. The absolute value of Qu is not enough to cover the Pu value, which makes its sustainability performance low. Most of the cities in this type are heavy industry cities, such as Suzhou, Zibo, Dongying, etc. Although their positive efforts to contribute to the city's GDP have resulted in a relatively good population, income and road infrastructure, the problems of environmental pollution caused by heavy industry are significant. This is especially true for excessive emissions of Nox and excessive energy consumption, which have resulted in these cities having higher Pu values than Qu values. For this type of city, a heavy industry is a double-edged sword for their sustainability performance. On the one hand, the industries have created more income and jobs and given the cities a higher Qu value. On the other hand, they have had a huge negative impact on the environment. Based on this, their priority should be to focus on the control of environmental pressure. For example, Zibo and Dongying require comprehensive industrial restructuring and upgrading, including reduction in the traditional fossil energy consumption and actively laying out new energy sources such as wind and solar power, etc.

(5) Lagging negative type: there are 19 cities in this type, including Shaoguan, Shijiazhuang, Huaibei, Heze, Xingtai, etc. The sustainability performance of this type of city is relevantly low, with the lowest Qu values and slightly higher Pu values than other types. These cities are mostly small or middle-sized cities and have for a long time lagged behind in their environmental governance capacity. They particularly lack urban attractiveness and infrastructure development, although the cities are of medium population, GDP and income. These cities also have high environmental pressures from Nox emissions and energy consumption, causing their future sustainable development to be of concern. As a result, cities with medium amounts of urban development have faced the same situation as some negative mega cities. They have similar Qu values to some positive cities, but it is not well controlled enough to

combat the high Pu values that make their sustainability performance overall negative. Therefore, the suggested solution would be to make proper policy for their city's industrial development, especially the heavy industry-based cities such as Shijiazhuang and Handan. It is heartening to see that some cities are embarking on more green industries, such as the city of Dezhou which has been working hard to develop the largest solar industry in recent years. This is the beginning of a good trend that will help turn negative performance into positive performance in the near future.

(6) Purely negative type: there are seven cities in this type including Tongling, Tangshan, Ma'anshan, Rizhao, Binzhou, Yancheng and Chengde. These cities have the highest Pu values and relatively low Qu values among all six types of cities, showing a remarkable state of unsustainable development. It can be found that most of these cities are heavy industrial cities with high energy consumption and high pollution. To produce higher GDP, urban population and income, they consume overly large amounts of energy, causing much higher Nox discharge and many other environmental problems. For the cities in this type, most of them rely on heavy-industry and natural resource consumption, such as Tongling, Tangshan, Ma'anshan and Binzhou. The coal industry and petroleum industry are the mainstay of their economies. However, they should be aware that once urban resources are depleted, they cannot be repaired, even at great cost. Therefore, a strict policy of energy saving, consumption reduction and tighter control of environmental pressures will work best for these cities. This will take sustained patience over several years to gradually transform into positive sustainability performance.

Table 4. List of cities by sustainability performance clustering classification.

No.	City Classification	Common Characteristics Description	Level of Urban Sustainability	List of Cities	N
I	**Advanced positive type**	Advanced economic cities with high urbanization and high-level administrative division	High sustainability with high quality	Shanghai, Beijing, Shenzhen, Guangzhou, Dongguan, Qingdao, Hangzhou	7
II	**Coordinated positive type**	Urban quality and environment are in a well-coordinated but relatively backward cities	Moderate sustainability with low pressure	Wenzhou, Zhoushan, Jieyang, Yunfu, Zhanjiang, Chaozhou, Fuyang, Meizhou, Suqian, Weifang, Heyuan, Maoming, Liu'an, Suuzhou, Hengshui, Baoding, Shantou	17
III	**Smart positive type**	Steady improvement cities with joint implementation of urban construction and environmental control strategies.	Basic sustainability with low quality	Tianjin, Xuzhou, Foshan, Ningbo, Linyi, Langfang, Tai'an, Jiaxing, Zhongshan, Nantong, Yangzhou, Taizhou, Zaozhuang, Lishui, Huai'an, Chuzhou, Shanwei, Anqing, Zhaoqing, Yangliang, Weihai, Huangshan, Zhuhai, Huizhou, Lianyungang, Yantai, Jiangmen, Taizhou, Shaoxing, Jinhua, Haozhou	31
IV	**Transitional negative type**	In the process of industrial transformation and upgrading but relatively developed cities	Slight unsustainability with high quality	Nanjing, Suzhou, Hefei, Jinan, Wuxi, Changzhou, Zibo, Dongying, Huzhou, Zhenjiang	10
V	**Lagging negative type**	Long-term multi-dimensional lagged behind cities with inadequate environmental governance capacity	Intermediate unsustainability with low quality	Shijiazhuang, Handan, Chizhou, Shaoguan, Huaibei, Huainan, Wuhu, Qinhuangdao, Quzhou, Cangzhou, Liaocheng, Heze, Bengbu, Dezhou, Qingyuan, Jining, Xuancheng, Zhangjiakou, Xingtai	19
VI	**Purely negative type**	Heavy industrial cities with high energy consumption and high pollution	Severe unsustainability with high pressure	Tangshan, Tongling, Ma'anshan, Yancheng, Rizhao, Binzhou, Chengde	7

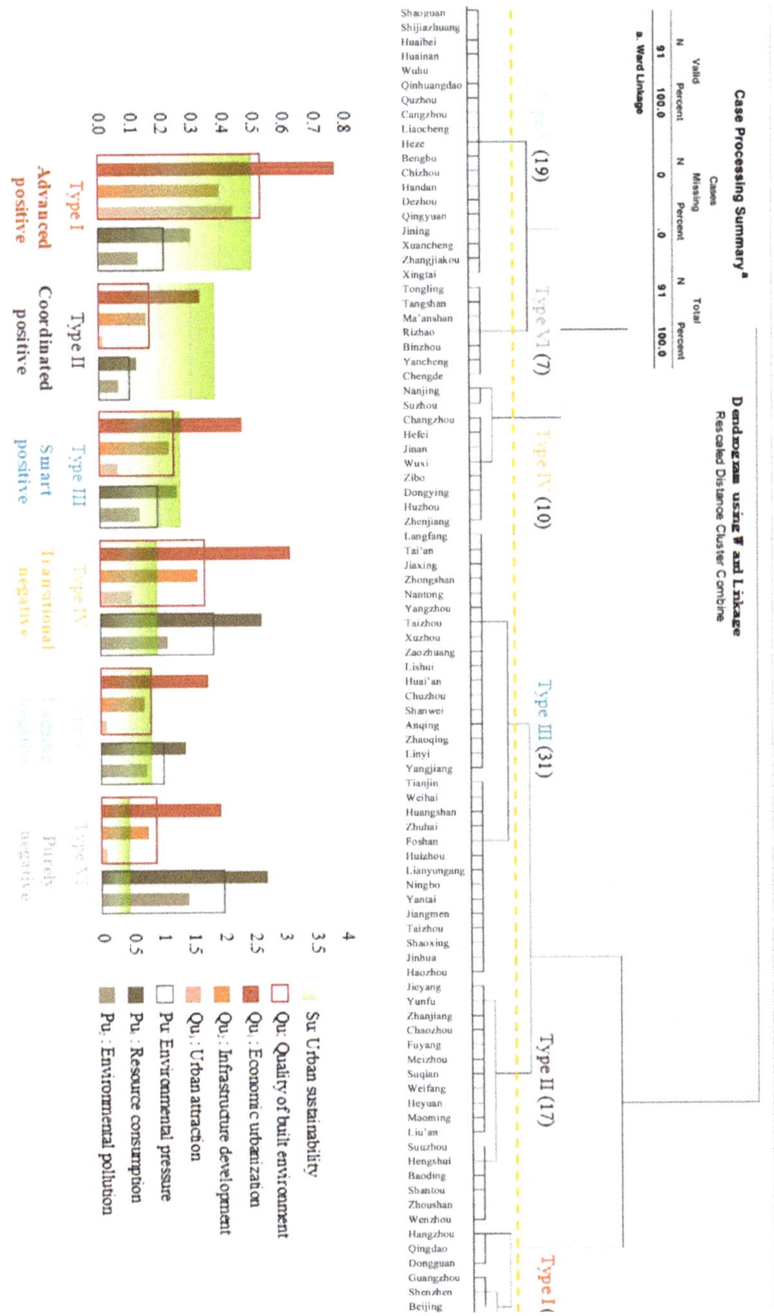

Figure 9. Result of hierarchical Clustering and comparison of the mean evaluation scores of each indicator categories among six types of cities in 2018.

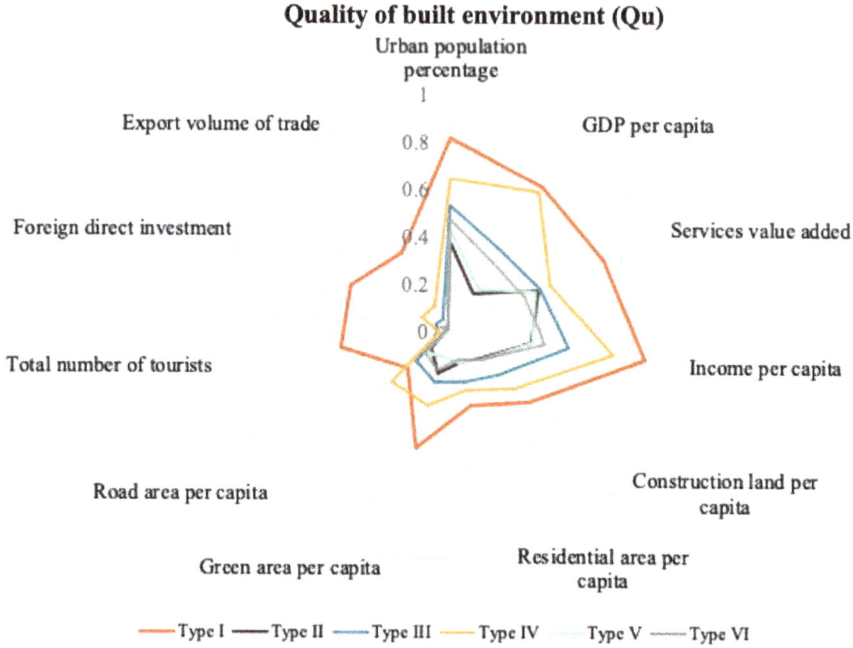

(**a**) The mean evaluation scores of Qu among six types of cities in 2018

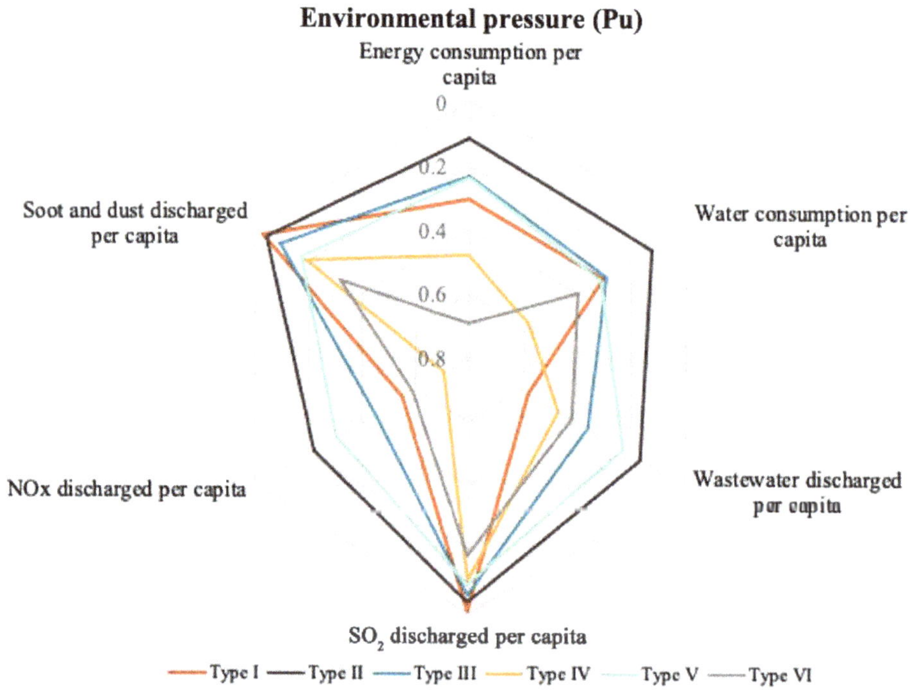

(**b**) The mean evaluation scores of Pu among six types of cities in 2018

Figure 10. Comparison of the mean evaluation scores among six types of cities in 2018.

5. Conclusions

By comparing the sustainability performance results of four major Chinese megacities, the following main conclusions are obtained:

From the comparative analysis of the data from 2010 to 2018, (1) all four megalopolises show continuously positive trends in terms of the quality of the built environment with a steady growth rate. The Pearl River Delta has the highest Qu score, followed by the Yangtze River Delta, Shandong Peninsula and JingJinJi. (2) Environmental pressure first experienced an upward and then a downward trend. The Pearl River Delta megalopolis has the least environmental pressure, followed by JingJinJi, Yangtze River Delta and the Shandong Peninsula in 2018. Comprehensively, (3) the Pearl River Delta has the highest level of sustainable development, ranking at the top for nine consecutive years. It is followed by JingJinJi, Yangtze River Delta and Shandong Peninsula. Additionally, the JingJinJi megalopolis has the fastest improvement of urban sustainability. In general, (1) the sustainability of each megalopolis is showing a positive trend of improvement (2) but due to the different speed of improvement, the gap between the urban agglomerations could potentially widen, which may cause imbalanced development of the entire country and is not conducive to the coordinated development in the future.

Based on the sustainability performance clustering, cities can be classified into six types. The cities of advanced positive, coordinated positive and smart positive type have an overall positive sustainability performance, whereas the remaining three types of transitional negative, lagging negative and purely negative type have negative sustainability performance with different degrees of environmental pressure. The corresponding characteristics and implications of the different types of cities mentioned above provide a reference for cities in similar situations. These cities can be informed by indicators that they need to pay more attention to in order to avoid unsustainable risks.

There is an urgent need to further expand the scope of urban sustainable development research to gradually cover more regions all over the world in the future, for example, by establishing interconnections between cities based on their complementarity. A visual diagnostic and evaluation system can also be established for each city to facilitate real-time self-examination and provide decision aids for city policymakers. In the next stage, as developing countries are prepared to enter a new status of deceleration stage or terminal stage, the evaluation systems need to be optimized and enriched as well. All of the above are prospects for our future research.

Author Contributions: Conceptualization, X.F. and W.G.; methodology, X.F. and X.S.; software, X.S. and T.K.P.; validation, X.F. and W.G.; formal analysis, T.K.P.; investigation, X.F. and X.S.; resources, X.S.; data curation, T.K.P.; writing—original draft preparation, X.F.; visualization, T.K.P.; supervision, W.G. All authors have read and agreed to the published version of the manuscript.

Funding: This research is funded by the Key Technology Research and Development Program of Shandong, grant number 2019GSF110004.

Acknowledgments: This study was supported by the Innovation Institute for Sustainable Maritime Architecture Research and Technology (iSMART) of Qingdao University of Technology. We would like to acknowledge and express our gratitude to all the people who helped complete this research and to the editors and reviewers for their suggestions for improving the manuscript.

Conflicts of Interest: The authors declare no conflict of interest.

References

1. Behera, S.R.; Dash, D.P. The effect of urbanization, energy consumption, and foreign direct investment on the carbon dioxide emission in the SSEA (South and Southeast Asian) region. *Renew. Sustain. Energy Rev.* **2017**, *70*, 96–106. [CrossRef]
2. United Nations | Department of Economic and Social Affairs Publications. 2018 Revision of World Urbanization Prospects. 2018. Available online: https://desapublications.un.org/publications/2018-revision-world-urbanization-prospects (accessed on 30 July 2022).
3. Arfanuzzaman, M.; Dahiya, B. Sustainable urbanization in Southeast Asia and beyond: Challenges of population growth, land use change, and environmental health. *Growth Change* **2019**, *50*, 725–744. [CrossRef]

4. UN. *UN-Habitat Sustainable Urbanization in Asia: A Sourcebook for Local Governments*; UN: Nairobi, Kenya, 2012.
5. Higo, M.; Klassen, T.R. Retirement in Japan and Korea in an era of rapid population aging. In *Retirement in Japan and South Korea: The Past, the Present and the Future of Mandatory Retirement*; Routledge: England, UK, 2015; pp. 1–29.
6. Kajimura, M. Changes in the Demographic Structure and Economic Growth in East and Southeast Asia. ISEAS Yusof Ishak Institute. 2020. Available online: http://hdl.handle.net/11540/12284 (accessed on 30 July 2022).
7. Nogimori, M.; Seki, S.; Kumagai, S. *The Uneven Economic Recovery in Asia*; Japan Research Institute: Tokyo, Japan, 2020; Volume 3.
8. Yang, Z.; Dunford, M. City shrinkage in China: Scalar processes of urban and hukou population losses. *Reg. Stud.* **2018**, *52*, 1111–1121. [CrossRef]
9. Dunford, M.; Weidong, L. *The Geographical Transformation of China*; Routledge: England, UK, 2014; ISBN 131761478X.
10. Li, H.; Mykhnenko, V. Urban shrinkage with Chinese characteristics. *Geogr. J.* **2018**, *184*, 398–412. [CrossRef]
11. Yang, Z.; Dunford, M. Cluster evolution and urban industrial dynamics in the transition from a planned to a socialist market economy: The case of Beijing. *Spat. Econ. Anal.* **2017**, *12*, 50–71. [CrossRef]
12. Liu, Z.; Liu, S.; Song, Y. Understanding urban shrinkage in China: Developing a multi-dimensional conceptual model and conducting empirical examination from 2000 to 2010. *Habitat Int.* **2020**, *104*, 102256. [CrossRef]
13. Gonzalez, E.T.; Mendoza, M.L. *Governance in Southeast Asia: Issues and Options*; PIDS Discussion Paper Series; Philippine Institute for Development Studies (PIDS): Makati City, Philippines, 2002.
14. Henderson, J.V.; Quigley, J.; Lim, E. *Urbanization in China: Policy Issues and Options*; Brown University: Providence, RI, USA, 2009; *unpublished manuscript*.
15. Kirkby, R.J.R. *Urbanization in China: Town and Country in a Developing Economy 1949–2000 AD*; Routledge: England, UK, 2018; Volume 3, ISBN 1351171623.
16. Lin, J.Y.; Liu, Z. Fiscal decentralization and economic growth in China. *Econ. Dev. Cult. Change* **2000**, *49*, 1–21. [CrossRef]
17. Zhang, L.; LeGates, R.; Zhao, M. *Understanding China's Urbanization: The Great Demographic, Spatial, Economic, and Social Transformation*; Edward Elgar Publishing: Cheltenham, UK, 2016; ISBN 1783474742.
18. Chaolin, G.U.; Liya, W.U.; Cook, I. Progress in research on Chinese urbanization. *Front. Archit. Res.* **2012**, *1*, 101–149. [CrossRef]
19. Henderson, J.V. Cities and development. *J. Reg. Sci.* **2010**, *50*, 515–540. [CrossRef]
20. Chan, K.W. Fundamentals of China's urbanization and policy. *China Rev.* **2010**, 63–93.
21. Fang, X.; Shi, X.; Phillips, T.K.; Du, P.; Gao, W. The coupling coordinated development of urban environment towards sustainable urbanization: An empirical study of Shandong Peninsula, China. *Ecol. Indic.* **2021**, *129*, 107864. [CrossRef]
22. Li, S.; Xiao, W.; Zhao, Y.; Xu, J.; Da, H.; Lv, X. Quantitative analysis of the ecological security pattern for regional sustainable development: Case study of Chaohu Basin in Eastern China. *J. Urban Plan. Dev.* **2019**, *145*, 4019009. [CrossRef]
23. Yi, S. *Research on Sustainable Development Strategy and Evaluation Model of the of Urban Wastewater Treatment System in China*; Shanghai Institute of Technology: Shanghai, China, 2010; pp. 4918–4922.
24. Cheshmehzangi, A.; Zhu, Y.; Li, B. Application of environmental performance analysis for urban design with Computational Fluid Dynamics (CFD) and EcoTect tools: The case of Cao Fei Dian eco-city, China. *Int. J. Sustain. Built Environ.* **2017**, *6*, 102–112. [CrossRef]
25. Fang, X.; Shi, X.; Gao, W. Measuring urban sustainability from the quality of the built environment and pressure on the natural environment in China: A case study of the Shandong Peninsula region. *J. Clean. Prod.* **2020**, *289*, 125145. [CrossRef]
26. Fang, C.; Li, G.; Wang, S. Changing and Differentiated Urban Landscape in China: Spatiotemporal Patterns and Driving Forces. *Environ. Sci. Technol.* **2016**, *50*, 2217–2227. [CrossRef] [PubMed]
27. Harrison, J.; Gu, H. Planning megaregional futures: Spatial imaginaries and megaregion formation in China. *Reg. Stud.* **2021**, *55*, 77–89. [CrossRef]
28. Meng, L.; Crijns-Graus, W.H.J.; Worrell, E.; Huang, B. Impacts of booming economic growth and urbanization on carbon dioxide emissions in Chinese megalopolises over 1985–2010: An index decomposition analysis. *Energy Effic.* **2018**, *11*, 203–223. [CrossRef]
29. Yu, J.; Zhou, K.; Yang, S. Land use efficiency and influencing factors of urban agglomerations in China. *Land Use Policy* **2019**, *88*, 104143. [CrossRef]
30. Wan, H.; Rieckmann, J.M.; Hou, G. Cluster of Financial Industry of Urban Agglomeration in China—A Panel Data Study. In Proceedings of the 2018 5th International Conference on Industrial Economics System and Industrial Security Engineering (IEIS), Beijing, China, 3–6 August 2018; pp. 1–9.
31. Guo, L.; Qu, Y.; Wu, C.; Wang, X. Identifying a pathway towards green growth of Chinese industrial regions based on a system dynamics approach. *Resour. Conserv. Recycl.* **2018**, *128*, 143–154. [CrossRef]
32. Guan, X.; Wei, H.; Lu, S.; Dai, Q.; Su, H. Assessment on the urbanization strategy in China: Achievements, challenges and reflections. *Habitat Int.* **2018**, *71*, 97–109. [CrossRef]
33. Fang, C.; Yu, D. *China's Urban Agglomerations*; Springer: Berlin/Heidelberg, Germany, 2020; ISBN 9811515514.
34. Cohen, B. Urbanization in developing countries: Current trends, future projections, and key challenges for sustainability. *Technol. Soc.* **2006**, *28*, 63–80. [CrossRef]
35. Fang, L.; Song, H. Notice of Retraction: Research on Differences of Capacity for Sustainable Development of Shandong Peninsula City Groups. In Proceedings of the 2011 International Conference on E-Business and E-Government (ICEE), Shanghai, China, 6–8 May 2011; pp. 1–3.

36. National Bureau of Statistics China Statistical Yearbook. Available online: http://www.stats.gov.cn/tjsj/ (accessed on 30 July 2022).
37. Borza, M. The Connection between Efficiency and Sustainability—A Theoretical Approach. *Procedia Econ. Financ.* **2014**, *15*, 1355–1363. [CrossRef]
38. Leibeistein, H. *General X-Efficiency Theory & Economic Development*; Oxford University Press: Oxford, UK, 1978; ISBN 019502379X.
39. State Environmental Protection Administration The Ecological Counties, Ecological Cities Construction Evaluation System (2003 Trial version). *Environmental Protection*; State Environmental Protection Administration The Ecological Counties. 2003. Available online: https://www.epd.gov.hk/epd/english/links/mainchina/link_mainchina.html (accessed on 30 July 2022).
40. Zhang, W. The concept and discussion of China's Livable Cities Scientific Evaluation Standards. *Urban Plan. Forum* **2007**, *3*, 30–34.
41. Shang, l.; Su, X.; Wang, M.; Wang, M. Research Progress on Urban Low-Carbon Index Evaluation. *Urban Rural Plan.* **2018**, *1*, 78–83.
42. Ministry of Housing and Urban-Rural Development; Ministry of Finance; National Development and Reform Commission. *Green and Low-carbon Cities Evaluation Indicators*; Ministry of Housing and Urban-Rural Development; Ministry of Finance; National Development and Reform Commission: Beijing, China, 2011.
43. Ministry of Housing and Urban-Rural Development; Ministry of Finance; National Development and Reform Commission. *National Environmental Protection Model City Assessment Index and Implementation Rules*; Ministry of Housing and Urban-Rural Development; Ministry of Finance; National Development and Reform Commission: Beijing, China, 2011.
44. Ministry of Housing and Urban-Rural Development; Ministry of Finance; National Development and Reform Commission. *National Ecological Civilization Pilot Policy Evaluation System*; Ministry of Housing and Urban-Rural Development; Ministry of Finance; National Development and Reform Commission: Beijing, China, 2013.
45. Ministry of Housing and Urban-Rural Development; Ministry of Finance; National Development and Reform Commission. *Technical Criterion for Ecosystem Status Evaluation*; Ministry of Housing and Urban-Rural Development; Ministry of Finance; National Development and Reform Commission: Beijing, China, 2015.
46. National Development and Reform Commission of PRC. *Evaluation System for the Construction of National Ecological Civilization Model City (Trial Version)*; National Development and Reform Commission of PRC: Beijing, China, 2016.
47. National Development and Reform Commission of PRC. *China Green Development Index System*; National Development and Reform Commission of PRC: Beijing, China, 2016.
48. National Development and Reform Commission of PRC. *Evaluation System for Ecological Civilization Construction*; National Development and Reform Commission of PRC: Beijing, China, 2016.
49. Zong, J.; Li, Y.; Lin, L.; Bao, W. Evaluation indicators for green cities. *IOP Conf. Ser. Earth Environ. Sci.* **2017**, *267*, 052009. [CrossRef]
50. Ministry of Housing and Urban-Rural Development; Ministry of Finance; National Development and Reform Commission. *Evaluation System of the Construction Target of the National Ecological Civilization Pilot Zone*; Ministry of Housing and Urban-Rural Development; Ministry of Finance; National Development and Reform Commission: Beijing, China, 2018.
51. Standardization Administration of China. *Indicators for City Services and Quality of Life*; Standardization Administration of China: Beijing, China, 2019.
52. Ministry of Housing and Urban-Rural Development; Ministry of Finance; National Development and Reform Commission. *New-Type Urbanization—Evaluation Index System of Quality City (Trial Version)*; Ministry of Housing and Urban-Rural Development; Ministry of Finance; National Development and Reform Commission: Beijing, China, 2020.
53. Wackernagel, M.; Rees, W. *Our Ecological Footprint: Reducing Human Impact on the Earth*; New Society Publishers: Gabriola Island, BC, Canada, 1998; Volume 9, ISBN 086571312X.
54. Huang, L.; Wu, J.; Yan, L. Defining and measuring urban sustainability: A review of indicators. *Landsc. Ecol.* **2015**, *30*, 1175–1193. [CrossRef]
55. Zhang, W.L. The application of principal component analysis in SPSS. *Mark. Res.* **2005**, *12*, 31–34.
56. Guo, H.; Yang, C.; Liu, X.; Li, Y.; Meng, Q. Simulation evaluation of urban low-carbon competitiveness of cities within Wuhan city circle in China. *Sustain. Cities Soc.* **2018**, *42*, 688–701. [CrossRef]
57. Adler, N.; Golany, B. Including principal component weights to improve discrimination in data envelopment analysis. *J. Oper. Res. Soc.* **2002**, *53*, 985–991. [CrossRef]
58. Dodge, Y.; Cox, D.; Commenges, D. *The Oxford Dictionary of Statistical Terms*; Oxford University Press on Demand: Oxford, UK, 2006; ISBN 0199206139.
59. Johnson, S.C. Hierarchical Clustering Schemes. *Psychometrika* **1967**, *32*, 241–254. [CrossRef]
60. Murtagh, F.; Contreras, P. Algorithms for hierarchical clustering: An overview. *Wiley Interdiscip. Rev. Data Min. Knowl. Discov.* **2012**, *2*, 86–97. [CrossRef]
61. Romesburg, C. *Cluster Analysis for Researchers*; Lulu Press: Morrisville, NC, USA, 2004; ISBN 1411606175.
62. Kaufman, L.; Rousseeuw, P.J. *Finding groups in Data: An Introduction to Cluster Analysis*; John Wiley & Sons: Hoboken, NJ, USA, 2009; Volume 344, ISBN 0470317485.
63. Gower, J.C. A comparison of some methods of cluster analysis. *Biometrics* **1967**, *23*, 623–637. [CrossRef] [PubMed]
64. Meng, W.; Hu, B.; He, M.; Liu, B.; Mo, X.; Li, H.; Wang, Z.; Zhang, Y. Temporal-spatial variations and driving factors analysis of coastal reclamation in China. *Estuar. Coast. Shelf Sci.* **2017**, *191*, 39–49. [CrossRef]

65. Fang, C.; Zhao, S. A comparative study of spatiotemporal patterns of urban expansion in six major cities of the Yangtze River Delta from 1980 to 2015. *Ecosyst. Health Sustain.* **2018**, *4*, 95–114. [CrossRef]
66. Shi, B.; Yang, H.; Wang, J.; Zhao, J. City green economy evaluation: Empirical evidence from 15 sub-provincial cities in China. *Sustainability* **2016**, *8*, 551. [CrossRef]
67. Wilkins, K.; Zurawski, A. Infrastructure Investment in China | Bulletin–June Quarter 2014. *Bulletin* **2014**. Available online: https://www.rba.gov.au/publications/bulletin/2014/jun/bu-0614-4a.html (accessed on 30 July 2022).
68. Ellsworth-Krebs, K. Implications of declining household sizes and expectations of home comfort for domestic energy demand. *Nat. Energy* **2020**, *5*, 20–25. [CrossRef]
69. Qian, Y.; Zhou, W.; Li, W.; Han, L. Understanding the dynamic of greenspace in the urbanized area of Beijing based on high resolution satellite images. *Urban For. Urban Green.* **2015**, *14*, 39–47. [CrossRef]
70. Wang, X.-J. Analysis of problems in urban green space system planning in China. *J. For. Res.* **2009**, *20*, 79–82. [CrossRef]
71. Zhou, W.; Wang, J.; Qian, Y.; Pickett, S.T.A.; Li, W.; Han, L. The rapid but "invisible" changes in urban greenspace: A comparative study of nine Chinese cities. *Sci. Total Environ.* **2018**, *627*, 1572–1584. [CrossRef]
72. Contractor, F.J.; Dangol, R.; Nuruzzaman, N.; Raghunath, S. How do country regulations and business environment impact foreign direct investment (FDI) inflows? *Int. Bus. Rev.* **2020**, *29*, 101640. [CrossRef]
73. Yeung, G. Foreign direct investment and investment environment in Dongguan Municipality of Southern China. *J. Contemp. China* **2001**, *10*, 125–154. [CrossRef]
74. Li, X.; Lu, Y.; Huang, R. Whether foreign direct investment can promote high-quality economic development under environmental regulation: Evidence from the Yangtze River Economic Belt, China. *Environ. Sci. Pollut. Res.* **2021**, *28*, 21674–21683. [CrossRef] [PubMed]
75. Li, T.; Wang, Y.; Zhao, D. Environmental Kuznets curve in China: New evidence from dynamic panel analysis. *Energy Policy* **2016**, *91*, 138–147. [CrossRef]
76. Babbar, K.; Martin, J.; Ruiz, J.; Parray, A.A.; Sommer, M. Menstrual health is a public health and human rights issue. *Lancet Public Health* **2022**, *7*, e10–e11. [CrossRef]
77. Swanson, T.; Lin, T. *Economic Growth and Environmental Regulation: China's Path to a Brighter Future*; Routledge: England, UK, 2009; Volume 20, ISBN 1135247471.
78. Liu, J.; Xue, J.; Yang, L.; Shi, B. Enhancing green public procurement practices in local governments: Chinese evidence based on a new research framework. *J. Clean. Prod.* **2019**, *211*, 842–854. [CrossRef]
79. Yang, J.; Ma, J.; Sun, Q.; Han, C.; Guo, Y.; Li, M. Health benefits by attaining the new WHO air quality guideline targets in China: A nationwide analysis. *Environ. Pollut.* **2022**, *308*, 119694. [CrossRef] [PubMed]
80. Dahiya, S.; Myllyvirta, L. *Global SO 2 Emission Hotspot Database*; Greenpeace Environment Trust: Beijing, China, 2019; pp. 1–3.
81. Ward, J.H. Hierarchical Grouping to Optimize an Objective Function. *J. Am. Stat. Assoc.* **1963**, *58*, 236–244. [CrossRef]

buildings

MDPI

Article

Transit-Oriented Development in New Towns: Identifying Its Association with Urban Function in Shanghai, China

Lingzhu Zhang [1,2], Peng Hou [1] and Dan Qiang [1,2,*]

[1] College of Architecture and Urban Planning, Tongji University, Shanghai 200092, China
[2] Key Laboratory of Ecology and Energy-Saving Study of Dense Habitat, Tongji University, Shanghai 200092, China
* Correspondence: 1932131@tongji.edu.cn

Abstract: Transit-oriented development (TOD) construction is considered critical for economic growth and the population's daily well-being in suburban development. However, there are few empirical evaluations of TOD performance and typology in suburban areas of high-density cities. In this study, we selected 23 metro stations in the five new towns in Shanghai as research objects to understand their TOD characteristics. By proposing a data-driven framework built on points of interest (PoIs) to characterize urban functions of metro stations in new towns, four thematic topic functions were extracted by implementing Latent Dirichlet Allocation (LDA) topic modeling. Five types of stations were revealed through a hierarchical cluster analysis based on their main functions. Then, an extended "Node-Place" model with a third "design" dimension was applied to classify TOD typologies. After establishing an evaluation framework by calculating the results of 15 indicators, five TOD topologies were identified through hierarchical cluster analysis. In addition, results from the ANOVA analysis showed that the classification according to thematic topics changes according to the "place" dimension indicators. Ultimately, the identified urban thematic function types and TOD types provide a useful tool for planners and governors to diagnose common problems and design targeted strategies in new towns.

Keywords: transit-oriented development (TOD); new town; urban function; multi-sourced urban data; metro station

1. Introduction

Chinese cities have experienced unprecedented suburban growth since 2000. New town development has been adopted as an important means to optimize the urban spatial structure of large cities and cope with urban growth. In Shanghai, there is a decreasing difference in population density between the inner city and the suburbs, and a growing trend in the multi-center population patterns [1]. Although the suburban new towns in Shanghai have become the main clusters for economic activities, the scope of employment in each new town is still concentrated, and the single-centered expansion remains the city's dominant spatial pattern [2]. Accordingly, rail transit has been adopted as an instrument for financing suburban development to facilitate growth, and a transit-oriented development (TOD) strategy has been proposed as an efficient approach to concentrate jobs, housing, and services around metro stations [3].

In Shanghai, the government has promoted the development of new towns to alleviate the pressure on the urban core. The strategy evolved from the "One City, Nine Towns Development Plan" of 2001 to the development of the "Seven New Cities" during the 12th Five-year Plan and the plan to build the "Five New Cities" in suburban Jiading, Qingpu, Songjiang, and Fengxian districts as well as Nanhui in the Pudong New Area during the 14th Five-Year Plan (2021–2025). The development of Shanghai's new towns is closely related to its urban railway development. In 2007, Songjiang New City was connected to the central urban area by metro line 9, becoming the first of the five new towns to operationalize the metro. Jiading was connected to Central Shanghai by Metro line 11 at the end of 2009. Metro line 9 benefited Songjiang, and line 11 promoted Jiading and Lingang [4]. In Shanghai's strategic plan for future development, the "Five New Cities" are the top priority. In addition, the TOD community and 15-min community life circles are expected to be the focus of metro construction in new towns with major developments planned for rail transit projects [5]. By 2035, each new town with over 100,000 residents is expected to have its metro stations. The catchment areas of the metro stations with a radius of 600 m covers 30% of the area of the new city, and the resident population and employment coverage ratios within a 600 m radius of the metro station in five new towns will reach 40%.

In this context, it is imperative to examine the TOD performance of metro station areas in new towns and to discuss how TOD should be adapted to suburban Shanghai. The rapid growth of multi-source geospatial big data and new technologies provides novel analysis methods for quantifying the characteristics of transit station areas [6–10]. Nevertheless, previous studies using non-traditional data have seldom investigated the existing urban functions and how they interact with built environments to shape vibrant metro stations areas and achieve their planning objectives [7]. By 2035, each new town is expected to have over one million people, which is comparable to most second-and third-tier cities in China. From this perspective, the TOD research on new towns is more universal than that on the metropolis. It is imperative to examine the TOD performance of metro station areas for new towns and to discuss how TOD should be adapted to suburban Shanghai. However, few such studies have examined TOD performance and typology for suburban cities. Accordingly, we attempt to address existing research gaps by exploring the following research questions:

By identifying the most prototypical thematic topics of urban functions for catchment areas, how do such thematic topics depict a metro station and each new town?

1. To what extent do existing conditions of TOD sites meet TOD standards?
2. Can TOD performance be measured and determined via thematic urban functions?

To answer the research questions, 23 metro station catchment areas of 600 m radius around 23 metro stations in the five new towns in Shanghai, China were examined in this study. All the stations are located 30–60 km from Shanghai's city center. This study aimed to investigate whether the implementation of TOD of metro stations in new towns aligns with their planning objectives, to quantitatively explore their TOD performance using multi-source urban data, and to observe their TOD typology classification by extending the classic "node-place" model with a third "design" dimension. These can capture the TOD characteristics of metro station areas, thus providing an efficient method for planners and governors to diagnose common problems. The analysis also generates discussions concerning how TOD should be adapted to new towns in a large and growing metropolis.

The remainder of the article is organized as follows. Section 2 reviews the literature on the development of TOD in China and the extension of the node-place model. Section 3 introduces the data and methodology. Section 4 presents the results of the thematic urban functions and TOD performances. Finally, Section 5 provides a discussion on urban planning implications and limitations.

2. Literature Review

2.1. The Growth of TOD in China

Since Calthorpe introduced the TOD concept [11], considerable research related to this subject has been conducted in the West. Although China has a relatively short history regarding this topic [12], extensive research has investigated how TOD has unfolded over time in China's metropolises. This is because urbanization in China has been increasing exponentially, with the urban population rising from 80 million in 1978 to 901.99 million in 2021 [13],; consequently, there were serious problems related to automobiles, pollution, and congestion [14]. TOD is one of the most appropriate strategies for sustainable land use, urban form, and transit [15]. Moreover, given that the Chinese urban periphery takes the form of high-density apartment buildings compared to the USA's low-density houses, the potential for TOD implementation in China is extremely high [16].

Therefore, scholars are committed to adapting TOD principles to the Chinese context. Considering the overall prospect of the TOD strategy in China, Yanping Chen stated that the development of transport-oriented land use forms is the key to solving urban traffic problems in the country [17]. Scholars in China and abroad have published various influential articles focusing on large Chinese cities with developed metro networks, such as Hong Kong, Taipei, Beijing, Shanghai, and Shenzhen. For example, Alice Chow believes Tseung Kwan O new town in Hong Kong achieved transport and urban sustainability through the TOD concept [18]. Kheir Al-Kodmany identified four types of Tall Buildings and Transit-Oriented Development (TB-TOD) models within different areas in Hong Kong, from the old city and city edge to the suburban and new areas [19]. Zidong Yu proposed a framework to describe four thematic urban functions of metro stations through points of interest (PoIs) in Hong Kong [7]. Additionally, Pengjun Zhao investigated the lifestyle, shopping, and entertainment choices of passengers around Beijing subway station areas, believing that in the process of suburbanization, land-use planning based on the TOD model can help shape sustainable mobility [20]. Moreover, taking Shanghai as an example, Bindong Sun analyzed the built environment's impact on commuting behavior and concluded that the TOD model can help shorten travel distances [21]. Some researchers explored variations and typology of TOD in Shanghai: Yao Chen [22] categorized them into five types with different degrees of congestion, putting forward optimizing suggestions for high congestion stations; Zekun Li [23] clustered them into four types based on the "node-place" model, offering optimization plan for the stations with low TOD index values. Furthermore, Jiawen Yang took Shenzhen as an example to analyze the impact of location selection on TOD planning [24]. Ming Zhang proposed a practical TOD model to build a sustainable transportation system that combined the experiences of Hong Kong, Taipei, and mainland Chinese cities [25].

TOD has been widely adopted in Chinese cities, and the local government has realized the importance of applying this concept. Nevertheless, studies exploring TOD areas around suburban metro stations remain scarce. This is because of China's unbalanced urban and rural development and the lack of access to detailed data [26]. Although Chinese urban planners typically prefer to adopt the TOD model when planning new neighborhoods, the current TOD implementation practices are still at a nascent stage. Many new towns in Hong Kong, acknowledged as the most successful TOD cases in China, are highlighted as the benchmark for implementing TOD in mainland China, even more widely in Asia [12,18]. Consequently, an increasing number of scholars have recognized the importance of metro stations in the urban periphery to conduct studies in megacities such as Beijing [20,27,28] and Nanjing [12], and uncover the logic of the TOD framework in small cities [29]. However, the existing investigations in Shanghai from this perspective were surprisingly limited. Haixiao Pan [30] examined two station areas in Songjiang, Shanghai, to assess the effectiveness of their TOD application and Jie Shen [3] examined the same cases to understand the social relations and processes of suburban infrastructures. However, these studies relied on traditional manually collected data, making it difficult to provide large-scale feedback on current planning practice. Thus, China still has a long way to go in this direction.

In short, though there have been numerous studies about TOD classification and evaluation in China, there is little quantitative evidence on estimating new town TOD. Although TOD has been proposed to serve Chinese new metro projects on the edges of cities, they have not been designed as vectors to shape new towns' urbanization [31]. Strategies and suggestions for the construction of TOD in new towns generally focus on features, such as scale, facilities, position and classification, and mainly include qualitative elaboration, they lack refined quantitative measurement. Furthermore, most research has barely combined new towns' comprehensive plans with TOD implementation on a large and fine scale. Therefore, this study considers all 23 metro stations in five new towns of Shanghai utilizing multi-source urban data and natural language processing techniques to define their main functions and more accurately explore their TOD performance.

2.2. Extension of "Node-Place" Model

The "node-place" model, typically used to classify TOD typologies, was established by Bertolini to describe various TOD forms in cities [32]. It focuses on the integration of nodes (the stations) and places (the catchment areas), projecting these two dimensions on a 2D coordinate system and further dividing five representative TOD typologies to depict the different combinations of development levels of one "node" and its surrounding "place". The simple equation line in the coordinate system is divided into three parts (Dependence, Balance, and Stress), leaving the other two areas (Unbalanced node and Unbalanced place) outside the spindle-shaped outline. The node and place values are both weak in the "dependent" areas, the bottom of the line. The "balanced" areas are found along the middle of the line, where the relative positions on nodes and places are almost equal. The "stressed" areas at the top of the line can become problematic as both the node and the place are overused. Moreover, the rest areas are two unbalanced conditions. The "unbalanced node" sets above the spindle, where the transport dimension is more advanced than the urban development. By contrast, the "unbalanced place" is located below the spindle, where the situation is the opposite. Accordingly, massive research has been conducted to identify thorough and accurate indicators to describe the performance of TOD projects more precisely. Features related to traffic, station, and road/lines are usually assigned to the "node" dimension, whereas those that reveal density, development, and population are considered part of the "place" dimension [33–35]. This framework presents an effective way to reveal the relative position of one metro station within an urban network.

Some researchers noticed limitations of the classical "node-place" theory when it was widespread [15,36]. They pointed out that the two-dimensional model was insufficient to capture diverse TOD manifestations [37,38]. Consequently, various three-dimensional extended node-place models were constructed to better measure and describe the existing characteristics of TOD. As for the third dimension, some scholars emphasized walking friendliness and corresponding indicators at an early stage; lately, others combined and discussed various aspects in the extended dimension (Table 1).

In previous years, researchers argued that evaluating walkability can assess the built environment dimensions which that influence traffic, as it greatly impacts the accessibility of TOD's catchment areas [39,40]. In the beginning, several elementary indices were chosen to depict walkability. Kamruzzaman [41] and Singh [42] tried to apply the "3 Ds" [43] theory (Density, Diversity, Design) to establish criteria. Both proposed using intersection density. However, Singh considered the streetscape quality for walking/cycling more important than Kamruzzaman's cul-de-sac density, which is similar to intersection density. Additionally, Monajem and Nosratian [44] utilized closeness and betweenness calculated by space syntax tools to measure pedestrian friendliness, whereas Vale [38] introduced "pedshed ratios," calculated by dividing pedestrian catchment areas with a circle with the same distance radius. Before long, Lyu [37] and Nigro [45] tried to add assorted indicators to the extended dimension. Nigro's research concentrated on TOD strategies in the context of low-density areas, whereas Lyu focused more attention on Chinese high-density areas, which are more relevant to our study.

More recently, "5 Ds" [46] (Density, Diversity, Design, Destination accessibility, and Distance to transit) or "3 Ds + A" [47] (Density, Diversity, Design, and Accessibility) have become increasingly popular in the extended node-place model. Researchers have begun to use a more comprehensive framework to amend the node-place model. For example, Loo and Verle [48] completely implied the 3Ds framework and selected 6–7 variables for each "D", where factors such as income, retail, road connectivity/density, covered walkways, and open spaces can represent the economy, walkability, and space quality. In contrast, other experts chose the indicators in a more unfettered way, even though they also considered the 3Ds theory as standard. Singh [49] used eight separate criteria, including walkability and cyclability, economic development, and user-friendliness of transit systems, to estimate TOD with different weights. Going one step further Zhou goes much further. It is worth mentioning that Zhou [9] exploits big and open data to examine 16 indicators of regional accessibility/centrality, density and diversity, and design. Furthermore, many scholars flexibly applied the 5Ds theory to improve the node-place model by adding a third dimension: design, functionality, orientation, or tie [8,23,50–52].

One of the newest and most representative studies in this approach considered serviceability, accessibility, and walkability in the extended dimension to compare TOD in five typical Chinese megacities and offer more targeted strategies [52]. Therefore, the current study chooses indicators based on this cutting-edge research. However, unlike [52], we combined local planning indicators into the framework, making it more comprehensive and in accord with the local context. Only in this way can the framework be truly practical when urban planners use it to verify whether the planning objectives are met. The details of how we add some of the new towns' planning guidelines to our extended "node-place" model is explained in Section 3.5.

Concerning the datasets, except for Zhou [9], Vale [50], and Qiang [8], most of the aforementioned studies used entropy or proportion of traditional land use data rather than PoIs to calculate land/functional mix. Nevertheless, only Vale [38,50] Nigro [45], Singh [49], and Qiang [8] utilized pedestrian network distance catchment areas; others used straight-line radius. Accordingly, this research used multi-sourced data to develop a 3D node-place model focusing on the TOD in Shanghai's five new towns.

Table 1. Indicators of various extended dimensions.

Extended Dimension	Name	Year	Subject/Title	Indicators of the Extended Dimension
Walking/Pedestrian Oriented	Kamruzzaman, Md. et al. [41]	2014	Advance TOD typology in Brisbane	Intersection density; Cul-de-sac density
	Singh Y.J. et al. [42]	2014	Measuring TOD for Arnhem and Nijmegen	Quality and suitability of Streetscape; Density of controlled intersections/street crossings
	Monajem, S. & Ekram Nosratian, F [44]	2015	The evaluation of the spatial integration via the node-place model	Two spatial indices (To-movement and through-movement)
	Vale, D.S. [38]	2015	Evaluating and classifying station areas in Lisbon	The pedestrian shed ratio (Pedshed ratio)
	Lyu, G. et al. [37]	2016	TOD typology for Beijing metro station areas	Block size; Distance from the station to jobs/residents; Length of paved footpath per acre; Intersection density; Walk scores
	Nigro, A et al. [45]	2019	Land use and public transport integration in small cities and towns	Feeder transport (Walking; Bike lanes; Expected traffic intensity. Road size/slope et al.)
Design/Tie (Function, Walking and economy et al.)	Loo, B.P.Y. & Verle, F. du [48]	2017	TOD toward a two-level sustainable mobility strategy	Diversity of land uses; Convenience of public transit; Retail; Road connectivity/density; Open space; Covered walkway; Exit system
	Singh Y.J. et al. [49]	2017	Measuring TOD around transit nodes	Land use diversity; Economic development; Walkability and Cyclability; User-friendliness of the transit system

Table 1. *Cont.*

Extended Dimension	Name	Year	Subject/Title	Indicators of the Extended Dimension
Design/Tie (Function, Walking and economy et al.)	Vale, D.S. et al. [50]	2018	The extended node-place model at the local scale	Variety/Number of PoIs; Degree of the functional mix; Pedshed ratio; Intersection density; Accessible network length
	Li Z. et al. [23]	2019	Typology, Optimization, and implications	Accessibility; Walkability
	Pezeshknejad, P. et al. [51]	2020	Evaluating BRT via extended node-place model	Functional mix; Streets integration/choice; Streets connectivity; Street density
	Zhou, J. et al. [9]	2020	Using Big and Open Data to Analyze TOD	Destination intensity of retail/entertainment/restaurant/residence; Simpson index; Walkability; Bikeability
	Su S. et al. [52]	2021	TOD typologies in China	Serviceability; Accessibility; Walkability
	Qiang, D. et al. [8]	2022	Evaluation of TOD Based on Multi-Source Data in Shanghai	Density of PoIs; Function mix; Density of road/pedestrian network; Accessibility of pedestrian network; Intersection density; Entrance

3. Data and Methodology

3.1. Analytical Framework

The four key steps of the present study were data collection, analysis, and classification, TOD evaluation, measurement, and guidance (Figure 1). First, information on metro lines and stations, 3D building information and PoIs, the third economic census, and the street network of Shanghai were gathered. The second step has two parts: (1) After cleaning, processing, and recategorizing the PoIs, we used Latent Dirichlet Allocation (LDA) and hierarchical cluster analysis to classify five groups of stations based on their main functions. (2) Other data types of data were imported to ArcGIS to be analyzed using sDNA(a spatial network analysis software) [53] and other toolsets. Moreover, the results in step one, key references for D1 indicators, were integrated into the "design" dimensions to calculate the weighted average walking route distance from the metro station to its representatively functional quarters. Third, three critical dimensions—node, place, and design—were presented to rate the TOD performances in new towns. Each dimension was evaluated by five sub-indicators, which were then summed to obtain the degree of every metro station's TOD performance; this was displayed in radar charts to show the main differences. The calculation details are discussed later. Finally, a few optimizing guidance points are provided to help TOD construction in Shanghai's new towns, according to their diverse TOD characteristics. This framework can diagnose problems, help determine whether the TOD development matches its new town policy and prospects, and more importantly, guide the future metro construction in new towns.

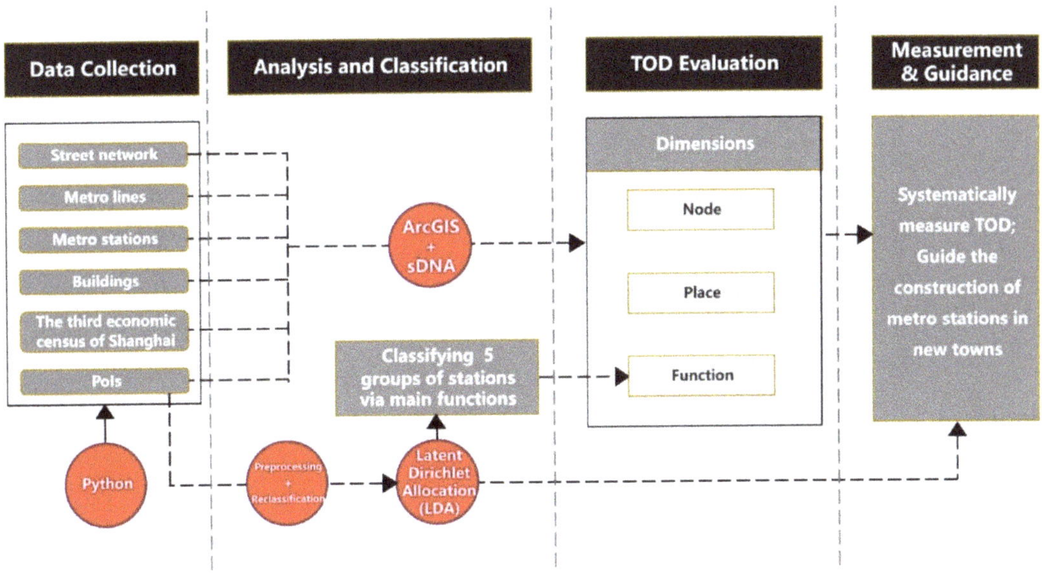

Figure 1. Analytic framework.

3.2. Study Area

This study applied the extended framework to the five new towns of Shanghai. The centers of these new towns cover a total area of 748 km^2, accounting for 11.8% of Shanghai's total area. These five sub-centers of Shanghai were chosen as the research area for two reasons. First, the metro railway system in Shanghai was established in 1993; As of December 2021, it had 508 metro stations (including two maglev stations) and 20 operating lines (including maglev lines), with a total operational mileage length of 831 km (including 29 km of maglev lines). Second, massive, solid evidence was presented to support the appearance of a typical TOD mode in Shanghai [8,22,54]. Nevertheless, there is little quantitative evidence on the relationship between the development of new towns in Shanghai and the development of their local TOD. However, according to the latest Shanghai Comprehensive Plan 2017–2035 (hereafter referred to as the 2035 Plan), the surrounding areas of new towns' metro stations should and will represent the new towns' characteristics. Although most stations on one suburb subway line were constructed simultaneously, the new towns are developing in stages. The catchments of some stations could not present their thematic function, except for the stations located in sub-centers. Therefore, we only used 23 stations in the core districts of five new towns as examples, as specified in the 2035 Plan.

Next, we used a 600 m walking distance based on a real pedestrian network to define the catchment area of each metro station. Previous investigations have suggested a radius of 400–600 m as the most appropriate distance for the catchment area of TOD precincts [37,38]. In our case, a catchment area with a 600 m walking distance was selected as the analysis boundary of each site, as this distance can cover more population and PoIs and better reflect the identical functions of each site; this is because the distance between two suburb stations is longer and the surrounding density is lower compared to stations in the city center. Nevertheless, the circular catchment calculated by a Euclidean-based measurement method cannot simulate real-world walking scenarios. Therefore, we calculated the 600 m service area of every station's exit using the Network Analyst Toolbox on ArcGIS. The final catchment area of a metro station consists of overlapping service areas of the station's exits (Figure 2).

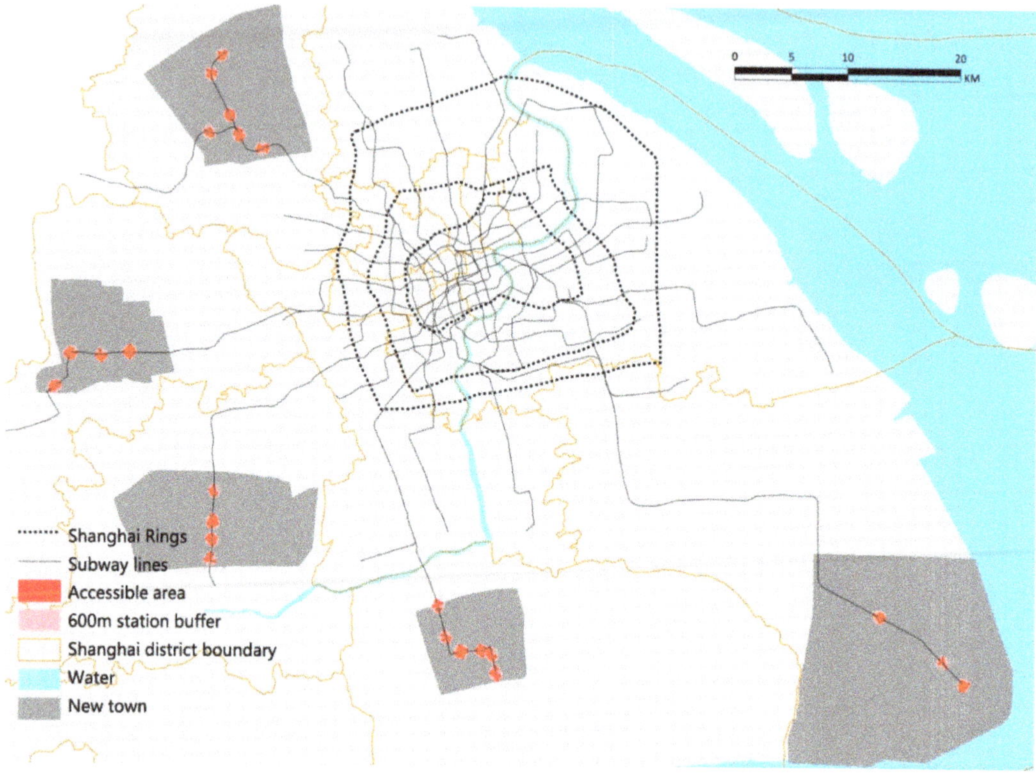

Figure 2. Study area.

3.3. Raw Data Pre-Processing

Multi-sourced urban data, including traditional and non-traditional data, were collected to represent the aforementioned indicators. For example, for non-traditional data, 77,724 street polylines from OpenStreetMap (OSM) (2019), 559,777 street polylines, 754,607 building plots data from Baidu Map API and Python (2019), and over 2.7 million PoI data from Gaode Maps (2019) were collected. Then, traditional data obtained from the Shanghai metro official website and the Shanghai census were used to represent the physical characteristics of metro stations and social attributes of surrounding neighborhoods.

The PoI data used in this study were provided by AutoNavi Map, a platform that has collected over 270,000 records of Shanghai. Three pre-processing steps were performed before PoI analysis:

First, we used Python to grab PoI data from the AutoNavi Map and imported it into ArcGIS for coordinate system transformation. The obtained PoI data included the coordinates, names, and three classification types: major class, medium class, and small class. In the subsequent analysis, "major class" was used as the analysis sample.

Second, abnormal PoI records with incomplete information were removed from the dataset, and PoI data of the type "auto repair, car sales, motorcycle repair" were merged into the "Life Services" category. A total of 15 major types of PoIs were obtained: catering, scenic spots, public facilities, corporate enterprises, shopping, transportation facilities services, financial insurance services, science, education and cultural services, commercial residences, life services, sports and leisure services, healthcare services, government agencies and social groups, and accommodation services.

Finally, the PoIs within the pre-defined TOD area were extracted on the ArcGIS platform and approximately 7500 pieces of data were obtained, providing a basis for classifying metro stations and measuring functional diversity for subsequent research.

3.4. Evaluating Thematic Functions via LDA Topic Modeling

Thematic functions of metro stations can be characterized by the abundance of urban functions in their catchment areas. First, 15 major types of PoIs were analyzed by LDA topic modeling to extract the functional topics of the metro station areas. Second, metro station areas were clustered by agglomerative hierarchical clustering according to the importance weights of thematic functional topics, such that the metro station areas with similar topics belong to the same type. The specific steps are as follows:

1. Establish two analysis matrices: PoI data were used as analysis keywords to form a thesaurus of each station catchment area. Two matrices for extracting topics were established, namely topic-to-PoIs and catchment-to-topic matrices. In the former, each PoI type was assigned a probability within each latent topic, and the semantic features corresponding to the topic were obtained. The latter represents the probability of each topic corresponding to different station areas.
2. Determine the analysis parameters: The number of potential topics n, Max_df, and Min_df are key parameters in the LDA model and need to be pre-defined. Here, we employed the perplexity score metric to determine the optimal number of topics. The smaller the value, the better the quality of the topic results. According to the semantic features corresponding to the topic, whether this parameter was reasonable could be determined. After the analysis, the number of potential topics was determined to be 4, the value of Max_df as 0.95, and the value of Min_df as 4. The semantic results generated by the topic reflected the characteristics of the site area function.
3. Based on the corresponding probabilities of topics within the metro station areas, we used a standard deviation measure to calculate the uniqueness of topics. The clusters with the most similar attributes were merged into a new cluster using the agglomerative hierarchical clustering method, and through multiple iterations, the classification results of metro stations that conform to the actual situation were finally obtained.

3.5. Measuring TOD Performance of Five New Towns in Shanghai

An extended "node-place" model, with the third dimension of "design", was used to measure Shanghai's five new towns' TOD performance. Each of these three dimensions has a different significant value: the "Node" dimension describes the transportation capacity of the metro station itself, while the "Place" dimension represents the degree of urban development, dynamics, and diversity of the surrounding area of the station. Meanwhile, the "Design" dimension identifies the quality of the environment and convenience of the whole network within the catchment area.

We selected indicators in accordance with: the logic of Su's "node-place–functionality" model [52] and Qiang's "Transportation (T)–Pedestrian–oriented (O)–Development (D)" framework [8]; the concrete TOD characteristics of Shanghai; the official reports of Comprehensive Plan and General Land-Use Plan, 2017–2035 of Five New Cities in Shanghai; data availability; and previous cases.

Each dimension was evaluated by five sub-indicators. Specific indicator descriptions, calculation methods, and data sources are listed in Table 2. The "node" dimension is portrayed by indicators of carrying capacity (metro frequency (N4) and passenger flow (N5)) and transit connectivity (accessibility of the metro station (N1), reachable metro stations within 20 min (N2) and accessible bus station number (N3)). The local metro company provided the original data on the metro network, metro frequency, and passenger flow. To describe the accessibility of metro stations, we referred to the betweenness centrality variable, a measure that reflects the through-movement potential of each link, calculated by sDNA [53] based on the official metro network. Reachable metro stations within 20 min were automatically measured via Baidu Map's API. The number of bus stops which are reachable in the catchment can be indicated by the corresponding PoIs from the AutoNavi's Map.

Table 2. Overview of TOD indicators.

Dimension	Indicator	Description/Calculation	Data Sources
Node	N1 Accessibility of the metro station	The average betweenness centrality value of the lines connecting to the station	Shanghai metro's official website
	N2 Reachable metro stations within 20 min	The number of metro stations that one station can reach within 20 min	Calculated on Baidu Map
	N3 Accessible bus station number	Number of bus stops within the catchment area	PoIs from AutoNavi's Map
	N4 Metro frequency	Train frequency during non-peak hours at the station on weekdays	Shanghai metro's official website
	N5 Passenger flow	The passenger flow on July 18 (weekday) and July 20 (weekend), 2019	Shanghai Shentong Metro Group Co., Ltd.
Place	P1 Density of PoIs	The Density of PoIs within the catchment	PoIs from AutoNavi's Map
	P2 Functional mixture	Shannon entropy of various categories of PoIs within the catchment area	PoIs from AutoNavi's Map
	P3 Floor area ratio (FAR)	The ratio of the gross floor area of buildings and the total buildable area	Map World (National Platform for Common Geospatial Information Services) https://www.tianditu.gov.cn/ (accessed on 12 July 2022)
	P4 Employment density	Employment density within the catchment	Third Economic Census in 2013
	P5 Population density	Resident population density within the catchment area of the station	https://www.swguancha.com/ (accessed on 12 July 2022)
Design	D1 Distance to representative function	The average walking distance from the nearest metro station exits to certain PoIs, which belonging to the station's representative functions obtained from LDA modeling analysis	Calculated on ArcGIS based on LDA results.
	D2 Green public area	Area of green space within the catchment of the station	AoIs from the AutoNavi Map
	D3 Density of pedestrian network	Pedestrian road network density within the catchment	Baidu network
	D4 Job-housing balance	The absolute value of the difference between the ratio of standardized population density against employment density and 1	https://www.swguancha.com/ (accessed on 12 July 2022)
	D5 Accessibility of pedestrian network	Average betweenness centrality value within the catchment	Baidu network

The "place" dimension was constructed from two aspects: multi-function degree and density. The former aspect includes indicators of the Density of PoIs (P1) and functional mixture (P2); the latter includes the floor area ratio (FAR) (P3), employment density (P4), and population density (P5). The calculation methods of P1, P3, and P4, provided in Table 2, are clear and uncomplicated. P2 was quantified by the Shannon entropy [55] of the 15 categories of PoIs obtained from Section 3.4. More importantly, P5 was the standard of both the 2035 Plan and the construction strength around TOD. It can represent the guideline of "the coverage rate of permanent residents within 600 m of metro stations" in "the 2035 Plan" of Five New Cities [56–61].

Concerning the "design" dimension, we established most of the indicators from the main indicator table in "the 2035 Plan" of Five New Cities in Shanghai [56–61]. All the new towns' 2035 plans have set regulations on the 15-min walk accessibility rate of comprehensive service facilities, per capita parks and green space, pedestrian network density (D3), and job–housing balance (D4). We retained the last two indicators and transformed the first two into the distance to representative function (D1) and green public area (D2) based on the data availability. Moreover, we added accessibility of the pedestrian network (D5), calculated by sDNA [53], to describe the pedestrian network topological connectivity. When calculating D1, we evaluated the average distance from each PoI that belongs to the station's dominant topics to the nearest station exit. The larger the D1 value, the farther the subway station was from the leading function, which meant more inconvenience. The accessible area of green public space within the catchment was obtained from areas of interest (AoIs) from the AutoNavi Map, while the original data for D3 were from a pedestrian network on the Baidu Map. However, the job–housing balance was a little perplexing. First, we generated a ratio of the number of jobs to the number of households. Then, we took the absolute value of the difference between one and this ratio, as job–housing is the most balanced when the ratio reaches one. Lastly, the final value negatively correlated with the state of the job–housing balance. The equation to calculate the value is as follows:

$$JHR = |1 - \frac{J_i}{H_i}| \quad (1)$$

where J_i = number of jobs of catchment i, H_i = number of households in the catchment i. The number of households is obtained by dividing the population by the average household size of Shanghai (Average Household Size: Shanghai data was reported at 2.32 Person in the Seventh National Population Census in May 2021. http://tjj.sh.gov.cn/, accessed on 12 July 2022).

The abovementioned subindices were given equal weight to join these 15 indicators into a final TOD index. Before integrating all indicators, they were separately normalized and standardized using the formula below:

$$\text{X\_index} = (\text{X} - \text{Min}(\text{X}))/(\text{Max}(\text{X}) - \text{Min}(\text{X})) \quad (2)$$

Except for D1 and D4, other indicators positively correlated with the TOD index. Therefore, D1 and D4 must be subtracted from 1 before adding up to obtain the "design" or TOD index.

4. Results

4.1. Thematic Topics of Urban Functions in Catchment Areas

Four thematic topics were extracted by implementing LDA topic modeling, indicating the characteristics of the urban function in rail transit station areas. Figure 3 shows the results of LDA topic modeling and the frequency of each keyword under each topic, showing that each topic had good independence. The main keywords of each topic were then displayed by using a comparison word cloud, as shown in Figure 4.

The standard deviation measure of each topic generated by LDA topic modeling results represented the different strengths of each catchment on different topics. As a complex and dense urban space, a catchment cannot be simply defined by a single urban function type. Thus, railway stations were divided into different clusters to study their corresponding urban functions. As shown in Figure 5, all metro stations were classified into five groups through hierarchical clustering. The dominant topic of cluster one was Topic 2, due to the low development degree of these three stations and because their functions were mainly public facilities built prior to urban development. Only two stations from group 2 showed a predominant interest in Topic zero. This revealed that many building material stores were in the catchment areas of the Zuibaichi and Qingpuxincheng stations. Stations from group 3 show strong interest in Topic 2, which is relevant to daily residential activities and accounts for 43% of metro stations in the five new towns. Notably, clusters four and five showed more than one interest in urban functions. Stations from group 4 showed relatively strong interest in Topics 1 and 3, because large businesses and residential areas surrounded such stations; thus, they showed the compound property of shopping and catering. Dishuihu station was the only station in cluster 5 and its leading topics were Topics 2 and 3, which were related to the goal of building a world-class cultural and sports tourism destination that combines business, culture, sports, tourism, and residence. Based on the LDA topic modeling results, the station catchment classification had a strong coupling relationship with the built environment, which matched the development objectives of the new towns.

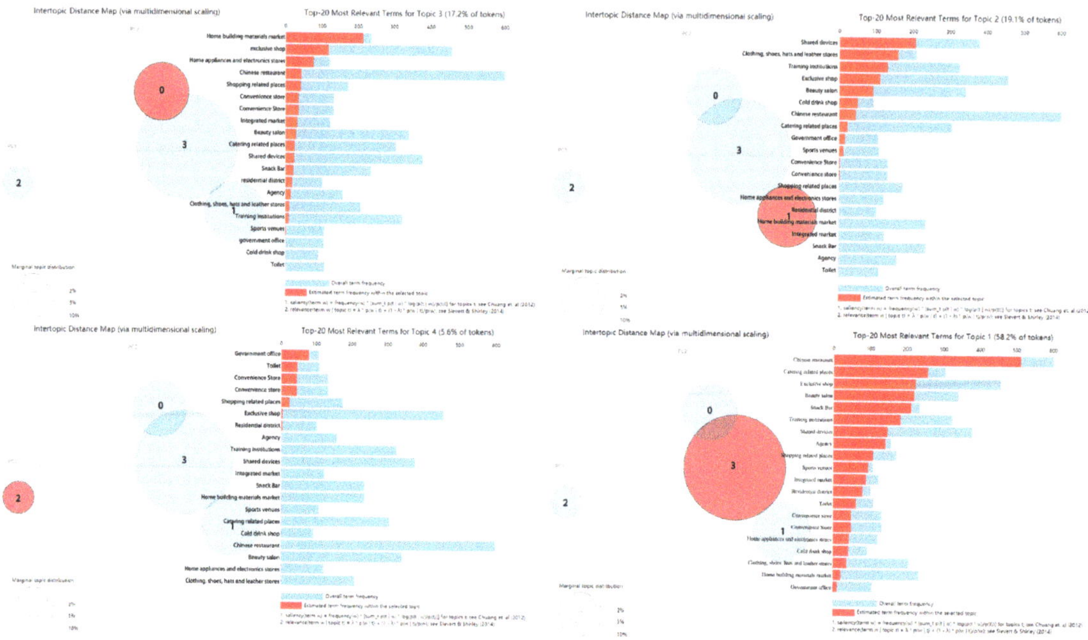

Figure 3. LDA results of LDA topic modeling.

Figure 4. LDA comparison word cloud.

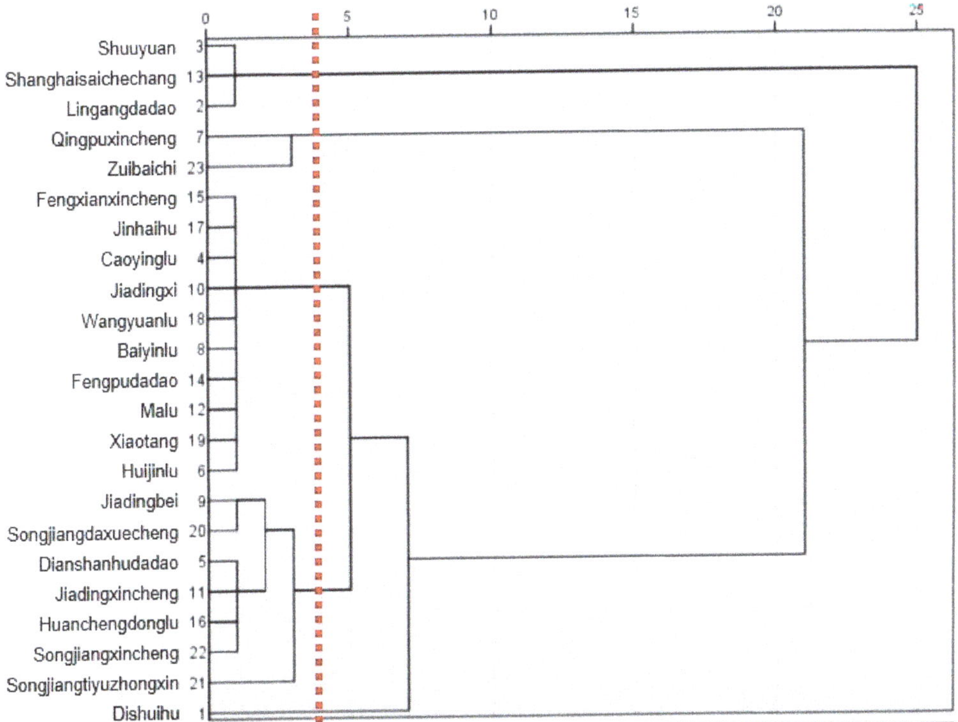

Cluster	Cluster 1	Cluster 2	Cluster 3	Cluster 4	Cluster 5
Station Name	Shuyuan Shanghaisaichechang Lingangdadao	Qingpuxincheng Zuibaichi	Fengxianxincheng Jinhaihu Caoyinglu Jiadingxi Wangyuanlu Baiyinlu Fengpudadao Malu Xiaotang Huijinlu	Jiadingbei Songjiangdaxuecheng Dianshanhudadao Jiadingxincheng Huanchengdonglu Songjiangxincheng Songjiangtiyuzhongxin	Dishuihu
Dominant Topic	Topic 2	Topic 0	Topic 2	Topic 1 Topic 3	Topic 2 Topic 3

Figure 5. Hierarchical clustering results of station based on standard deviation metric.

From a geospatial perspective, we discovered the clustering patterns shown in Figure 6, using distinct colors to label stations from different groups. We found that although the same new town often included different clusters, the dominant topics of stations in Nanhui, Fengxian, Songjiang, and Jiading New City were relatively uninformed. By contrast, the dominant topics of Qingpu New City were more diversified, which is related to the functional positioning of innovation research, business trade, and tourism.

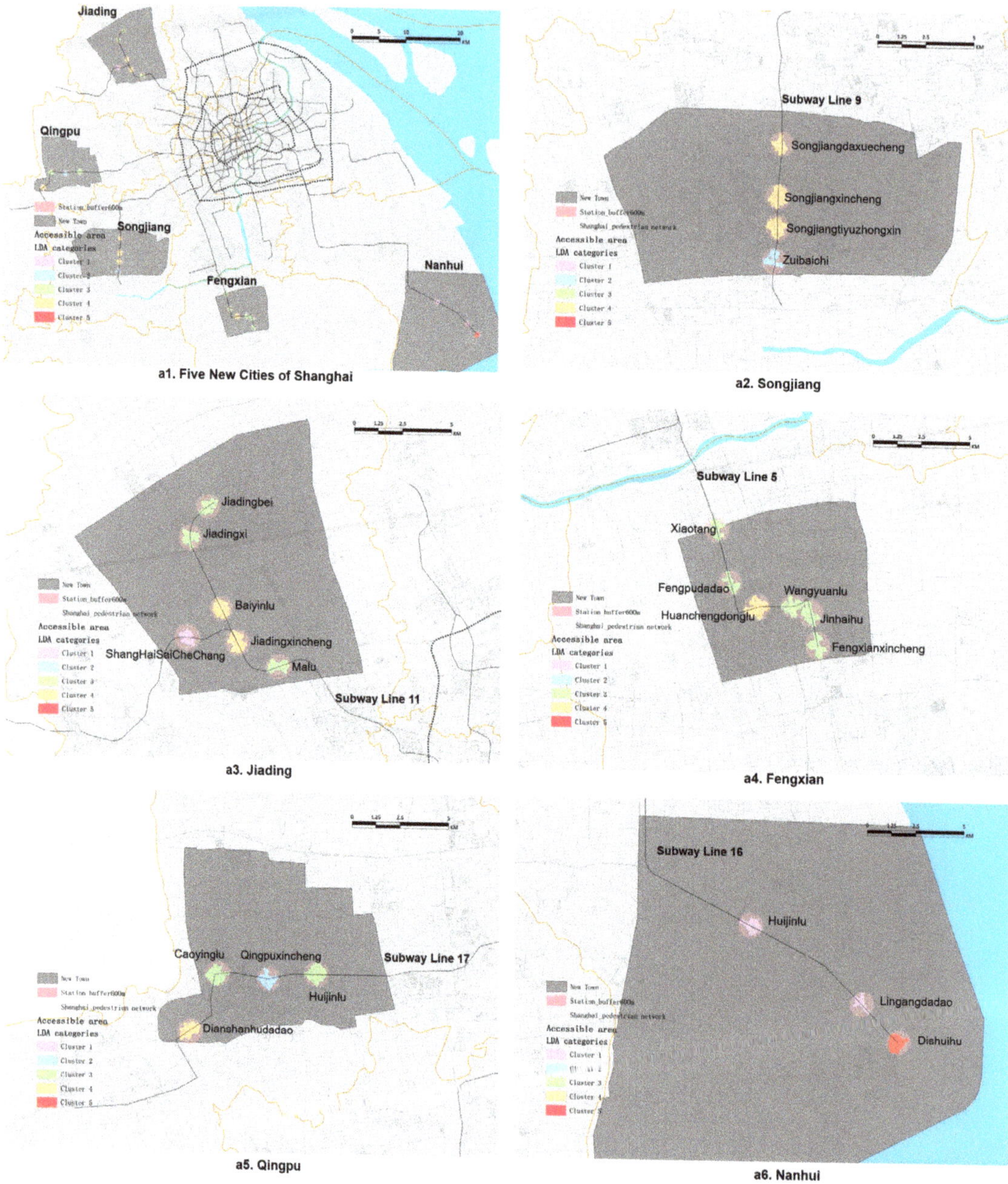

Figure 6. The spatial distribution of aggregated metro stations regarding their thematic topic compositions.

4.2. Node, Place, and Design Indices and TOD Performance in Catchment Areas

Figure 7 shows that the combined TOD indices in Songjiang New City rank the highest, and those of Nanhui New City stand the lowest. The TOD in Jiading New City has the largest variation, accounting for four out of five types, with a high degree of inconsistency. The TOD indices of Qingpu New City and Fengxian New City were relatively close, with only two types. Moreover, except for Fengxian and Nanhui New Cities, the closer the metro station was to downtown Shanghai, the higher the TOD index. This could be because Fengxian New City has the Huangpu river as a natural barrier, and Nanhui New City is far from the central area.

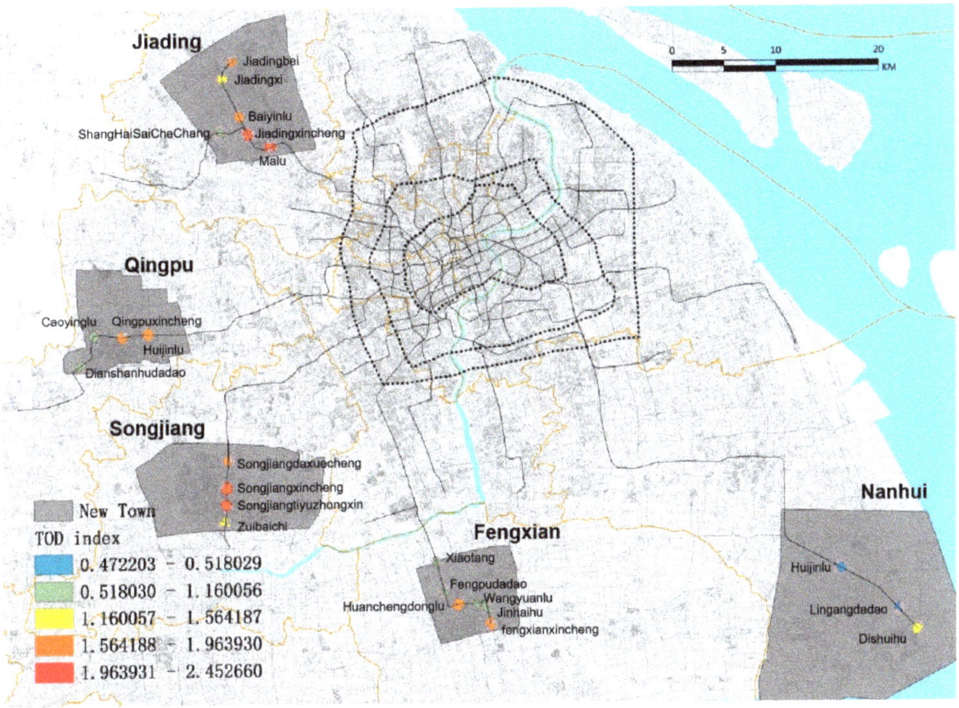

Figure 7. The spatial distribution of metro stations regarding their TOD indices.

We performed a separate diagnosis of each dimension of metro stations in the Five New Cities (Figure 8). First, the overall performance of the three dimensions of Songjiang and Jiangding New Cities looked excellent but deeply uncoordinated. For example, Songjiang-daxuecheng station in the north of Songjiang New City featured the highest "node" and "place" degree, but the lowest "design" index. This was similar to the Malu station in Jiading New City. By contrast, Songjiangxincheng and Songjiangtiyuzhongxin stations in the center of Songjiang New City had the highest "place" and "design" and the lowest "node" scores. Additionally, Jiadingxincheng station had superior scores in all three dimensions, whereas Shanghaisaichechang was far from satisfactory. Therefore, the two new towns had various problems with different metro stations. The government should take appropriate action in response to these concerns. Second, the TOD area of Qingpu New City had issues of high density and functional mixture, but there were fewer people and employment positions. This indicates that the real estate in Qingpu New City was over-developed, and the surrounding supporting greening and public facilities were insufficient, resulting in problems such as an insufficient permanent resident population. Third, the "place" and "design" were fairly good in stations of Fengxian New City, but the "node" was surprisingly unacceptable, especially in the nearest two stations closest to Fengxianxincheng station. In this case, the core area could not promote a synergy effect between the regions to enhance its sub-center status. Hence, they urgently needed to increase the capacity and connectivity of the stations. Lastly, the stations in Nanhui New City had the worst "node" and "place" degrees. Only the "design" index of Dishuihu station made people comfortable, as it is close to a huge lake and many museums, parks, and hotels. Therefore, Nanhui New City, has a long way to go in applying TOD.

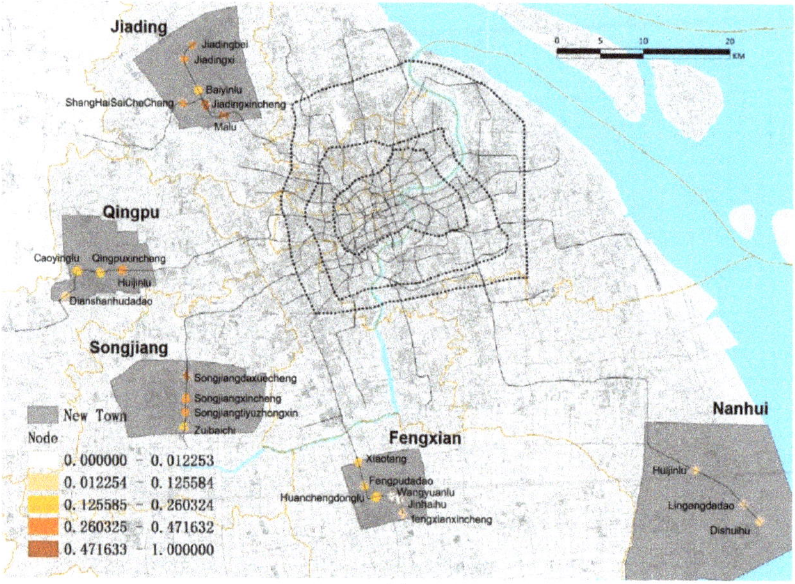

(a)

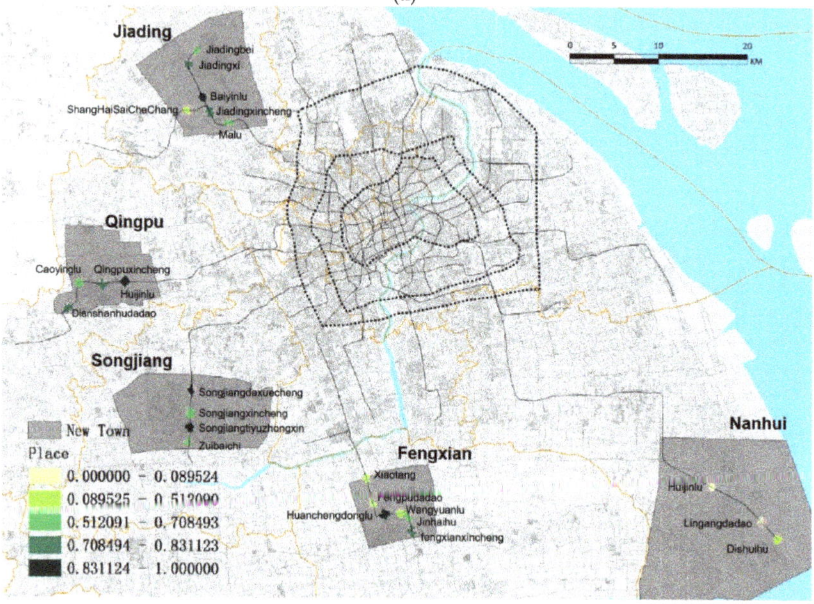

(b)

Figure 8. *Cont.*

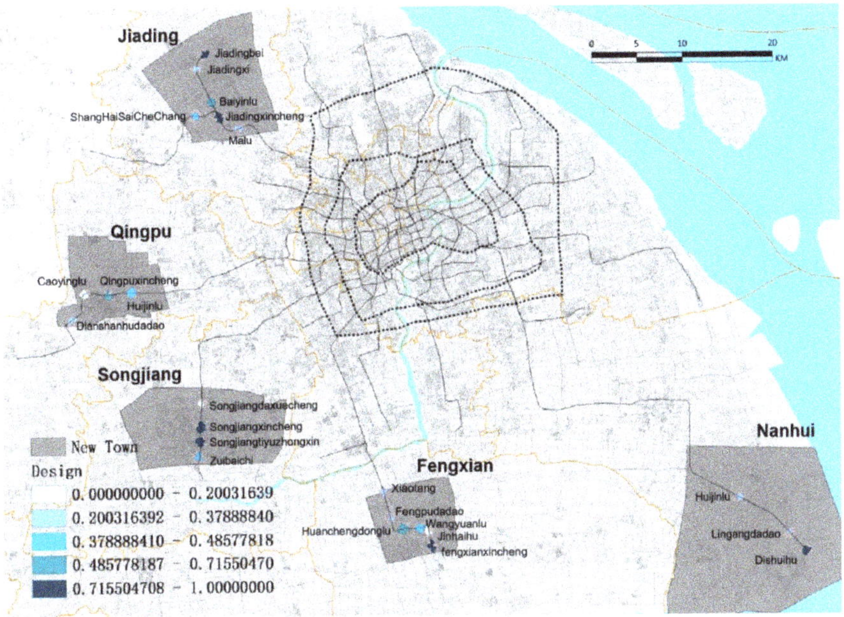

(c)

Figure 8. The spatial distribution of metro stations regarding Node, Place, and Design indices. (**a**): The spatial distribution of metro stations regarding the Node index. (**b**): The spatial distribution of metro stations regarding the Place index. (**c**): The spatial distribution of metro stations regarding the Design index.

Additionally, most of the indicators of "design" were abstracted from "the 2035 Plan". For the "design" aspect (Figure 8c), the overall values were considerably high in Jiading and Songjiang New Cities; however, but the values of the stations in each new town varied greatly. The "design" values in other new towns were fairly low, although they were all equipped with one or two stations with high "design" values. Interestingly, the stations with the highest degree of "design" were the central stations of the five new towns, namely Songjiangxincheng, Songjiangtiyuzhongxin, Jiadingxincheng, Fengxianxincheng, and Dishuihu (Dishuihu station was formerly known as Lingangxincheng Station in planning guidance in 2009, and Lingang New City was formerly known as Nanhui New City). Furthermore, the station with the highest "design" index in Qingpu New City was also Qingpuxincheng station, although its score only ranked second. As can be observed, the "design" score supported the planning guidance well. The construction of the central areas of new towns led to the implementation of the planning indicators, which meant that the goal of developing Five New Cities to become sub-centers of Shanghai in "the 2035 Plan" took a huge step forward. However, the stations with the highest "design" values did not have very high "node" or "place" indices, indicating that the TOD concept is still not well-integrated with planning objectives. Urban planners and policymakers should thus focus on combining TOD and comprehensive urban planning.

4.3. TOD Typologies in Catchment Areas

We used a hierarchical cluster analysis method to classify 23 metro stations into four categories based on three dimensions (Figure 9), focusing on the geographical distribution and characteristics of each type. The radar charts in Table 3 visualized every dimension of each type for easy comparison. Figure 10 shows the spatial location distribution of the four types. Some conclusions can be drawn when we combine Figure 10 and Table 3:

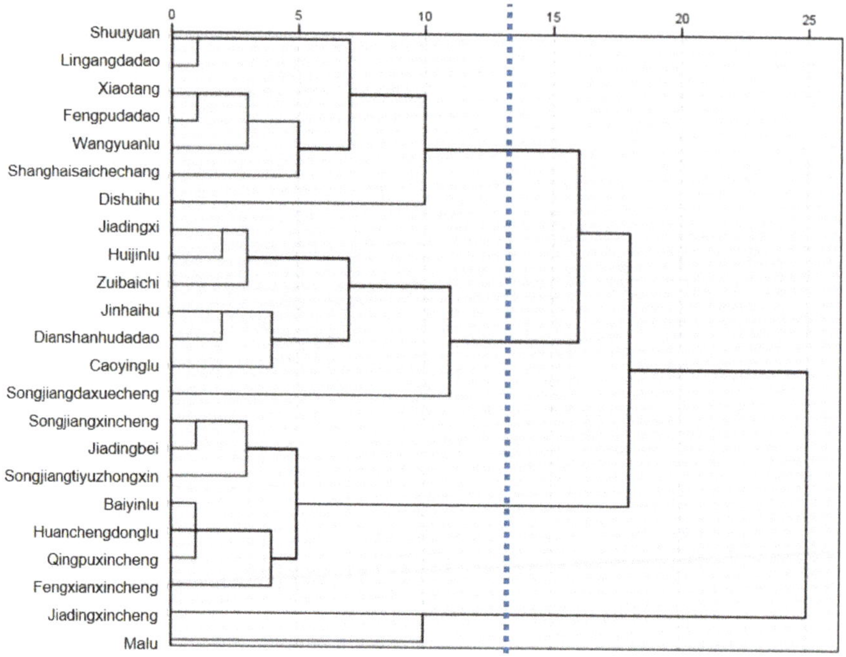

Figure 9. Hierarchical cluster analysis based on three dimensions.

Table 3. Rader charts for each dimension by types.

Type 1	Type 2
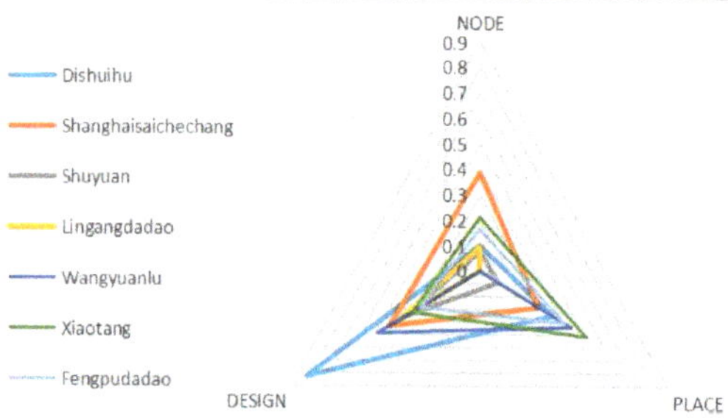	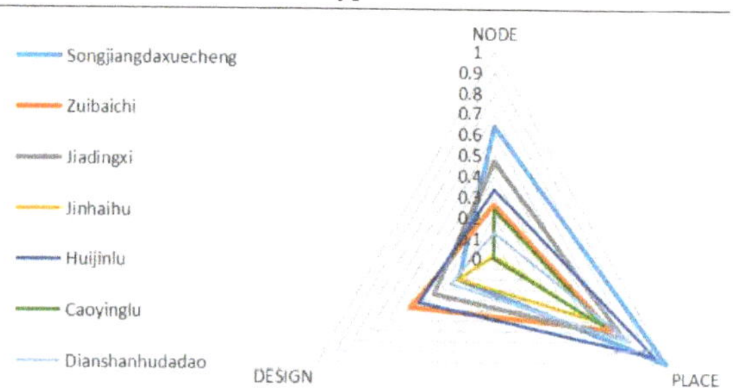
Type 3	**Type 4**
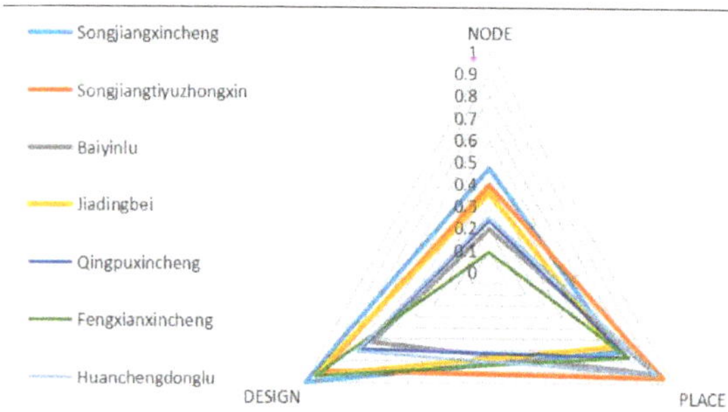	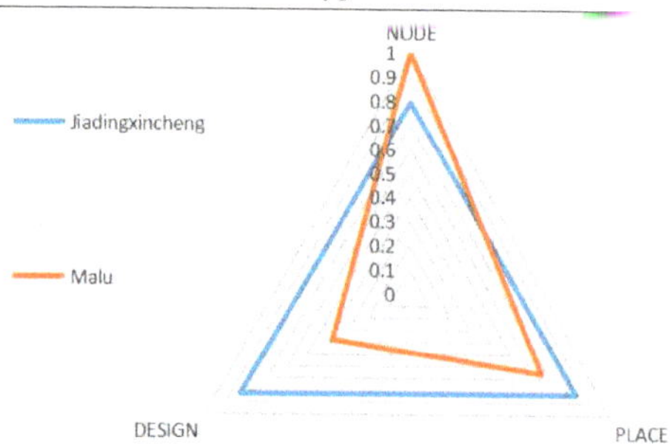

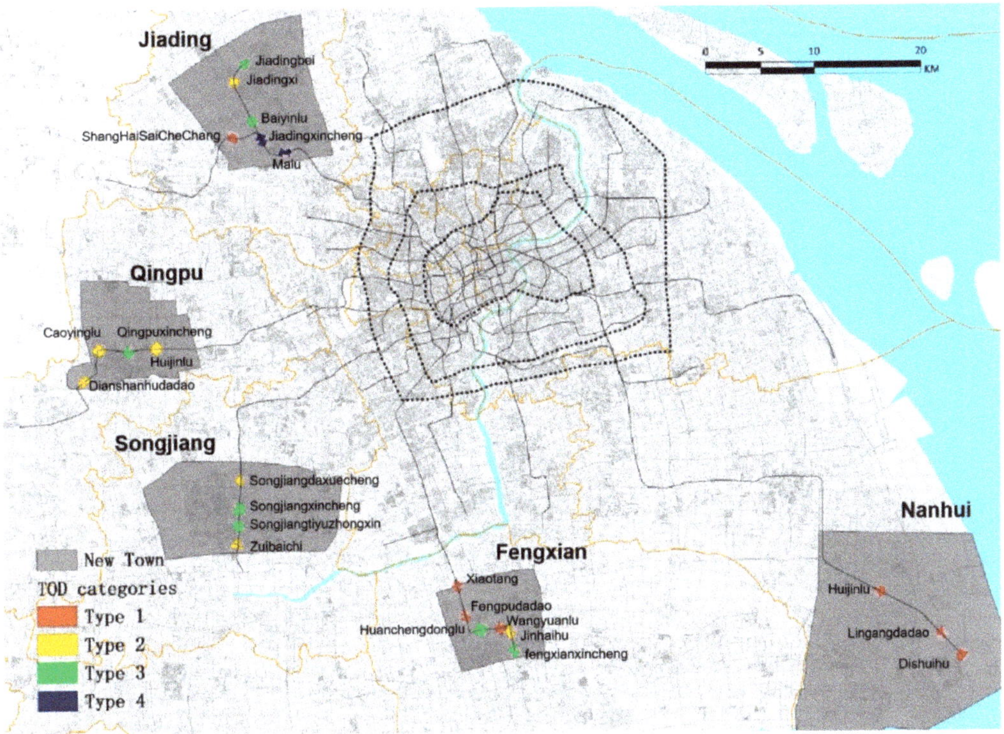

Figure 10. The spatial distribution of clustered metro stations regarding their TOD typologies.

Type 1—Unsustained: The implementation of TOD in this type of station is at a very early stage. All dimension values of the stations belonging to this type were below 0.5, except for the "design" variable of the Dishuihu station. Nevertheless, despite the unsatisfactory performance, the three indicators of these stations were relatively balanced. Most stations were concentrated in Nanhui and Fengxian New Cities, suggesting that the development of TOD in these areas is very slow. For this type of metro station, the priority of the local government is improving the urban development to attract more citizens to settle down in the new town.

Type 2—Unbalanced place: This type of station's performance is extremely poor in living quality. Their environments cannot support the overcrowded density. This type was characterized by extremely high "place" values but relatively low "node" and "design" values. This suggested that the population density and land use development around metro stations far outweighed the potential demand created by transportation. Most stations of this type were around the central areas of Qingpu, Songjiang, and Jiading New Cities. The density of real estate development near these stations was similar to that in the new town centers; however, but the quality of transport services and the environment was lower. This can lead to an unbalanced place for such stations. All of them happen to be close to the central station. Consequently, strengthening their "node" and "design" dimensions can help the Five New Cities achieve the goals of "the 2035 Plan" faster and more thoroughly.

Type 3—Unbalanced, lack of nodes: The transportation capability of stations in the third cluster is far from enough. They comprised a comparable balance between "design" and "place" and intermediate "node". This indicated that urban design and development provided a well-equipped and highly accessible neighborhood to satisfy the needs of an important transportation hub. The only problem is that the operational attributes of the stations were too moderate to match the first two features. The vast majority of Type

3 stations are located in the center of new towns, such as Qingpuxincheng, Songjiangxincheng, and Fengxianxincheng stations, as well as their adjacency, such as Baiyinglu and Songjiangtiyuzhongxin stations. Practically, all of them were built before "the 2035 Plan" was proposed. Therefore, the design of these stations cannot fully meet the rapid development of the surrounding area and needs to be optimized in the future. Local planners need to concentrate more on renovating and upgrading the outmoded stations.

Type 4—Balanced: This type includes two metro stations, Jiangdingxincheng and Malu stations, which can be considered the highest standard of TOD in Shanghai's new towns, both of which belong to Jiading New City. Malu station was quite balanced but was under stress threat since the "node" was slightly superior to the "place" and the "node" dimensions. By contrast, Jiadingxincheng station was the most balanced station among the 23 new town stations. As shown in Table A1, only the bus station number, green public areas, and accessibility of the pedestrian network were weak and needed to be optimized. Nevertheless, it should be a role model for other stations in developing an outstanding TOD project. For them, further optimization is the punctual enhancement of targeted indicators.

4.4. Analysis of Variance

A one-way analysis of variance (ANOVA) was conducted to evaluate the relationship between the TOD indicators and urban functional topics. We see that the significance value for the five place dimension variables was 0.002, 0.000, 0.035, 0.065, and 0.001, respectively (Table A2), revealing a statistically significant difference in indicators of the "place" dimension in the classification of thematic topics. That is, the classification according to thematic topics changes according to the Density of PoIs, Population Density, Function Mixture, and Plot Ratio. Furthermore, except for the green public area (D2, with a p-value of 0.05), the remaining indicators did not statistically significantly affect the thematic topics' classification.

5. Discussion and Conclusions

Many cities in China have adopted TOD planning around metro stations. TOD construction is considered critical for economic growth and the population's daily well-being in suburban developments. By 2035, each new town in Shanghai is expected to have a population of over one million people. From this point of view, the TOD research on new towns is more comprehensive than that on the metropolis. It is imperative to examine the TOD performance of metro station areas for new towns and to discuss how TOD should be adapted to suburban Shanghai. Several studies have attempted to investigate the effectiveness of the TOD application at the urban peripheral locations in a large metropolitan region [3,30]. However, few such studies have quantitatively examined TOD performance and TOD typology for suburban areas.

This research aimed to investigate the TOD implementation of the catchment areas of metro stations in new towns and examine how the TOD typology can be used as a decision-support tool for planners by introducing a data-driven analytical framework. Unlike prior research, this study provides a comprehensive and realistic evaluation approach for measuring TOD performance by incorporating the "design" dimension in an extended "node–design–place" framework and employing relevant indicators supporting planning objectives.

The differences in the thematic topic were visualized by identifying the most prototypical thematic topics of urban functions for catchment areas. Five types of stations were identified via hierarchy clustering. We observed that most metro stations only had one thematic topic. In contrast, eight of the 23 metro stations studied were labeled with multiple thematic topics, revealing that these station areas have started to develop a multifunctional system. Additionally, the TOD index of each catchment area is based on an extended "node-place" model. The study revealed that there is uneven development of new towns in Shanghai. The TOD typologies further distinguished in this study can provide valuable insights for TOD planning. Planners should design individualized strategies for each TOD type based on their function and planning targets.

A few limitations need to be pointed out. First, a set of indicators was selected based on a literature review, five new towns' comprehensive plans, and data availability. This procedure can cause limitations in subjective bias and knowledge scope. In the future, planners and governors should verify the validity of the chosen indicators and determine their proportions. Second, although PoIs are positively correlated with urban vitality, it is worthwhile to collect and analyze more user-generated datasets to provide a more accurate and comprehensive insight into local activities. Third, this study focuses on the current conditions of TOD sites. Finally, there is an absence of planning data for estimating the future condition of TOD sites. Follow-up research should further explore information such as projected or assumed population and employment density in 2035. An evaluation of TOD performance based on more in-depth information and planning guidelines could help planners and governors to diagnose and correct problems in the planning stage, and to guide the planning process.

Author Contributions: Conceptualization, L.Z. and D.Q.; methodology, L.Z. and D.Q.; formal analysis, P.H. and D.Q.; writing—review and editing, DQ., L.Z. and P.H.; funding acquisition, L.Z. All authors have read and agreed to the published version of the manuscript.

Funding: This research was funded by the National Natural Science Foundation of China (52008297); Shanghai Pujiang Program (21PJC114).

Institutional Review Board Statement: Not applicable.

Informed Consent Statement: Not applicable.

Data Availability Statement: The datasets analyzed for this study are included in the article/table, further inquiries can be directed to the corresponding authors.

Conflicts of Interest: The authors declare no conflict of interest.

Appendix A

Table A1. Rader charts for each indicator by types.

Type 1	Type 2
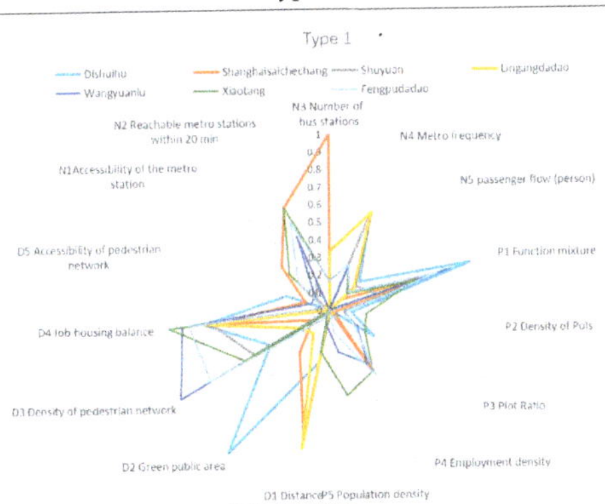	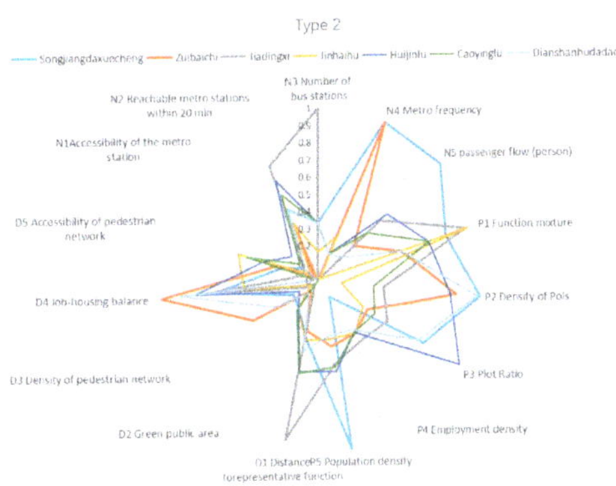

Table A1. *Cont.*

Type 3

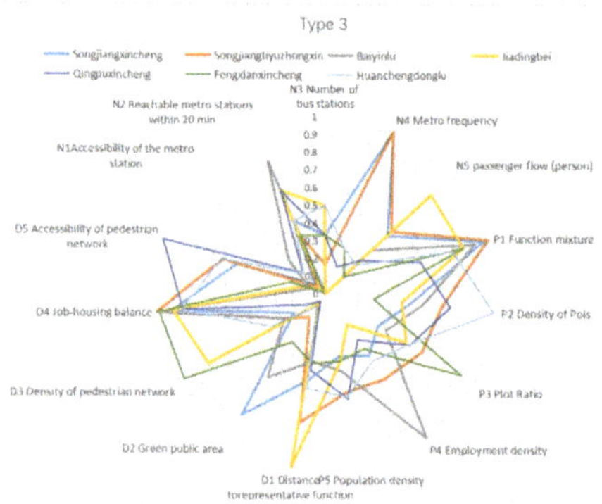

Type 4

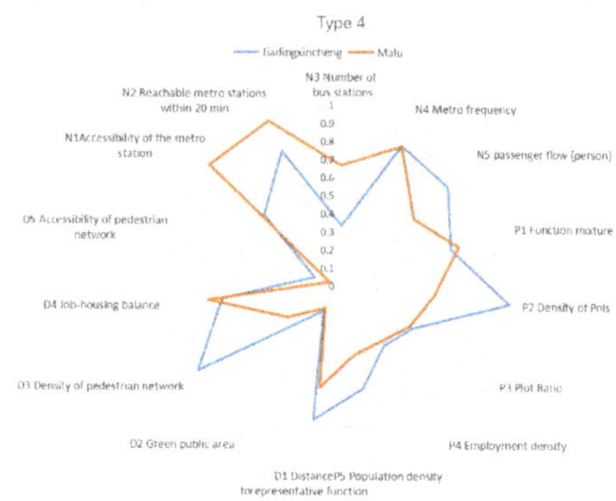

Table A2. One-way ANOVA result.

		Sum of Squares	Df	Mean Squares	F	*p*
N1Accessibility of the metro station	Between groups	0.065	4	0.016	0.284	0.885
	Within groups	1.035	18	0.057		
	Total	1.100	22			
N2 Reachable metro stations within 20 min	Between groups	0.502	4	0.125	2.877	0.053
	Within groups	0.785	18	0.044		
	Total	1.287	22			
N3 Number of bus stations	Between groups	0.120	4	0.030	0.361	0.833
	Within groups	1.492	18	0.083		
	Total	1.611	22			
N4 Metro frequency	Between groups	0.608	4	0.152	1.204	0.343
	Within groups	2.272	18	0.126		
	Total	2.880	22			
N5 passenger flow (person)	Between groups	0.334	4	0.084	1.212	0.340
	Within groups	1.240	18	0.069		
	Total	1.575	22			
P1 Function mixture	Between groups	0.706	4	0.177	6.315	0.002
	Within groups	0.503	18	0.028		
	Total	1.209	22			
P2 Density of PoIs	Between groups	1.929	4	0.482	11.614	0.000
	Within groups	0.748	18	0.042		
	Total	2.677	22			
P3 Plot Ratio	Between groups	0.661	4	0.165	3.285	0.035
	Within groups	0.905	18	0.050		
	Total	1.566	22			
P4 Employment density	Between groups	0.344	4	0.086	2.683	0.065
	Within groups	0.577	18	0.032		
	Total	0.920	22			
P5 Population density	Between groups	0.805	4	0.201	7.655	0.001
	Within groups	0.473	18	0.026		
	Total	1.279	22			

Table A2. *Cont.*

		Sum of Squares	Df	Mean Squares	F	p
D1 Distance to representative function	Between groups	0.242	4	0.061	0.946	0.460
	Within groups	1.152	18	0.064		
	Total	1.394	22			
D2 Green public area	Between groups	0.800	4	0.200	5.498	0.005
	Within groups	0.655	18	0.036		
	Total	1.455	22			
D3 Density of pedestrian network	Between groups	0.133	4	0.033	0.305	0.871
	Within groups	1.970	18	0.109		
	Total	2.103	22			
D4 Job-housing balance	Between groups	0.113	4	0.028	0.559	0.695
	Within groups	0.906	18	0.050		
	Total	1.018	22			
D5 Accessibility of pedestrian network	Between groups	0.479	4	0.120	2.195	0.111
	Within groups	0.983	18	0.055		
	Total	1.462	22			

References

1. Wang, C.; Yang, S. Dynamic Analysis on Population Suburbanization and New Town Development in Shanghai. *City Plan. Rev.* **2016**, *25*, 70–75.
2. Niu, X.; Ding, L.; Song, X. Analyzing Suburban New Town Development in Shanghai from the Perspective of Jobs-Housing Spatial Relationship. *City Plan. Rev.* **2017**, *41*, 47–53.
3. Shen, J.; Wu, F. Paving the Way to Growth: Transit-Oriented Development as a Financing Instrument for Shanghai's Post-Suburbanization. *Urban Geogr.* **2020**, *41*, 1010–1032. [CrossRef]
4. Ximing, L.; Xiang, W. Regional Rapid Rail Transit Planning in Shanghai Metropolitan Area. *Urban Transp. China* **2014**, *6*, 8–17.
5. Shanghai Municipal People's Government. Shanghai Urban Master Plan. Available online: https://www.supdri.com/2035/public/ebook07.pdf (accessed on 12 July 2022).
6. Zhou, J.; Yang, Y.; Gu, P.; Yin, L.; Zhang, F.; Zhang, F.; Li, D. Can TODness Improve (Expected) Performances of TODs? An Exploration Facilitated by Non-Traditional Data. *Transp. Res. Part D Transp. Environ.* **2019**, *74*, 28–47. [CrossRef]
7. Yu, Z.; Zhu, X.; Liu, X. Characterizing Metro Stations via Urban Function: Thematic Evidence from Transit-Oriented Development (TOD) in Hong Kong. *J. Transp. Geogr.* **2022**, *99*, 103299. [CrossRef]
8. Qiang, D.; Zhang, L.; Huang, X. Quantitative Evaluation of TOD Performance Based on Multi-Source Data: A Case Study of Shanghai. *Front. Public Health* **2022**, *10*, 820694. [CrossRef]
9. Zhou, J.; Yang, Y.; Webster, C. Using Big and Open Data to Analyze Transit-Oriented Development. *J. Am. Plan. Assoc.* **2020**, *86*, 364–376. [CrossRef]
10. Gu, P.; He, D.; Chen, Y.; Christopher Zegras, P.; Jiang, Y. Transit-Oriented Development and Air Quality in Chinese Cities: A City-Level Examination. *Transp. Res. Part D Transp. Environ.* **2019**, *68*, 10–25. [CrossRef]
11. Calthorpe, P. *The Next American Metropolis: Ecology, Community, and the American Dream*; Princeton Architectural Press: New York, NY, USA, 1993; ISBN 978-1-878271-68-6.
12. Bruce, C. *Transit Oriented Development in China: Designing a New Transit Oriented Neighbourhood in Hexi New Town, Nanjing, Based on Hong Kong Case Studies*; Blekinge Institute of Technology & Nanjing Forestry University: Nanjing, China, 2012.
13. National Bureau of Statistics of China. The Main Data of the Seventh National Census. Available online: http://www.stats.gov.cn/tjsj/zxfb/202105/t20210510_1817176.html (accessed on 10 July 2022).
14. Campanella, T.J. *The Concrete Dragon: China's Urban Revolution and What It Means for the World*; Princeton Architectural Press: New York, NY, USA, 2008; ISBN 978-1-56898-627-2.
15. Ibraeva, A.; de Almeida Correia, G.H.; Silva, C.; Antunes, A.P. Transit-Oriented Development: A Review of Research Achievements and Challenges. *Transp. Res. Part A Policy Pract.* **2020**, *132*, 110–130. [CrossRef]
16. Cervero, R.; Murakami, J. Rail and Property Development in Hong Kong: Experiences and Extensions. *Urban Stud.* **2009**, *46*, 2019–2043. [CrossRef]
17. Chen, Y. The Public Transport Community and The Urban Land Use Form Guided by The Public Transport. *City Plan. Rev.* **2000**, *3*, 10–14+64.
18. Chow, A.S.Y. Urban Design, Transport Sustainability and Residents' Perceived Sustainability: A Case Study of Transit-Oriented Development in Hong Kong. *J. Comp. Asian Dev.* **2014**, *13*, 73–104. [CrossRef]
19. Al-Kodmany, K.; Xue, Q.; Sun, C. Reconfiguring Vertical Urbanism: The Example of Tall Buildings and Transit-Oriented Development (TB-TOD) in Hong Kong. *Buildings* **2022**, *12*, 197. [CrossRef]

20. Zhao, P.; Li, S. Suburbanization, Land Use of TOD and Lifestyle Mobility in the Suburbs: An Examination of Passengers' Choice to Live, Shop and Entertain in the Metro Station Areas of Beijing. *J. Transp. Land Use* **2018**, *11*, 195–215.
21. Sun, B.; Ermagun, A.; Dan, B. Built Environmental Impacts on Commuting Mode Choice and Distance: Evidence from Shanghai. *Transp. Res. Part D Transp. Environ.* **2017**, *52*, 441–453. [CrossRef]
22. Chen, Y.; Yao, M.; Cai, Z. Research on the Classification of Urban Rail Transit Stations—Taking Shanghai Metro as an Example. In Proceedings of the 2018 15th International Conference on Service Systems and Service Management (ICSSSM), Hangzhou, China, 21–22 July 2018; pp. 1–6.
23. Li, Z.; Han, Z.; Xin, J.; Luo, X.; Su, S.; Weng, M. Transit Oriented Development among Metro Station Areas in Shanghai, China: Variations, Typology, Optimization and Implications for Land Use Planning. *Land Use Policy* **2019**, *82*, 269–282. [CrossRef]
24. Yang, J.; Chen, J.; Le, X.; Zhang, Q. Density-Oriented versus Development-Oriented Transit Investment: Decoding Metro Station Location Selection in Shenzhen. *Transp. Policy* **2016**, *51*, 93–102. [CrossRef]
25. Zhang, M. Chinese Edition of Transit-Oriented Development. *Transp. Res. Rec.* **2007**, *2038*, 120–127. [CrossRef]
26. Spear, J. Urban Transport Themes in China and Lessons from International Experience. In Proceedings of the European Transport Conference, Strasbourg, France, 18–20 September 2006; pp. 1–23.
27. Gu, Y.; Zheng, S. The Impacts of Rail Transit on Housing Prices and Land Development Intensity: The Case of No.13 Line of Beijing. *SSRN Electron. J.* **2008**, *40*, 124–133. [CrossRef]
28. Deng, T.; Nelson, J.D. The Impact of Bus Rapid Transit on Land Development: A Case Study of Beijing, China. *Int. J. Humanit. Soc. Sci.* **2010**, *4*, 1169–1179.
29. Liang, Y.; Du, M.; Wang, X.; Xu, X. Planning for Urban Life: A New Approach of Sustainable Land Use Plan Based on Transit-Oriented Development. *Eval. Program Plan.* **2020**, *80*, 101811. [CrossRef] [PubMed]
30. Pan, H.; Shen, Q.; Liu, C. Transit-Oriented Development at the Urban Periphery: Insights from a Case Study in Shanghai, China. *Transp. Res. Rec.* **2011**, *2245*, 95–102. [CrossRef]
31. Doulet, J.-F.; Delpirou, A.; Delaunay, T. Taking Advantage of a Historic Opportunity? A Critical Review of the Literature on TOD in China. *J. Transp. Land Use* **2017**, *10*, 77–92.
32. Bertolini, L. Spatial Development Patterns and Public Transport: The Application of an Analytical Model in the Netherlands. *Plan. Pract. Res.* **1999**, *14*, 199–210. [CrossRef]
33. Reusser, D.E.; Loukopoulos, P.; Stauffacher, M.; Scholz, R.W. Classifying Railway Stations for Sustainable Transitions—Balancing Node and Place Functions. *J. Transp. Geogr.* **2008**, *16*, 191–202. [CrossRef]
34. Chorus, P.; Bertolini, L. An Application of the Node Place Model to Explore the Spatial Development Dynamics of Station Areas in Tokyo. *J. Transp. Land Use* **2011**, *4*, 45–58.
35. Zemp, S.; Stauffacher, M.; Lang, D.J.; Scholz, R.W. Classifying Railway Stations for Strategic Transport and Land Use Planning: Context Matters! *J. Transp. Geogr.* **2011**, *19*, 670–679. [CrossRef]
36. Phani Kumar, P.; Ravi Sekhar, C.; Parida, M. Identification of Neighborhood Typology for Potential Transit-Oriented Development. *Transp. Res. Part D Transp. Environ.* **2020**, *78*, 102186. [CrossRef]
37. Lyu, G.; Bertolini, L.; Pfeffer, K. Developing a TOD Typology for Beijing Metro Station Areas. *J. Transp. Geogr.* **2016**, *55*, 40–50. [CrossRef]
38. Vale, D.S. Transit-Oriented Development, Integration of Land Use and Transport, and Pedestrian Accessibility: Combining Node-Place Model with Pedestrian Shed Ratio to Evaluate and Classify Station Areas in Lisbon. *J. Transp. Geogr.* **2015**, *45*, 70–80. [CrossRef]
39. Park, S.; Choi, K.; Lee, J.S. To Walk or Not to Walk: Testing the Effect of Path Walkability on Transit Users' Access Mode Choices to the Station. *Int. J. Sustain. Transp.* **2015**, *9*, 529–541. [CrossRef]
40. Nawrocki, J.; Nakagawa, D.; Matsunaka, R.; Oba, T. Measuring Walkability and Its Effect on Light Rail Usage: A Comparative Study of the USA and Japan. *Urban Transp. XX* **2014**, *138*, 10–2495.
41. Kamruzzaman, M.; Baker, D.; Washington, S.; Turrell, G. Advance Transit Oriented Development Typology: Case Study in Brisbane, Australia. *J. Transp. Geogr.* **2014**, *34*, 54–70. [CrossRef]
42. Singh, Y.J.; Fard, P.; Zuidgeest, M.; Brussel, M.; van Maarseveen, M. Measuring Transit Oriented Development: A Spatial Multi Criteria Assessment Approach for the City Region Arnhem and Nijmegen. *J. Transp. Geogr.* **2014**, *35*, 130–143. [CrossRef]
43. Cervero, R.; Kockelman, K. Travel Demand and the 3Ds: Density, Diversity, and Design. *Transp. Res. Part D Transp. Environ.* **1997**, *2*, 199–219. [CrossRef]
44. Monajem, S.; Ekram Nosratian, F. The Evaluation of the Spatial Integration of Station Areas via the Node Place Model; an Application to Subway Station Areas in Tehran. *Transp. Res. Part D Transp. Environ.* **2015**, *40*, 14–27. [CrossRef]
45. Nigro, A.; Bertolini, L.; Moccia, F.D. Land Use and Public Transport Integration in Small Cities and Towns: Assessment Methodology and Application. *J. Transp. Geogr.* **2019**, *74*, 110–124. [CrossRef]
46. Ewing, R.; Cervero, R. Travel and the Built Environment: A Meta-Analysis. *J. Am. Plan. Assoc.* **2010**, *76*, 265–294. [CrossRef]
47. Vale, D.S.; Pereira, M. Influence on Pedestrian Commuting Behavior of the Built Environment Surrounding Destinations: A Structural Equations Modeling Approach. *Int. J. Sustain. Transp.* **2016**, *10*, 730–741. [CrossRef]
48. Loo, B.P.Y.; Verle, F. du Transit-Oriented Development in Future Cities: Towards a Two-Level Sustainable Mobility Strategy. *Int. J. Urban Sci.* **2017**, *21*, 54–67. [CrossRef]

49. Singh, Y.J.; Lukman, A.; Flacke, J.; Zuidgeest, M.; Van Maarseveen, M.F.A.M. Measuring TOD around Transit Nodes—Towards TOD Policy. *Transp. Policy* **2017**, *56*, 96–111. [CrossRef]
50. Vale, D.S.; Viana, C.M.; Pereira, M. The Extended Node-Place Model at the Local Scale: Evaluating the Integration of Land Use and Transport for Lisbon's Subway Network. *J. Transp. Geogr.* **2018**, *69*, 282–293. [CrossRef]
51. Pezeshknejad, P.; Monajem, S.; Mozafari, H. Evaluating Sustainability and Land Use Integration of BRT Stations via Extended Node Place Model, an Application on BRT Stations of Tehran. *J. Transp. Geogr.* **2020**, *82*, 102626. [CrossRef]
52. Su, S.; Zhang, H.; Wang, M.; Weng, M.; Kang, M. Transit-Oriented Development (TOD) Typologies around Metro Station Areas in Urban China: A Comparative Analysis of Five Typical Megacities for Planning Implications. *J. Transp. Geogr.* **2021**, *90*, 102939. [CrossRef]
53. Cooper, C.H.; Chiaradia, A.J. SDNA: 3-d Spatial Network Analysis for GIS, CAD, Command Line & Python. *SoftwareX* **2020**, *12*, 100525.
54. Dou, M.; Wang, Y.; Dong, S. Integrating Network Centrality and Node-Place Model to Evaluate and Classify Station Areas in Shanghai. *ISPRS Int. J. Geo-Inf.* **2021**, *10*, 414. [CrossRef]
55. van Eggermond, M.A.B.; Erath, A.L. Quantifying Diversity: An Assessment of Diversity Indices and an Application to Singapore. *FCL Mag. Spec. Issue—Urban Breed. Grounds* **2016**, 30–37. [CrossRef]
56. Shanghai Pingpu District People's Government. *Comprehensive Plan and General Land-Use Plan of Qingpu District, Shanghai, 2017–2035*; Shanghai Pingpu District People's Government: Shanghai, China, 2019.
57. Shanghai Songjiang District People's Government. *Shanghai Municipal Bureau of Planning and Natural Resources, Comprehensive Plan and General Land-Use Plan of Songjiang District, Shanghai, 2017–2035*; Shanghai Songjiang District People's Government: Shanghai, China, 2019.
58. Shanghai Fengxian District People's Government. *Shanghai Municipal Bureau of Planning and Natural Resources, Comprehensive Plan and General Land-Use Plan of Fengxian District, Shanghai, 2017–2035*; Shanghai Fengxian District People's Government: Shanghai, China, 2019.
59. Shanghai Jiading District People's Government. *Shanghai Municipal Bureau of Planning and Natural Resources, Comprehensive Plan and General Land-Use Plan of Jiading District, Shanghai, 2017–2035*; Shanghai Jiading District People's Government: Shanghai, China, 2019.
60. Shanghai Pudong New Area People's Government. *Shanghai Municipal Bureau of Planning and Natural Resources, Comprehensive Plan and General Land-Use Plan of Pudong New Area, Shanghai, 2017–2035*; Shanghai Pudong New Area People's Government: Shanghai, China, 2018.
61. Shanghai Municipal People's Government. *Territory Spatial Master Plan of China (Shanghai) Pilot Free Trade Zone Lin-Gang Special Area (2019–2035)*; Shanghai Municipal People's Government: Shanghai, China, 2020.

buildings

MDPI

Article

Influence of Vegetation on Outdoor Thermal Comfort in a High-Altitude Tropical Megacity: Climate Change and Variability Scenarios

Angélica María Bustamante-Zapata [1], Carlos Alfonso Zafra-Mejía [2,*] and Hugo Alexander Rondón-Quintana [3]

1 Grupo de Investigación INDESOS, Facultad del Medio Ambiente y Recursos Naturales, Universidad Distrital Francisco José de Caldas, Carrera 5 Este #15-82, Bogotá E-111711, DC, Colombia; ambustamantez@correo.udistrital.edu.co

2 Grupo de Investigación en Ingeniería Ambiental—GIIAUD, Facultad del Medio Ambiente y Recursos Naturales, Universidad Distrital Francisco José de Caldas, Carrera 5 Este #15-82, Bogotá E-111711, DC, Colombia

3 Ingeniería Topográfica, Facultad del Medio Ambiente y Recursos Naturales, Universidad Distrital Francisco José de Caldas, Carrera 5 Este #15-82, Bogotá E-111711, DC, Colombia; harondonq@udistrital.edu.co

* Correspondence: czafra@udistrital.edu.co; Tel.: +57-6013239300 (ext. 4040); Fax: +57-6012841658

Abstract: The objective of this paper is to show a study on the influence of vegetation on the outdoor thermal comfort (OTC) of a high-altitude tropical megacity. The OTC is evaluated by the PET (Physiological Equivalent Temperature) index and by establishing three simulation scenarios: (i) Current OTC, (ii) OTC under RCPs 4.5 and 8.5 (Representative Concentration Pathway), and (iii) OTC under RCPs and ENSO (El Niño–Southern Oscillation). The results show that the hourly variation range of the current OTC in urban areas with vegetation is greater (+3.15 °C) compared to impermeable areas. Outdoor thermal stress due to cold in vegetated areas is 1.29 °C lower compared to impervious areas. The effect of vegetated coverage on the improvement of urban OTC increases as the phenomenon of global warming intensifies. On average, in the current, RCP4.5, and RCP8.5 scenarios for each 10% increase in urban vegetation coverage, an increase of 0.22, 0.24, and 0.28 °C in OTC is obtained, respectively. The hourly variation range of the PET index increases during the ENSO scenario (vegetated areas: +16.7%; impervious areas: +22.7%). In the context of climate change and variability, this study provides a reference point for decision-makers to assess possible planning options for improving OTC in megacities.

Keywords: ENSO; outdoor thermal comfort; physiological equivalent temperature; representative concentration pathway; urban coverage

1. Introduction

Worldwide, it was estimated that since pre-industrial times the temperature of the planet increased by approximately 1.0 °C and that this increase was progressive during the 21st century [1]. The increase in temperature associated with climate change could contribute to the occurrence of warmer days, mainly in low latitude areas, and in the manifestation of more heat waves, with greater incidence in cities [2,3]. This environmental problem has also been intensified in cities because urban growth was one of the main causes of natural ecosystem degradation and therefore of decrease in outdoor thermal comfort (OTC) [4]. The previous urban trend also increased the loss of environmental quality and well-being of the inhabitants [5]. Thus, the accelerated growth, reduction of natural areas, and low percentage of vegetated areas were aspects of concern in relation to the temperature increase in urban areas [6].

The effects of temperature increase in urban areas were also intensified due to the increase in energy demand from cooling systems, CO_2 emission by electric power plants, and emissions from the use of fossil fuels and refrigerant gases in air conditioners and refrigerators [7,8]. Urban adaptation strategies to the possible impacts of climate change included increased vegetation, use of cold roofs that could be vegetated or with greater solar reflectance, and increasing cold pavements [9,10]. Therefore, it is necessary to develop studies on the influence of vegetation on OTC in urban areas. These types of studies will be useful for decision-makers in public and private institutions, who are continually looking for tools to guide the planning of their cities. The increase in temperature is one of the planning challenges that cities must continue to address in the coming years, as well as its possible intensification due to phenomena such as El Niño—Southern Oscillation (ENSO) and heat islands.

In the IPCC's Fifth Assessment Report (AR5), climate change scenarios were based on total anthropogenic radioactive forcing by the end of the 21st century. Four representative concentration pathways (RCPs) for greenhouse gases (GHGs) were defined [11]. The four RCPs incorporated a strict mitigation scenario (RCP2.6), two intermediate scenarios (RCP4.5 and RCP6.0), and a scenario with extremely high GHG emissions (RCP8.5) [12]. These scenarios evolve based on the driving force: population growth, socioeconomic development, and GHGs [13]. In this study, the RCP4.5 and RCP8.5 scenarios are considered to analyze the influence of vegetation on the urban OTC. The RCP4.5 scenario forecasts climate under the assumption that the current level of CO_2 emissions is maintained. The RCP 8.5 scenario forecasts climate under the assumption that the current level of CO_2 emissions changes dramatically [14]. In the global warming context, climate variation generates significant worldwide interest given its direct and indirect regional effects on the economy, public health, and ecosystems [15]. Due to possible teleconnections [16], the ENSO phenomenon is considered a significant factor in predicting regional climate variations in the study megacity. The ENSO phenomenon can be classified into three phases: El Niño (warm phase), La Niña (cold phase), and Neutral [17]. ENSO cycle lasts between 12–18 months, in periods of two to seven years, although some El Niño and La Niña events can last longer than 24 months. This phenomenon occurs along the equator between latitudes 5° N and 5° S [18].

The OTC was defined as the relationship between the climate of a site and the population perception [19]. According to psychometric tables, OTC was observed in temperature ranges between 20–26.7 °C and relative humidity between 20–80% [20,21]. However, OTC was a subjective variable and depended on factors such as the person's age, whether they are female or male, body shape, skin color, eating habits, and health status [22]. Other studies also reported clothing type and metabolic activity as influencing factors. These last two factors were conditioned by environmental variables such as radiant temperature and airflow rate [23]. Although people could not control the climate conditions of a given site, they could adapt to them and the environmental conditions of their surroundings through clothing and housing [24]. The normal development of daily activities and habitability in cities required the existence of several hours of OTC, between 2–9 h [25,26]. In addition, other factors were reported to be considered, such as the importance of the cultural condition on OTC. That is, if it came from a region different from the site where it was inhabited, the thermal memory related to the weather of the last days, the adaptation to the environment, and the expectations of the population [20,27].

Urban vegetation was recognized as an outdoor thermal adaptation strategy for its benefits over temperature regulation, contribution with the efficient use of electrical energy, the contribution of shade, decrease in surface temperature, and influence of evapotranspiration in cities of temperate and subtropical climates [9,28]. The study city (Bogotá, Colombia) is in a tropical zone, at 2600 m.a.s.l., and shows hourly temperature variations on the same day of up to 10 °C. The city is also part of one of the Latin American regions of high altitude most susceptible to climate variations (tropical mountain climate—cold), and before any of the representative concentration pathway (RCP) of climate change, a temperature increase of up to 2.0 °C is forecast between 2041–2070 [29]. These temperature increases could generate a greater impact on the city because of the effects associated with the El Niño—Southern Oscillation (ENSO) phenomenon [30]. Moreover, in the city, there is evidence of the occurrence of the heat island phenomenon [31], which together with temperature variations due to climate change, the ENSO phenomenon, and its high altitude, can become adjuvants of discomfort or outdoor thermal stress in the city [32].

The objective of this paper is to show a study on the influence of vegetation on OTC in a tropical megacity of high altitude (Bogotá, Colombia). The effect of vegetation on OTC is studied from its spatial variation in relation to other types of urban coverage. The predominant coverage in the study megacity is that of the impervious type. Thus, monitoring stations with a predominance of vegetated coverage and monitoring stations with a predominance of impervious coverage were selected. Within the context of OTC, this study is relevant for the following aspects: (i) analyzes the influence of vegetation in tropical urban areas of high altitude, (ii) evaluates the effect of the variation of urban surface coverage, and (iii) studies the influences of climate change and the ENSO phenomenon on OTC.

2. Materials and Methods

2.1. Study Site

This study was conducted in the high-altitude tropical megacity of Bogotá, Colombia (4°35′57″ N; 74°04′51″ W). Four monitoring stations distributed throughout the megacity were established. The elevation of the monitoring stations was between 2577–2590 m.a.s.l. In this study, two monitoring stations were established where the predominant urban coverage was vegetated: Guaymaral = GU and Parque Simón Bolívar = PSB. In contrast, two monitoring stations were also established where the predominant urban coverage was impervious: Puente Aranda = PA and Kennedy = KE (Figure 1). On average, the annual climate characteristics of the study sites were as follows: Rainfall = 1091 mm, temperature = 12.9 °C, wind speed = 9.60 km/h, and relative humidity = 77.0%.

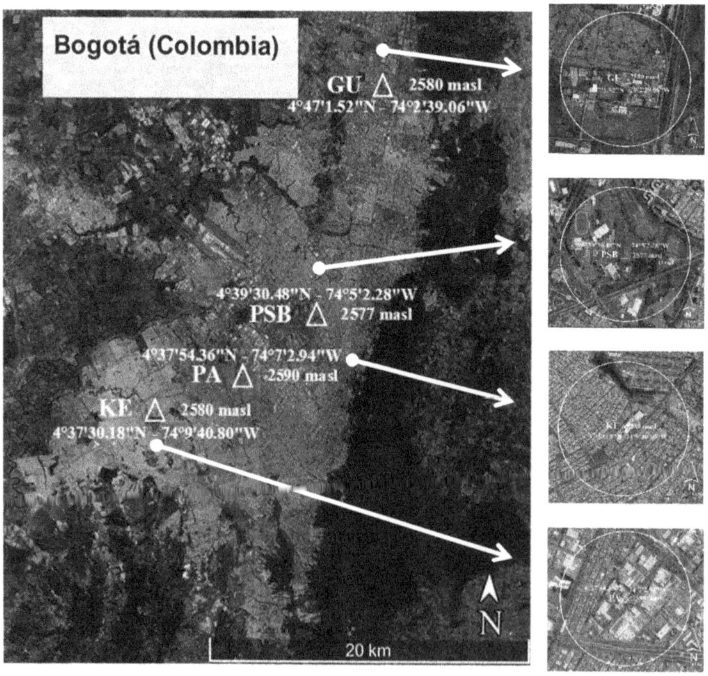

Figure 1. Location of monitoring stations in the megacity of Bogotá, Colombia (Google Earth Pro, 2022).

2.2. Information Collection

Orthomosaics from 2010–2017 were used. These were downloaded from the Spatial Data Infrastructure Platform for the Capital District of Bogotá, Colombia [33]. Subsequently, maps were generated at each monitoring station on a scale of 1:5000. The maps were generated within a radius of 500 m in relation to the physical location of each monitoring station [34,35]. Within this radius of influence, the following types of urban coverage were identified: impervious coverage—IC (pavement, concrete, and buildings), vegetated coverage—VC (grass, uncovered soil, and trees), and water coverage—WC (rivers, lakes, and wetlands). The types of urban coverage were quantified using ArcGis software (Version 10.3, Esri, San Diego, CA, USA). In addition, climate information was collected at each monitoring station during the study period: 1 January 2010–31 December 2017. The climate information was downloaded from the Platform of the Air Quality Monitoring Network of Bogotá, Colombia [36]. The climate variables considered were the following: air temperature (T_a), wind speed (V_a), and relative humidity (RH).

2.3. Information Analysis

Initially, a pre-processing of the climate information obtained at each monitoring station was conducted. The missing information in the time series of climate variables was completed using the normal ratio method [37]. Before completing the missing information, it was verified that the time series had more than 75% of the information from the study period considered. Subsequently, the existence of anomalies in the time series was analyzed using the double-mass method [38]. In relation to temperature, the months of greatest hourly variation during the same day were identified, taking as a reference their monthly average at the multi-year level. From the time series of climate variables, the mean radiant temperature (TMR) was calculated. This corresponded to the amount of radiation absorbed by the pedestrian during the day, where the sources were all the surfaces that surrounded him and the view factor that existed between the sources and the pedestrian [39]. The TMR considered both shortwave (direct, diffuse, and reflected) and longwave radiation (surrounding surfaces and from the atmosphere). The equations used for the calculation of TMR (°C) considered the guidelines of Sen and Nag [40]:

$$T_d = T_a - \left(\frac{100 - RH}{5}\right) \tag{1}$$

$$T_g = (0.811 \times T_d) + 9.93 \tag{2}$$

$$TMR = \left(1 + 0.22 \times V_a^{0.5}\right) \times (T_g - T_a) + T_a \tag{3}$$

where, T_d is the dew point temperature (°C), T_a is the air temperature (°C), RH is the relative humidity (%), T_g is the globe temperature (°C), V_a is the wind speed (m/s), and TMR is the mean radiant temperature (°C).

From the calculated TMR, the OTC in each monitoring station was analyzed using the Physiological Equivalent Temperature (PET) index. This is due to its wide use to evaluate the effect of vegetated coverage on OTC (e.g., [34,41]). The PET index calculation was performed with RayMan Pro Beta software (Version 3.1, Meteorological Institute, University Freiburg, Baden-Württemberg, Germany) [42] and based on TMR. This software was selected for its wide use in OTC studies on low and non-seasonal latitudes, such as India [40] and Singapore [43]. The following simulation conditions proposed by Matzarakis and Amelung [44] were used: physiological profile of a man, 35 years old, 1.75 m in height, 75 kg in weight, with 0.90 of clothing (dressed without a coat), standing, and 80 W of metabolic activity (standing and resting). In addition, it was assumed that the influencing factors in anthropogenic heat were uniform at all monitoring stations. Namely, no variations in OTC associated with factors such as heating and cooling systems, traffic, and industries were considered. Validation of the calculated PET index using RayMan Pro Beta software (Version 3.1, Meteorological Institute, University Freiburg, Baden-Württemberg,

Germany) was performed in two ways. Initially, the Universal Thermal Comfort Index (UTCI) calculated using RayMan Pro Beta software (Version 3.1, Meteorological Institute, University Freiburg, Baden-Württemberg, Germany) was compared with that calculated using the UTCI calculator [45]. Thus, differences in the calculation of the PET index were compared using these two free-use tools. Subsequently, linear correlation analyses were performed between the PET index calculated using RayMan Pro Beta software (Version 3.1, Meteorological Institute, University Freiburg, Baden-Württemberg, Germany) and the temperature values observed at the monitoring stations. A strong relationship between these two variables suggested adequate performance for the software used [40,42].

The current OTC was estimated for each study area during the 24 h and in the months detected with the greatest range of hourly temperature variation for the same day. The PET index was classified according to the following ranges [46]: <4 °C = Extreme cold stress, 4–8 °C = Strong cold stress, 8–13 °C = Moderate cold stress, 13–18 °C = Light cold stress, 18–23 °C = No thermal stress, and 23–29 °C = light heat stress. This was based on the PET range that was historically observed in the monitoring stations selected in the study megacity. Spatiotemporal analysis of the influence of vegetated coverage on the current OTC (PET index) was carried out based on the guidelines of Sun et al. [47]. The influence of the two types of predominant urban coverage (VC: vegetated coverage and IC: impervious coverage) on OTC was analyzed (Figure 1). This analysis was developed during the study period and at the following hours: 1:00, 7:00, 13:00, and 19:00 h.

The study of the influence of urban coverage under RCP scenarios was carried out for the month detected as the one with the greatest hourly variation in temperature for the same day, and with a simulation horizon to 2100. Future OTC was calculated from the T_a projection for the RCPs 4.5 and 8.5. In this analysis, the guidelines proposed by Aminipouri et al. [48] were followed and considering the increase in T_a predicted for the Bogotá city (Colombia). This temperature increase corresponded to 2.20 °C and 4.90 °C under the RCPs 4.5 and 8.5, respectively [30]. Thus, V_a and RH were also projected under the previous RCPs in all monitoring stations and from their correlation with T_a (linear regression models). Then, TMR with Equations (1)–(3), and OTC through the PET index were estimated for these RCP scenarios. All these calculations were executed with RayMan Pro Beta software (Version 3.1, Meteorological Institute, University Freiburg, Baden-Württemberg, Germany). Lastly, forecasts were also made based on average values of minimum and maximum T_a observed during the study period.

Base on OTC calculated for scenarios RCP 4.5 and 8.5, influence of the ENSO phenomenon was simulated (ENSO scenario). For this purpose, average PET index during the study period (2010–2017) was determined and compared with the years in which the occurrence of the ENSO phenomenon was reported. Hourly differences observed in the PET index between these two time periods (average and occurrence of ENSO) were considered to analyze the influence of the ENSO phenomenon on OTC under the RCPs considered. During the study period, the occurrence of a strong ENSO was reported for 2010 (Oceanic Niño Index—ONI = 1.50 °C) [49]. This was the reference year to study the future influence of the ENSO phenomenon under the RCP scenarios considered. Therefore, in this study, three scenarios of information simulation and analysis were considered: (i) Current OTC, (ii) OTC under RCPs 4.5 and 8.5, and (iii) OTC under RCPs 4.5 and 8.5, and ENSO (El Niño). Lastly, the following statistical tests were applied for the analysis of the information: Kolmogov-Smirnov normality test (p-values > 0.050), Pearson correlation coefficient, t-Student test, and linear regression models. All statistical tests were applied using IBM-SPSS software (Version 19.0, IBM, New York, NY, USA) and with 95% confidence.

3. Results and Discussion

3.1. Urban Surface Coverage

The coverage analysis during the study period (2010–2017) allowed us to identify variations in each monitoring station (Table 1). For example, the results showed that in 2017 the monitoring stations with the highest vegetated coverage were GU (VC = 75.9%) and PSB

(VC = 72.3%). In contrast, the PA and KE stations were those with the highest impervious coverage (IC: 83.1% and 75.8%) and the lowest vegetated coverage (VC: 16.9% and 24.2%), respectively. Only two monitoring stations showed water coverage (WC: GU = 4.40% and PSB = 3.90%), which was constant during the study period. In this study, there were two monitoring stations with a predominance of vegetated coverage and two stations with a predominance of impervious coverage to analyze the variation of urban OTC.

Table 1. Spatial variation of urban surface coverage during the study period (2010–2017).

Coverage Type	Monitoring Station	Year		
		2010	2014	2017
VC (%)	GU	79.3	78.9	75.9
	PSB	74.0	73.0	72.3
	PA	13.7	22.1	16.9
	KE	20.7	24.3	24.2
IC (%)	GU	16.3	16.8	19.6
	PSB	22.1	23.1	23.8
	PA	86.3	77.9	83.1
	KE	79.3	75.7	75.8
WC (%)	GU	4.40	4.40	4.40
	PSB	3.90	3.90	3.90

Note. VC = Vegetated coverage, IC = Impervious coverage, and WC = Water coverage.

3.2. Current OTC

On average, the results showed during the study period that the months with the greatest range of hourly temperature variation for the same day were in order of importance January, February, and December. For example, at the GU station, the maximum hourly temperature variation in January, February, and December was 12.9 °C, 11.5 °C, and 11.3 °C, respectively. These months were prioritized to analyze the OTC of all the monitoring stations under study. The Institute of Hydrology, Meteorology, and Environmental Studies of Colombia also reported equivalent results [29]. They reported hourly temperature variations for the same day exceeding 10 °C at the GU and PSB stations. The analysis of OTC in the monitoring stations and for the most critical month (January) showed that the time interval of the highest thermal stress due to strong cold (PET: 4–8 °C) was between 1:00–7:00 h (Table 2). The OTC was improved as the day progressed. Thus, at 13:00 h, OTC was of stress due to light cold (PET: 13–18 °C). However, at night, a reduction in OTC was again observed. This was of moderate cold thermal stress (PET: 8–13 °C). The results suggested that during the month of greatest temperature variation (January) there was no time interval without outdoor thermal stress. This trend was also observed for the other months and throughout the study period in this high-altitude megacity (tropical mountain climate—cold). During the other months, there were fewer hourly variations in the PET index for the same day. Though, cold thermal stress remained.

Table 2. Average values of PET index (°C) for January.

Hour	Year	ENSO		Monitoring Station			
		ONI	Category	GU	PSB	KE	PA
1:00	2010	+1.50	El Niño—Strong	7.80	8.50	9.00	8.90
	2014	−0.40	La Niña—Weak	10.5	10.1	9.60	9.30
	2017	−0.30	La Niña—Weak	12.0	10.4	10.0	10.1
7:00	2010	+1.50	El Niño—Strong	7.00	7.00	7.00	7.50
	2014	−0.40	La Niña—Weak	9.20	8.80	8.40	9.30
	2017	−0.30	La Niña—Weak	11.0	9.70	8.80	10.0
13:00	2010	+1.50	El Niño—Strong	16.0	17.0	13.7	13.5
	2014	−0.40	La Niña—Weak	15.8	15.6	12.6	12.3
	2017	−0.30	La Niña—Weak	15.3	15.5	12.2	11.7
19:00	2010	+1.50	El Niño—Strong	11.5	11.8	10.8	10.6
	2014	−0.40	La Niña—Weak	11.5	11.5	10.7	9.80
	2017	−0.30	La Niña—Weak	11.3	12.6	10.4	10.2
			Min.	7.00	7.00	7.00	7.50
			Max.	16.0	17.0	13.7	13.5
			Range	9.00	10.0	6.70	6.00
			Median	11.6	11.5	10.3	10.3

Note. ONI = Oceanic Niño Index. PET: 4–8 °C = Stress/strong cold, 8–13 °C = Stress/moderate cold, 13–18 °C = Stress/light cold, and 18–23 °C = No thermal stress.

A comparative analysis was conducted between 2010, 2014, and 2017 in relation to the occurrence of the ENSO phenomenon and its influence on OTC. This is for the month with the greatest hourly variation of temperature on the same day (January). The results showed the occurrence of El Niño for 2010 (ONI = 1.50 °C, strong), La Niña for 2014 (ONI = −0.40 °C, weak), and La Niña for 2017 (ONI = −0.30 °C, weak) [49]. When analyzing the possible influence of the ENSO phenomenon, the results showed that during its occurrence in 2010, OTC tended to decrease during the most critical hour (7:00 h). On average, thermal stress due to strong cold increased between 25.2–38.6% (Table 2). However, OTC tended to increase slightly during the least critical hour (13:00 h). On average, light cold thermal stress decreased between 2.45–8.07% during this hour. In the study megacity, the ENSO phenomenon historically manifested itself as a period of dry weather (decrease in rainfall) and an increase in temperature [49]. Therefore, the results suggested for the high-altitude megacity under study (tropical mountain climate—cold) that the ENSO phenomenon tended to increase outdoor thermal stress more compared to outdoor thermal comfort. This is for the identified hours of lowest (7:00 h) and highest (13:00 h) OTC during the day, respectively. In the study sites, as the ENSO phenomenon is associated with a greater number of days of dry weather and increased temperature, it is likely that the above findings will be intensified in the future due to the global warming phenomenon. In other words, there will possibly be a greater hourly range of temperature variation and, therefore, greater outdoor thermal stress during the day.

Overall, OTC at study sites ranged from strong cold stress (PET: 4–8 °C) to moderate cold stress (PET: 13–18 °C) (Table 2). A *t*-Student test showed the existence of significant differences in the PET index between monitoring stations with different urban surface coverage: GU–PA (p-value = 0.005), GU–KE (p-value = 0.010), PSB–PA (p-value = 0.007), and PSB–KE (p-value = 0.210). As mentioned, the GU and PSB stations showed predominantly vegetated coverage (GU = 78.0% and PSB = 73.1%) compared to the PA and KE stations, where the predominant coverage was impervious (PA = 82.4% and KE = 76.9%). Therefore, the results suggested that OTC was influenced by the type of surface coverage predominant in the study stations. Obiakor et al. [50] reported similar results in urban areas. The radius of influence analyzed in our study was 500 m, in relation to the physical location of the selected monitoring stations.

On average, the results suggested comparatively that an increase between 55.0–60.5% in vegetated coverage generated an increase between 12.4–12.7% (PET: 1.28–1.31 °C) in urban OTC. This trend was more marked during the hours of greater OTC (13:00 h). In this case, OTC increased between 22.3–28.3% (2.87–3.53 °C). In contrast, this trend was less marked during the hours of lower OTC (7:00 h). At this time, OTC increased only between 1.49–12.4% (0.13–1.00 °C) (Figure 2). The results suggested for the high-altitude megacity under study that the increase in vegetated coverage tended to improve thermal comfort during the hours of lower thermal stress (13:00 h) compared to the hours of greater thermal stress (7:00 h). The difference observed in the improvement of OTC between these two hours was between 9.90–26.8% (1.87–3.40 °C). The previous results were consistent with those reported by Chow et al. [43] in Singapore, where at noon the highest values of OTC (PET) were observed in areas with a predominance of vegetated coverage, and the lowest values were observed in areas with a predominance of impervious coverage. In our study, the average ratio between the percentage increase in vegetated coverage and the percentage increase in OTC was 4.60. Namely, for every 10.0% increase in urban vegetation coverage, an increase of 0.22 °C in OTC was obtained.

Additionally, the results showed, on average, that the hourly variation range of OTC in monitoring stations with a predominance of vegetated coverage was greater (PET range = 9.50 °C) compared to monitoring stations with a predominance of impervious coverage (PET range = 6.35 °C) (Figure 2). These findings indicated that monitoring stations with a predominance of impervious coverage tended to be more frequently in conditions of higher cold thermal stress (PET: 7.25–13.6 °C) compared to monitoring stations located in areas of predominance vegetated coverage (PET: 7.00–16.5 °C). During the study period, the outdoor thermal stress in the monitoring stations with a predominance of vegetated coverage (VC between 73.1–78.0%) was 1.29 °C lower compared to the monitoring stations with a predominance of impervious coverage (IC between 82.4–76.9%). Song and Wang [51] reported similar results in the City of Phoenix (USA).

As expected, there were no significant differences between the GU and PSB stations in relation to the PET index (p-value = 0.894). This is possibly due to the same type of urban surface coverage predominant in these two study stations: vegetated coverage (GU = 78.0% and PSB = 73.1%) (Figure 2). Likewise, between the PA and KE stations, there were no significant differences in relation to the PET index (p-value = 1.000). This trend was possibly associated with the predominant impervious coverage at these two monitoring stations (PA = 82.4% and KE = 76.9%). Therefore, consistency in the findings was observed from the analysis of the influence of urban surface coverage on OTC.

The results consistently showed that the climate variable that best correlated with the PET index at all monitoring stations was T_a. The relationship detected was direct and very strong (average r-Pearson = 0.955). This was the main climate variable that conditioned the PET index (Figure 3). However, the results showed differences for the other climate variables considered from the predominant urban surface coverage. In the monitoring stations with a predominance of vegetated coverage, the order of importance in the correlations between the PET index and the other climate variables was as follows: RH (average r-Pearson = −0.923) > V_a (average r-Pearson = 0.784). Conversely, in stations with a predominance of impervious coverage the order of importance was as follows: V_a (average r-Pearson = 0.890) > RH (average r-Pearson = −0.857). Therefore, the results suggested that in areas with a predominance of vegetated coverage (GU and PSB), RH had a greater influence on the PET index compared to V_a. Conversely, in stations with a predominance of impervious coverage (KE and PA), V_a had a greater influence compared to RH.

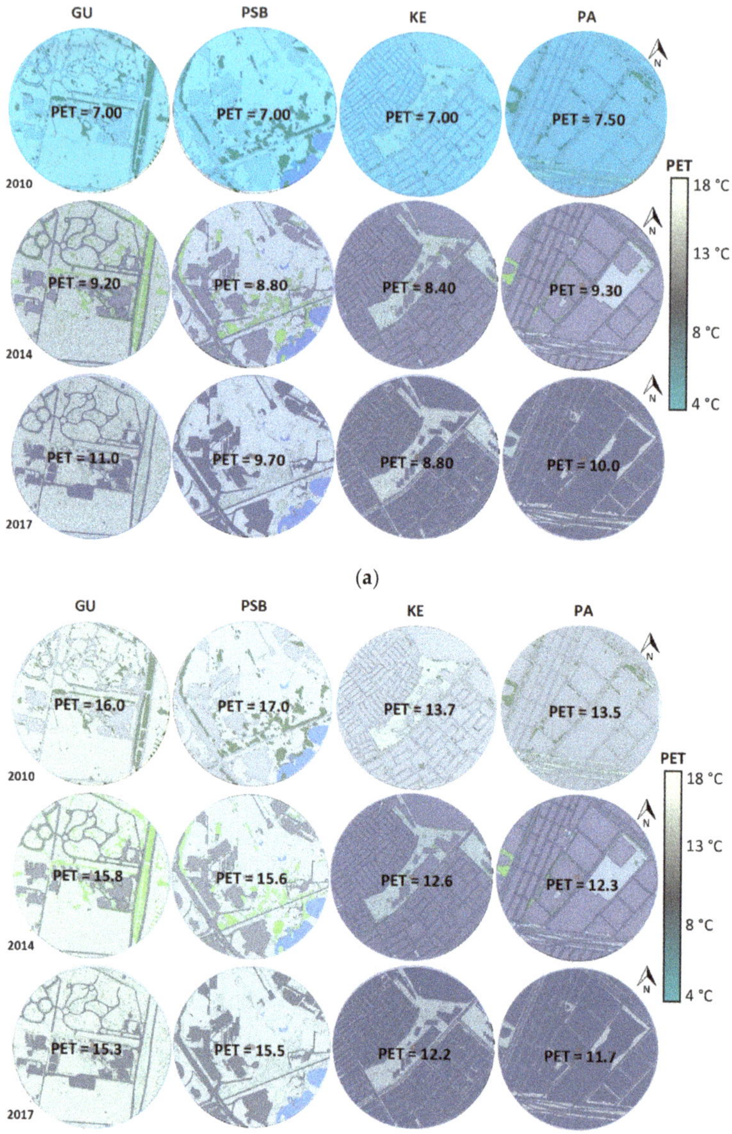

Figure 2. Average PET index for (**a**) 7:00 h and (**b**) 13:00 h during January. PET: 4–8 °C = Stress/strong cold, 8–13 °C = Stress/moderate cold, 13–18 °C = Stress/light cold, and 18–23 °C = No thermal stress.

Under the climate conditions of the study (tropical mountain climate—cold), the results showed that when T_a increased, the outdoor thermal stress due to cold decreased (PET) in all the monitoring stations considered (Figure 3). On average, in the monitoring stations with a predominance of vegetated coverage, for each increase of 1.0 °C in T_a, the OTC increased by 7.27% (0.84 °C). In the monitoring stations with a predominance of impervious coverage, for every increase of 1.0 °C in T_a, the OTC increased by 7.00% (0.72 °C). Comparatively, as T_a increased, OTC increased more rapidly in monitoring stations with vegetated coverage rather than in monitoring stations with impervious coverage. The findings also hinted that when RH decreased, OTC was increased in the high-altitude megacity under study. On average, in monitoring stations with a predominance of vegetated coverage, for every 10.0% decrease in RH, the OTC increased by 14.8% (1.71 °C). In monitoring stations with a predominance of impervious coverage, for every 10.0% decrease in RH, the OTC increased by 15.7% (1.61 °C).

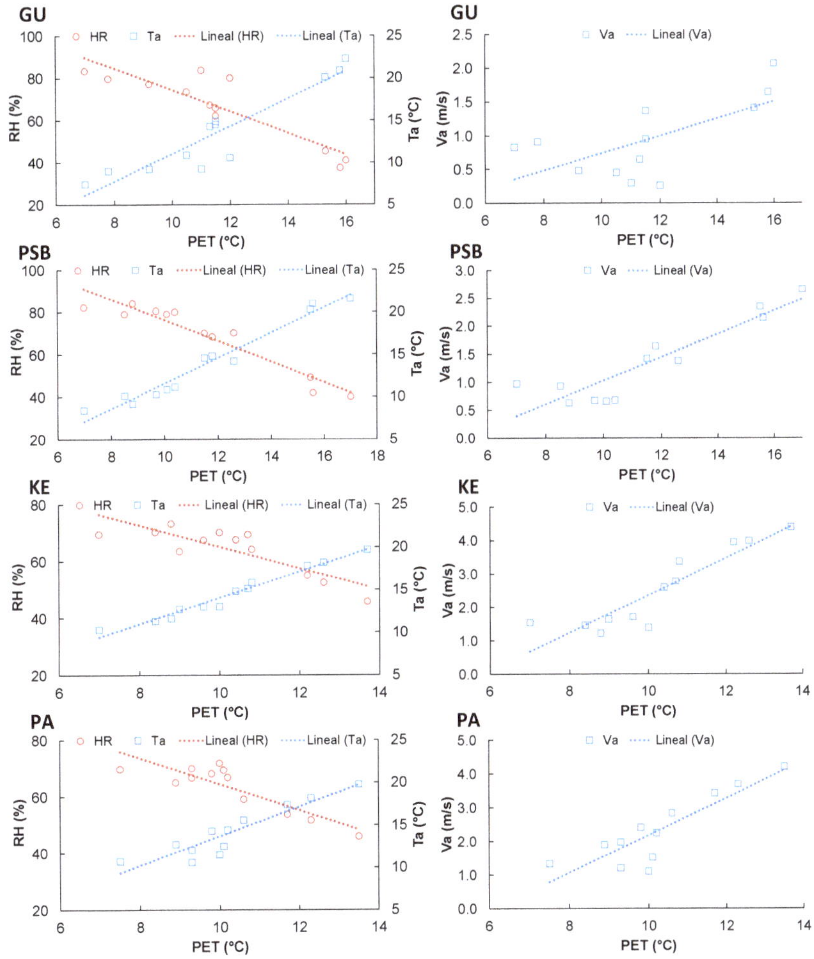

Figure 3. Average trend of climate variables in relation to the PET index. T_a = Temperature, RH = Relative humidity, and V_a = Wind speed.

The results showed that during the hours in which the highest wind speeds were observed, the greatest OTC was also observed. However, thinking in this study that when V_a increased OTC also increased could be a mistake. These increases in V_a towards the hour of greater OTC (13:00 h) were possibly related to atmospheric instability conditions in the high-altitude megacity under study [52]. If this trend had not occurred, OTC would have been greater during this time. When analyzing the trend of the main variables to determine OTC (T_a, RH, and V_a), it was observed that, although there was a high T_a (18.0 °C) and a low RH (between 37.0–52%), an increase in V_a (between 0.60 m/s–1.70 m/s) could reduce OTC, and therefore increase the outdoor thermal stress due to cold. A multiple linear regression analysis between the PET index (dependent variable) and the climate and coverage variables (independent variables) showed that there were no significant models for the study sites. In this study, the PET index could not be explained statistically from the type of urban coverage considered. This index could only be explained statistically from the three climate variables considered for its calculation ($R^2 \geq 0.916$, p-value < 0.001) (Figure 3).

Finally, during the validation of the selected software (RayMan Pro Beta software, Version 3.1, Meteorological Institute, University Freiburg, Ba-den-Württemberg, Germany)), a very strong linear correlation was evidenced between the calculated PET index and the temperature values observed in all monitoring stations (r-Pearson $\geq$ 0.920, p-value < 0.001). In addition, the results showed that the differences in the calculation of UTCI using the RayMan Pro Beta software (Version 3.1, Meteorological Institute, University Freiburg, Baden-Württemberg, Germany) and UTCI calculator [45] were between 0.26–2.62 °C. Therefore, in this study, the findings suggested good performance of the selected software.

3.3. RCP Scenarios

Overall, the results showed that OTC predicted under RCP scenarios (2070–2100) had a similar hourly trend compared to that observed during the baseline study period (2010–2017). This is based on the increases in the PET index forecast under the two RCP scenarios considered (RCPs 4.5 and 8.5). The results showed that the months with the greatest hourly temperature variation for the same day were in order of importance January, February, and December. Thus, OTC under RCP scenarios was evaluated for the most critical month (January). Figure 4 shows the observed average PET values, and the minimums and maximums forecast under the RCP scenarios considered for the most critical month.

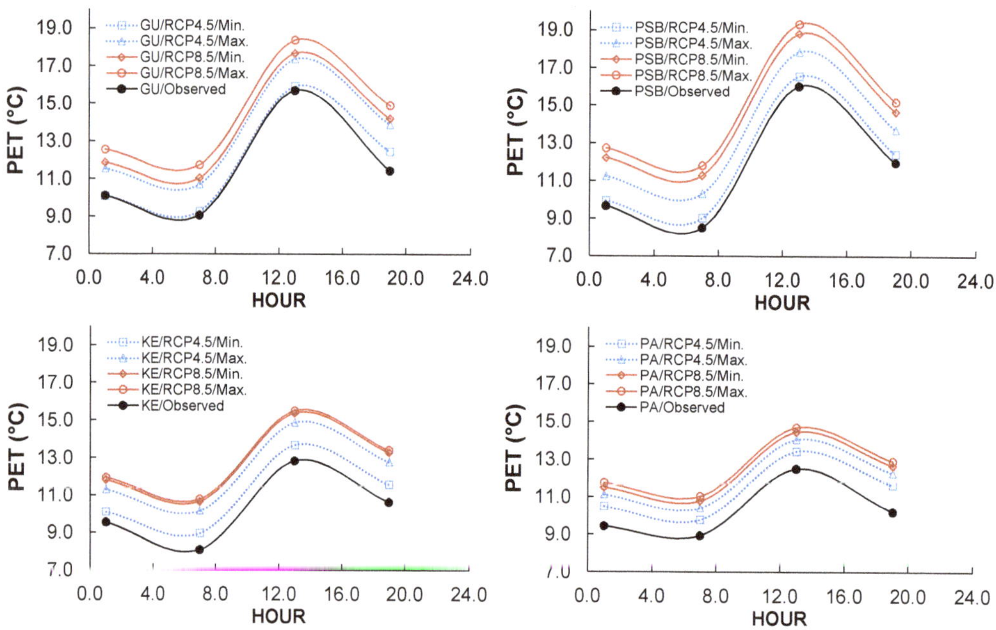

Figure 4. Observed average PET values, and the minimums and maximums forecast under RCP scenarios at monitoring stations. PET: 4–8 °C = Stress/strong cold, 8–13 °C = Stress/moderate cold, 13–18 °C = Stress/light cold, and 18–23 °C = No thermal stress.

A t-Student test showed the existence of significant differences in the predicted PET index between monitoring stations with different urban surface coverage: GU–PA (p-values < 0.037), GU–KE (p-values < 0.004), PSB–PA (p-values < 0.027), and PSB–KE

(p-values < 0.003). The results hinted at differences in predicted OTC from the type of coverage predominant in the study stations. As mentioned, vegetated coverage predominated at GU and PSB stations, and impervious coverage predominated at KE and PA stations. On average, greater differences were observed between monitoring stations with different coverage when considering the predicted maximum PET values (difference = 12.2%, 1.52 °C), compared to the predicted minimum PET values (difference = 9.12%, 1.09 °C). This is under each of the RCP scenarios considered (Figure 4). Therefore, the findings suggested that the differences in OTC between urban areas with a predominance of vegetated and impervious coverage were greater when maximum temperatures were observed under the RCP scenarios. The results also showed that the global warming phenomenon had a greater effect during the hour of lower OTC (7:00 h) compared to the hour of greater OTC (13:00 h). On average, regardless of the coverage type, the increase in PET at 7:00 h was 21.4% and at 13:00 h was 13.2%. The previous trend suggested that the global warming phenomenon had a lower impact during the hours of greater OTC (13:00 h) compared to the hours of lower OTC (7:00 h). This is under the particular high-altitude climate conditions of the tropical megacity under study.

On average, for the RCP4.5 scenario, the results suggested comparatively that an increase between 55.0–60.5% in vegetated coverage generated an increase between 8.36–13.5% (PET: 1.06–1.71 °C) in urban OTC. Namely, a decrease in moderate cold stress was observed. Under this scenario, during the hour of greatest OTC (13:00 h) it was observed that in the monitoring stations with a predominance of vegetated coverage the thermal stress was light cold. In monitoring stations with a predominance of impervious coverage, the outdoor thermal stress was similar (Figure 4). On average, for the RCP8.5 scenario, the findings comparatively hinted that an increase between 55.0–60.5% in vegetated coverage generated an increase between 8.57–14.0% (PET: 1.20–2.04 °C) in urban OTC. That is, cold stress decreased from moderate to light. During the hour of greatest OTC (13:00 h), it was observed that in the monitoring stations with a predominance of vegetated coverage the thermal stress was between light cold and without thermal stress. While in the monitoring stations with a predominance of impervious coverage there was always stress due to light cold. In the high-altitude tropical megacity under study, the results suggested that the effect of vegetated coverage on improving OTC increased as the phenomenon of global warming (RCP8.5) intensified. This compared to the urban coverage of impervious type. The global warming phenomenon under the scenarios RCPs 4.5 and 8.5 caused cold stress to decrease in the megacity under study.

Finally, under the RCP4.5 scenario, for each 10.0% increase in urban vegetation coverage, an average increase of 0.24 °C in OTC was obtained. Under the RCP8.5 scenario, for each 10.0% increase in urban vegetation coverage, an average increase of 0.28 °C in OTC was obtained. Without climate change scenarios, the average increase observed was 0.22 °C for each 10.0% increase in vegetated coverage. Therefore, the results suggested that the positive effect of vegetation on OTC increased as the global warming phenomenon intensified.

3.4. ENSO Scenario

In this study, the effect of climate change (RCPs 4.5 and 8.5) during the ENSO phenomenon (2070–2100) was considered. Under this scenario, the results showed that the months with the highest hourly temperature variation for the same day were in order of importance January, February, and December. Thus, OTC under the ENSO scenario was evaluated for that month of greatest variation in the PET index (January). Figure 5 shows the observed average PET values and those forecast under the ENSO scenario for January. A t-Student test showed the existence of significant differences in the predicted PET index between monitoring stations with different urban coverage: GU–PA (p-values < 0.019), GU–KE (p-values < 0.044), PSB–PA (p-values < 0.021), and PSB–KE (p-values < 0.040). As reported, vegetated coverage predominated at GU and PSB stations and impervious coverage predominated at KE and PA stations. On average, greater differences were observed between monitoring stations with different coverage when considering the predicted maximum PET values (difference = 12.2%, 0.53 °C), compared to the predicted minimum PET values (difference = 1.36%, 0.16 °C). This considers the ENSO scenario and each of the RCP scenarios. The results suggested that the difference in OTC between monitoring stations with different types of coverage (vegetated and impervious) was greater when maximum temperatures were observed under the ENSO scenario.

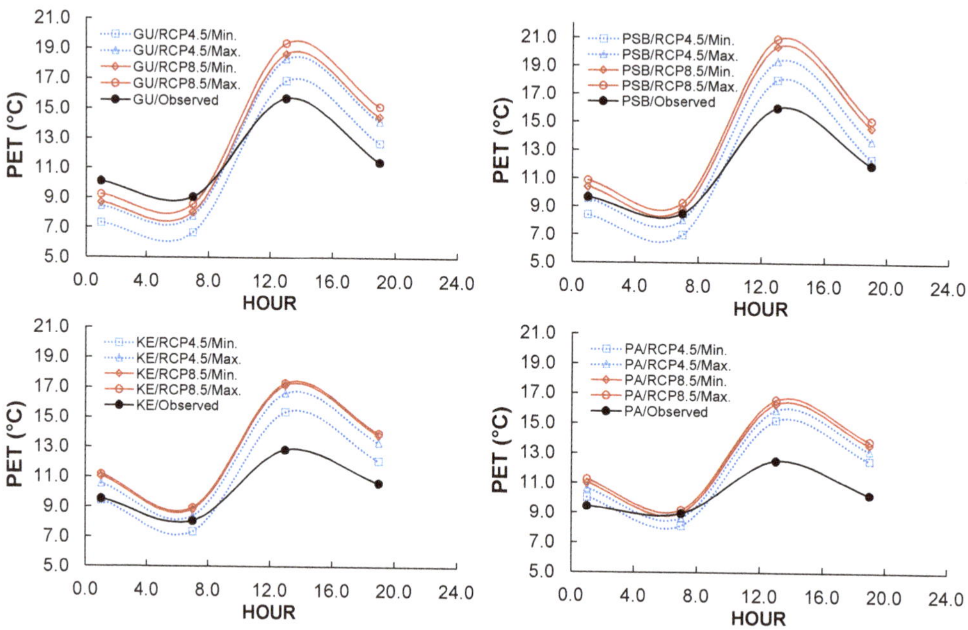

Figure 5. Observed average PET values and the minimums and maximums forecast under the ENSO scenario at monitoring stations. PET: 4–8 °C = Stress/strong cold, 8–13 °C = Stress/moderate cold, 13–18 °C = Stress/light cold, and 18–23 °C = No thermal stress.

The results showed that under the ENSO scenario, OTC had a different behavior than that predicted for the RCP scenarios. In other words, cold stress increased during the hour of lower OTC (7:00 h) (Figure 5). This trend was most evident for monitoring stations with a predominance of vegetated coverage. On average, at the GU and PSB stations, OTC decreased by 16.6% (PET: 1.29 °C) and 2.74% (PET: 0.23 °C), respectively. In contrast, in the monitoring stations with a predominance of impervious coverage, no trend was appreciable during this hour. In the KE station, a slight increase was observed (3.89%, PET: 0.31 °C), and in the PA station, a slight decrease in OTC was observed (2.39%, PET: 0.21 °C). At this time, the outdoor thermal stress was between strong cold and moderate cold. During the hour of greatest OTC (13:00 h), in all monitoring stations, an increase in the PET index was observed regardless of the coverage type. However, this increase in OTC was stronger for monitoring stations with a predominance of impervious coverage. In the KE and PA stations (impervious coverage), the increase in OTC was 29.5% (PET: 3.79 °C) and 28.0% (PET: 3.50 °C), respectively. In the GU and PSB stations (vegetated coverage), the increase in OTC was 16.7% (PET: 2.62 °C) and 22.7% (PET: 3.64 °C), respectively. During this time, the outdoor thermal stress was between light cold and no thermal stress. Therefore, the results suggested that the ENSO phenomenon had a greater impact on the monitoring stations with a predominance of vegetated coverage during the hours of lower OTC (7:00 h). Conversely, the ENSO phenomenon had a greater impact on monitoring stations with a predominance of impervious coverage during the hours of greatest OTC (13:00 h).

On average, during the ENSO scenario, it was observed that the hourly variation range of OTC in monitoring stations with a predominance of vegetated coverage was greater (PET range = 11.0 °C) compared to monitoring stations with a predominance of impervious coverage (PET range = 7.80 °C) (Figure 5). The hourly ranges observed without ENSO and RCP scenarios were 1.16 and 1.23 times lower for vegetated and impermeable coverage stations, respectively (Figure 4). The results suggested that the hourly variation range of the PET index increased during the occurrence of the ENSO phenomenon. In other words, the ENSO phenomenon increased the range of outdoor thermal stress during the day in the high-altitude tropical megacity under study. Lastly, under the ENSO scenario, for each 10.0% increase in urban vegetated coverage, an average increase of 0.28 °C was obtained in OTC.

4. Conclusions

The results of this study on the influence of vegetation on the outdoor thermal comfort (PET) of a high-altitude tropical megacity (cold mountain climate) allow us to visualize the following conclusions:

- Hourly variation range of the current OTC in urban vegetated areas is greater (+3.15 °C) compared to impervious areas. Thus, the outdoor thermal stress due to cold in vegetated areas is 1.29 °C lower compared to impervious areas.
- The effect of vegetated coverage on the improvement of urban OTC increases as the phenomenon of global warming intensifies. On average, in the current, RCP4.5, and RCP8.5 scenarios, for each 10.0% increase in urban vegetation coverage, an increase of 0.22, 0.24, and 0.28 °C in OTC is obtained, respectively.
- ENSO scenario has a greater negative impact on vegetated areas during the hours of lower OTC (7:00 h). Conversely, the ENSO scenario has a greater positive impact on impervious areas during the hours of greater OTC (13:00 h).
- Hourly variation range of the PET index increases during the ENSO scenario: vegetated areas = +16.7% and impervious areas = +22.7%.
- On average, in the ENSO scenario, for each 10.0% increase in urban vegetated coverage, an increase of 0.28 °C in OTC is obtained. This trend is like that reported in the RCP8.5 scenario.

The following detected limitations should be considered to visualize future research lines: (i) the influence of anthropogenic heat (e.g., heating and cooling, transport, and industrial processes) was not considered in the simulations performed, and (ii) the analysis of coverage was simplified, i.e., only urban areas with a predominance of vegetated or impervious coverages were considered. Indeed, a more detailed study on OTC requires consideration of all types of coverage existing in a megacity. Lastly, in the context of climate change and variability, this study constitutes a reference point for decision-makers in public and private organizations to evaluate possible planning options for improving OTC in megacities.

Author Contributions: Conceptualization, A.M.B.-Z. and C.A.Z.-M.; methodology, A.M.B.-Z., C.A.Z.-M. and H.A.R.-Q.; validation and formal analysis, A.M.B.-Z., C.A.Z.-M. and H.A.R.-Q.; resources, A.M.B.-Z., C.A.Z.-M. and H.A.R.-Q.; writing—original draft preparation, A.M.B.-Z. and C.A.Z.-M.; writing—review and editing, A.M.B.-Z., C.A.Z.-M. and H.A.R.-Q.; funding acquisition, A.M.B.-Z., C.A.Z.-M. and H.A.R.-Q. All authors have read and agreed to the published version of the manuscript.

Funding: This research received no external funding.

Institutional Review Board Statement: Not applicable.

Informed Consent Statement: Not applicable.

Data Availability Statement: Not applicable.

Acknowledgments: The authors appreciate the academic support of the Universidad Distrital Francisco José de Caldas (Colombia) and of the INDESOS and GIIAUD research groups.

Conflicts of Interest: The authors declare no conflict of interest.

References

1. Brown, S.; Nicholls, R.; Lázár, A.N.; Hornby, D.; Hill, C.; Hazra, S.; Addo, K.A.; Haque, A.; Caesar, J.; Tompkins, E.L. What are the implications of sea-level rise for a 1.5, 2 and 3 °C rise in global mean temperatures in the Ganges-Brahmaputra-Meghna and other vulnerable deltas? *Reg. Environ. Chang.* **2018**, *18*, 1829–1842. [CrossRef]
2. Matthews, T. Humid heat and climate change. *Prog. Phys. Geogr. Earth Environ.* **2018**, *42*, 391–405. [CrossRef]
3. McMichael, A.J.; Woodruff, R.E.; Hales, S. Climate change and human health: Present and future risks. *Lancet* **2006**, *367*, 859–869. [CrossRef]
4. Najafzadeh, F.; Mohammadzadeh, A.; Ghorbanian, A.; Jamali, S. Spatial and Temporal Analysis of Surface Urban Heat Island and Thermal Comfort Using Landsat Satellite Images between 1989 and 2019: A Case Study in Tehran. *Remote Sens.* **2021**, *13*, 4469. [CrossRef]
5. Pigliautile, I.; Pisello, A.; Bou-Zeid, E. Humans in the city: Representing outdoor thermal comfort in urban canopy models. *Renew. Sustain. Energy Rev.* **2020**, *133*, 110103. [CrossRef]
6. Silva, J.S.; da Silva, R.M.; Santos, C.A.G. Spatiotemporal impact of land use/land cover changes on urban heat islands: A case study of Paço do Lumiar, Brazil. *Build. Environ.* **2018**, *136*, 279–292. [CrossRef]
7. Solano–Olivares, K.; Romero, R.; Santoyo, E.; Herrera, I.; Galindo–Luna, Y.; Rodríguez–Martínez, A.; Santoyo-Castelazo, E.; Cerezo, J. Life cycle assessment of a solar absorption air-conditioning system. *J. Clean. Prod.* **2019**, *240*, 118206. [CrossRef]
8. Crutzen, P.J.; Mosier, A.R.; Smith, K.A.; Winiwarter, W. N_2O Release from Agro-biofuel Production Negates Global Warming Reduction by Replacing Fossil Fuels. In *A Pioneer on Atmospheric Chemistry and Climate Change in the Anthropocene*; Crutzen, P.J., Brauch, H.G., Eds.; Springer International Publishing: Cham, Switzerland, 2016; pp. 227–238. [CrossRef]
9. Tan, M.L.; Juneng, L.; Tangang, F.T.; Chung, J.X.; Firdaus, R.B.R. Changes in temperature extremes and their relationship with ENSO in Malaysia from 1985 to 2018. *Int. J. Clim.* **2021**, *41*, E2564–E2580. [CrossRef]
10. Taleghani, M. Outdoor thermal comfort by different heat mitigation strategies–A review. *Renew. Sustain. Energy Rev.* **2018**, *81*, 2011–2018. [CrossRef]
11. IPCC. *Climate Change 2013—The Physical Science Basis: Working Group I; Contribution to the Fifth Assessment Report of the Intergovernmental Panel on Climate Change*; Intergovernmental Panel on Climate Change, Ed.; Cambridge University Press: Cambridge, UK, 2014. [CrossRef]
12. Bienvenido-Huertas, D.; Rubio-Bellido, C.; Marín-García, D.; Canivell, J. Influence of the Representative Concentration Pathways (RCP) scenarios on the bioclimatic design strategies of the built environment. *Sustain. Cities Soc.* **2021**, *72*, 103042. [CrossRef]
13. Nilawar, A.P.; Waikar, M.L. Impacts of climate change on streamflow and sediment concentration under RCP 4.5 and 8.5: A case study in Purna river basin, India. *Sci. Total Environ.* **2019**, *650*, 2685–2696. [CrossRef]
14. Kim, H.G.; Lee, D.K.; Park, C.; Kil, S.; Son, Y.; Park, J.H. Evaluating landslide hazards using RCP 4.5 and 8.5 scenarios. *Environ. Earth Sci.* **2015**, *73*, 1385–1400. [CrossRef]
15. Zhou, P.; Liu, Z.; Cheng, L. An alternative approach for quantitatively estimating climate variability over China under the effects of ENSO events. *Atmos. Res.* **2020**, *238*, 104897. [CrossRef]
16. Diniz, F.R.; Iwabe, C.M.N.; Piacenti-Silva, M. Valuation of the human thermal discomfort index for the five Brazilian regions in the period of El Niño-Southern Oscillation (ENSO). *Int. J. Biometeorol.* **2019**, *63*, 1507–1516. [CrossRef] [PubMed]
17. Ren, H.-L.; Zheng, F.; Luo, J.-J.; Wang, R.; Liu, M.; Zhang, W.; Zhou, T.; Zhou, G. A Review of Research on Tropical Air-Sea Interaction, ENSO Dynamics, and ENSO Prediction in China. *J. Meteorol. Res.* **2019**, *34*, 43–62. [CrossRef]
18. McGregor, G.R.; Ebi, K. El Niño Southern Oscillation (ENSO) and Health: An Overview for Climate and Health Researchers. *Atmosphere* **2018**, *9*, 282. [CrossRef]
19. Johansson, E.; Thorsson, S.; Emmanuel, R.; Krüger, E. Instruments and methods in outdoor thermal comfort studies–The need for standardization. In Proceedings of the Urban Climate, ICUC8: The 8th International Conference on Urban Climate and the 10th Symposium on the Urban Environment, Dublin, Ireland, 1 January 2014; Volume 10, pp. 346–366. [CrossRef]
20. Shooshtarian, S.; Lam, C.K.C.; Kenawy, I. Outdoor thermal comfort assessment: A review on thermal comfort research in Australia. *Build. Environ.* **2020**, *177*, 106917. [CrossRef]
21. Tahbaz, M. Psychrometric chart as a basis for outdoor thermal analysis. *Iran Univ. Sci. Technol.* **2011**, *21*, 95–109.
22. Tung, C.-H.; Chen, C.-P.; Tsai, K.-T.; Kántor, N.; Hwang, R.-L.; Matzarakis, A.; Lin, T.-P. Outdoor thermal comfort characteristics in the hot and humid region from a gender perspective. *Int. J. Biometeorol.* **2014**, *58*, 1927–1939. [CrossRef] [PubMed]
23. Haldi, F.; Robinson, D. Modelling occupants' personal characteristics for thermal comfort prediction. *Int. J. Biometeorol.* **2011**, *55*, 681–694. [CrossRef] [PubMed]
24. Patiño, E.D.L.; Vakalis, D.; Touchie, M.; Tzekova, E.; Siegel, J. Thermal comfort in multi-unit social housing buildings. *Build. Environ.* **2018**, *144*, 230–237. [CrossRef]
25. Golasi, I.; Salata, F.; Vollaro, E.D.L.; Coppi, M.; Vollaro, A.D.L. Thermal Perception in the Mediterranean Area: Comparing the Mediterranean Outdoor Comfort Index (MOCI) to Other Outdoor Thermal Comfort Indices. *Energies* **2016**, *9*, 550. [CrossRef]
26. Zain, Z.M.; Taib, M.N.; Baki, S.M.S. Hot and humid climate: Prospect for thermal comfort in residential building. In Proceedings of the Ninth Arab International Conference on Solar Energy (AICSE-9), Manama, Bahrain, 30 April 2007; Volume 209, pp. 261–268. [CrossRef]

27. Aljawabra, F.; Nikolopoulou, M. Influence of hot arid climate on the use of outdoor urban spaces and thermal com-fort: Do cultural and social backgrounds matter? *Intell. Build. Int.* **2010**, *2*, 198–217. [CrossRef]
28. Lai, D.; Liu, W.; Gan, T.; Liu, K.; Chen, Q. A review of mitigating strategies to improve the thermal environment and thermal comfort in urban outdoor spaces. *Sci. Total Environ.* **2019**, *661*, 337–353. [CrossRef]
29. IDEAM. *La Variabilidad Climática y el Cambio Climático en Colombia;* Instituto de Hidrología, Me-Teorología y Estudios Ambientales: Bogotá, Colombia, 2018.
30. IDEAM. *Modelo Institucional del IDEAM Sobre el Efecto Climático de los Fenómenos El Niño y La Niña en Colombia (No. 063–2007);* Instituto de Hidrología, Meteorología y Estudios Ambientales: Bogotá, Colombia, 2007.
31. Ramírez-Aguilar, E.A.; Souza, L.C.L. Urban form and population density: Influences on Urban Heat Island intensities in Bogotá, Colombia. *Urban Clim.* **2019**, *29*, 100497. [CrossRef]
32. Anselm, N.; Rojas, O.; Brokamp, G.; Schütt, B. Spatiotemporal Variability of Precipitation and Its Statistical Relations to ENSO in the High Andean Rio Bogotá Watershed, Colombia. *Earth Interact.* **2020**, *24*, 1–17. [CrossRef]
33. Ideca: Infraestructura de Datos Espaciales para el Distrito Capital. Available online: https://www.ideca.gov.co/ (accessed on 15 January 2018).
34. Meili, N.; Acero, J.A.; Peleg, N.; Manoli, G.; Burlando, P.; Fatichi, S. Vegetation cover and plant-trait effects on outdoor thermal comfort in a tropical city. *Build. Environ.* **2021**, *195*, 107733. [CrossRef]
35. Zafra, C.; Ángel, Y.; Torres, E. ARIMA analysis of the effect of land surface coverage on PM 10 concentrations in a high-altitude megacity. *Atmospheric Pollut. Res.* **2017**, *8*, 660–668. [CrossRef]
36. RMCAB: Red de Monitoreo de Calidad del Aire de Bogotá. Available online: http://201.245.192.252:81/Report/stationreport (accessed on 15 January 2018).
37. Tang, W.; Kassim, A.; Abubakar, S. Comparative studies of various missing data treatment methods–Malaysian experience. *Atmos. Res.* **1996**, *42*, 247–262. [CrossRef]
38. Li, D.; Lu, X.X.; Yang, X.; Chen, L.; Lin, L. Sediment load responses to climate variation and cascade reservoirs in the Yangtze River: A case study of the Jinsha River. *Geomorphology* **2018**, *322*, 41–52. [CrossRef]
39. Krüger, E.L.; Minella, F.O.; Matzarakis, A. Comparison of different methods of estimating the mean radiant temperature in outdoor thermal comfort studies. *Int. J. Biometeorol.* **2014**, *58*, 1727–1737. [CrossRef]
40. Sen, J.; Nag, P.K. Effectiveness of human-thermal indices: Spatio–temporal trend of human warmth in tropical India. *Urban Clim.* **2019**, *27*, 351–371. [CrossRef]
41. Farhadi, H.; Faizi, M.; Sanaieian, H. Mitigating the urban heat island in a residential area in Tehran: Investigating the role of vegetation, materials, and orientation of buildings. *Sustain. Cities Soc.* **2019**, *46*, 101448. [CrossRef]
42. Matzarakis, A.; Rutz, F.; Mayer, H. Modelling radiation fluxes in simple and complex environments: Basics of the RayMan model. *Int. J. Biometeorol.* **2010**, *54*, 131–139. [CrossRef]
43. Chow, W.T.; Akbar, S.N.B.A.; Heng, S.L.; Roth, M. Assessment of measured and perceived microclimates within a tropical urban forest. *Urban For. Urban Green.* **2016**, *16*, 62–75. [CrossRef]
44. Matzarakis, A.; Amelung, B. Physiological Equivalent Temperature as Indicator for Impacts of Climate Change on Thermal Comfort of Humans. In *Seasonal Forecasts, Climatic Change and Human Health;* Health and Climate, Advances in Global Change Research; Thomson, M.C., Garcia-Herrera, R., Beniston, M., Eds.; Springer: Dordrecht, The Netherlands, 2008; pp. 161–172. [CrossRef]
45. IfADo: Leibniz Research Centre for Working Environment and Human Factors. Available online: http://www.utci.org/utcineu/utcineu.php (accessed on 15 February 2019).
46. Zare, S.; Hasheminejad, N.; ElahiShirvan, H.; Hemmatjo, R.; Sarebanzadeh, K.; Ahmadi, S.; Zare, S.; Hasheminejad, N.; ElahiShirvan, H.; Hemmatjo, R.; et al. Comparing Universal Thermal Climate Index (UTCI) with selected thermal indices/environmental parameters during 12 months of the year. *Weather Clim. Extrem.* **2018**, *19*, 49–57. [CrossRef]
47. Sun, S.; Xu, X.; Lao, Z.; Liu, W.; Li, Z.; Garcia, E.H.; He, L.; Zhu, J. Evaluating the impact of urban green space and landscape design parameters on thermal comfort in hot summer by numerical simulation. *Build. Environ.* **2017**, *123*, 277–288. [CrossRef]
48. Aminipouri, M.; Rayner, D.; Lindberg, F.; Thorsson, S.; Knudby, A.J.; Zickfeld, K.; Middel, A.; Krayenhoff, E.S. Urban tree planting to maintain outdoor thermal comfort under climate change: The case of Vancouver's local climate zones. *Build. Environ.* **2019**, *158*, 226–236. [CrossRef]
49. NOAA. NOAA's Climate Prediction Center. Available online: https://origin.cpc.ncep.noaa.gov/products/analysis_monitoring/ensostuff/ONI_v5.php (accessed on 22 March 2022).
50. Obiakor, M.O.; Ezeonyejiaku, C.D.; Mogbo, T.C. Effects of Vegetated and Synthetic (Impervious) Surfaces on the Microclimate of Urban Area. *J. Appl. Sci. Environ. Manag.* **2012**, *16*, 85–94. [CrossRef]
51. Song, J.; Wang, Z.-H. Impacts of mesic and xeric urban vegetation on outdoor thermal comfort and microclimate in Phoenix, AZ. *Build. Environ.* **2015**, *94*, 558–568. [CrossRef]
52. Zafra, C.; Suárez, J.; Pachón, J. Public Health Considerations for PM_{10} in a High-Pollution Megacity: Influences of Atmospheric Condition and Land Coverage. *Atmosphere* **2021**, *12*, 118. [CrossRef]

buildings

MDPI

Article

Spatial Distribution and Accessibility Measurements for Elderly Day Care Centers in China's Urban Built-up Area: The Case of Tianjin Nankai District

Da Wan [1,2], Hui Liu [1], Jiaxing Guo [1,3], Lian Guo [1], Dongchen Qi [1], Sheng Zhang [1,*], Pengbo Li [1] and Hiroatsu Fukuda [2]

1 School of Architecture, Tianjin Chengjian University, Tianjin 300380, China
2 Department of Architecture, Faculty of Environmental Engineering, The University of Kitakyushu, Kitakyushu 808-0135, Japan
3 Tianjin Architecture Design Institute Co., Ltd., Tianjin 300074, China
* Correspondence: zs_057@163.com

Abstract: The elderly community day care model is an emergent solution to the aging problem influenced by the Eastern perspective of family in China. Due to the structural problem of spatial disorder in most of China's urban built-up areas, the planning and construction of elderly day care centers (EDCCs) is facing great challenges. This study aims to comprehensively compare the spatial distribution and accessibility measurement methods for elderly residents and EDCCs in typical Chinese urban built-up areas based on the accessibility theory and spatial analysis methods from the community living circle perspective. The results show that different spatial distribution analysis methods have their own emphases and limitations, requiring comprehensive application in practice. The potential model method is most suitable for the accessibility measurement in this scenario. The threshold setting of service distance for the urban built-up areas public service facilities in the current Chinese standard needs to be further optimized. The existing EDCCs suffer from serious quantity deficiencies and misplaced supplies in the region. These findings can reveal the EDCCs distribution characteristics of typical Chinese urban built-up areas and provide new insights for urban planners and policy makers who are assessing the equity and efficiency of public service facilities.

Keywords: elderly care; public facility planning; spatial allocation; service distance; ArcGIS

1. Introduction

According to the report of the Ministry of Civil Affairs of China in 2020, there are 264 million people aged 60 and over in China, accounting for 18.7% of the total population. The population aged 65 and over is 191 million, accounting for 13.5%, which is expected to reach 14% in 2025. The increasing number of elderly groups, such as empty nesters, singletons, and semi-disabled will pose a significant challenge to China's elderly service system. Many outstanding elderly care systems can be observed around the world, especially in developed countries. Examples include the foster care programs and self-helping elderly living communities in Sweden, Belgium, and Denmark; the self and mutually supporting elderly care services in the United States; the family retirement insurance program in Singapore; the community-based integrated care system in Japan; and the community care and assistance services and residential care services for the elderly in Hong Kong [1,2].

In China, despite the growth of smaller and more nucleated families due to urbanization, the traditional family-based concept still leans toward a "family-centered" elderly care model, with a certain degree of clannishness and exclusivity. To address these challenges, the community elderly day care mode allows the elderly who are independent or semi-independent to access day care services, such as meals, personal care, physical examinations, and social recreation through the community, with the family still as the core.

This is an intermediate model between home care and institutional care, preserving the traditional form of home care while fully utilizing personal, family, community, and social resources to provide close and convenient services to the elderly. This not only satisfies the traditional psychological and material demands of the elderly, but also reduces the daily care load for their children.

The previous community planning system of the "per 1000 population model" in China focused on supply-side planning. It measured the floor area and land area of each public service facility per 1000 residents as the control index for urban public facilities. This is a top-down allocation process from the supply-side perspective. The implementation path is a one-way path from supply planning to users, without considering the feedback mechanism from users to the supply side. However, the growing multi-level demands of residents are not easily met by this system. The mismatch between the supply and demand structure is a critical issue. To cope with it, the latest Standard for Urban Residential Area Planning and Design (GB 50180-2018) [3] (hereinafter referred to as "the Standard") changed the classification model of residential areas based on the number of residents to the community living circle classified by pedestrian scale. The theoretical basis of this model is the spatial accessibility, which was first introduced by Hansen in 1959 as the magnitude of the interactive impact of each node in the road networks [4].

In recent years, with the development of computer technology and social science research, the study on accessibility has been gradually intensified. Several scholars have reviewed accessibility research in terms of road networks' evaluation, transportation efficiency comparison, and geographic region research [5–9]. Regarding the spatial distribution of public service facilities, the main research objects involve high demand for physical proximity of urban public service facilities, such as schools, hospitals, and greenspaces [10–15]. Some studies focused on the accessibility of low-income workers, the elderly, and disabled people [16–18]. Currently, spatial accessibility is one of the most effective tools to measure the spatial allocation rationality of public service resources. It is widely used in the evaluation of emergency services, transportation, education, medical care, and other public facilities arrangements [19–23]. Based on these studies, the pedestrian scale community classification model developed in China guides the distribution of service facilities with accessibility to meet the requirements of residents who can reach the corresponding facilities within an appropriate walking timeframe. Public service facilities will progressively develop a distribution strategy that is constrained from both the supply side and the demand side.

However, due to social and historical conditions, the early construction of urban areas failed to anticipate the rapid aging of China's population. Therefore, the number of community elderly facilities allocated in the early urban construction was few, and the distribution was not discussed in a scientific manner. There is still limited study on the spatial distribution and accessibility of EDCCs in China's urban built-up areas. In addition, how to carry out an effective arrangement of elderly facilities in built-up areas with high density and how to promote the construction of such facilities are still two of the difficulties to the researchers.

This study aims to provide a theoretical basis and methodological support for the planning and renovation of EDCCs in China's urban built-up areas by examining the spatial distribution and accessibility measurement methods. Firstly, we investigated the EDCCs for the elderly in Nankai District, Tianjin, and obtained the basic construction data of EDCCs and the data of residences in Nankai District and the surrounding communities. After that, we combined the road network data to construct the database of "EDCCs—residences". Then, the standard deviation ellipse (SDE), spatial autocorrelation (SA), average nearest neighbor (ANN), kernel density estimation (KDE), and buffer zone methods were used to comprehensively evaluate the spatial distribution characteristics of EDCCs in the region. The accessibility was analyzed and compared by the proportional method, the nearest distance method, the two-step floating catchment area method (2SFCA), the inverted two-step floating catchment area method (i2SFCA), and the potential model method. The results show the emphases of different spatial distribution algorithms and their limitations. Regarding 2SFCA and the potential model, the latter demonstrated more advantages for the analysis of this scenario. The threshold setting of public facilities' service distances for the urban built-up areas in the current standard requires further exploration. The analysis results also confirmed that the number of existing EDCCs in Nankai District is severely small, the distribution is random, and a spatially mismatch in supplies with the distribution of the elderly population exists. Finally, the principles and strategies for planning and optimization of EDCC are presented based on the analysis (Figure 1). These findings of this study may provide new insights for urban planners and policy makers in assessing the equity and efficiency of the accessibility and spatial distribution of public service facilities.

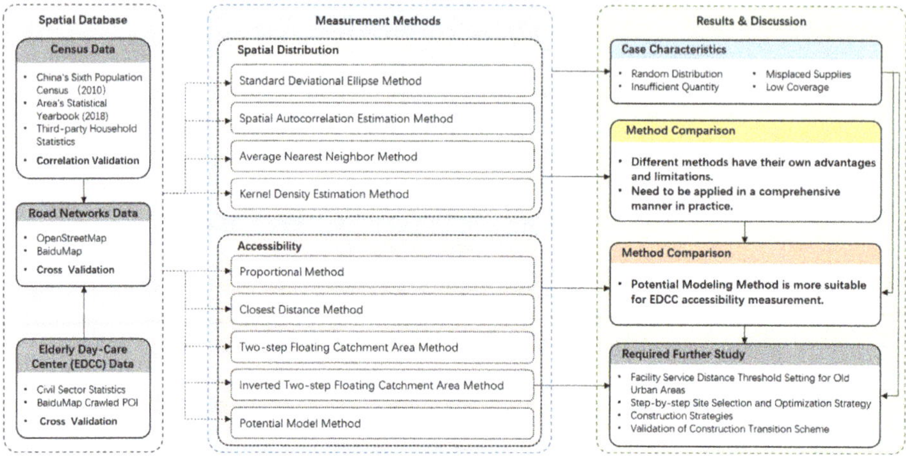

Figure 1. Study framework.

2. Materials and Methods

2.1. Study Area

According to the 6th Census in 2010, Nankai District has a population of 1,018,196, of which 16.78% (170,885 people) are 60 years old and above, which is similar to the national data. The district has existed since the 14th century. It includes 12 sub-districts and 171 communities (Table 1). It was selected as the second batch of national experimental areas for reforming home and community-based elderly care services in 2017 and is typical case of elderly service development recommended by the Ministry of Civil Affairs (Figure 2).

Table 1. Basic information of Nankai sub-districts.

Name	Area (km^2) [a]	Number of Community [b]	Aging Rate [b] (Aged 60 and Over, %)	Elderly Population Density (Persons/km^2)
Gulou	2.117	9	7.55	1061
Xingnan	1.7	9	16.9	5102
Guangkai	1.7	13	15.1	6458
Changhong	2.635	13	14.09	3930
Xiangyanglu	10	19	16.72	5962
Jialingdao	3.672	16	19.29	5048
Wanxing	3.54	22	18.66	7697
Xuefu	4.7	14	14.23	3384
Shuishanggongyuan	5.633	10	13.92	1717
Wangdingdi	5.016	23	21.19	5542
Tiyuzhongxin	6.467	12	14.08	1601
Huayuan	1.92	11	19.79	5049

Data resource: [a] Nankai District Statistical Yearbook 2018; [b] 6th Census of China.

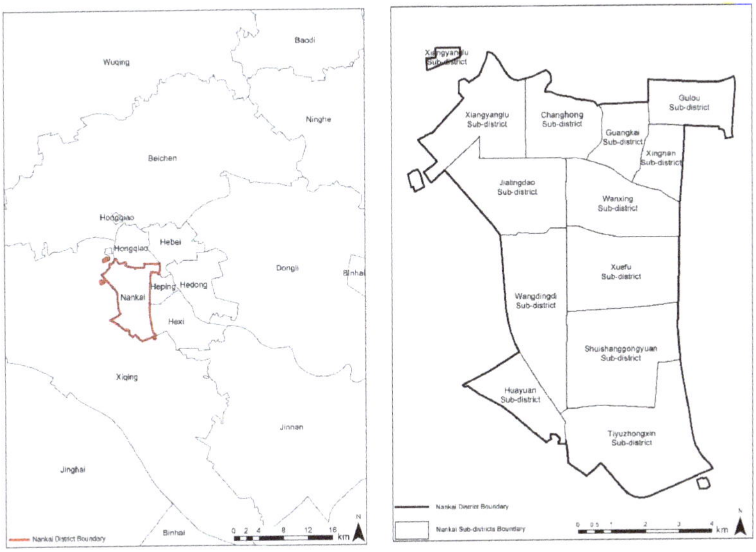

Figure 2. Study area. Left: Location of Nankai District. Right: Sub-districts of Nankai District.

2.2. Related Standard

The Standard, implemented in 2018, is the main authority for the allocation of community public service facilities. There is a fundamental change in the classification of residential areas. From the classifying of residential groups, blocks, and areas based on the number of households, it has changed to the living circles system based on the pedestrian scale. This change is combined with residents' travel behavior to guide the arrangement of facilities (Table 2). The Standard stipulates that a 5 min pedestrian scale neighborhood should include EDCCs, and the services' distances should not exceed 300 m.

Table 2. Classification model of residential areas.

Item	15-Min Pedestrian-Scale Neighborhood	10-Min Pedestrian-Scale Neighborhood	5-Min Pedestrian-Scale Neighborhood	Neighborhood Block
Walking Distance (m)	800~1000	500	300	-
Resident Population	50,000~100,000	15,000~25,000	5000~12,000	1000~3000
Number of Households	17,000~32,000	5000~8000	1500~4000	300~1000

Data resource: Urban Residential Area Planning and Design Standard (GB 50180-2018).

2.3. Data Collection and Spatial Database

1. Road Networks

Urban road networks are important to know about when performing spatial distribution and accessibility analysis. For this study, road network data were provided by Urban Data Party (Tianjin Road Networks, based on Baidu Maps) as the base map and validated with Open Street Map (OpenStreetMap).

2. Elderly Population

The source of the sub-district population data was the 6th China Census (2010), applied through the Nankai District Information Disclosure Platform (Table 3).

Table 3. Elderly populations of Nankai sub-districts in 2010.

Name	Total Population	Aged 65 and Over	Aged 60 and Over	Aging Rate (%)
Gulou	28,788	1264	2174	7.55%
Xingnan	43,744	4922	7392	16.90%
Guangkai	73,299	7357	11,071	15.10%
Changhong	73,346	7298	10,337	14.09%
Xiangyanglu	145,263	16,425	24,293	16.72%
Jialingdao	102,405	13,767	19,757	19.29%
Wanxing	140,684	18,900	26,253	18.66%
Xuefu	103,319	11,137	14,699	14.23%
Shuishanggongyuan	64,853	6389	9029	13.92%
Wangdingdi	122,599	18,368	25,975	21.19%
Tiyuzhongxin	66,948	6292	9429	14.08%
Huayuan	52,948	7304	10,476	19.79%

Data resource: 6th Census of China.

3. Population Variables

It has been more than a decade since the 6th Census in China. To ensure the timeliness of the study, the total number of households in each sub-district was counted using the ArcGIS overlay analysis tool, and a Pearson correlation test was conducted with the total population and the number of elderly populations, respectively. By means of data crawling, the household data of Nankai District was obtained from the Lianjia (a Chinese real-estate brokerage company: lianjia.com, accessed on 23 July 2022). Data from five adjacent districts were used for joint analysis to avoid boundary effects. The results show that the correlation coefficients are 0.777 and 0.796, respectively, and both are significantly correlated at the 0.01 level (2-tailed). This proves that the elderly population shows a strong correlation with the number of households in the area (Table 4). Combining the above correlation analysis and the household capacity criteria of the community living circle, the number of households was used as a population variable in this study for analysis and calculation.

Table 4. Correlation test between the number of households and population in 12 sub-districts.

	Correlation	Population	Population Aged 60 and Over
Households	Pearson Correlation	0.777	0.796
	Sig. (2-tailed)	0.003 [1]	0.002 [1]

[1] At the 0.01 level (2-tailed), the correlation is significant.

4. EDCC Data

The EDCC data were retrieved from the Civil Affairs Department of Nankai District and compiled by online and offline investigation. The data include the locating sub-district, name, detailed address, building area, number of staff, number of beds, type of services, etc. The geographical coordinates were crawled by Baidu Maps API (lbsyun.baidu.com/jsdemo.htm#wAddressParseGroup, accessed on 23 July 2022) and then rectified.

5. Spatial Database

The spatial database was constructed by ArcGIS. Only walking-distance EDCCs were considered in this study. Therefore, the original design speed of the road was not considered, and the traffic cost was calculated by the length of the road network. The population data were set as surface elements, containing the total population, the elderly population aged over 60, and the aging rate of the sub-district.

2.4. Spatial Distribution Analysis Methods

Number of facilities and density data can be used in a preliminary statistical analysis of the arrangement but fail to reflect the spatial distribution characteristics. Therefore, this study conducted comparative spatial distribution analysis by using the SDE, SA, an ANN, a KDE, and the buffer zone method in ArcGIS. These could identify the relationship between EDCCs and elderly distribution from various aspects.

1. Standard deviation ellipse method (SDE)

SDE can analyze the center of spatial distribution and the trend of distribution. The long axis of the ellipse indicates the direction with the most spatially distributed points, and the short axis indicates the direction with the least spatially distributed points [24,25]. The center of the ellipse is the average coordinates of all points.

In this study, SDE was conducted separately for Nankai District residences weighted with the elderly population, and Nankai District EDCCs with the facility area as the weighting value. The standard deviation level was set to one, and the proportion of data inclusion was 68%.

2. Spatial autocorrelation method (SA)

SA (Global Moran's I) is an important research method in spatial statistics to study the phenomenon of objects' spatial distribution, or whether there is a correlation between an attribute of the object and its spatial distribution [26,27]. It is mainly used to test whether a certain attribute of the object has a high-to-high adjacency or high-to-low misallocation.

In this study, the correlation between the spatial distribution of EDCCs and their area of location was analyzed by ArcGIS using the global Moran's I (the value is generally between −1 and 1). The closer the absolute value of Moran's I is to 1, the stronger the correlation is. When the index is 0, it indicates that the study objects are randomly distributed in space; when close to +1, it indicates the existence of spatial aggregation; and when close to −1, it indicates spatially discrete distribution. Thus, the spatial clustering model of facilities could be expressed [28]. In this study, the elderly population and the area of EDCCs in the residences were analyzed separately.

3. Average nearest neighbor method (ANN)

An ANN can be used to describe the spatial distribution of a large number of points in a specific region, including three types of dispersed, random, and clustered distributions [29]. In this study, ANN analysis was conducted for the residences and EDCCs in Nankai District, respectively.

In ArcGIS, when the resulting R-value (ratio of ANN) is 1, the facilities are randomly distributed. If it is less than 1, the facilities are relatively close to each other in space and have a spatially clustered distribution. If it is greater than 1, the spatial distribution shows a dispersed or competing distribution pattern.

The returned Z-value and p-value were used to determine the statistical significance. Z-value represents the multiple of standard deviation (positive or negative). The higher the Z-value, the higher the degree of aggregation, and the closer to zero, the more random the distribution. p-value is the confidence interval, i.e., the probability of the observed spatial distribution pattern, which corresponds to the Z-value. When the absolute Z-value is larger, the corresponding p-value is smaller, and the facility is less likely to be randomly distributed in the space, and vice versa.

4. Kernel density estimation method (KDE) and buffer zone method

A KDE is a spatial analysis in which the densities of elements in their domains are calculated and visualized. The method calculates the density of elements in the analysis of points around a single pixel, each of which has a smooth surface. The value is highest at the position of the point of interest (POI). As the distance from the POI location increases, the value gradually decreases and the connection with the facility decreases. The density distribution within the study area is then obtained by superimposing the densities at the same locations and generating a gridded graph. Within the search radius h, the distance to other study objects is inversely proportional to the distance. KDE assumes that points in space are interconnected, and the strength of the connection is inversely proportional to the distance. This property is the same as the reality that the attractiveness of public service facilities decreases with distance [30].

KDE was performed for EDCCs at 300, 500, and 1000 m, respectively. Based on this, the proximity problem was solved by buffer zone analysis, which was used to explain the proximity or coverage of the facilities.

2.5. Accessibility Measurements

Most common measures of accessibility are evaluated from a regional or spatial perspective, with urban planning, transportation, and economic benefits as the objects of study. The different methods of measuring accessibility have their own applicable facilities and different usage situations.

1. Proportional method

The proportional method treats the study area as a "container," which can be a district, a sub-district, a community, a neighborhood, or a census unit, or can be subjectively defined according to the needs of the study. This method is often used to calculate the ratio of total service supply to total population in demand; higher ratios indicate better accessibility. The ratio method assumes that all facilities in the study area are only available to residents in the area. To make this assumption valid, the study area needs to be large enough to ensure that residents in the area would not choose facilities outside the area, and that residents outside the area would not choose facilities inside the area.

2. Nearest distance method

The nearest distance method, i.e., measuring the distance from the demand point to the nearest facility point, generally uses the Euclidean distance or road distance. This method assumes that residents will certainly choose the closest facility. It emphasizes the influence of distance on residents' activity decisions; the closer the distance, the better the accessibility. This method ignores the influence of other facilities with different levels, sizes, and types [31]. EDCCs serve elderly people, and the distance factor has a greater influence on their facility choice. Therefore, the nearest distance method has a certain reference value in the study. We obtained the accessibility values by applying the nearest facility point analysis of the Network Analyst module in ArcGIS. The accessibility expression was performed by the spatial interpolation method.

3. Two-step floating catchment area method (2SFCA)

The 2SFCA is essentially a combination of the floating catchment area method (FCA) and the gravity model, first proposed by Professors John Radke and Lan Mu at UCB and then improved by Luo and Wang [32,33]. The so-called "two-step floating" refers to two "movements" in the search area between the demand and supply points.

Step 1: Centering on supply point j, set the distance threshold d_0 to determine the supply point search area, search the set of demand points k within the established search area, and calculate the supply-to-demand ratio R_j for supply point j.

Step 2: With demand point i as the center and d_0 as the radius, calculate the accessibility index R_j of demand point i in the search area and find the sum of supply and demand ratio A_i.

$$A_i = \sum_{j\in(d_{ij}\leq d_0)} R_j = \sum_{j\in(d_{ij}\leq d_0)} \frac{S_j}{\sum_{k\in(d_{kj}\leq d_0)} P_k} \quad (1)$$

where A_i is the accessibility of demand point i. R_j is the supply-to-demand ratio of supply point j in the range of travel threshold d_0, i.e., the service capacity it provides. S_j represents the supply scale of supply point j, which was quantified by floor area in this study. d_{kj} is the distance between supply point j and demand point k. d_{ij} is the distance between point i

and point j. P_k is the number of demand population in the search range. The accessibility of demand point i is positively correlated with the value of A_i.

Although the basic 2SFCA considers the supply–demand relationship between supply and demand points, it fails to address the effect of distance decay on accessibility. Therefore, to simulate a more realistic spatial relationship between supply and demand points, researchers have proposed to extend the 2SFCA with an additional distance decay function within the radius of the 2SFCA, whose general equation is:

$$A_i = \sum_{n=1}^{n} \frac{S_j f(d_{ij})}{\sum_{k=1}^{m} P_k f(d_{ij})} \text{ where } f(d_{ij}) = \begin{cases} f(d_{ij}),\ d_{ij} < d_0 \\ 0,\ d_{ij} > d_0 \end{cases} \quad (2)$$

After that, enhanced, gravity, kernel density, and Gaussian 2SFCA have been proposed and applied successively. The core operation is to use segmental assigned weights, distance decay function of gravity model, kernel density function, Gaussian function, etc., as distance decay function [34–37].

Referring to relevant Chinese studies on walking distance decay functions for the elderly in China, we chose the gravity 2SFCA (G2SFCA) as the measurement method:

$$f(d_{ij}) = d_{ij}{}^{-\beta},\ d_{ij} \le d_0 \quad (3)$$

where β is the decay parameter that takes a value generally between 1 and 2, which varies with the service type of the facility and the range of the study. EDCC is a non-emergency demand type facility, so it is generally taken as 1.

In this study, the "Origin–Destination (OD)" cost matrix was established in ArcGIS. The starting point is the residential site of the community (usually at the community gate or center of mass), the EDCC is the destination, the impedance is set as the length of the road, the limit travel distance is 300 m, the scale of the facility is the area of the EDCC, and the scale of the demand point is the number of households in the community.

When calculating the OD matrix, coordinate points overlap occurs because some EDCCs are built directly inside residential areas or transformed first floors of residential buildings. This overlap phenomenon is the ideal EDCC distribution pattern for a residential area. However, the OD distance score for such cases is 0, which cannot be calculated due to the issue of road network accuracy. To highlight the accessibility advantage of such situations, this study considered the field investigation and the accuracy of the road network, and set the OD cost distance of such situations as 10 m uniformly, which is the nearest walking distance from the residential unit to the facility.

4. Inverted 2SFCA (i2SFCA)

Wang designed a reverse 2SFCA method for evaluating facility congestion and supply coordination or saturation in 2017 [38]. By means of this method, this study tried to explain the saturation of EDCCs, i.e., the usage efficiency.

5. Potential model method

The potential model is an approach that introduces the law of gravity to urban economics and geography, revealing the relationship between economy and space in cities. The term "potential" refers to the potential energy of the relationship among objects in space. This energy is defined in the study of spatial accessibility as the potential for development, which decreases with distance. The potential model was first proposed and introduced for use by Hansen in 1959 with the following general expression for the model [4].

The original potential model only considers the potential effects of supply and demand points and ignores the countervailing effects of demand points. For example, population competition for the usage of facilities can cause the accessibility dilution. Alun E Joseph added the size factor V_j to take into account the competing relationship between demand points [13].

$$A_i^p = \sum_{j=1}^{n} \frac{S_j}{d_{ij}^{\beta} V_j} \text{ where } V_j = \sum_{k=1}^{m} \frac{P_k}{d_{kj}^{\beta}} \quad (4)$$

where A_i^p is the spatial accessibility from demand point i to all facilities j. S_j represents the scale of supply point j. d_{ij} is the traffic impedance between node i and node j, and β is the friction coefficient, which takes the same value as above. P_k is the size of the demand point, which was set as the population of the residence in this study. n and m denote the numbers of supply points j and demand points k, respectively. d_{kj}^{β} is the traffic impedance between demand point k and supply point j.

Based on the characteristics of EDCCs and the research of other Chinese scholars, we added the service distance of elderly facilities to the potential model method for calculating the accessibility of elderly facilities. The facility service threshold distance was set to 300 m. The facility area was used as the size factor V_j. The transportation cost was the same as in 2SFCA, which was obtained through the OD cost matrix, and the value of β was also taken as 1.

3. Results

3.1. Basic Statistics

3.1.1. EDCC Statistics

The results of the EDCC data in Nankai District are shown in Table 5. It was found that the number of EDCCs in Nankai District is seriously insufficient, and there is no EDCC in some sub-districts. In addition, there are some discrepancies between the number of EDCCs in each sub-district and the distribution of the elderly population. For instance, the number of EDCCs in Shuishanggongyuan sub-district accounts for 21.74% of the total, but only 5.28% of the district's elderly requires an EDCC. Gulou sub-district, which has 5.42% of the district's residents and 1.27% of the elderly population, does not yet have an EDCC.

Table 5. Population ratio and EDCC data of Nankai sub-districts in 2018.

Name	Number of EDCC	Ratio of EDCC (%)	Ratio of Elderly (%)	Ratio of Residence (%)
Gulou	0	0.00%	1.27%	5.42%
Xingnan	1	4.35%	4.33%	4.80%
Guangkai	1	4.35%	6.48%	8.13%
Changhong	1	4.35%	6.05%	5.58%
Xiangyanglu	3	13.04%	14.22%	11.11%
Jialingdao	2	8.70%	11.56%	8.57%
Wanxing	3	13.04%	15.36%	13.55%
Xuefu	2	8.70%	8.60%	7.58%
Shuishanggongyuan	5	21.74%	5.28%	5.04%
Wangdingdi	2	8.70%	15.20%	14.65%
Tiyuzhongxin	1	4.35%	5.52%	11.73%
Huayuan	2	8.70%	6.13%	3.85%
Total	23	100.00%	100.00%	100.00%

The "ratio" in each column refers to the ratio between their respective sub-district's value and the value in Nankai District.

3.1.2. EDCC Density Statistics

According to the Nankai District Statistical Yearbook and the number of EDCCs in each sub-district, the density and proportion of EDCCs in each sub-district were calculated as shown in Table 6. The analysis results show that the number of EDCCs in each sub-district is seriously insufficient in terms of the capacity of households in 5 min pedestrian scale neighborhoods. The best situation is only 0.86 units per 4000 households in the Shuishanggongyuan sub-district. At the same time, there is a significant problem with the density and the number of facilities per 1000 elderly in each sub-district. Except for the Gulou sub-district, the sub-districts with relatively decent facility density are the Huayuan sub-district and the Shuishanggongyuan sub-district. The results were only 0.19 and 0.55 per 1000 elderly people, respectively. The highest average number of facilities in the community was only 0.5 per community in the Shuishanggongyuan sub-district. The rest of the sub-districts had values of around 0.1.

Table 6. EDCC density statistics of Nankai sub-districts.

Name	Density of EDCC (/km^2)	Average Number of EDCC (/4000 Households)	Average Number of EDCC (/1000 Elderly Over 60)	Average Number of EDCC (/Community)
Gulou	0.00	0.00	0.00	0.00
Xingnan	0.59	0.18	0.14	0.11
Guangkai	0.59	0.11	0.09	0.08
Changhong	0.38	0.16	0.10	0.08
Xiangyanglu	0.30	0.23	0.12	0.16
Jialingdao	0.54	0.20	0.10	0.13
Wanxing	0.85	0.19	0.11	0.14
Xuefu	0.43	0.23	0.14	0.14
Shuishanggongyuan	0.89	0.86	0.55	0.50
Wangdingdi	0.40	0.12	0.08	0.09
Tiyuzhongxin	0.15	0.07	0.11	0.08
Huayuan	1.04	0.45	0.19	0.18

3.2. Results of Spatial Distribution Analysis

1. SDE

The corresponding SDE parameters (Table 7) obtained based on Section 2.4 are plotted and discussed in Section 4.

Table 7. SDE parameters of EDCCs and residences.

Item	Center X	Center Y	X StdDist	Y StdDist	Rotation	Area (km^2)	Perimeter (km)
SDE of Residences	117.144213	39.114371	0.031713	0.022796	162.990528	36.053588	22.544944
SDE of EDCCs	117.143797	39.107079	0.02796	0.020537	173.313796	28.634609	20.086912

2. SA

The results showed that the Z-value of residents and elderly population in Nankai District exceeded the critical value of 1.65 (rejecting the zero-dummy threshold). The p-value was less than 0.05, showing a significant positive autocorrelation, and a clustered distribution was identified. The Z-value of EDCC showed a negative autocorrelation. The p-value is greater than 0.1, indicating a high probability of a random distribution. This shows that the population spatial distribution effect was not considered in the preliminary planning, and there is a misplaced supply phenomenon (Figure 3).

3. ANN

The R-value of the community residences is less than 1, the absolute value of Z-value is greater than 1.65, and the p-value is less than 0.01. This indicates that their spatial distribution is relatively highly clustered, statistically significantly so (Figure 4a).

In contrast, the R-value of EDCC is greater than 1, the absolute value of Z-value is less than 1.65, and the p-value is greater than 0.01. It indicates that its spatial distribution presents a random distribution (Figure 4b).

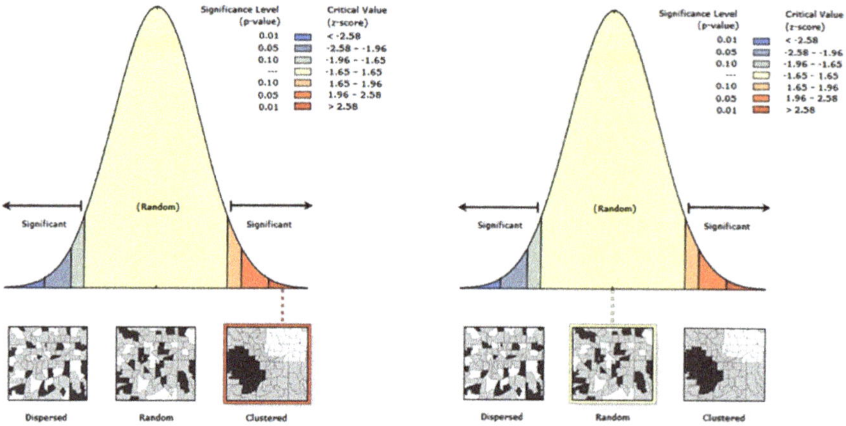

Moran's I	0.149140
Variance	0.001852
Z-value	3.497965
P-value	0.000469

(**a**)

Moran's I	−0.142380
Variance	0.008734
Z-value	−1.189126
P-value	0.234390

(**b**)

Figure 3. SA result: (**a**) elderly distribution; (**b**) EDCC distribution.

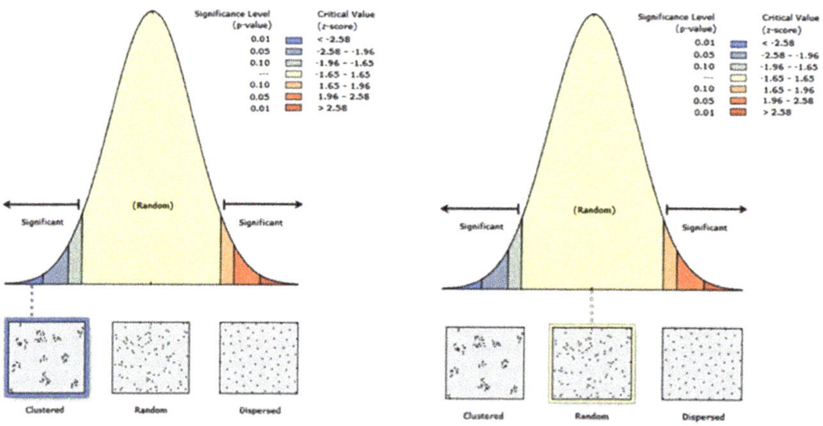

Average Observation Distance	165.2189 m
Expected Average Distance	196.7422 m
R-value	0.839774
Z-value	−8.173338
P-value	0.000000

(**a**)

Average Observation Distance	907.1162 m
Expected Average Distance	786.4417 m
R-value	1.153444
Z-value	1.407810
P-value	0.159187

(**b**)

Figure 4. ANN result: (**a**) distribution of residences; (**b**) EDCC distribution.

4. KDE and buffer zone method

Comparing the kernel density heat maps of the elderly population obtained at different search distances (Figure 5), we can see that the distribution of EDCCs in Nankai District presents an irregular point distribution. There is no significant relationship between planning and the distribution of elderly population.

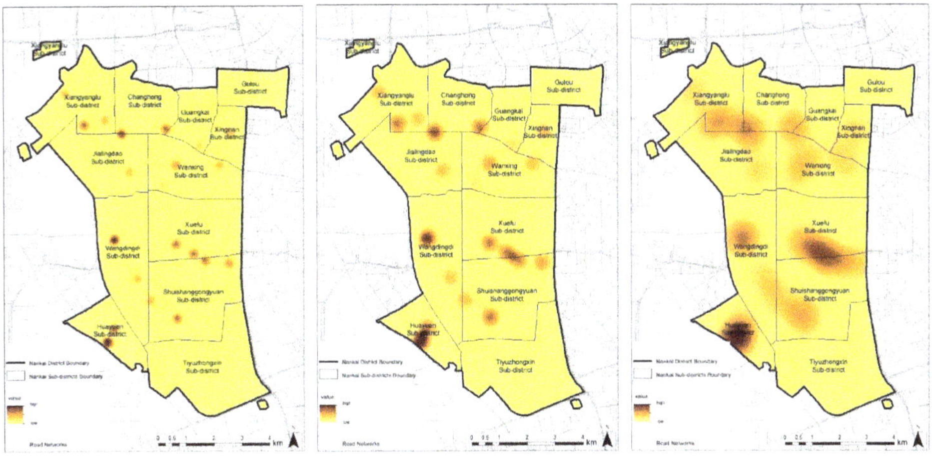

Figure 5. KDE of EDCCs: (**a**) 300 m radius; (**b**) 500 m radius; (**c**) 1000 m radius.

The areas with high densities of EDCCs are the Huayuan sub-district, the Wangdingdi sub-districts, and the junction of Shuishanggongyuan and Xuefu sub-districts. Most of the remaining sub-districts have missing facilities or very low densities. Regarding the existing results, the overall lack and unreasonable spatial distribution of EDCCs in Nankai District are the main problems.

The buffer zone analysis of the 23 existing EDCCs in Nankai District showed that none of community coverage, household coverage, and elderly population coverage exceeded 10%, and the facility coverage was extremely low (Figure 6).

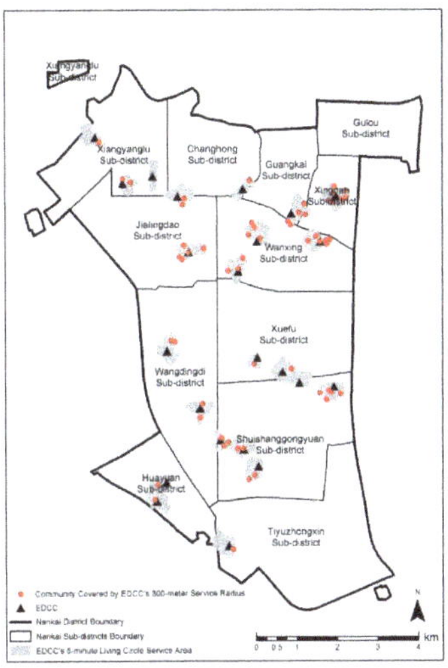

Figure 6. EDCCs in 300 m coverage area.

3.3. Results of Accessibility Measurements

1. Proportional Method

The proportional method's accessibility values were obtained from the total service supply of the EDCCs in each sub-district divided by the number of elderly individuals in the sub-district. Subsequently, the data were prepared for visual display using the ArcGIS tool (Figure 7).

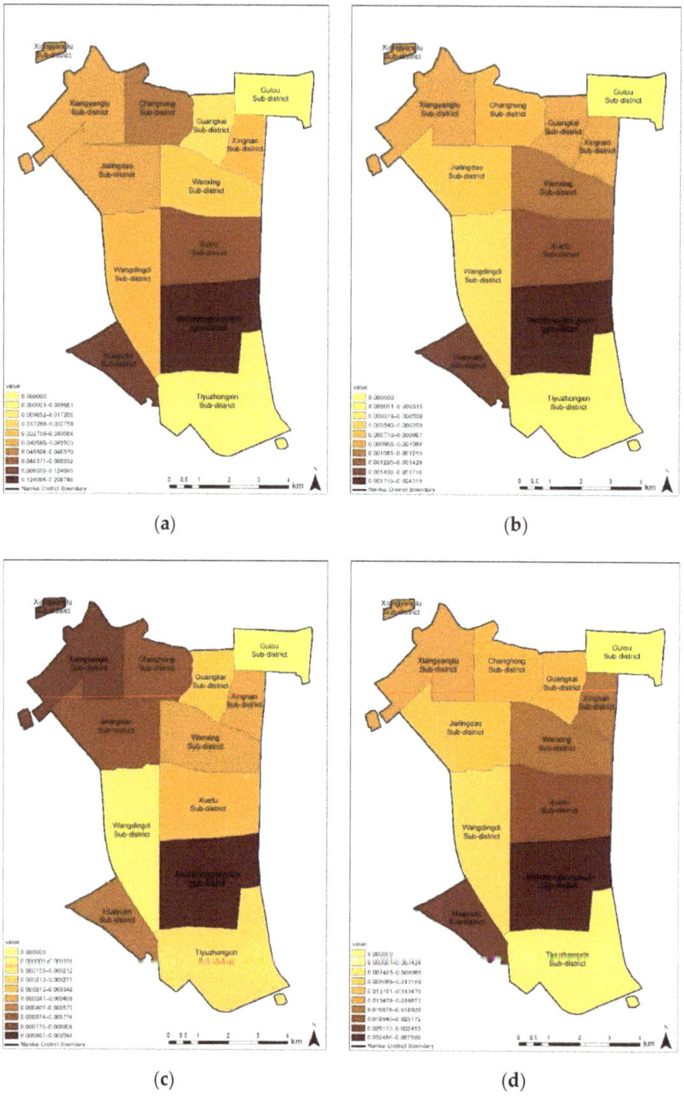

Figure 7. Proportional method results. (**a**) Facility area. (**b**) Number of beds. (**c**) Number of service staff. (**d**) Score of service.

2. Nearest Distance Method

The results showed that 717 residences were generally far from the nearest EDCC (Figure 8). Only 49 residences were within 300 m (walking distance) of an EDCC. The farthest distance among the 717 residences was 3493.17 m (Figure 9a). The shortest was the EDCC built directly within the block (Figure 9b).

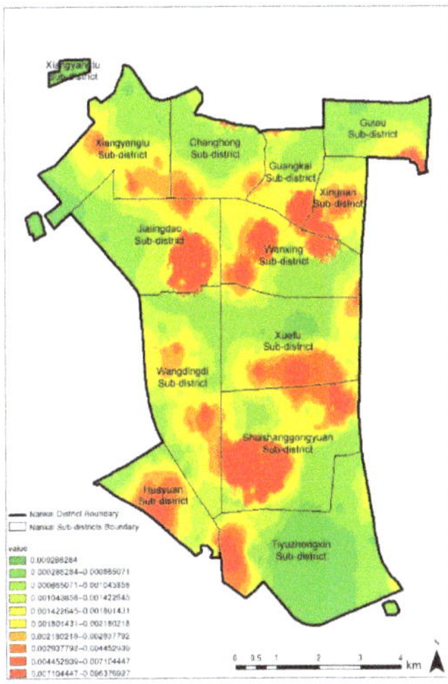

Figure 8. Nearest distance result.

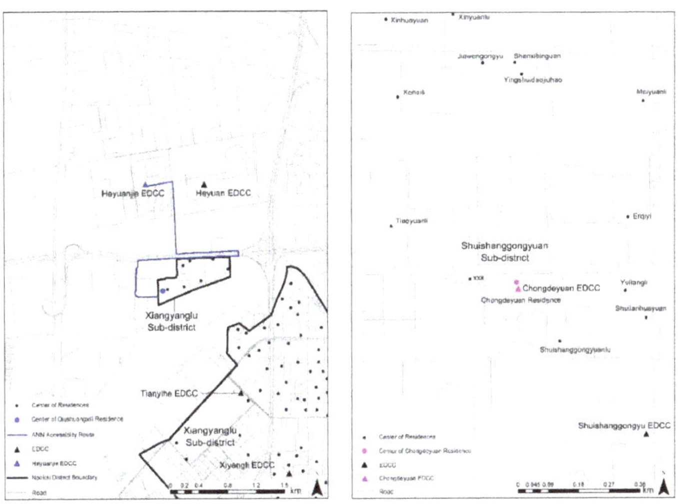

(a) (b)

Figure 9. Two extreme cases of the nearest distance method. (**a**) Longest distance (3493.17 m from Qiushuangxili residence to Heyuanjie EDCC). (**b**) Shortest distance (Chongdeyuan EDCC built in Chongdeyuan residence).

3. i2SFCA

According to the residential area classification standards from the "Standard", the 5 min pedestrian scale neighborhood residential area is limited to a walking distance of 300 m. The number of dwellings is 1500–4000 units. The EDCC area was taken as the lower limit of 350 m^2. The number of dwellings was used as the upper and lower limits of the demand scale to calculate the EDCC saturation scores. The results showed the following EDCC saturation criteria: low saturation value of 4.28 and high value of 11.43.

Except for Santandongli community and Yangguangyibai community, the saturation was at the lowest value, i.e., in an inefficient supply state (Figure 10).

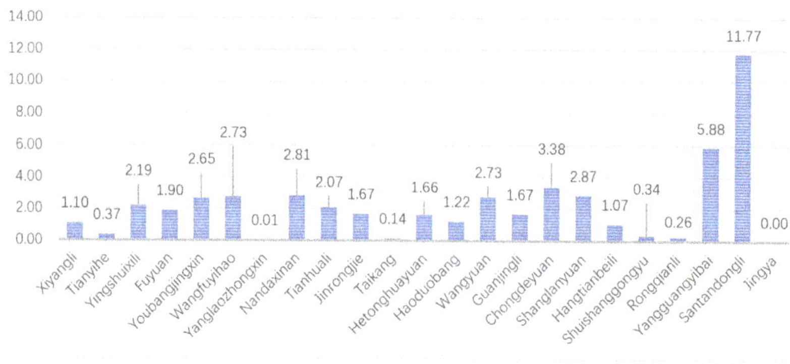

Figure 10. Saturation statistics of existing EDCC facilities in Nankai District.

4. G2SFCA and Potential Model Method

The results obtained by the G2SFCA and potential model methods are visually represented (Figure 11). The results show that the accessibility distributions presented by the two methods are very different.

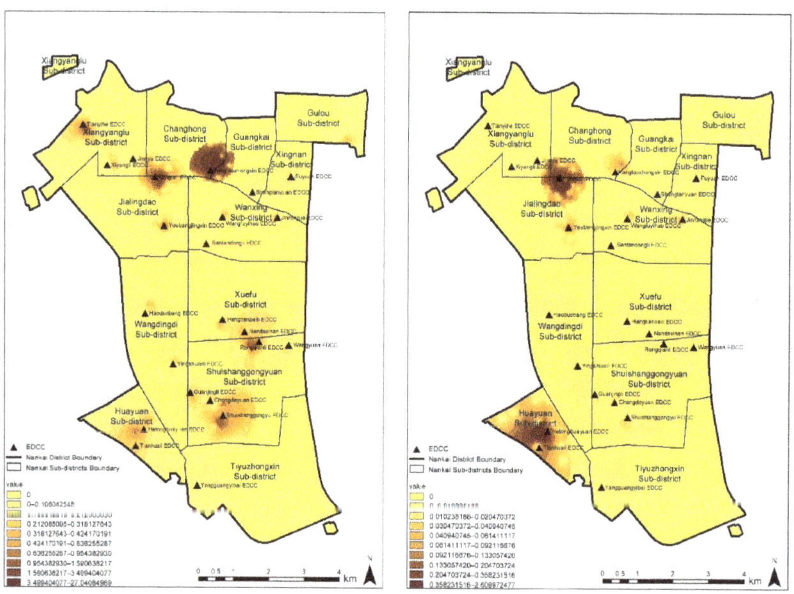

(a) (b)

Figure 11. Comparison of differences in results. (**a**) Based on G2SFCA. (**b**) Based on potential model.

In the accessibility distribution map obtained by G2SFCA, the facilities with high accessibility are mainly concentrated in the Yanglaozhongxin, Kangtai, Shuishanggongyu, and Tianyihe sub-districts. Among them, the Yanglaozhongxin EDCC has the highest facility accessibility.

In the accessibility distribution map obtained by the potential model method with added service thresholds, the communities with higher accessibility are mainly concentrated in the Huayuan sub-district and the vicinity of the Kangtai EDCC. Among them, the accessibility of the neighborhoods in the Huayuan sub-district is the highest.

4. Discussion

Based on the results above, the following points should be considered in terms of the spatial distribution and accessibility of EDCCs in Nankai District. This could serve as advice for other Chinese urban built-up areas.

1. Significant quantity shortage

The number of existing EDCCs in Nankai District is seriously low. The existing 23 EDCCs are far from the minimum number of allocations required by the Standard. Most of the communities are not served by an EDCC. The accessibility measurement shows that only the residences around the EDCCs have accessibility values. Less than 10% of the community has EDCC coverage.

The analysis results show that the accessibility of Shuishanggongyuan and Huayuan sub-districts is higher than that of other sub-districts. Gulou sub-district has the lowest accessibility in the entire district. Other sub-districts showed no significant changes in the accessibility results for the four indicators during the analysis. Xiangyanglu and Jialingdao sub-districts showed increased accessibility in the measurement of the population served. The reason for this is that the EDCCs in these two sub-districts are mostly installed in nursing homes, and the number of served is calculated as in institutional nursing facilities. This is one of the advantages of the joint construction of EDCC and institutional elderly care centers.

According to the field investigation, two other phenomena were found to exist. On the one hand, there are several EDCCs under construction or undergoing validation. To ensure the quality of facility services, some private facilities with EDCC functions have not yet been accepted for entry into the relevant subsidy and management lists. The improvement of these facilities may meet a part of the demand in the future. On the other hand, the demand for "partial functions" of the existing EDCCs has already been satisfied. For example, the number of elderly canteens in Nankai District has reached 185, with an official claim of 100% community coverage. However, according to the data of Lianjia, there are about 460,000 households in Nankai District. At least 112 EDCCs are needed for every 4000 households. Therefore, the existing facilities can rarely meet the demand both in coverage and quantity.

2. Spatial misplacement of supply

The SDE results showed that the spatial distribution of EDCC in Nankai District presented spatial misplacement with the distribution of elderly population. By comparing the SDE parameter tables generated by the calculation (Table 7), the SDE flat rate of EDCCs is larger than that of residences. The direction of EDCCs distribution is obvious, and the degree of dispersion is larger. By comparing the SDE areas of them, the area of residences is larger than that of EDCCs, indicating that the distribution of facilities does not cover all residences (Figure 12).

Therefore, it could be presumed that Nankai District missed the preliminary spatial distribution planning for EDCCs. Through interviews with the person in charge of the Elderly Services Division of the Tianjin Civil Affairs Bureau, we learned that the existing EDCC construction model in Tianjin is that the sub-district proposes construction requirements, the District Civil Affairs Bureau approves and submits them, and finally, the Municipal Civil Affairs Bureau allocates funds or subsidizes the construction. In terms of the construction of facilities, the idea of construction is mostly a matter of: "Where could it be available? Where can it be built? Where must it be built?" Due to the historical reasons for urban built-up area, the construction of facilities was preferred to the financial power, land property, and the nature of the residence (whether it is a high-class neighborhood or a retirement community), with less consideration for the spatial coupling with the elderly.

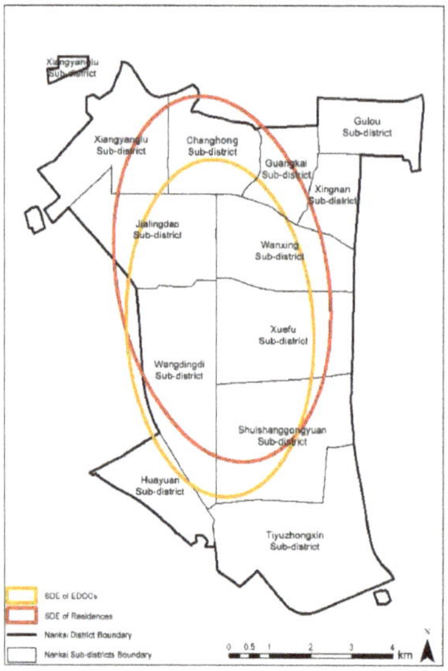

Figure 12. SDE comparison of EDCC and elderly population spatial distribution.

This will inevitably pose both equity and efficiency deficits to the operation of EDCCs. In terms of equity, the accessibility analysis shows that most communities are in an inaccessible state, and some communities are over-accessible (Figure 11). In terms of efficiency, the saturation analysis shows that the saturation of EDCC usage appears to be over-saturated in some cases at 11.43, while in others it appears to be unavailable at 0, as the distance threshold is exceeded (Figure 10). The above indicate that: the accessibility of facility distribution is uneven, i.e., inequitable; the saturation of facility usage varies greatly, resulting in inefficiency of facility usage.

3. Service distance requires further verification

According to the Standard, the saturation of existing EDCC usage analyzed by i2SFCA is significantly lower than the target under the 300 m service distance threshold. The attainment rate is less than 10%, and most of the EDCC is "unreachable". This result is significantly different from the actual investigation. Therefore, it is necessary to further analyze the reasons behind this, and re-calculate the appropriate service distance thresholds considering the maximum efficiency of facility use and the walkability of the elderly.

4. The potential model method is more suitable for EDCC accessibility analysis

When comparing the equations of G2SFCA and the potential model, it can be found that the distance decay function $f(d_{ij})$ is included in both the numerator and denominator of the G2SFCA equation, whereas the potential model equation only includes the transportation cost factor ${d_{ij}}^{\beta}$ in the denominator. Due to the small distance threshold and small number of existing EDCC services in the urban built-up areas of China, this result is a one-to-one correspondence between the supply and demand points in most cases (Figure 13). This then leads to the G2SFCA equation being approximately reduced into:

$$A_i = \frac{S_j}{P_k} \tag{5}$$

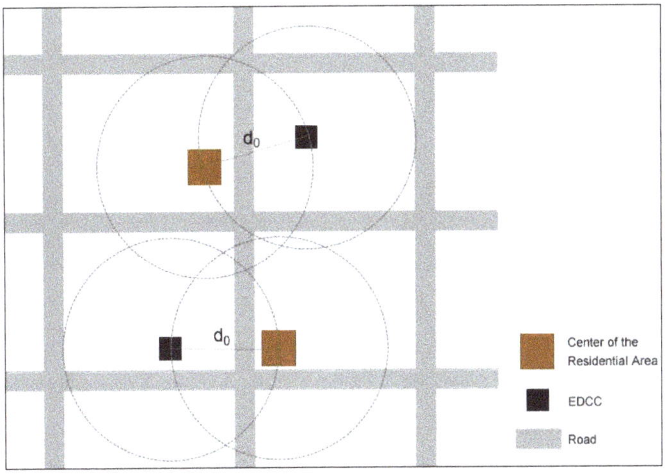

Figure 13. The one-to-one correspondence between supply and demand points within 2SFCA.

In this case, the equation no longer considers the facility traffic cost, but only the supply and demand. The potential model method, on the other hand, still considers the distance decay of transportation costs in such cases. This leads to differences in the calculation results of the two methods for the accessibility measure. According to the results of the field study, the potential model method with distance threshold can more accurately simulate the real travel and the usage of the elderly in the calculation of the accessibility of EDCCs in China's urban built-up areas.

5. The principles of planning and optimizing for EDCCs

Through the previous analysis of the construction process and distribution characteristics of EDCCs in urban built-up areas, the following strategies and objectives should be considered in further planning and distribution optimization.

(1) Aiming for community-wide accessible.

The deteriorating mobility habits and abilities of the elderly dictate that their activities are concentrated in community areas. This is particularly evident in built-up urban areas with high building density. Therefore, EDCCs should be within pedestrian accessible range of the elderly in the corresponding communities to ensure their equity.

(2) Indicating with the demand of residents.

The scale of EDCC should be coupling verified according to the elderly population within its service distance. This will improve the balance and efficiency of its spatial distribution and enable the integrated and coordinated development of the regional community elderly service system.

(3) Optimizing by progressive phased construction.

Firstly, the reasonable service distance threshold of EDCC in the region should be determined based on data analysis and simulation to maximize its utility. Secondly, the service distance can be gradually reduced, such as from 500 to 300 m, to guide the progressive phased construction with comprehensive consideration of cost and requirements.

(4) Composing based on the existing buildings.

Insufficient dedicated facilities and land constraints are commonly found in built-up areas of Chinese cities. When carrying out EDCC allocation in such areas, existing buildings, such as elderly institutions, community councils, community health service stations, or residences, can be considered for composite construction methods. These can be used as the candidate points for phased optimization and construction.

6. Limitations

This study attempted to interpret the arrangement of EDCCs in the urban built-up area through spatial distribution and accessibility analysis, relying only on the corresponding standards of community living circle. With this aim, we tried to propose a planning philosophy for the allocation planning and construction of EDCC in the area. However, the spatial arrangement is only one of the important factors affecting the construction of community elderly service system and the actual utilization of the elderly. In practice, constrained by social, economic, cultural, and medical factors in the built-up area, further evidence from a wider scope is still needed, while also relying on constant improvement and optimization in operation.

In addition, the development of advanced technologies such as smart wear and service-oriented robots has also led to more possibilities and alterations in architecture design and planning arrangement. Through shared services, machine supervised services, and others, the physical space requirements and accessibility of EDCCs may produce a new definition, which also bring more opportunities and challenges for future research.

5. Conclusions

This study took 12 sub-districts, 171 communities, 717 residences, and 23 EDCCs in Nankai District of Tianjin as the research objects and evaluated the spatial distribution and accessibility status of EDCCs in this region through calculations by relevant technologies and algorithms of the ArcGIS platform. The main conclusions of the study are as follows.

1. Different spatial distribution analysis methods have their own emphases and limitations. In practice, multiple methods need to be applied to obtain the spatial distribution characteristics of facilities in a comprehensive manner. The Moran's I can be used to analyze the dispersion, randomness, and clustering of the spatial distribution of facilities, but the degree of clustering or dispersion of facilities needs to be analyzed by the ANN. The ANN can only analyze the degree of clustering of facilities in space, but cannot accurately analyze the distribution area and form of facilities. The KDE can form a smooth space of discrete degree facility distribution and visualize the density analysis map, thereby making up for the limitation of the ANN.
2. The goal of conventional accessibility models is to illustrate the ratio between facility supply capacity and demand, and the spatial relationship between them. This can be used to evaluate the accessibility of public services to residents. The i2SFCA can be used to capture the "crowding" of facilities, i.e., the scarcity of facilities or the intensity of competition. However, the potential modeling approach is better suited to simulating the actual accessibility of such facilities that are limited in number, insufficient in supply, and have a small service distance, and where distance has a greater impact on residents' choices.
3. The Standard implemented in 2018 proposes that the service distance of an EDCC should not exceed 300 m. Through actual investigation, simulation, and calculation, it was found that there is a low utilization rate of facilities, wastage of resources, substandard saturation, and even no residences in part of the service distance, according to the limitation of a 300 m service distance threshold. It is necessary to further study whether the service distance can be set to 500 m or others, as in a transitional construction scheme.
4. Through spatial visualization and statistics of the 6th Census data and residence data, it was found that: the regional distribution of the elderly population in Nankai District varies significantly in each sub-district and has the characteristics of clustering distribution. The distribution of the elderly population shows a trend of high in the west and north, and low in the east and south. The EDCCs in the region have the problems of random spatial distributed, mismatched supply, and insufficient quantity. The problem of low saturation of facility services within the service distance exists also.

The problem of community aging in place has plagued China and countries around the world for a long time. The construction of a sophisticated elderly service system involves social, economic, cultural, medical, urban planning, and architectural factors. This study only analyzed the corresponding spatial distribution and accessibility from the perspectives of architecture and urban planning. However, architecture and spatial arrangement are only one of the factors that influence the construction of community elderly service systems and the actual decisions of elderly people to use them. In the practical planning, multi-disciplinary, multi-contextual, and multi-angle argumentation should be performed. At the same time, it needs to be constantly improved and optimized in use. Only then can the EDCC be truly built to the satisfaction of the elderly and the acceptance of the residents.

Based on this study, we will further explore the EDCC service distance thresholds that are suitable for urban built-up areas by combining relevant standards, population data, and elderly travel modes. This will be used to identify construction strategies and candidate sites. Secondly, a stepwise optimization of site selection for new EDCCs in Nankai District will be carried out by using the "regional allocation model." After that, the new sites will be evaluated and optimized mainly by the potential model method. Finally, the feasibility of the proposal will be evaluated with the practical investigation.

Author Contributions: Conceptualization, D.W. and H.F.; methodology, S.Z. and H.L.; software, J.G.; validation, D.W., L.G. and P.L.; formal analysis, D.W.; investigation, J.G. and D.Q.; resources, S.Z. and H.L.; data curation, D.W. and J.G.; writing—original draft preparation, D.W. and J.G; writing—review and editing, D.W., D.Q. and H.F.; visualization, J.G. and L.G.; supervision, S.Z.; project administration, D.W. and H.F.; funding acquisition, H.L. and P.L. All authors have read and agreed to the published version of the manuscript.

Funding: This work was supported by the National Key Research and Development Program of China (grant number 2019YFD1100402) and the Key Technology Research and Development Program of Shandong (grant number 2019GSF110004).

Institutional Review Board Statement: Not applicable.

Informed Consent Statement: Not applicable.

Data Availability Statement: Data available on request due to restrictions, e.g., privacy or ethics.

Acknowledgments: We are very grateful to Yu Miao (Department of Civil and Environmental Engineering, University of Los Angeles, CA, USA) for his generous advice and help in the writing process, and the DAMlab (Digital Architecture and Manufacture Laboratory) Tianjin Sector at Tianjin Chengjian University for providing excellent equipment support during this study.

Conflicts of Interest: The authors declare no conflict of interest.

References

1. Sudo, K.; Kobayashi, J.; Noda, S.; Fukuda, Y.; Takahashi, K. Japan's Healthcare Policy for the Elderly through the Concepts of Self-Help (Ji-Jo), Mutual Aid (Go-Jo), Social Solidarity Care (Kyo-Jo), and Governmental Care (Ko-Jo). *Biosci. Trends* **2018**, *12*, 7–11. [CrossRef]
2. Social Welfare Department of Hong Kong. Hong Kong: The Fact-Social Welfare. Available online: https://www.gov.hk/en/about/abouthk/factsheets/docs/social_welfare.pdf (accessed on 24 August 2022).
3. Ministry of Housing and Urban-Rural Development of the People's Republic of China. Standard for Urban Residential Area Planning and Design. Available online: https://www.mohurd.gov.cn/gongkai/fdzdgknr/tzgg/201811/20181130_238590.html (accessed on 24 August 2022).
4. Hansen, W.G. How Accessibility Shapes Land Use. *J. Am. Inst. Plann.* **1959**, *25*, 73–76. [CrossRef]
5. Higgs, G. A Literature Review of the Use of GIS-Based Measures of Access to Health Care Services. *Health Serv. Outcomes Res. Methodol.* **2004**, *5*, 119–139. [CrossRef]
6. Liu, S.; Zhu, X. Accessibility Analyst: An Integrated GIS Tool for Accessibility Analysis in Urban Transportation Planning. *Environ. Plan. B Plan. Des.* **2004**, *31*, 105–124. [CrossRef]
7. Vickerman, R.W. Accessibility, Attraction, and Potential: A Review of Some Concepts and Their Use in Determining Mobility. *Environ. Plan. A Econ. Sp.* **1974**, *6*, 675–691. [CrossRef]
8. Sakkas, N.; Pérez, J. Elaborating Metrics for the Accessibility of Buildings. *Comput. Environ. Urban Syst.* **2006**, *30*, 661–685. [CrossRef]

9. Kwan, M.-P.; Murray, A.T.; O'Kelly, M.E.; Tiefelsdorf, M. Recent Advances in Accessibility Research: Representation, Methodology and Applications. *J. Geogr. Syst.* **2003**, *5*, 129–138. [CrossRef]
10. Talen, E.; Anselin, L. Assessing Spatial Equity: An Evaluation of Measures of Accessibility to Public Playgrounds. *Environ. Plan. A Econ. Sp.* **1998**, *30*, 595–613. [CrossRef]
11. Stern, E.; Michlis, M. Redefining High School Catchment Areas with Varying Effects of Achievement Equality. *Appl. Geogr.* **1986**, *6*, 297–308. [CrossRef]
12. Pacione, M. Access to Urban Services—The Case of Secondary Schools in Glasgow. *Scott. Geogr. Mag.* **1989**, *105*, 12–18. [CrossRef]
13. Joseph, A.E.; Bantock, P.R. Measuring Potential Physical Accessibility to General Practitioners in Rural Areas: A Method and Case Study. *Soc. Sci. Med.* **1982**, *16*, 85–90. [CrossRef]
14. Song, Y.; Huang, B.; Cai, J.; Chen, B. Dynamic Assessments of Population Exposure to Urban Greenspace Using Multi-Source Big Data. *Sci. Total Environ.* **2018**, *634*, 1315–1325. [CrossRef]
15. Song, Y.; Chen, B.; Kwan, M.-P. How Does Urban Expansion Impact People's Exposure to Green Environments? A Comparative Study of 290 Chinese Cities. *J. Clean. Prod.* **2020**, *246*, 119018. [CrossRef]
16. Lau, J.C.Y.; Chiu, C.C.H. Accessibility of Low-Income Workers in Hong Kong. *Cities* **2003**, *20*, 197–204. [CrossRef]
17. Alsnih, R.; Hensher, D.A. The Mobility and Accessibility Expectations of Seniors in an Aging Population. *Transp. Res. Part A Policy Pract.* **2003**, *37*, 903–916. [CrossRef]
18. Church, R.L.; Marston, J.R. Measuring Accessibility for People with a Disability. *Geogr. Anal.* **2003**, *35*, 83–96. [CrossRef]
19. Tsou, K.-W.; Hung, Y.-T.; Chang, Y.-L. An Accessibility-Based Integrated Measure of Relative Spatial Equity in Urban Public Facilities. *Cities* **2005**, *22*, 424–435. [CrossRef]
20. Jones, A.P.; Bentham, G. Emergency Medical Service Accessibility and Outcomefrom Road Traffic Accidents. *Public Health* **1995**, *109*, 169–177. [CrossRef]
21. Murray, A.T.; Wu, X. Accessibility Tradeoffs in Public Transit Planning. *J. Geogr. Syst.* **2003**, *5*, 93–107. [CrossRef]
22. Talen, E. School, Community, and Spatial Equity: An Empirical Investigation of Access to Elementary Schools in West Virginia. *Ann. Assoc. Am. Geogr.* **2001**, *91*, 465–486. [CrossRef]
23. Guagliardo, M.F. Spatial Accessibility of Primary Care: Concepts, Methods and Challenges. *Int. J. Health Geogr.* **2004**, *3*, 3. [CrossRef] [PubMed]
24. Chew, V. Confidence, Prediction, and Tolerance Regions for the Multivariate Normal Distribution. *J. Am. Stat. Assoc.* **1966**, *61*, 605–617. [CrossRef]
25. Wang, B.; Shi, W.; Miao, Z. Confidence Analysis of Standard Deviational Ellipse and Its Extension into Higher Dimensional Euclidean Space. *PLoS ONE* **2015**, *10*, e0118537. [CrossRef]
26. Griffith, D.A. Spatial Autocorrelation. In *Encyclopedia of Social Measurement*; Kempf-Leonard, Ed.; Elsevier: New York, NY, USA, 2005; pp. 581–590, ISBN 978-0-12-369398-3.
27. Getis, A. Spatial Pattern Analysis. In *Encyclopedia of Social Measurement*; Kempf-Leonard, Ed.; Elsevier: New York, NY, USA, 2005; pp. 627–632, ISBN 978-0-12-369398-3.
28. Getis, A.; Ord, J.K. The Analysis of Spatial Association by Use of Distance Statistics. *Geogr. Anal.* **2010**, *24*, 189–206. [CrossRef]
29. Clark, P.J.; Evans, F.C. Distance to Nearest Neighbor as a Measure of Spatial Relationships in Populations. *Ecology* **1954**, *35*, 445–453. [CrossRef]
30. Okabe, A.; Satoh, T.; Sugihara, K. A Kernel Density Estimation Method for Networks, Its Computational Method and a GIS-based Tool. *Int. J. Geogr. Inf. Sci.* **2009**, *23*, 7–32. [CrossRef]
31. Madge, C. Public Parks and the Geography of Fear. *Tijdschr. Econ. Soc. Geogr.* **1997**, *88*, 237–250. [CrossRef]
32. Radke, J.; Mu, L. Spatial Decompositions, Modeling and Mapping Service Regions to Predict Access to Social Programs. *Ann. GIS* **2000**, *6*, 105–112. [CrossRef]
33. Luo, W.; Wang, F. Measures of Spatial Accessibility to Health Care in a GIS Environment: Synthesis and a Case Study in the Chicago Region. *Environ. Plan. B Plan. Des.* **2003**, *30*, 865–884. [CrossRef]
34. Luo, W.; Qi, Y. An Enhanced Two-Step Floating Catchment Area (E2SFCA) Method for Measuring Spatial Accessibility to Primary Care Physicians. *Health Place* **2009**, *15*, 1100–1107. [CrossRef] [PubMed]
35. Wang, F.; Tang, Q. Planning toward Equal Accessibility to Services: A Quadratic Programming Approach. *Environ. Plan. B Plan. Des.* **2013**, *40*, 195–212. [CrossRef]
36. Dai, D.; Wang, F. Geographic Disparities in Accessibility to Food Stores in Southwest Mississippi. *Environ. Plan. B Plan. Des.* **2011**, *38*, 659–677. [CrossRef]
37. Dai, D. Racial/Ethnic and Socioeconomic Disparities in Urban Green Space Accessibility: Where to Intervene? *Landsc. Urban Plan.* **2011**, *102*, 234–244. [CrossRef]
38. Wang, F. Inverted Two-Step Floating Catchment Area Method for Measuring Facility Crowdedness. *Prof. Geogr.* **2018**, *70*, 251–260. [CrossRef]

buildings

MDPI

Article

A 3D Spatial Diagnostic Framework of Sustainable Historic and Cultural District Preservation: A Case Study in Henan, China

Man Zhang [1], Yaoxin Zhang [1], Xue Fang [2,*] and Xiaoqi Wang [3]

[1] School of Architecture and Urban Planning, Beijing University of Civil Engineering and Architecture, Beijing 100044, China
[2] Research Institute for Science and Technology, Tokyo University of Science, Tokyo 162-8601, Japan
[3] Graduate School of Environmental Engineering, The University of Kitakyushu, Fukuoka 802-8577, Japan
* Correspondence: fangxue@rs.tus.ac.jp

Abstract: Three-dimensional (3D) information technology has become an important technical support in digital heritage preservation practice. However, due to the lack of systematic quantitative research, it is difficult to form a comprehensive understanding of the historic and cultural districts, from macro to micro. Our study aimed to establish a systematic 3D spatial diagnostic framework combining 3D scanning and SPSS data descriptive analysis and regression analysis for historic and cultural districts to promote sustainable historic and cultural area preservation. Taking Zhongshan Street in Qi County as an example, data statistical analysis was carried out on morphological feature data from the macro level of the district, the meso level of architecture, and the micro level of elements. The research conclusion shows that at the macro level the street form continues the main features of a traditional alley spatial skyline, height–width ratio, and sectional symbol language. At the meso level, the architecture reveals various periods of style in terms of the facade width and mathematical relationship between traditional architectural facades. At the micro level, architectural detailing explains the main reasons for the recent new construction being inconsistent with the historic and cultural district appearance. This quantitative diagnostic method can accurately analyze the current characteristics of historic and cultural districts and easily provide effective suggestions for follow-up preservation methods.

Keywords: historic and cultural district; spatial diagnostic framework; 3D laser scanning; sustainable preservation

1. Introduction

Historic and cultural districts are defined as areas with a high concentration of historical buildings and abundant cultural relics that can reflect traditional patterns and historical styles in a relatively integrated and authentic manner. Therefore, they are usually important areas that carry out functions related to residency, the economy, and culture. In the space where tradition and modernity coexist, old and new buildings are interwoven. At the same time, they also face the contradiction between preservation and development. Historical and cultural districts need to explore a balance between the spatial changes caused by economic development and the protection needs that limit the material environment [1].

As an important region where an urban sense of place and identity is concentrated, historic and cultural districts are typically located in the city center. Thus, their revitalization is closely related to urban revitalization [1]. The basic elements of historical and cultural districts are cultural lineage and morphology, including spatial contours, urban texture, street patterns, overall cityscape features, buildings, and structures [2]. Since the 1990s, China has entered a period of rapid modernization, urbanization, and globalization, as well as a special period when social and economic transitions overlap. With rapid changes in the internal structure and expansion of external contours, historic districts in cities face the predicament of preservation and development, which involves the gradual disappearance of historical heritage and the continuous deterioration of the living environment.

The Charter for the Conservation of Historic Towns and Urban Areas established in 1987 protects the street patterns and spatial forms and formal appearance (scale, size, style, construction, materials, colors, and decoration) in historic towns and historic urban areas. To achieve the effective protection and sustainable development of historical and cultural districts, researchers focus on the spatial form, street density, street continuity, height–width ratio [3–6], skyline [7–9], and store density [10]. Architectural features focus on architectural style, material, roof form, roof material, gable form [11–13], height, and width [14,15]. The architectural details focus on traditional façade elements, window and door ratios, and facade opening degrees [10,16,17]. Steven Tiesdell emphasizes that new buildings should respect not only the spatial form of historical districts, including the scale and size of buildings, but also the architectural features, including vertical and horizontal rhythms, facade emptiness and fullness, and recognizable materials and practices [1].

In addition, with the development of technology, 3D scanning technology is an important technical means of heritage preservation. Three-dimensional laser scanning technology has several advantages, including non-contact data acquisition, a high data sampling rate, high resolution, high precision, and multi-directional data acquisition. These capabilities enable the technology to efficiently capture 3D information about the structure, details, and materials of a given object or site. However, it is important to note that while 3D laser scanning can quickly obtain accurate and detailed data, the process of creating 3D models from these data can still be time-consuming, and additional post-processing may be required. It has been widely used in data collection, archiving, data analysis, status investigation and monitoring, and digital display [18–20]. The study not only involves cultural heritage, such as ancient sites [21], ancient buildings [22], and cave temples [23], but also includes larger-scale historical and cultural cities [24] and historical and cultural districts [25]. In addition, with the development of computer technology, 3D scanning technology can be combined with GIS (Geographic Information System) and BIM (Building Information Modeling). Murphy Maurice proposed a new concept of HBIM (Heritage Building Information Modelling), which includes 3D laser scanning, point cloud data processing, parametric logic design, parametric modeling, and HBIM model value assessment [26]. The combination of HBIM and GIS can extend the research scope from the building scale to the city scale [27]. BIM and HBIM can provide parametric modeling for buildings, automatic score and 3D point cloud semantics, and spatial analysis in combination with GIS thematic mapping, which ensures sustainable preservation and management of each life cycle [28,29].

Traditional manual surveying has limitations in efficiency and accuracy because historic and cultural districts have overlapped elements from different historical periods, with a large number of buildings on both sides that are diverse in style and structure. Additionally, with a mixed pattern of commercial and residential, the store signs conceal a large amount of building information. The intervention of digital preservation provides important technical support for the protection of historical districts. The 3D model of a historic district constructed by UAV (unmanned aerial vehicle) tilt photogrammetry technology can be used to analyze the posting rate, spatial enclosure, and visibility of the current street space and make suggestions for the design of public space renewal in the historic district [30]. The combination of 3D laser scanning and photogrammetry techniques can create high-precision 3D models [31]. Traditional manual surveying can be time-consuming and labor-intensive compared to 3D laser scanning, even though scanning large scales can also be time-consuming; however, manual surveying is prone to omissions and errors, which can be avoided with scanning technology. Through 3D laser scanning, it is possible to quickly obtain information about various features, including facades, second contours, commercial spaces, open space structures within the hidden fabric, skyline perception, and the evolution of old street skyline profiles. Furthermore, point cloud datasets obtained through 3D laser scanning can be used to perform spatial analysis, providing a more substantiated understanding of the site [32].

To sum up, the research on the preservation of the space form of historic and cultural districts is mainly based on the qualitative and quantitative analysis of the street space features or architectural form features [33–35]. Due to the lack of systematic research, it is difficult to form a macro-to-micro understanding of historic and cultural districts. At the same time, in recent years, there has been a trend to use digital technology to analyze spatial characteristics but relatively few studies have combined it with the spatial form of the historic and cultural district, which needs to be strengthened. Therefore, this study aims to create a complete set of analysis methods for the 3D characteristics of the historic and cultural district. We hope to provide some quantitative indexes for the extraction, protection and renewal of street culture features. The 3D diagnostic framework enables efficient and accurate data collection of historical and cultural districts. It outperforms traditional manual methods by reducing labor consumption and enhancing accuracy. The indicator establishment stage involved a literature review to find commonly used quantitative indicators of spatial characteristics. During the analysis stage, quantitative data analysis with SPSS can provide an effective description of street characteristics and control building data for updated designs. The framework has broad application prospects. It can be used to compare pre- and post-design outcomes, guide and evaluate designs, and provide technical support for sustainable preservation. By combining with new technology such as HBIM and GIS, the framework can achieve full life cycle management and protection in the future.

2. Materials and Methods

2.1. Study Area

The 3D diagnosis approach proposed in this paper takes Zhongshan Street, Qi County, Hebi City, Henan Province as an example. Qi County was once the capital of four emperors in the Shang and Yin dynasties, as well as the capital of the Principality of Wey in the Zhou Dynasty [36]. Zhongshan Street is the political and economic center of Qi County, which is reflected in its name. During the Republic of China, a central street of the political and economic center of a city was usually renamed Zhongshan Street. As a significant form of space commemoration and as a political project, there were about 532 streets with the name Zhongshan in the Republic of China. As a Chinese political landscape that venerates Sun Yat-sen, Zhongshan Street is one of the world's most influential places of personal memory [37] (Figure 1).

In history, the whole process of transformation of Zhongshan Street, from a commercial center in the Ming and Qing dynasties to an administrative center in the new China period, presents an evolving history and culture through a spatial language of continuous evolution. From the Ming dynasty to the Republic of China, Zhongshan Street is the commercial center of the county seat. According to the book Qi County Zhi, there was a street market in the middle of the county (now Zhongshan Street) [38]. The early years of the Republic of China witnessed a brief commercial boom on Zhongshan Street. There were 119 stores in the county, most of which were located along the street [39]. After 1949, the economy gradually recovered and developed in Qi County. The administrative function status of Zhongshan Street has gradually become prominent. Influenced by the ideological shift, cultural and commercial buildings with the PRC-style (new Chinese architectural style after the founding of the People's Republic of China) have been emerging on Zhongshan Street. In the 1960s and 1970s, a large number of buildings with the PRC-style were built, including the county government building, the Xinhua Bookstore, the office building of the People's Bank, and the Post Office. Zhongshan Street has gradually evolved into a new and unique street space where traditional and PRC-style buildings are equally divided. The most obvious phenomenon is that there is both the traditional street style with sloping roofs and green bricks and gray tiles as the main features; the PRC-style with vertical line division, facade decoration symbolizing socialist culture; and water-brushed stone, green bricks, and red walls as the main features.

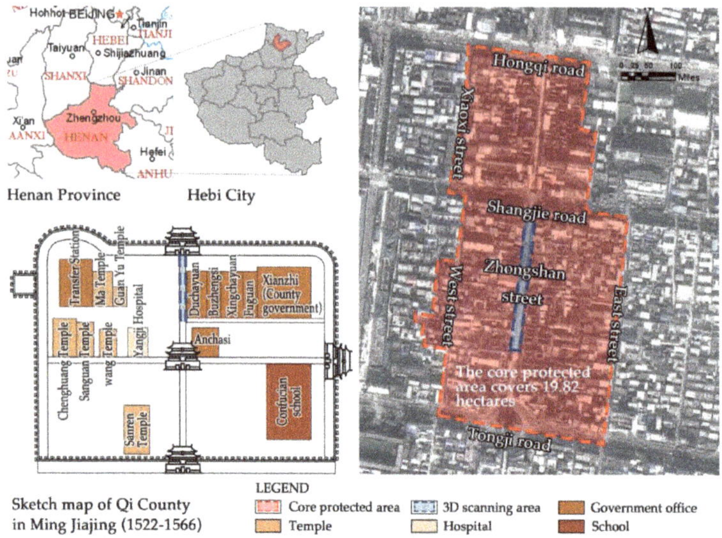

Figure 1. The core protected area of Zhongshan Street Historical and cultural district (note: the historical map of Qi County Zhi from Chinese Local Chronicles of Henan Prefecture County Integrated 29 Series Qi County Zhi in Qing Shunzhi [38]).

China's historic and cultural districts are facing many problems, such as the continuous fragmentation of spatial structure, decay of the living environment, continuous disorder of the spatial appearance, commercial overdevelopment, and the loss of cultural authenticity [40,41]. In order to achieve sustainable preservation of a district, it is vital to refer to current practices. These include preserving the texture, historic architecture, and traditional appearance of the district, as well as fostering public participation through the establishment of a robust slow-walking system. Additionally, repairing the material spatial entities and recasting the immaterial human environment are vital components of this effort. By implementing these measures, the historical value of the district can be preserved, the cultural lineage can continue to thrive, the living environment can be improved, and a sense of place and identity can be cultivated [42–44]. To strike a balance between safeguarding and developing historical and cultural districts, addressing the issue of vanishing traditional culture and architectural features, and preserving their traditional appearance, we need to find effective solutions. The study chose Zhongshan Street in Qi County for the following reasons. On the one hand, it is a sample to study the dynamic relationship of urban space; the architectural style and functions change dramatically in Zhongshan Street where traditional buildings, PRC-style buildings and modern buildings coexist. On the other hand, Zhongshan Street is facing the same urgent problems as other historic and cultural districts. Although it has a prominent political and economic status in history, there is a big gap between the serious damage to the streetscape and its prominence, so the quality of the space needs to be improved. The study aims to help people quickly comprehend the spatial features of historical and cultural districts. It also offers practical solutions and quantitative indicators to ensure the preservation of traditional appearance.

2.2. Indicator System Building

Traditional architecture is an essential asset in cities. The preserving and reusing not only supports sustainable land use [45] but also contributes to cultural sustainability, in the context of urban renewal. Historic and cultural districts usually contain a large number of traditional buildings from different periods. Their preservation and renovation require continuity of the historical context and identifiable spatial features over time. Therefore, it is crucial to extract the street space and architectural features.

In the preservation of historic and cultural districts, attention is given to street interfaces, including bottom, side, and top interfaces [46]. The bottom interface is usually focused on ground paving which creates a sense of place. The side interface considers spatial scale and the psychological feelings it can evoke. Historic and cultural districts are different from modern streets in that they have a pleasant and safe scale, and the street width-to-height ratio proposed by Yoshinobu Ashihara is commonly used in studying street scale [4–6]. The top interface is typically interested in skyline fluctuation [7,8]. Controlling height through fluctuation calculation ensures harmony in the visual outline and prevents confusion and destruction of the urban silhouette by buildings and facilities. The study selected the skyline fluctuation (W) index of the top interface and the street width-to-height ratio (D_h) of the side interface. The ground paving of historic and cultural districts in China is relatively uniform and challenging to quantify, so it was not included in the index system.

Extracting architectural features typically involves identifying typical styles from different eras, reflecting the building's massing in terms of its width and height, and analyzing its structural elements, all of which are important in shaping the cultural identity of a city. Therefore, we have analyzed the architectural styles (S) from different eras and the width (W_i) of buildings in our study of architectural features [11–15,47]. We also attempted to use linear regression to obtain the mathematical relationship (Y) of traditional Chinese architecture between the three-parts, as proposed by Yu Hao in the ancient book "Mu Jin", which can help us understand the proportionality of traditional Chinese buildings and provide a reference for updating design.

However, during the process of urbanization, not all buildings can maintain their traditional character during urbanization. User modifications, facade additions, and the installation of doors and windows have become important factors that can damage traditional architectural aesthetics [48]. Quantifying data on door and window density (D_i), fragmentation degree (C_i), facade addition density (P_s), and addition height (A_h) can help us effectively control these factors during planning and design [17].

2.3. Dataset, Tool and Technical Framework

The proposed framework for sustainable historical and cultural district preservation includes a systematic 3D spatial diagnosis framework (Figure 2), an indicator system (Table 1), and research tools (Table 2). The framework aims to explore the street space characteristics and provide information and technical support for future renewal designs. It can better preserve and display the historic and cultural districts' characteristics, thereby continuing the historical context and traditional street space experience.

The study consists of three main steps. The first step of the study is data collection and preprocessing. To record the information on Zhongshan Street, a survey was conducted in December 2018 to gather 3D data before the construction work began in 2019. The survey used a Trimble TX8 3D laser scanning device, which is capable of acquiring high-precision data at a speed of 1 million points/second within a range of 120 m. This survey covered 260 m, from the Xianwei Hutongs to the Shangjie Street Road (Figure 1), including 20 traditional buildings, 13 PRC-style buildings, and 6 modern buildings. By setting one station per 5 m to ensure 30% overlap of adjacent stations. In the processing stage, point cloud data were aligned, denoised, cut, and exported to the las file format using Trimble Realworks software. Since the 3D model coordinates of each station are different, the point clouds need to be aligned during the integration process. The alignment accuracy in this study is within 3 mm. The data were then converted to a rcs file format by Autodesk Recap software. AutoCAD was utilized to slice the point cloud model, resulting in 1 street plan, 2 street elevations, and 24 street sections at 10 m intervals (Figure 3).

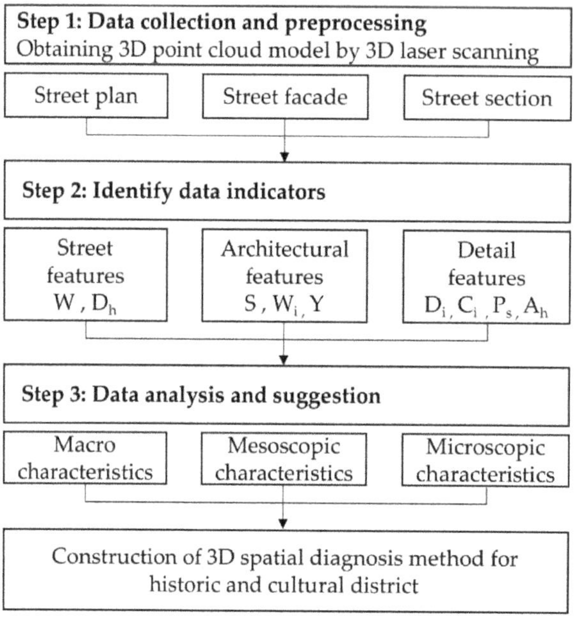

Figure 2. Technical framework of the 3D diagnosis approach.

Table 1. Data indicators of the 3D diagnosis approach.

Indicator	Definition	Formula
W [1]	Reflects the degree of the skyline fluctuation.	$W = \frac{\sum_{i=1}^{n} \Delta h_i}{n}$ (m)
D_h [2]	Street width-to-height ratio in section, reflecting the spatial scale of the street.	$D_h = \frac{D}{H}$
S	Architecture style.	-
W_i	Architecture width.	- (m)
Y [3]	The height mathematical relationship between the upper section (roof), the middle section (body) and the lower section (base).	$Y = b_0 + b_1 x_1 + b_2 x_2 + \varepsilon$
D_i [4]	The density of the doors' and windows' holes to the total area of the facade.	$D_i = \frac{\sum_{i=1}^{n} a_i}{A_i} \times 100\%$
C_i [5]	The percentage of fragmentation of the facade.	$C_i = \frac{N_i}{A_i} \times 100\%$
P_s [6]	The density of facade area covered by additions.	$P_s = \frac{S_i}{A_i} \times 100\%$
A_h [7]	Average height of the additions from the ground.	$A_h = \frac{\sum_{i=1}^{n} h_i}{n}$ (m)

* [1] W—the degree of the skyline fluctuation; Δh_i—the absolute value of the height difference between two adjacent buildings; n—building number. [2] Dh—street width to height ratio; D—the width of the street; H—the height of the street. [3] Y—the upper section height (the dependent variable); b_0—the constant; b_1—slope (regression coefficient) for variable x_1; x_1—the middle section height (independent variable); b_2—slope for variable x_2; x_2—the lower section height (independent variable); ε—error (or residual) value. [4] D_i—degree of facade openings; a_i—total area of holes in doors and windows; A_i—total area of the facade. [5] C_i—degree of facade fragmentation; N_i—the number of doors and windows in the No.i facade. [6] P_S—degree of facades additions; S_i—area of the additions. [7] A_h—the average height of the additions from the ground; h_i—the height of the additions from the ground; n—additions number.

Table 2. Hardware and software.

	Hardware	Software	Input Data File Format	Output Data File Format	Quality/Accuracy of the Data
Step 1	Trimble TX8 3D laser scanning	-	-	TZF	Camera: 10 megapixel resolution, full field of view
	-	Trimble Realworks	TZF	LAS	-
	-	Autodesk Recap	LAS	RCS	-
	-	Autodesk AutoCAD	RCS	DWG/PDF	High quality
Step 3	-	SPSS descriptive analysis and regression analysis	XLSX	JPG	Centimeter

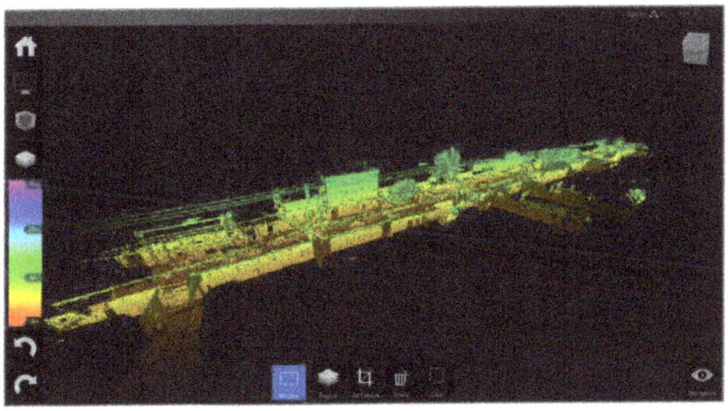

Figure 3. The point cloud data in Autodesk Recap.

In the second step, the study identified data indicators based on three levels, street features, architectural features, and detail features, which were identified through the literature review. There are two common indicators at the street level, including the building skyline and street width-to-height ratio. At the architectural level, we choose the architectural style, width, and three parts of Chinese traditional buildings. The density and fragmentation degree of windows and doors are commonly used indicators in architectural details. In order to obtain the distribution characteristics of the facade additions, we added indicators of the density and height of the additions attached to the wall or the street facing. Combining the above indicators, we identified nine indicators for the 3D spatial diagnosis framework.

The third step is data analysis and suggestion, which includes analyzing macro characteristics, mesoscopic characteristics, and microscopic characteristics. The skyline fluctuation degree and street width-to-height ratio were used to analyze the macro characteristics. The data of them calculated by the formula are used to analyze whether the street skyline and street scale have better maintained the traditional characteristics or have been destroyed during urban evolution. Mesoscopic characteristics (architectural features) can be concluded by the proportion of architectural style, the width of the architectural surface, and the mathematical relationship of architectural facades. The architectural style reflects the typical characteristics of different styles of architecture. The use of 3D laser scanning allows for the rapid quantification of the proportion of traditional elements that need to be preserved, as well as the identification of those that need to be updated that are not in keeping with the traditional style. By SPSS descriptive analysis and regression analysis on the facade width and three parts (roof, body, and base), the data characteristics of traditional buildings on the facade can be obtained. This supports the control of the dimensions and proportions of the new building in the updated design to ensure harmony with the traditional building. Microscopic characteristics (detail features) can be summed by the density and fragmentation degree of doors and windows, as well as the density and height of the additions. Through SPSS descriptive analysis of the dimensional data of windows, doors, and additions of different styles of buildings, it can guide how to control the area and number of windows and doors, as well as the area and height limit of additions in the updated design.

3. Results

3.1. Macro Characteristics

3.1.1. Street Skyline Fluctuation Degree

As the first impression of a city, the skyline plays an important role in shaping the image of the city. Its fluctuation is the vertical height variation between buildings, reflecting the vertical sequence rhythm [9]. In Zhongshan Street, the roofs of the buildings constituting the skyline are composed of 46% Yingshan roofs and 54% flat roofs.

The skyline fluctuation degree within 30 m is considered a smooth contour, while the opposite is a fluctuation contour [49]. Taking 20 m as the calculation unit, the study obtains the building height and calculates the skyline fluctuation degree in the 3D point cloud model. The average height of the buildings in the scanning area is 6.62 m, with the average height of the buildings on the east side being 6.60 m and the average height of the buildings on the west side being 6.63 m, both of which are similar in mean value. The maximum value of skyline fluctuation in the street is 8.51 m, the minimum value is 0.64 m, and the average value is 2.09 m, with the building fluctuation on the east side being 1.53 m and the building fluctuation on the west side being 2.65 m (Figure 4). Therefore, it indicates that the skyline of Zhongshan Street is smooth, and the building facades on both sides of the street are relatively stable in height (Figure 5).

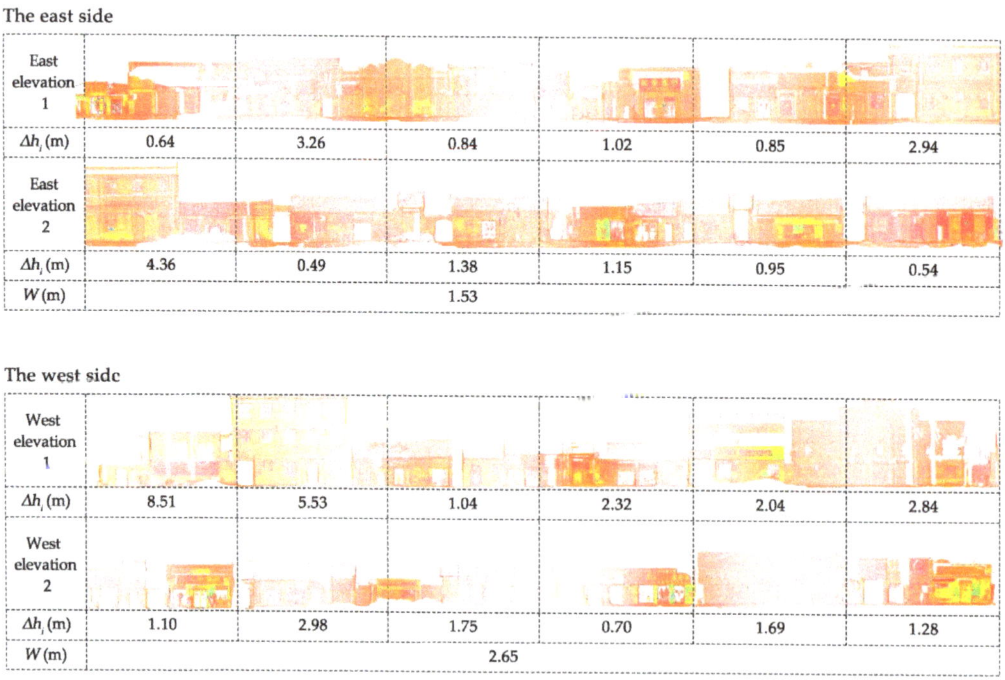

Figure 4. The smooth skyline fluctuation of Zhongshan Street on the east and west facades.

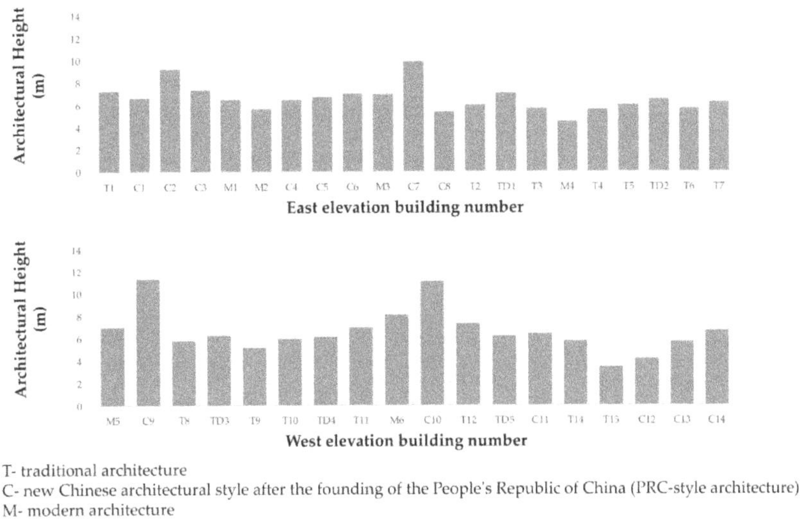

Figure 5. Schematic diagram of the east and west elevation skyline of Zhongshan Street.

In future renovation design, the building should be kept within a small fluctuation degree of the skyline at no more than two floors in order to ensure a smooth and orderly street skyline.

3.1.2. Street Section

In order to understand the scale and spatial shape of the streets, based on the 3D laser scanning data, 24 street profiles were generated by generating a section every 10 m. It can be calculated that the typical street scale D_h of Zhongshan Street is concentrated in 1–2 (Figure 6). The proportion of harmony and comfort means that Zhongshan Street has continued the traditional commercial street scale and street space form [50].

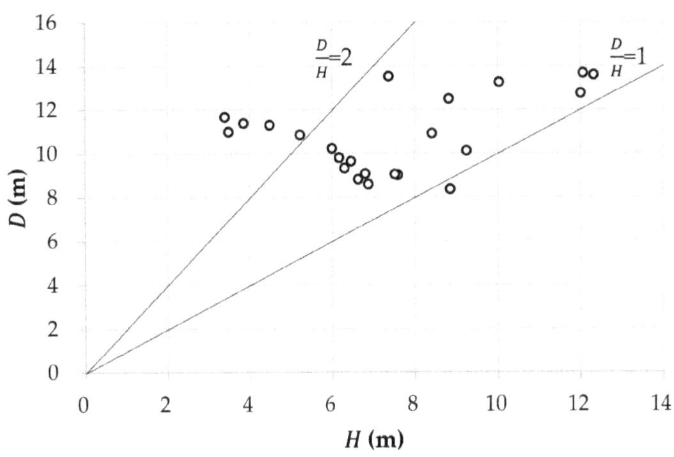

Figure 6. Scatterplot of Street D_h.

The street space of Zhongshan Street is shaped by various historical layers that have accumulated over time. Building walls, roads, greenery, and building appendages come together to form the street space, which is characterized by three distinct vocabularies

( ) that create spatial diversity. The first vocabulary ( ) represents the traditional form of Zhongshan Street, where both sides of the street have traditional Yingshan roofs. The second and third vocabularies (and) are characterized by flat-roofed buildings on one or both sides of the street. Together, these three vocabularies shape the spatial diversity of the street section (Figure 7).

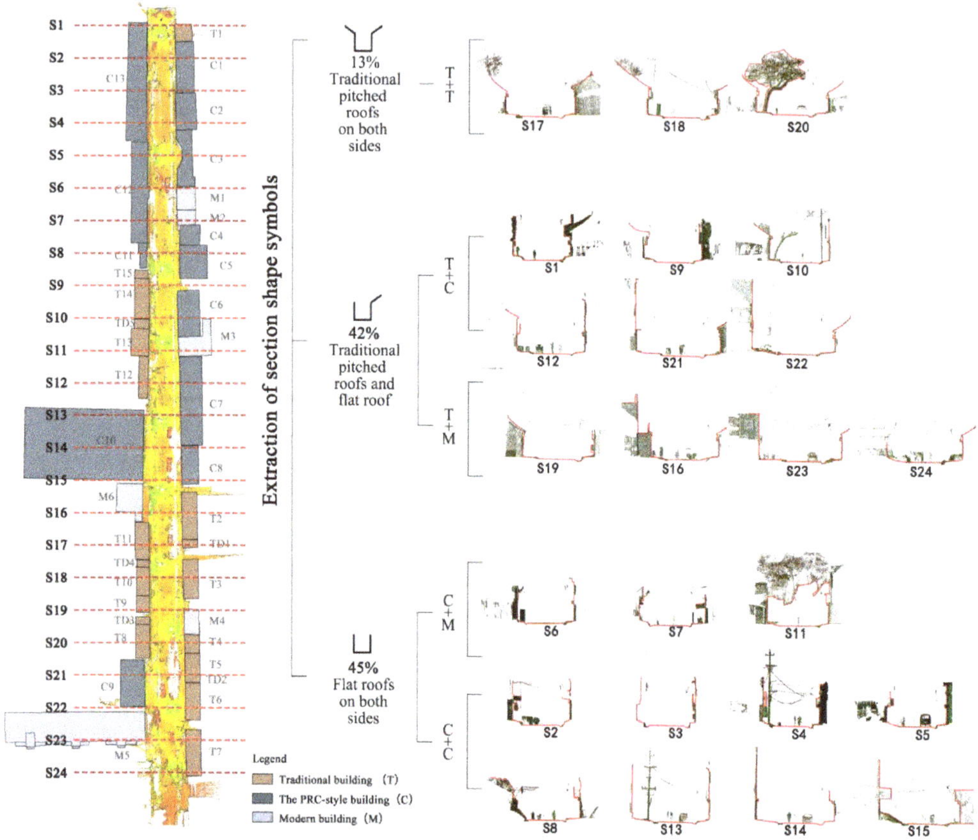

Figure 7. Extraction and classification of street morphological characteristic vocabularies.

The street section is characterized by three distinct vocabularies that reveal its long history of evolution. The dynamic change process from traditional pitched roofs to the PRC-style and new buildings' flat roofs reflects the characteristic shape of Zhongshan Street, marked by the coexistence of various kinds from different time periods and multiple spatial forms. The traditional street scale and the diversity of characteristic vocabularies should be continued in the future.

3.2. Mesoscopic Characteristics

3.2.1. Architecture Style

Within the range of 3D laser scanning data, there are 39 street-facing buildings on both sides. According to the architectural features, it can be divided into 20 traditional buildings, 13 PRC-style buildings, and 6 recent new buildings (Figure 8).

Figure 8. Three types of architectural styles and table chart of preservation and renewal sections.

Traditional buildings on Zhongshan Street continue the three-part composition of Chinese traditional architecture, including the roof, body, and foundation. The roof is a Yingshan roof with a tile and brick ridge. The body of the building is a one- to two-floor structure with between one and three openings. The walls are made of brick pillar adobe walls or bricks. The PRC-style buildings mainly adopt brick or brick and mixed structures and are single-story with walls made of red or green bricks. The facades of those buildings place emphasis on a symmetrical composition, with vertical lines dividing them, and the inclusion of red five-star elements symbolizing socialist culture [51]. These features highlight the commemorative nature of PRC architecture and reflect the idea that during the revolutionary era, architectural culture and art were utilized to serve the revolutionary struggle of workers and farmers [52]. The new buildings are all one to two floors and are more similar to the first two in terms of opening separation scale. Still, the overall architectural style is incongruous with the traditional and new Chinese period architectural styles due to the large shading of architectural additions, especially advertising signs.

3.2.2. Architecture Width

Based on the width data of 20 traditional buildings, 13 PRC-style buildings, and 6 modern buildings, the buildings' width is from 3.43 m to 7.30 m for traditional buildings, from 5.7 m to 37.88 m for the PRC-style buildings, and from 4.90 m to 12.86 m for modern buildings (Table 3). Through the standard deviation and variance of the width data, it can be observed that traditional buildings have the largest number of buildings, the least variation and the most stable form with commercial and residential functions. On the other hand, the PRC-style buildings have the largest range of sizes and the most diverse original functions of the building including office, cultural, commercial, and residential functions. To a certain extent, this reflects the relationship between the function and width of the building (Figure 9).

Table 3. Width statistics of three styles of buildings.

	N	Minimum (m)	Maximum (m)	Mean (m)	Std. Deviation	Variance Statistic
Traditional building	20	3.43	7.31	6.08	0.851	0.725
PRC-style building	13	5.70	37.88	17.83	9.848	9.698
Modern building	6	4.90	12.86	8.63	2.832	8.023

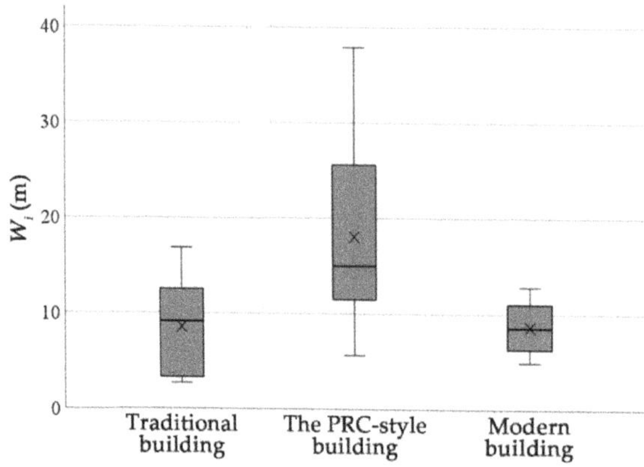

Figure 9. Width boxplot of three styles of buildings.

3.2.3. The Mathematical Relationship in Traditional Building

These three parts are referred to as the upper section, the middle section, and the lower section which specifies the features of the architectural configuration. The study obtained the facade height data for each of the 18 traditional Yingshan roof buildings in 3 parts. The maximum values of the upper section were 1.58 m and 3.24 m with a mean value of 2.40 m, the maximum values of the middle section were 2.92 m and 4.77 m with a mean value of 3.53 m, and the maximum values of the lower section were 0.06 m and 0.57 m with a mean value of 0.28 m.

The multiple linear regression model enabled the establishment of a mathematical relationship between the three components. The model (Table 4) illustrates that the middle and lower section height can effectively explain 63.4% of the variation in the upper section height and that the regression equation has a high degree of fit. The final equation is $Y = 2.374 - 0.154x_1 + 2.001x_2$ (Table 5). The result indicates that the traditional building modulus is not only reflected in the relationship between the column diameter or Doukou and other component dimensions but may also have a mathematical relationship in the three parts. Therefore, in renewal and design projects, we should follow traditional architectural mathematical relationships to avoid architectural imbalance in proportions.

Table 4. Linear regression model of three parts of the traditional buildings.

R	R^2	Adjusted R^2	R^2 Change	F Change	Sig.F Change	Durbin–Watson
0.796	0.634	0.585	0.634	12.968	0.000	2.107

Table 5. Table of regression coefficients of three parts of the traditional buildings.

	Unstandardized Coefficients		Standardized Coefficients	t	Sig.	Collinearity Statistics	
	B	Std. Error	Beta			Tolerance	VIF
Constant (b0)	2.374	0.503		4.720	0.000		
the middle sec tion height (x_1)	−0.154	0.134	−0.182	−1.154	0.267	0.985	1.015
the lower sec tion height (x_2)	2.001	0.419	0.753	4.781	0.000	0.985	1.015

3.3. *Microscopic Characteristics*

3.3.1. Windows and Doors

The degree of facade opening took values ranging from 4.39% to 48.80%, with a mean value of 22.36%. The degree of facade fragmentation took values in the range of 1.54–12.09%, with a mean value of 5.76% (Table 6). Both degrees of modern buildings are higher than those of traditional and PRC-style buildings (Figures 10 and 11), which indicates that the facade integrity of the modern buildings is poor. Thus, it is necessary to pay attention to the coordination between the doors and windows of modern buildings and the traditional style in urban renewal.

Table 6. Statistical table of D_i and C_i.

	T1	T2	T3	T4	T5	T6	T7	T8	T9	T10
D_i (%)	22.84	11.44	11.67	22.10	18.07	13.70	16.59	13.48	23.16	22.04
C_i (%)	4.65	2.27	4.13	5.71	4.97	5.70	4.76	5.19	7.31	12.09
N_i	2	2	3	2	2	4	5	3	2	6
S_i (m^2)	9.82	10.08	8.47	7.74	7.26	9.61	17.43	7.79	6.33	10.93

Table 6. *Cont.*

	T11	T12	T13	T14	T15	TD1	TD2	TD3	TD4	TD5
D_i (%)	24.26	16.89	20.13	32.66	45.06	19.02	17.18	24.67	25.82	17.58
C_i (%)	5.76	7.86	5.33	5.43	10.65	4.05	4.76	4.87	5.04	4.90
N_i	4	6	3	4	1	1	1	1	1	1
S_i (m^2)	16.84	12.89	11.33	24.05	4.23	4.70	3.61	5.06	5.12	3.59
	C1	C2	C3	C4	C5	C6	C7	C8	C9	C10
D_i (%)	30.56	16.17	16.81	21.57	20.02	23.88	21.37	37.99	25.87	15.54
C_i (%)	4.56	7.25	3.07	4.75	5.47	5.23	5.63	5.70	8.83	5.78
N_i	5	8	3	2	4	5	16	4	15	12
S_i (m^2)	33.50	17.83	16.44	9.09	14.63	22.84	60.73	26.68	43.93	32.28
	C11	C12	C13	M1	M2	M3	M4	M5	M6	
D_i (%)	11.68	32.31	25.54	48.80	25.22	17.51	27.91	4.39	30.44	
C_i (%)	2.92	6.27	5.37	11.91	7.09	4.42	5.56	1.54	7.70	
N_i	1	9	12	8	2	2	2	1	12	
S_i (m^2)	3.99	46.40	57.05	32.78	7.12	7.92	10.05	2.86	31.63	

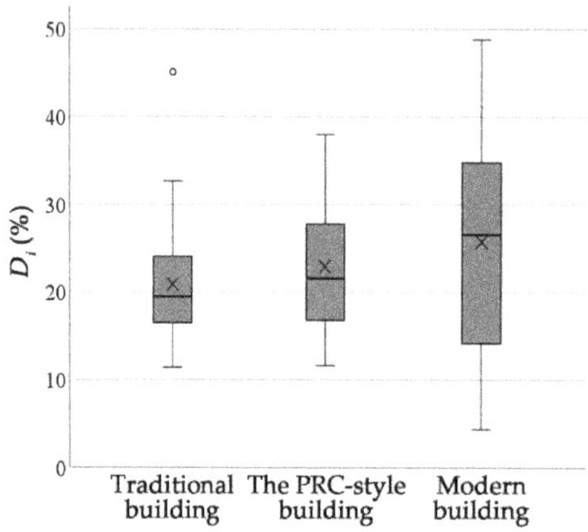

Figure 10. Degree of facade openings. (Note: circle represents outliers; × represents the average value.)

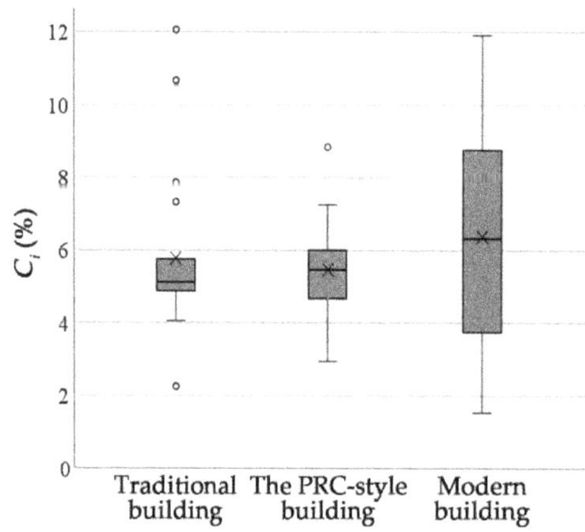

Figure 11. Degree of facade fragmentation.

3.3.2. The Second Contour Line

The second contour line refers to the fixed elements and temporary installations on the building facade. The First Contour Line refers to the original appearance of the building with a sense of order and clear structure [3]. For the street, the lower the disturbance of the second contour line, the more beautiful the image of the district will be while leaving a stronger impression. Therefore, in the preservation of Zhongshan Street, it is necessary to consider both the aesthetics of the street and the second contour line due to commerce. It is found that the shading of PRC-style buildings and modern buildings on Zhongshan Street is larger than that of traditional buildings in terms of the facade percentage, the average height of the facade and data dispersion. Among them, modern buildings have the most serious situation of building facade shading (Figure 12 and Table 7).

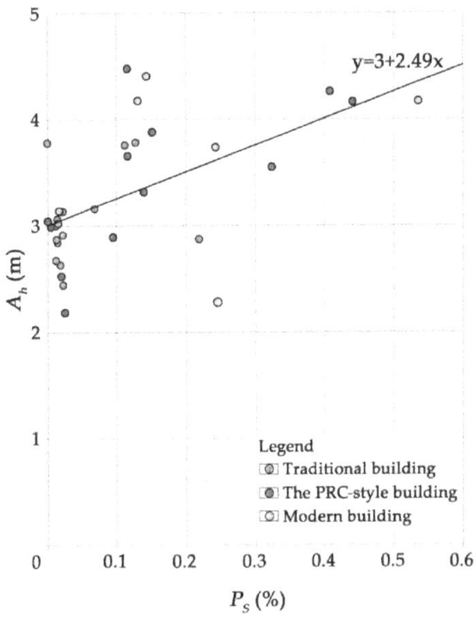

Figure 12. The relationship between the degree of facades additions and the average height of additions from the ground.

Table 7. The degree of facade additions and the average height of additions from the ground in 3 styles of building.

	N	Minimum (m)	Maximum (m)	Mean (m)	Std. Deviation	Variance Statistic
P_s (T)	15	0.00	0.22	0.05	0.06	0.00
A_h (T)	15	2.45	3.78	3.07	0.41	0.17
P_s (C)	13	0.00	0.44	0.14	0.15	0.02
A_h (C)	13	0.00	4.48	3.15	1.16	1.36
P_s (M)	6	0.02	0.54	0.22	0.18	0.03
A_h (M)	6	2.28	4.40	3.64	0.81	0.65

4. Conclusions

This study constructs a systematic 3D spatial quantitative diagnosis method, covering the whole process, from 3D data acquisition to index establishment, and then to data analysis and suggestions. The method is applied to the historical and cultural district of Zhongshan Street in Qi County. The main conclusions are as follows:

- This study involved three stages of analysis. The first stage focused on efficient and accurate data collection and processing of historical blocks using 3D laser scanning equipment. The resulting point cloud model could be quickly viewed through AutoCAD for arbitrary data acquisition and output. The point-cloud-based data management system ensured the permanent preservation of information data and facilitated future visual management of the historical and cultural blocks in the study area. This system also aided in the dynamic monitoring of changes in historical and cultural blocks over different periods.
- In the second stage, we reviewed the literature on street, building, and detailing features and extracted nine commonly used indicators. With these indicators, we constructed a system of diagnostic framework from the macro to micro level. This system allowed us to quickly obtain spatial characteristics.
- In the third stage, data analysis and suggestions were provided based on the 3D diagnosis approach's data indicators. Using Zhongshan Street in Qi County as an example, we quantitatively analyzed and summarized factors. At the macroscopic level, we evaluated the traditional characteristics of street spaces through the calculation of W and D_h. At the meso architectural level, we analyzed S, W_i, and Y to determine which buildings along the old street require preservation or updating and how to control the facade width and height to achieve harmony with the traditional architectural scale. At the micro detail level, we analyzed facade windows, doors, and the second contour line to understand the problems of modern buildings that do not conform to traditional styles. We addressed these issues by referring to the corresponding data of traditional buildings through the analysis of the diagnostic index system.

This study provides an effective and general approach for other historic districts that have multi-period architectural styles or multiple functions coexisting under the urgent need for spatial quality improvement. The method has broad application prospects. Applying the 3D quantitative diagnosis method with the analysis of the current situation characteristics of the district can scientifically reveal the spatial characteristics of streets in order to better protect and present the historical and cultural characteristics of Zhongshan Street, continue the historical heritage and the spatial experience of traditional streets and alleys, and provide theoretical support for the preservation and sustainable development of the district. Thus, the unique characteristics of the historic and cultural district can be better preserved and presented through the continuation of the historical lineage and the spatial experience of the traditional streets.

Compared with other methods in the technical field, this 3D diagnostic framework is more efficient than the traditional manual measurement, and it can grasp high-precision actual measurement data and reduce the difficulty of mapping historical buildings. In the index field, this study establishes a macro-to-micro index system based on the synthesis of the building facade features commonly mentioned in the existing studies, so it not only grasps the typical characteristics of the district more comprehensively and quickly but also can make effective suggestions from more dimensions. However, this study still has some limitations for the establishment of complex feature indicators, which need further improvement. In the future, BIM and HBIM can be introduced on the basis of research to provide sustainable protection and management for all of the life cycle of buildings [53]. In addition, based on the spatial analysis of the GIS platform, a thematic map could be formed to realize the integration and analysis of qualitative and quantitative data [54].

Author Contributions: Conceptualization, M.Z., Y.Z. and X.F.; methodology, M.Z.; software, Y.Z. and X.W.; validation, M.Z., Y.Z. and X.F.; formal analysis, Y.Z.; investigation, M.Z. and X.W.; resources, M.Z. and X.W.; data curation, Y.Z.; writing—original draft preparation, Y.Z.; writing—review and editing, M.Z., Y.Z., X.F. and X.W.; visualization, X.F.; supervision, X.F.; project administration,

M.Z.; funding acquisition, M.Z. All authors have read and agreed to the published version of the manuscript.

Funding: This research was funded by the Major Program of National Natural Science Foundation of China, grant number 51938002, the General Program of Beijing Association of Higher Education, grant number YB202196, and Education and Teaching Research Projects of China Association of Construction Education, grant number 2019006.

Data Availability Statement: Not applicable.

Acknowledgments: The authors would like to thank all the anonymous reviewers and editors who contributed their time and knowledge to this study. The authors also thank Liu X. for scanning technology support.

Conflicts of Interest: The authors declare no conflict of interest.

References

1. Tiesdell, S.; Heath, T.; Oc, T. *Revitalising Historic Urban Quarters*; Routledge: London, UK, 2013; ISBN 978-1-136-02074-2.
2. Cao, C.Z. The stoatlgy and conservation situation of historical and cultural city, town and village profoundly. *China Anc. City* **2011**, *114*, 20–30. [CrossRef]
3. Yoshinobu, A. *The Aesthetic Townscape*; The MIT Press: Cambridge, MA, USA, 1984.
4. Hata, H.; Sternberg, E. Framing the beholder's visual experience: An investigation of perspectival thinking for urban design. *J. Urban. Int. Res. Placemaking Urban Sustain.* **2021**, *16*, 242–265. [CrossRef]
5. Balasubramanian, S.; Irulappan, C.; Kitchley, J.L. Aesthetics of urban commercial streets from the perspective of cognitive memory and user behavior in urban environments. *Front. Arch. Res.* **2022**, *11*, 949–962. [CrossRef]
6. Zou, H.; Liu, R.; Cheng, W.; Lei, J.; Ge, J. The Association between Street Built Environment and Street Vitality Based on Quantitative Analysis in Historic Areas: A Case Study of Wuhan, China. *Sustainability* **2023**, *15*, 1732. [CrossRef]
7. Ying, X.; Gao, J.; Liu, Z.; Qin, X.; Chen, J.; Shen, L.; Han, X. Investigation of Pedestrian-Level Wind Environment with Skyline Quantitative Factors. *Buildings* **2022**, *12*, 792. [CrossRef]
8. Shih, N.-J.; Diao, P.-H.; Hsu, W.-T. A Section-based Illustration of Old Street Fabric and 2nd Contour Using 3D Scan Data. *Comput. -Aided Des. Appl.* **2019**, *17*, 598–606. [CrossRef]
9. Zhang, P.F. Spatial fluctuation of urban architecture. *J. Arid. Land Resour. Environ.* **2016**, *30*, 51–57.
10. Zhang, Z.; Fang, K.; Wang, X.; Chen, L.; Zhang, W.; Furuya, N. Characteristics, correlations of traditional street space elements and tourist density following spontaneous renovation: A case study on Suzhou's Shantang Street. *J. Asian Arch. Build. Eng.* **2021**, *20*, 29–43. [CrossRef]
11. Spence, C. Senses of place: Architectural design for the multisensory mind. *Cogn. Res. Princ. Implic.* **2020**, *5*, 46. [CrossRef]
12. Yi, Y.K.; Zhang, Y.; Myung, J. House style recognition using deep convolutional neural network. *Autom. Constr.* **2020**, *118*, 103307. [CrossRef]
13. Fu, J.; Zhou, J.; Deng, Y. Heritage values of ancient vernacular residences in traditional villages in Western Hunan, China: Spatial patterns and influencing factors. *Build. Environ.* **2021**, *188*, 107473. [CrossRef]
14. Che, E.; Jung, J.; Olsen, M.J. Object Recognition, Segmentation, and Classification of Mobile Laser Scanning Point Clouds: A State of the Art Review. *Sensors* **2019**, *19*, 810. [CrossRef] [PubMed]
15. Usui, H. Automatic measurement of building setbacks and streetscape widths and their spatial variability along streets and in plots: Integration of streetscape skeletons and plot geometry. *Int. J. Geogr. Inf. Sci.* **2023**, *37*, 810–838. [CrossRef]
16. Klimkowska, A.; Cavazzi, S.; Leach, R.; Grebby, S. Detailed Three-Dimensional Building Façade Reconstruction: A Review on Applications, Data and Technologies. *Remote Sens.* **2022**, *14*, 2579. [CrossRef]
17. Hu, Y. Quantitative Research on Similarity of Architectural Facade in Historic Setting. Ph.D. thesis, Tianjin University, Tianjin, China, 2018.
18. Walmsley, A.P.; Kersten, T.P. The Imperial Cathedral in Königslutter (Germany) as an Immersive Experience in Virtual Reality with Integrated 360° Panoramic Photography. *Appl. Sci.* **2020**, *10*, 1517. [CrossRef]
19. Alshawabkeh, Y.; El-Khalili, M.; Almasri, E.; Bala'awi, F.; Al-Massarweh, A. Heritage documentation using laser scanner and photogrammetry. The case study of Qasr Al-Abidit, Jordan. *Digit. Appl. Archaeol. Cult. Herit.* **2020**, *16*, e00133. [CrossRef]
20. Kushwaha, S.K.P.; Dayal, K.R.; Sachchidanand;Raghavendra, S.; Pande, H.; Tiwari, P.S.; Agrawal, S.; Srivastava, S.K. 3D Digital Documentation of a Cultural Heritage Site Using Terrestrial Laser Scanner—A Case Study. In *Applications of Geomatics in Civil Engineering: Select Proceedings of ICGCE 2018*; Ghosh, J.K., da Silva, I., Eds.; Springer: Singapore, 2020; pp. 49–58.
21. Li, J.; Li, K.; Zhao, F.; Feng, X.; Yu, J.; Li, Y.; Chao, X.; Wang, J.; Mai, B.; Cao, J. Three-Dimensional Laser Scanning Technology Assisted Investigation and Extraction of Human Bone Information in Archaeological Sites at Shenna Ruins, China. *Coatings* **2022**, *12*, 1507. [CrossRef]
22. Masciotta, M.G.; Sanchez-Aparicio, L.J.; Oliveira, D.V.; Gonzalez-Aguilera, D. Integration of Laser Scanning Technologies and 360° Photography for the Digital Documentation and Management of Cultural Heritage Buildings. *Int. J. Arch. Herit.* **2023**, *17*, 56–75. [CrossRef]

23. Elbshbeshi, A.; Gomaa, A.; Mohamed, A.; Othman, A.; Ibraheem, I.M.; Ghazala, H. Applying Geomatics Techniques for Documenting Heritage Buildings in Aswan Region, Egypt: A Case Study of the Temple of Abu Simbel. *Heritage* **2023**, *6*, 742–761. [CrossRef]
24. Noor, N.M.; Ibrahim, I.; Abdullah, A.; Abdullah, A.A.A. Information Fusion for Cultural Heritage Three-Dimensional Modeling of Malay Cities. *ISPRS Int. J. Geo-Inf.* **2020**, *9*, 177. [CrossRef]
25. Shih, N.-J.; Lee, C.-Y.; Jhan, S.-W.; Wang, G.-S.; Jhao, Y.-F. Digital preservation of a Taiwanese historical settlement: Using 3D post-construction scan to develop an application framework and reference for Beipu Township. *Cities* **2011**, *28*, 193–205. [CrossRef]
26. Murphy, M. Historic Building Information Modelling (HBIM): For Recording and Documenting Classical Architecture in Dublin 1700 to 1830. Ph.D. Thesis, Department of Civil, Structural and Environmental Engineering, Trinity College, Dublin, Ireland, 2012.
27. Colucci, E.; De Ruvo, V.; Lingua, A.; Matrone, F.; Rizzo, G. HBIM-GIS Integration: From IFC to CityGML Standard for Damaged Cultural Heritage in a Multiscale 3D GIS. *Appl. Sci.* **2020**, *10*, 1356. [CrossRef]
28. Yang, X.; Grussenmeyer, P.; Koehl, M.; Macher, H.; Murtiyoso, A.; Landes, T. Review of built heritage modelling: Integration of HBIM and other information techniques. *J. Cult. Herit.* **2020**, *46*, 350–360. [CrossRef]
29. Pepe, M.; Costantino, D.; Alfio, V.S.; Restuccia, A.G.; Papalino, N.M. Scan to BIM for the digital management and representation in 3D GIS environment of cultural heritage site. *J. Cult. Herit.* **2021**, *50*, 115–125. [CrossRef]
30. Yi, X.; Zhai, F.; Huang, S.C.; Chen, Y.J. Urban Renewal and Cultural Value Activation Design in Historic Blocks based on Digital Technology: A Case Study of the Historic area a of Nanputing, Nanjing. *China Anc. City* **2021**, *35*, 38–47.
31. El-Din Fawzy, H. 3D laser scanning and close-range photogrammetry for buildings documentation: A hybrid technique towards a better accuracy. *Alex. Eng. J.* **2019**, *58*, 1191–1204. [CrossRef]
32. Shih, N.J.; Hsu, W.T.; Diao, P.H. Point Cloud-Oriented Inspection of Old Street's Sustainable Transformation from the Ceramic Industry to Cultural Tourism: A Case Study of Yingge, a Ceramic Town in Taiwan. *Sustainability* **2019**, *11*, 4749. [CrossRef]
33. Kou, H.; Zhou, J.; Chen, J.; Zhang, S. Conservation for Sustainable Development: The Sustainability Evaluation of the Xijie Historic District, Dujiangyan City, China. *Sustainability* **2018**, *10*, 4645. [CrossRef]
34. Li, G.; Li, Y.; Wang, L. Evaluation on Spatial Adaptability of Historic Urban Blocks for Commercial Regeneration. *J. Build. Constr. Plan. Res.* **2020**, *8*, 203–216. [CrossRef]
35. Zhao, Y.; Liu, J.; Zheng, Y. Preservation and Renewal: A Study on Visual Evaluation of Urban Historical and Cultural Street Landscape in Quanzhou. *Sustainability* **2022**, *14*, 8775. [CrossRef]
36. Yan, Z.A. Historical study of Chaoge, the capital of Shang Dynasty. In Proceedings of the Research on China's Ancient Capitals (16th Series)—Proceedings of the 16th Annual Meeting of the Chinese Society for Ancient Capital Studies and Ju Cultural Seminar; the Chinese Society for Ancient Capital Studies, Ju County, China, 28 October 1999; pp. 356–358.
37. Chen, Y.X. Zhongshan Roads of Republican China and the Routinization of Ideology. *J. Hist. Sci.* **2007**, *326*, 108–117. [CrossRef]
38. Shanghai Bookstore Publishing House. *Chinese Local Chronicles of Henan Prefecture County Integrated 29 Series*; Shanghai Bookstore Publishing House: Shanghai, China, 2013; ISBN 7-5458-0672-4.
39. Chinese People's Political Consultative Conference Committee of Qixian County, Henan Province Cultural and Historical Materials Committee. *Qi County Literature and History Data Volume 4*; Chinese People's Political Consultative Conference Committee of Qixian County, Henan Province Cultural and Historical Materials Committee: Hebi, China, 1992; Volume 4.
40. Yang, L.; Tang, F.F. Research and Consideration on the Implementation Model of Historic and Cultural Districts Renewal in China. *Urban Dev. Stud.* **2019**, *26*, 32–38.
41. Bian, L.; Wu, Q.; Shi, Y. Discrimination and Thinking of the Conservation of Historic and Cultura Area in Beijing Old City. *Beijing Plan. Rev.* **2019**, 34–41. Available online: https://kns.cnki.net/kcms2/article/abstract?v=3uoqIhG8C44YLTlOAiTRKibYlV5Vjs7i8oRR1PAr7RxjuAJk4dHXou8mJpoAjHOR2REVAV8WEYVo8o56PQISruLkEuAuDDBo&uniplatform=NZKPT (accessed on 10 May 2023).
42. Xie, Y. Study on the Protection and Sustainable Utilization of Historical and Cultural Blocks from the Perspective of Urban Design: A Case of Baihuazhou Historical and Cultural District in Jinan. *Urban. Archit.* **2021**, *18*, 25–28. [CrossRef]
43. Li, R.; Li, C.; Rui, G. Compilation Method of Historical and Cultural District from the Perspective of HUL: A Case in Guangzhou. *Planners* **2020**, *36*, 66–72.
44. Ma, B. Research on Conservation Strategies of Historic Conservation Area Based on the Concept of Urban Repair: A Case Study of Dengfeng Historic Conservation Area. *Archit. Cult.* **2019**, *183*, 116–119. [CrossRef]
45. Zheng, H.W.; Shen, G.Q.; Wang, H. A review of recent studies on sustainable urban renewal. *Habitat Int.* **2014**, *41*, 272–279. [CrossRef]
46. Wang, L.; Xu, D. Study on the Landscape Elements and Protection Strategies of Street Interface in Historical Districts: Taking the Historical and Cultural District of Yushan in Qingdao as an Example. *J. Green Sci. Technol.* **2020**, *5*, 37. [CrossRef]
47. Liu, J.; Wu, Z.-K. Rule-Based Generation of Ancient Chinese Architecture from the Song Dynasty. *J. Comput. Cult. Herit.* **2016**, *9*, 1–22. [CrossRef]
48. Wang, J. Exploring the Approaches of Adaptive Conservation, Reconstruction and Historic Districts a Case Study of Gunanjie Street in Dingshu, Yixing. *Archit. J.* **2021**, *631*, 1–7. [CrossRef]
49. Peng, Q.X. *Research on Evaluation and Control Methods of Urban Skyline: A Case Study on the Lakeshore New District in Hefei*; Hefei University of Technology: Hefei, China, 2015.

50. Xia, Z.W. Spatial Form of Traditional Commerce Street. *Chongqing Archit.* **2010**, *9*, 42–45. Available online: https://kns.cnki.net/kcms2/article/abstract?v=3uoqIhG8C44YLTlOAiTRKgchrJ08w1e7_IFawAif0mwOIHgfNHDSrvZUG63M69NIlUWrHXPl1hmFJ71kCha80mAocTkeBLUS&uniplatform=NZKPT (accessed on 10 May 2023).
51. Guo, W.B. Renovation policies for renovation of modern building in the earlier new period of China. *Archicreation* **2010**, *133*, 104–112.
52. Han, G.F.; Xu, M.Y. Discussion on the process of characteristics evolution and developmental tendency of revolutionary commemorative building in China. *Urban. Archit.* **2008**, *48*, 27–30.
53. Liu, J.; Xu, D.; Hyyppa, J.; Liang, Y. A Survey of Applications With Combined BIM and 3D Laser Scanning in the Life Cycle of Buildings. *IEEE J. Sel. Top. Appl. Earth Obs. Remote Sens.* **2021**, *14*, 5627–5637. [CrossRef]
54. Tsilimantou, E.; Delegou, E.T.; Nikitakos, I.A.; Ioannidis, C.; Moropoulou, A. GIS and BIM as Integrated Digital Environments for Modeling and Monitoring of Historic Buildings. *Appl. Sci.* **2020**, *10*, 1078. [CrossRef]

buildings

MDPI

Article

Study on the Correlation Analysis between Urban Morphological Factors and Microclimate Based on Empirical Methods on a University Campus in a Hot-Summer–Cold-Winter Region

Ruihan Wei [1], Chang Xu [1], Dexuan Song [2], Hui Tong [1,*] and Zhengshu Chen [1]

[1] College of Architecture and Urban Planning, Shandong Jianzhu University, Jinan 250101, China; weiruihan19@sdjzu.edu.cn (R.W.)
[2] College of Architecture and Urban Planning, Tongji University, Shanghai 200092, China
* Correspondence: rover@sdjzu.edu.cn

Abstract: With the continuous progress of urbanization, contemporary scholars are increasingly focusing their attention on the study of urban microclimates. This research aims to investigate the relationship between urban morphology and microclimates in regions characterized by hot summers and cold winters. Environmental meteorological data, including temperature, solar radiation, and wind speed during the spring and summer seasons, were collected through on-site monitoring in a specific area of Shanghai. Various methods, such as on-site surveys and drone photography, were employed to obtain and analyze a range of urban morphological parameters, including floor area ratio (FAR) and greenery ratio (GnPR). The sky view factor (SVF) was determined using a fisheye lens technique. Subsequently, the collected datasets were analyzed to assess the varying impact of distinct urban morphological factors on microclimates. The findings indicate a significant correlation between different building morphology factors and temperature and solar radiation during spring, while their associations with wind speed become more prominent in summer. Notably, greenery and pavement demonstrate stronger correlations with temperature and radiation in summer, aligning with the seasonal growth patterns of plants. These research findings provide both a theoretical foundation and practical guidelines for the urban design of environmental microclimates.

Keywords: urban morphological factors; microclimates; correlation mechanisms; empirical methods; hot-summer–cold-winter regions

1. Introduction

Based on the primary data from the seventh National Population Census conducted by the National Bureau of Statistics, as of May 2021, China's urbanization rate, reflecting the proportion of permanent residents living in urban areas, reached 63.89%. This signifies the emergence of an urban-oriented social structure, representing a profound societal transformation. In an environment where various resources, particularly land, are scarce, high-density development not only characterizes the present state of urban development in our country but also embodies a pivotal future trend that distinguishes it from foreign urban development models. While we reap the benefits of the conveniences afforded by the high-density urbanization process, this developmental paradigm imposes heightened demands and challenges on the comprehensive carrying capacity of the urban environment. Numerous environmental and climatic issues linked to high-density development, including the urban heat island effect, persistent and unresolved haze, extreme weather events, and inadequate urban ventilation, significantly compromise urban comfort, elevate energy consumption, and exacerbate pollution accumulation. Consequently, these factors exert substantial impacts on both social development and residents' well-being.

The atmosphere encompasses a wide range of spatial and temporal phenomena. Climatic issues at different scales are distinguished by their unique representations of time and space for urban surfaces, giving rise to distinct equilibrium equations [1]. Recent years have witnessed significant advancements in the study of urban climate systems across multiple scales, including the urban and regional scale (meteorological mesoscale), neighborhood scale (meteorological local scale), and block scale (meteorological microscale). The meteorological mesoscale, covering several to hundreds of kilometers, captures weather phenomena such as storms. At the meteorological local scale, calculations and simulations typically operate within a range of hundreds of meters to a few kilometers, incorporating specific parameters such as surface reflectivity and roughness length to characterize the surface attributes of urban areas [2]. Meanwhile, the microscale level of meteorology involves precise computations of physical processes within a range spanning from a few meters to nearly 1 km.

With the advancement of research on urban microclimates, scholars have conducted extensive investigations into the impact of diverse urban forms on urban climates. The morphology and configuration of urban buildings play a direct role in shaping urban microclimates [3]. Descriptions of the urban thermal environment can be achieved through six urban form indicators, namely the sky view factor, pavement cover ratio, vehicle traffic density, green plot ratio, frontal area index, and proximity to heat sinks. In situ meteorological measurements enable the examination of spatial variations in microclimate parameters [4]. Different urban forms exert significant influence on the spatiotemporal dimensions of temperature [5,6]. Notably, various types of street trees have a profound impact on enhancing the thermal environment of road spaces and improving pedestrian thermal comfort [7]. Trees exhibit different effects under varying wind speeds, air conditions, solar radiation, and stomatal resistance. Higher wind speeds diminish the cooling effect of trees, while the regulation of stomatal resistance by trees can mitigate the impact of environmental changes on leaf surface temperature [8]. The representation and impact of vegetation in microclimate models provide a basis for assessing the thermal environment and comfort of outdoor microclimates in urban settings [9]. In terms of buildings, factors such as street interface density, size and location of open spaces, and the ratio of street width to building height significantly influence thermal comfort. Leveraging building shadows and creating ventilation corridors are crucial for improving thermal comfort [10]. Researchers have analyzed the correlation between urban form indicators and street-level wind environments [11], leading to the proposal of tailored planning strategies to address surface heat island effects based on different types of urban spatial forms [12].

In recent years, research has undergone a shift towards quantification, leading to a greater diversification of research methods and data sources. In terms of methodology, linear regression has predominantly been employed to analyze the correlation between urban morphology and temperature [13–21]. Factors such as proximity to the sea, sky view factor (SVF), vegetation cover, and impervious surfaces have been examined through linear regression analysis to evaluate temperature variations in Gothenburg, Sweden [17]. Additionally, stepwise linear regression has been utilized to investigate urban canopy temperatures [18]. Regarding data sources, remote sensing satellite imagery, mesoscale meteorological simulation software, traffic flow measurements, and GIS platform technologies are commonly utilized at the urban scale. The WRF/UCM system has been utilized to simulate urban heat islands and explore the relationship between urban form indicators and heat island intensity [22–24]. Moreover, digital technologies including RS and GIS have played a crucial role in assessing the climate regulation effects of green open spaces [25], as well as enabling spatial identification by integrating satellite remote sensing data with meteorological observations [26]. At the community level, research primarily involves computer numerical simulations and wind tunnel experiments. Drawing upon the local climate zone (LCZ) method, studies have examined the spatial distribution of urban heat islands and their association with urban morphology [27–29]. Low-altitude thermal infrared remote sensing and thermal climate wind tunnel experiments offer distinct advantages for studying outdoor environments, allowing for comprehensive investigations into thermal comfort by integrating these experimental approaches with traditional measurement methods [30–32]. At the block scale, common research methods include point measurements and questionnaire surveys. For instance, the relationship between thermal comfort and urban microclimate was analyzed by observing the outdoor activities of university students in extremely cold regions [33]. Likewise, correlation analyses were conducted to explore the link between thermal comfort in a commercial pedestrian street in cold regions and crowd behavior activities, utilizing on-site microclimate measurements, crowd behavior annotations, questionnaire surveys, and SPSS statistical analysis [34]. Scholars have also dedicated attention to the development of corresponding numerical calculation methods for urban climates at various scales [35–38].

In contrast to climate scale classification, this research addresses climate-related issues in urban planning and design, focusing on three distinct scales: city scale, block scale, and building scale. At the city scale, investigating "ventilation corridors" that incorporate urban geography and climate characteristics can enhance rational urban planning and green space system design. Ventilation corridors should align with prevailing wind directions and, where feasible, channel airflow towards developed areas, while considering natural airflows such as ocean, land, and valley winds [39]. Research teams in cities such as Stuttgart, Tokyo, and Hong Kong [40] have effectively employed "environmental climate maps" to propose measures for controlling and guiding ventilation corridors within urban environments. At the block scale, an example from Tianjin Eco-City (Figure 1, left) illustrates the use of 400 m × 400 m blocks with 20 m wide green belts as pedestrian spaces. Nevertheless, actual implementation faced challenges related to the application of public slow traffic systems due to encroachments from community roads, parking, and green spaces [41]. Adopting smaller 100 m × 100 m block units, coupled with ecological considerations such as vegetation and rivers within the block, and adjusting the road network based on daylighting standards, can lead to more sensible block module determination (Figure 1, right). The development of large blocks in Chinese new urban areas often leads to car-oriented layouts, resulting in increased energy consumption and carbon emissions, while blocks with overly small scales may suffer from ventilation and lighting issues. Therefore, block scale determination should encompass comprehensive factors such as daylighting, ventilation, and pedestrian travel time. At the building scale, higher floor area ratios contribute to a greater reduction in the sky view factor (SVF), and variations in building forms have a more pronounced impact on SVF, particularly in sites with high density requirements. Consequently, buildings in such areas should adopt simplified and compact designs to increase SVF, as observed in locations such as Central in Hong Kong and Lujiazui in Shanghai. Introducing concepts such as "aerial clusters", "multiple ground layers", and "green return rates" in building design can enhance the economic and ecological sustainability of vertical cities [42]. It is crucial to consider and appropriately adapt the concept of green return rates to local climates in architectural design practices.

In conclusion, the urban climate is a distinct climate influenced by human activities, particularly urbanization, and it directly affects the comfort of urban residents. The main causes of urban climate can be attributed to diverse underlying surfaces and emissions resulting from human activities. Extensive research has convincingly demonstrated the significant influence of urban morphology on the climatic conditions of urban spaces. While scholars have analyzed the climate factors in specific regions such as Harbin, Dalian, Beijing, Shanghai, Chengdu, and Guangzhou, further validation of these findings is necessary to ascertain their applicability to other areas within Shanghai and cities of different scales. In this study, a micro-scale campus located in Shanghai was selected as the research site, comprising 56 measurement points. Through on-site measurements and questionnaire surveys, data on local urban morphological parameters and environmental climate were collected for the spring and summer seasons. By conducting a comparative analysis of the varying impacts of different urban morphological factors on microclimates during different seasons, this study aims to establish a basis for environmentally conscious urban design in the Shanghai area.

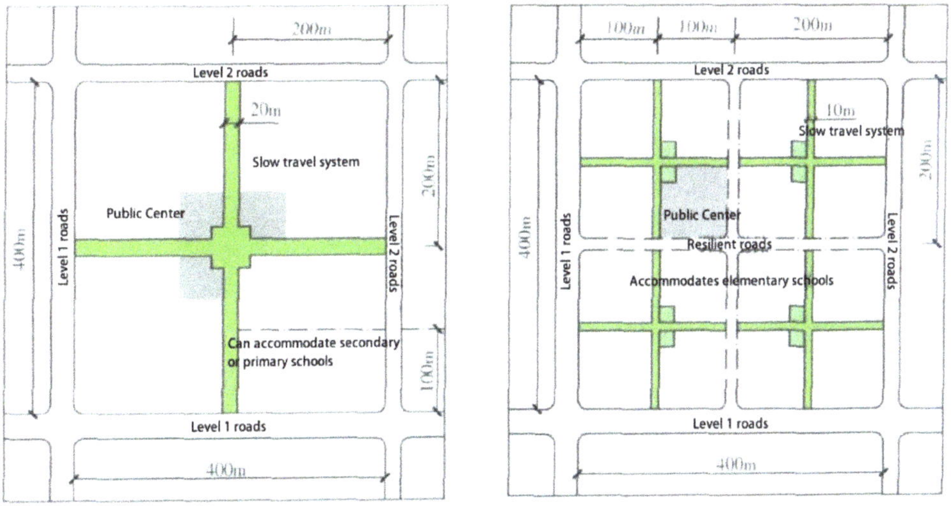

Figure 1. Tianjin Eco-City's basic unit model (**left**) and optimization mode (**right**).

2. Methodology

Field surveys were conducted to gather two types of data: (1) climate data, encompassing air temperature, humidity, solar radiation, wind speed, and wind direction; and (2) urban morphological factors, which encompass parameters associated with urban building density and form. The collected data underwent processing and analysis to derive results pertaining to urban morphological factors. Subsequently, the correlation between urban morphological factors and urban microclimate was examined and analyzed. Ultimately, conclusions were drawn to identify the factors that influence urban microclimate, providing a theoretical foundation and practical guidance for urban design.

2.1. Study Area

The Tongji University campus was chosen as the research site for this study due to its unique advantages. From an urban functionality perspective, the campus encompasses diverse functional zones, including residential areas, academic areas, administrative areas, and green spaces, each serving distinct purposes. The campus exhibits a varied mix of building heights, ranging from high-rise to mid-rise and low-rise structures, contributing to architectural diversity within its premises. The campus transportation network comprises different road types, such as carriageways and pedestrian walkways, featuring varying pavement materials. Furthermore, the campus boasts a rich variety of underlying surface materials, including buildings, paved areas, lawns, street trees, small forests, and water bodies, akin to a miniature urban model.

Located in Shanghai's Yangpu District, the Siping Campus of Tongji University spans approximately 900 acres, with a length of around 856 m and a width of about 693 m. Shanghai falls within the subtropical monsoon climate zone, characterized by hot summers and cold winters, resulting in longer summer and winter seasons and shorter spring and autumn seasons. Summers are marked by high temperatures and humidity, while winters are damp and cold, with minimal diurnal temperature fluctuations. The region experiences substantial annual precipitation and limited sunshine duration. The late spring to early summer period corresponds to the rainy season, frequently accompanied by cloudy and wet weather, including heavy rain and storms. The coastal and Yangtze River regions are susceptible to tropical storms and typhoons during the summer and autumn, resulting in intense rainfall and strong winds [43].

2.2. Site Research

For this study, 56 points were chosen, including 55 measurement points (Figure 2) and one reference point. The field survey aimed to measure and analyze urban morphological factors at these locations, including sky view factor (SVF), greenery ratio (GnPR), paved area (PAVE), building height-to-area ratio (HBDG), and others. The field survey encompassed the collection of sky hemisphere photographs, as well as surveys focusing on greenery and pavement conditions.

Figure 2. Distribution map of measurement points.

Sky view factor (SVF) photographs were captured at the measurement points using a horizontally positioned fisheye camera aimed at the sky. Two sets of photographs were taken, one during spring and the other during summer. The SVF was subsequently computed using Solweig(solar and longwave environmental irradiance geometry) software. The disparity in SVF between the two seasons primarily arises from variations in leaf quantity and density. Changes in leaf growth and senescence exert a notable influence on SVF, especially at measurement points with abundant vegetation. The greenery survey aimed to assess the vegetation surrounding the measurement points, including the types, quantities, and areas of trees, shrubs, and grasslands, which were measured and mapped. The pavement survey focused on identifying pavement materials and mapping the paved areas.

2.3. Field Measurement

To capture the significant influence of urban morphology on air temperature during clear and calm weather, measurement days were carefully selected to ensure clear conditions with weak winds. The study spanned 8 days, with 5 days allocated for spring measurements and 3 days for summer measurements. The influence of urban morphology on microclimates is most evident under specific conditions, such as clear skies, calm winds, no rain, and limited cloud cover. During the spring season in Shanghai, the temperature range varies significantly, ranging from 13 to 24 degrees Celsius. Recognizing the human body's heightened sensitivity to temperature fluctuations, we deliberately increased the number of measurements during the spring season using this unbalanced design to ensure focused and pertinent data collection. Each day, measurements were conducted from

8:00 AM to 20:00 PM, encompassing fixed-point measurements at the reference location as well as mobile measurements. The reference point employed the HOBO weather station U30, while the measurement points utilized the HOBO micro weather station H21, supplemented by two portable Kestrel data loggers. A team of 7 members was divided into 4 groups to assist with the measurements. One member was assigned to conduct fixed-point measurements at the reference point, overseeing the weather station and conducting comfort-related questionnaire interviews. The mobile measurements were conducted by 3 groups, each consisting of 2 members. One member was responsible for transporting the weather station, while the other conducted the questionnaire interviews. The mobile measurement points on the campus were divided into 4 routes, with 3 routes surveyed each day. By selecting different routes over the 3 days, each route was measured at least twice. Measurements were conducted in two time periods: morning (8:00 AM to 14:00 PM) and afternoon (14:30 PM to 20:00 PM). At 8:00 AM, the four weather stations were automatically initiated, and the three mobile groups commenced their measurements simultaneously. Upon reaching a measurement point, a timed interval of 8–10 min began, aligned with the arrival time of the last group. Subsequently, all groups departed the point simultaneously. This ensured that each measurement point had a minimum of 8 min of synchronized data, facilitating comparative analysis. While continuous fixed-point measurements would have been ideal, budget constraints and equipment safety considerations made this approach impractical. Hence, the study adopted a combination of fixed-point measurements at the reference location and mobile measurements. We made a conscious effort to carry out the measurements primarily during weekends to maximize the number of questionnaire responses. Despite the measurement period being limited to just eight days, we set up 56 measurement points and took two measurements at each point every day. This approach allowed us to obtain a robust and credible dataset consisting of 490 data entries.

The selection of measurement points aimed to represent diverse urban morphologies, vegetation forms, pavement types, and sky view variations. Functional areas, including residential, administrative, and high-rise areas, were deliberately chosen. Pavement types consisted of grass lawns, cement tiles, and asphalt roads. Vegetation types included lawns, trees, and shrubs, emphasizing the interaction and combination of these elements. Weather station sensors were uniformly positioned 2 m above the ground. The reference point was situated on the main lawn of the campus. Each weather station measured parameters such as air temperature (Ta), humidity (RH), globe temperature (GT), solar radiation (Solar), wind speed (WV), and wind direction (WD). Measurements were recorded at 10 s intervals. All equipment (refer to Table 1) was newly acquired and underwent factory calibration prior to testing. Calibration was conducted on all devices to ensure temperature discrepancies between weather stations were within a tolerance of 0.5 °C or an acceptable range. The meteorological data collected at the reference point served as the benchmark for establishing prediction models.

Table 1. List of device parameters.

	Device Model	Measurement Parameters	Accuracy Error	Range	Photograph
1	H21-002 Micro weather station data logger	Recorded data	First data point 0 to 2 s at 25 °C (77 °F)	/	
2	S-THB-M002 Temperature and humidity sensor (placed inside a white ventilated sunshade)	Air Temperature/ Relative Humidity	Temperature: ±0.2 °C Humidity: ±2.5%	−40 °C to 75 °C; 0 to 100%	

Table 1. *Cont.*

	Device Model	Measurement Parameters	Accuracy Error	Range	Photograph
3	S-LIB-M003 Total radiation sensor	Total solar radiation	±10 W/m^2 or ±5%	0 to 1280 W/m^2	
4	S-WSET-B Wind speed and direction sensors	Wind Speed/Direction	±1.1 m/s	0 to 45 m/s	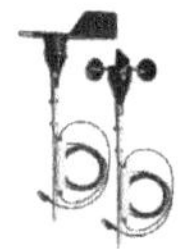
5	U23-001 Temperature and humidity recorder	Air Temperature/ Relative Humidity	Temperature: ±0.18 °C Humidity: ±2.5%	−40 °C to 70 °C; 0 to 100%	
6	UX100-014M Temperature logger	Temperature	±0.6 °C	−260 °C to 950 °C	

(Data source: Onset HOBO).

2.4. *Calculation of Urban Morphological Factors*

Urban form encompasses the spatial distribution and composition of different elements within a city [44]. Professor Meta Berhauser Pont from Delft University of Technology in the Netherlands has introduced an indicator system known as "Spacemate" [45]. This system integrates the floor space index (FSI), ground space index (GSI), average number of floors (L), and open space ratio (OSR) to create a chart that assesses the correlation between building density and urban form.

2.4.1. Methods for Calculating Urban Form Factors

Urban form factors can be classified based on the objects they describe. Factors associated with open space and visual perception include the sky view factor (SVF) and the open space ratio (OSR). Factors related to building form encompass the floor area ratio (FAR), wall surface area (WALL), average number of floors (L), building space coverage (BSC), and height-to-building area ratio (HBDG). Factors concerning vegetation encompass the greenery provision rate (GnPR). Factors related to pavement comprise the paving ratio (PAVE).

(1) Open Space Factors

SVF: The sky view factor is a numerical representation of the openness of urban three-dimensional space. It is a dimensionless value ranging from 0 to 1 [46], indicating the ratio of the sky area visible from a measurement point to the total area of the hemisphere. The fisheye photograph method is the most traditional approach for calculating the SVF, involving capturing a 180-degree fisheye photograph of the sky from a tripod at a height of 1 m and then using algorithms to compute the SVF based on the sky area in the fisheye photograph [47].

$$\Psi_s = \frac{1}{\pi R^2}\int_0^{S_v} \cos\theta dS \quad (1)$$

In the equation, S_v represents a small interface of the visible sky sphere; θ is the angle between S_v and the vertex of the sphere; R is the radius of the hemisphere.

OSR: open space ratio refers to the ratio of the total built-up area to the undeveloped open land area within a given site.

$$OSR = (1 - BSC_x)/FAR_x \quad (2)$$

In the equation, OSR represents the open space ratio, FAR_x denotes the floor area ratio of the land, and BSC_x refers to the building space coverage of the land.

(2) Building Form Factors

FAR: the floor area ratio, also known as the gross floor area ratio, represents the ratio of the total built-up area of a development to the land area.

$$FAR = F_x/A_x \tag{3}$$

In the equation, F_x represents the gross floor area in square meters (m^2), and Ax represents the area of aggregation, which refers to the land area in square meters (m^2).

WALL: the wall surface area represents the total surface area of building walls and, in another aspect, indicates urban building density.

L: the average number of floors refers to the ratio of the total floor area of buildings to the footprint area of the land parcel.

$$L = FAR_x/BSC_x \tag{4}$$

In the equation, AR_x represents the floor area ratio of parcel x, and BSC_x represents the building space coverage of parcel x.

BSC: Building site coverage refers to the ratio of the total footprint area of buildings to the total land area within a given range, expressed as a percentage. It reflects the open space ratio and the degree of building density within a specific land area.

$$BSC = B/A \tag{5}$$

In the equation, B represents the gross building footprint area in square meters (m^2), and A represents the land area of aggregation.

HBDG: height-to-building area ratio refers to the average ratio of building height to building area within a site, representing the thermal capacity of the buildings.

(3) Vegetation Factor

GnPR: Greenery provision rate is a parameter proposed by Ong [48] which calculates the average vegetation within a site using the leaf area index (LAI). It is determined by the sum of the products of each vegetation type and its corresponding LAI, divided by the total site area.

$$GnPR = \Sigma(n_n A_n \times LAI_n)/\text{Site Area} \tag{6}$$

In the equation, LAI_n represents the leaf area index for vegetation type n, A_n represents the canopy area for vegetation type n, and n_n represents the identification number of vegetation type n within the site.

(4) Paving Factors

PAVE: paving ratio refers to the proportion of hard-surfaced paving within a certain radius.

2.4.2. Results of Urban Form Factors

We conducted a comprehensive data collection of architectural, landscape, vegetation, and water body maps, complemented by on-site observations to enhance the accuracy of the time-delayed information in the maps. Relevant data, such as green space area, building area, and paved area, were directly extracted from the maps. Subsequently, we established a data processing workflow using FME (Feature Manipulate Engine) to calculate and extract urban geometric data for 55 designated measurement points.

A 50 m radius is considered optimal for influencing air temperature [5,49,50], and we developed a geographic information system (GIS) model that represents the urban form encompassing these 55 points. This model incorporates comprehensive information regarding buildings, roads, pavement, water bodies, and vegetation. By utilizing FME, a specialized data processing workflow tool was created based on this model to extract the urban form factors specific to each of the 55 measurement points, subsequently generating corresponding maps (Figures 3–5).

Figure 3. Green space ratio (GnPR) for each measurement point.

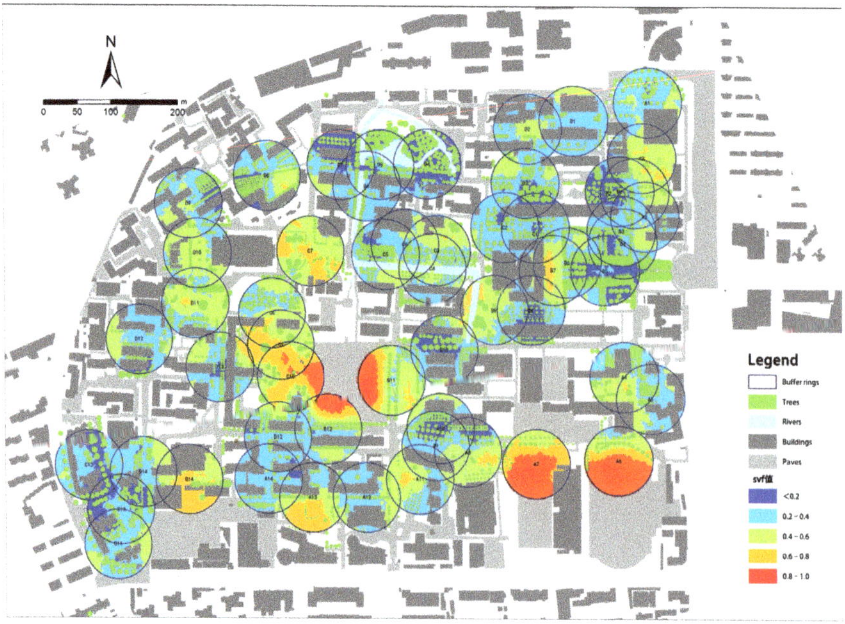

Figure 4. Sky view factor (SVF) for each measurement point.

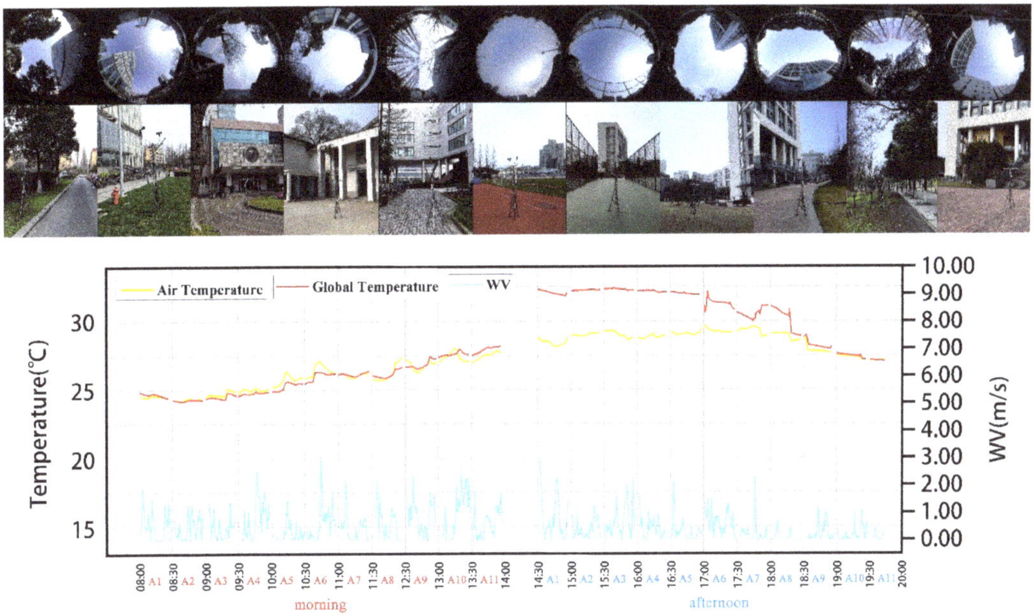

Figure 5. A1–A11 of fisheye lens images, on-site measurement point images, and meteorological data throughout a day.

3. Correlation Study between Urban Morphology and Microclimate

This section focuses on examining the correlation between meteorological parameters (such as temperature, solar radiation, and wind speed) at the measurement points and both environmental meteorological factors and urban morphology factors. We assume that these meteorological parameters are influenced by both the environmental meteorological factors and the urban morphology factors. To address equipment limitations, this study employed mobile measurements, resulting in non-simultaneous data collection for the meteorological parameters at the measurement points. The temporal variations in the meteorological environmental factors have a significant impact on the monitoring data at the measurement points. Therefore, we utilize partial correlation analysis to mitigate the interference of environmental meteorological parameters on the meteorological parameters at the measurement points. Moreover, we compare the impact of controlling environmental meteorological parameters (reference point meteorological parameters) and the meteorological parameters at the measurement points on the correlation with urban morphology factors.

3.1. Correlation Analysis between Urban Morphology and Temperature

Temperature plays a pivotal role in influencing human thermal comfort. The thermal environment, a vital metric for assessing urban environmental quality, has a profound impact on individuals' health and activity patterns. A pleasant outdoor thermal environment promotes mobility and activities, whereas excessively high urban temperatures can significantly affect urban development. When exploring the correlation between urban morphology factors and temperature, it is crucial to consider the influence of environmental meteorological parameters on urban morphology factors. To account for this, a pairwise correlation analysis was conducted, controlling for relevant variables. The findings are presented in Table 2.

Table 2. Correlation between temperature at measurement points and urban morphology factors during spring and summer with controlled variables.

Control Variables			PAVE	BSC	FAR	L	OSR	SVF	GnPR	Wall	HBDG
RefTa, RefRH, RefSolar, and RefWV	Spring	Correlation	−0.028	−0.175	−0.393	−0.319	0.109	0.264	0.020	−0.298	0.080
		Significance	0.668	0.007	0.000	0.000	0.097	0.000	0.761	0.000	0.222
		N	232	232	232	232	232	232	232	232	232
RefTa, RefRH, RefSolar, and RefWV	Summer	Correlation	0.189	0.008	−0.027	−0.058	0.199	0.467	−0.355	0.011	−0.044
		Significance	0.004	0.904	0.682	0.378	0.002	0.000	0.000	0.874	0.510
		N	228	228	228	228	228	228	228	228	228

The cells contain zero-order (Pearson) correlations.

After controlling for the influence of reference point meteorological parameters, the data in Table 2 indicate that, in both spring and summer, the bivariate Pearson correlation analysis between sky view factor (SVF) and temperature (Ta) yields a significance level (two-tailed) of less than 0.05. The correlation coefficients are 0.264 and 0.467, respectively, signifying a significant positive correlation between SVF and Ta. In spring, there is no significant correlation between green naturalness percentage (GnPR), paved ratio (PAVE), open space ratio (OSR), and Ta. However, in summer, GnPR exhibits a significant negative correlation with Ta, while PAVE and OSR display a significant positive correlation. The correlation between building shape compactness (BSC), floor area ratio (FAR), building height (L), wall ratio (WALL), and Ta transitions from a significant negative correlation in spring to a non-significant correlation in summer.

Based on the aforementioned analysis, it is evident that the correlation between building morphology factors and environmental temperature is more significant in spring, whereas in summer, the correlation between pavement, vegetation, and environmental temperature becomes more significant. The correlation between vegetation parameters and temperature shifts from non-significant in spring to significant in summer, aligning with the seasonal growth and decline of plants. The correlation between pavement ratio, openness index, and temperature changes from non-significant in spring to significant in summer, likely attributable to the increased solar radiation during summer. This results in the absorption of heat by pavements and surrounding building materials, as well as the release of long-wave radiation, leading to a significant increase in near-surface temperature.

3.2. Correlation Analysis between Urban Morphology and Solar Radiation

Solar radiation plays a crucial role in determining human thermal comfort. Urban surfaces absorb a significant amount of solar radiation, which is primarily used for heating the surrounding air and solid materials. The thermal response of different surface types to solar radiation varies, leading to significant impacts on the urban thermal environment. When examining the correlation between urban morphology factors and solar radiation, it is essential to consider the influence of environmental meteorological parameters on these factors. To address this issue, ordinary pairwise correlation analysis was performed with controlled variables, and the results are presented in Table 3.

Table 3. Correlation between solar radiation at measurement points and urban morphology factors in spring and summer with controlled variables.

Control Variables			PAVE	BSC	FAR	L	OSR	SVF	GnPR	Wall	HBDG
RefTa, RefRH, RefSolar, and RefWV	Spring	Correlation	0.138	−00.128	−00.228	−00.165	0.163	0.497	−00.074	−00.209	0.026
		Significance	0.033	0.047	0.000	0.012	0.013	0.000	0.256	0.001	0.697
		N	238	238	231	231	231	238	238	231	231
RefTa, RefRH, RefSolar, and RefWV	Summer	Correlation	0.081	−00.046	−00.131	−00.143	0.172	0.588	−00.276	−00.109	0.125
		Significance	0.262	0.523	0.070	0.048	0.017	0.000	0.000	0.133	0.084
		N	190	190	190	190	190	190	190	190	190

The cells contain zero-order (Pearson) correlations.

After controlling for the influence of reference point meteorological parameters, the table data reveal significant positive correlations between the open space ratio (OSR) and solar radiation, as well as between the sky view factor (SVF) and solar radiation, during both spring and summer. Conversely, a significant negative correlation is observed between building height (L) and solar radiation. In spring, no significant correlation is found between the green naturalness percentage (GnPR) and solar radiation, whereas in summer, a significant negative correlation exists between these two variables. The paved ratio (PAVE) demonstrates a significant positive correlation with solar radiation in spring, while building shape compactness (BSC), floor area ratio (FAR), and wall ratio (WALL) exhibit significant negative correlations. However, during summer, these factors do not show a clear correlation with solar radiation.

Based on the analysis above, it can be observed that the correlation between building morphology factors and solar radiation is more significant in spring, whereas in summer, the correlation between vegetation and solar radiation becomes more prominent. The correlation between urban morphology factors related to buildings and pavements and solar radiation weakens from being significant in spring to non-significant in summer. This can be attributed to the dense tree canopy during summer, which exhibits a significant correlation with solar radiation at the measurement points, resulting in a diminished correlation between factors related to buildings and pavements and solar radiation.

3.3. Correlation Analysis between Urban Morphology and Wind Speed

Urban ventilation has a direct impact on human thermal comfort and indirectly affects factors such as temperature and humidity. Efficient ventilation plays a crucial role in mitigating the urban heat island effect. When analyzing the correlation between urban morphology factors and wind speed, the influence of environmental meteorological parameters on urban morphology is relatively limited. After comparing the results of ordinary pairwise correlation analysis conducted prior to controlling for variables, the findings are presented in Table 4.

Table 4. Correlation between wind speed and urban morphology factors in spring and summer without controlling variables.

Control Variables			PAVE	BSC	FAR	L	OSR	SVF	GnPR	Wall	HBDG
None	Spring	Correlation	0.105	−0.014	0.362	0.370	0.058	0.072	−0.132	0.155	−0.238
		Significance	0.103	0.830	0.000	0.000	0.370	0.260	0.039	0.017	0.000
		N	242	242	235	235	235	242	242	235	235
None	Summer	Correlation	0.053	−0.186	0.249	0.346	0.059	0.297	−0.065	0.142	−0.123
		Significance	0.461	0.009	0.000	0.000	0.410	0.000	0.363	0.048	0.087
		N	194	194	194	194	194	194	194	194	194

The cells contain zero-order (Pearson) correlations.

In the absence of controlled variables, the data in the table reveal significant positive correlations between floor area ratio (FAR), building height (L), and wall ratio (WALL) with wind velocity (WV). No significant correlation is found between other morphology parameters such as paved ratio (PAVE) and open space ratio (OSR) with WV. Additionally, in spring, there is no significant correlation between building shape compactness (BSC) and sky view factor (SVF) with WV. However, in summer, BSC exhibits a significant negative correlation with WV, while SVF shows a significant positive correlation. The correlation between green naturalness percentage (GnPR) and height-based density gradient (HBDG) with WV changes from a significant negative correlation in spring to a non-significant correlation in summer.

Based on the above analysis, it is evident that the correlation between vegetation and wind velocity is more pronounced in spring, whereas the relationship between building morphology factors and wind velocity becomes more significant in summer. The correlation between sky view factor and wind velocity transitions from non-significant in spring to significant in summer. This can be attributed to the increased obstruction caused by dense vegetation during summer, resulting in a stronger association with wind velocity. In contrast to environmental temperature and solar radiation, wind velocity is less affected by seasonal variations. Moreover, controlling environmental meteorological parameters does not significantly influence the correlation between wind velocity and morphology factors, unlike temperature and solar radiation.

4. Multivariate Linear Regression Analysis of Urban Morphology and Microclimate

This section aims to investigate the potential impact of parameters that are significantly correlated with the meteorological variables at the measurement points on those variables and explore the feasibility of developing effective regression models based on these relationships. The dependent variables considered in this analysis are "Measurement Point Air Temperature (Ta)", "Measurement Point Solar Radiation (Solar)", and "Measurement Point Wind Velocity (WV)". The independent variables comprise meteorological parameters at the reference point, namely "Reference Point Air Temperature (RefTa)", "Reference Point Solar Radiation (RefSolar)", and "Reference Point Humidity (RefRH)", along with urban morphology factors that exhibit significant correlations with the respective meteorological variables. Prior to conducting the multivariate linear regression analysis, it is essential to examine the collinearity among the independent variables. In this study, correlation analysis is employed, with variables displaying correlation coefficients equal to or greater than 0.75 being considered as indicative of collinearity.

4.1. Multivariate Linear Regression Analysis of Urban Morphology and Temperature

In the multivariate linear regression analysis of urban morphology and temperature, a strong linear relationship was observed among the variables FAR, SVF, and GnPR, as indicated by Pearson correlation coefficients exceeding 0.75. The dependent variable in this analysis is Ta (measurement point air temperature), while the independent variables include reference point meteorological parameters (RefTa, RefRH, and RefSolar) and urban morphology factors (FAR, SVF, and GnPR). The results for both spring and summer seasons are summarized in Table 5.

Table 5. Multivariate regression models of urban morphology factors and temperature for spring and summer seasons.

	Regression Equation							R-Side	Variance	Significance
	Constants	**RefTa**	**SVF**	**FAR**	**GnPR**	**RefSolar**	**RefRH**			
Spring	−2.820	1.149	1.549	−0.257	/	/	/	0.900	338.536	0.000
Summer	0.260	0.992	1.800	/	−0.438	/	/	0.957	1417.137	0.000

SVF significantly impacts Ta in both spring and summer seasons. In spring, there is a significant influence of FAR on Ta, with a higher floor area ratio (FAR) correlating with lower air temperatures. This relationship can be attributed to the reduction in solar radiation within the street canyon due to building obstructions, leading to a decrease in temperature. In summer, GnPR has a significant effect on Ta, with higher green naturalness percentage (GnPR) associated with lower air temperatures. Consequently, it can be concluded that reducing FAR can increase environmental air temperature and enhance comfort in early spring, while increasing green area, particularly tree coverage, helps lower environmental air temperature and improve comfort during summer.

4.2. Multivariate Linear Regression Analysis of Urban Morphology and Solar Radiation

In the study of multivariate linear regression analysis on urban morphology and solar radiation, it was found that SVF, L, and "Tree Canopy Area" exhibited Pearson correlation coefficients exceeding 0.75, indicating a strong collinearity among these variables. After considering different combinations of independent variables, solar radiation was chosen as the dependent variable, while reference point meteorological parameters (RefSolar, RefRH, and RefWV) and urban morphology factors (SVF, L, and Tree) were selected as independent variables for the multivariate linear regression analysis. The results for both spring and summer seasons are presented in Table 6.

Table 6. Multivariate regression models of urban morphology factors and solar radiation in spring and summer seasons.

	Regression Equation							R-Side	Variance	Significance
	Constants	**RefSolar**	**SVF**	**L**	**RefRH**	**RefWV**	**Tree**			
Spring	−170.22	0.456	622.648	/	/	/	/	0.492	20.27	0.000
Summer	−114.825	0.63	649.127	−11.202	/	/	/	0.590	38.585	0.000

In both spring and summer, the sky view factor (SVF) emerges as the dominant urban morphology factor influencing solar radiation. A higher SVF corresponds to increased solar radiation received at the measurement points. In summer, the significance test confirms the influence of the parameter L, indicating that taller buildings result in reduced solar radiation reaching the measurement points. Additionally, although the green parameter Tree exhibits a significant correlation with solar radiation during both seasons, it fails to meet the significance test. This discrepancy may be attributed to the direct impact of tree canopies directly above the specific measurement points, outweighing the influence of the tree canopy area within a 50 m radius. Consequently, SVF encompasses the obstruction information arising from buildings and trees at the measurement points. Hence, the solar radiation models for both spring and summer do not incorporate the representation of green parameters.

4.3. Multivariate Linear Regression Analysis of Urban Morphology and Wind Velocity

The present study conducted a multivariate linear regression analysis to investigate the relationship between urban morphology and wind velocity. Wind velocity (WV) was selected as the dependent variable, while the independent variables consisted of RefRH, RefWV, and urban morphology factors HBDG, L, Tree, SVF, and BSC. These variables exhibited significant correlations with the wind speed at the measurement points. The obtained dataset was then used for the multivariate linear regression analysis. The results for both spring and summer seasons are summarized in Table 7.

Table 7. Multivariate regression models of urban morphology factors and wind velocity in spring and summer seasons.

	Regression Equation							R-Side	Variance	Significance
	Constants	**RefRH**	**RefWV**	**HBDG**	**L**	**Tree**	**SVF**	**BSC**			
Spring	0.112	0.026	0.303	−0.460	0.066	-1.88×10^{-4}	/	/	0.377	19.234	0.000
Summer	−0.115	0.008	/	−0.242	0.049	/	0.580	−0.667	0.457	26.526	0.000

In both spring and summer seasons, RefRH had a significant impact on wind velocity among the environmental meteorological factors (under calm wind conditions). Among the urban morphology factors, L exerted the greatest influence on wind velocity at the measurement points in both seasons. Additionally, HBDG had a significant impact on wind velocity during both seasons. Moreover, in spring, Tree had a significant influence on wind velocity, indicating a negative relationship between larger tree canopy areas and wind speeds. In summer, both SVF and BSC demonstrated a significant impact on wind velocity at the measurement points.

4.4. Validation of the Multivariate Linear Regression Model

4.4.1. Spring Air Temperature Regression Verification

Upon obtaining the aforementioned regression models, a rigorous evaluation of their goodness of fit becomes imperative. Linear regression necessitates the consideration of various assumptions, such as error normality and homoscedasticity, warranting a thorough investigation to ascertain their conformity with the observed data.

Assumption 1: In multiple linear regression, a fundamental assumption pertains to the residuals adhering to a normal distribution. The histogram of residuals for the linear regression model is illustrated in Figure 6, revealing a striking resemblance between the frequency distribution of residuals and the standard normal curve. Consequently, Assumption 1 is deemed satisfied for this multiple linear regression model.

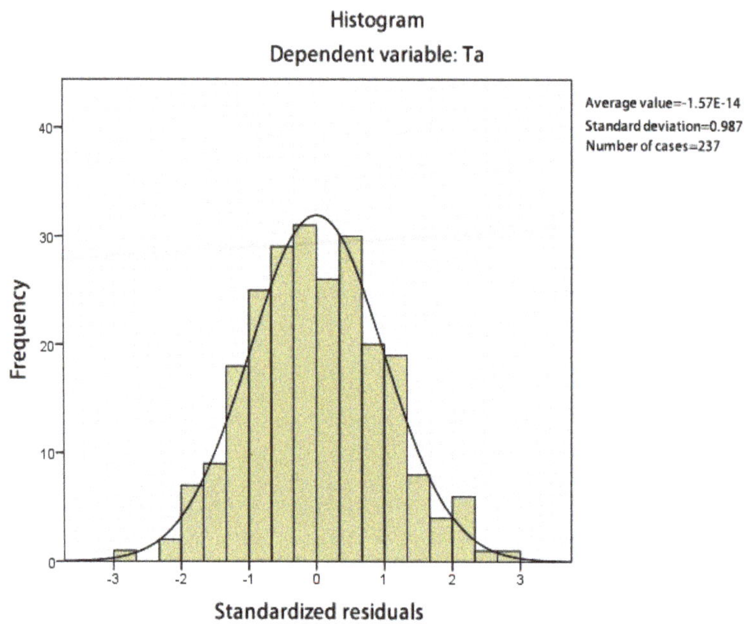

Figure 6. Residual histogram.

Assumption 2: In multiple linear regression, it is postulated that the residuals exhibit homoscedasticity, indicating consistent residual variance. Conversely, the presence of heterogeneous residual variance is referred to as "heteroscedasticity". To diagnose the existence of heteroscedasticity in the linear regression model, a scatter plot of residuals is examined (refer to Figure 7). The scatter plot showcases a relatively uniform distribution of residuals, providing evidence of the absence of significant heteroscedasticity. Thus, Assumption 2 is met.

Based on the comprehensive diagnostic analysis of the linear regression model, we can confidently conclude that the results obtained from this model are valid, and the derived linear regression equation holds practical value.

4.4.2. Summer Air Temperature Regression Verification

In multiple linear regression models, the assumption is made that the residuals conform to a normal distribution. As depicted in Figure 8, the residual histogram of this particular linear regression model exhibits a close resemblance to the standard normal curve, validating this assumption. Another critical assumption in multiple linear regression models pertains to the homoscedasticity of residuals. The scatter plot for this equation

(Figure 9) reveals a relatively uniform distribution of residuals, indicating the absence of significant heteroscedasticity. Furthermore, independence of residuals is also assumed in multiple linear regression models. Although there is a slight presence of autocorrelation in this model, it does not exert a substantial impact on the parameter estimates of the regression model. Upon conducting a thorough diagnostic assessment of the linear regression model, we can confidently assert the validity of the results, signifying the practical value of the derived linear regression equation.

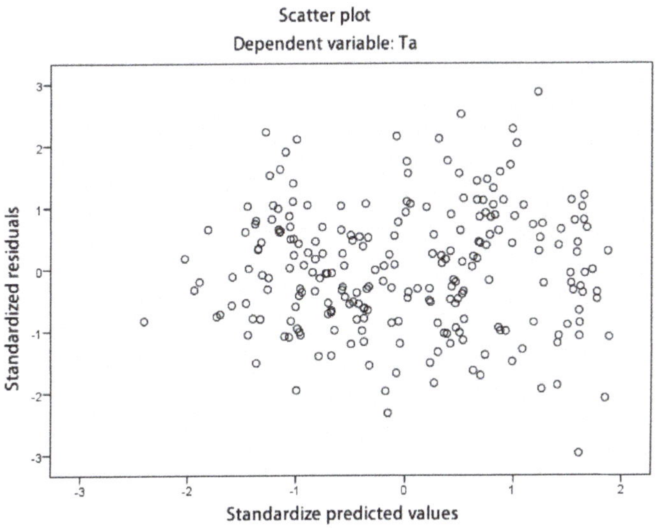

Figure 7. Residual scatter plot.

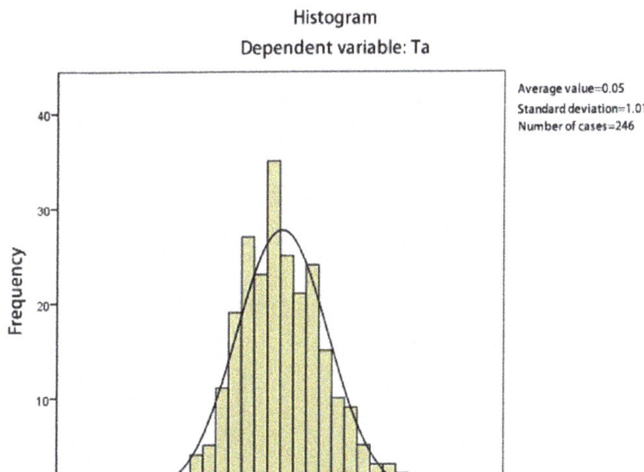

Figure 8. Residual histogram.

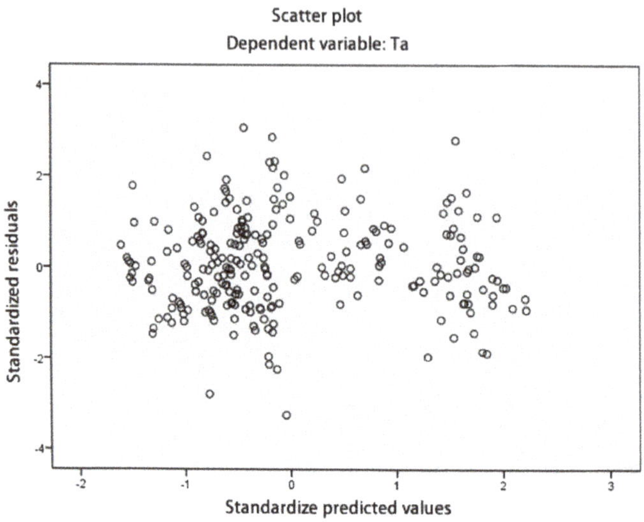

Figure 9. Residual histogram.

4.4.3. Spring Solar Radiation Regression Verification

Figure 10 presents the residual histogram of this linear regression model, displaying a clear alignment between the frequency distribution of residuals and the standard normal curve. The scatter plot of the regression equation (Figure 11) reveals a relatively uniform distribution of residuals, indicating the absence of significant heteroscedasticity. Additionally, there is no observed autocorrelation in this model. Based on a thorough diagnostic evaluation of the linear regression model, we can confidently affirm the validity of the results, thus emphasizing the practical significance of the derived linear regression equation.

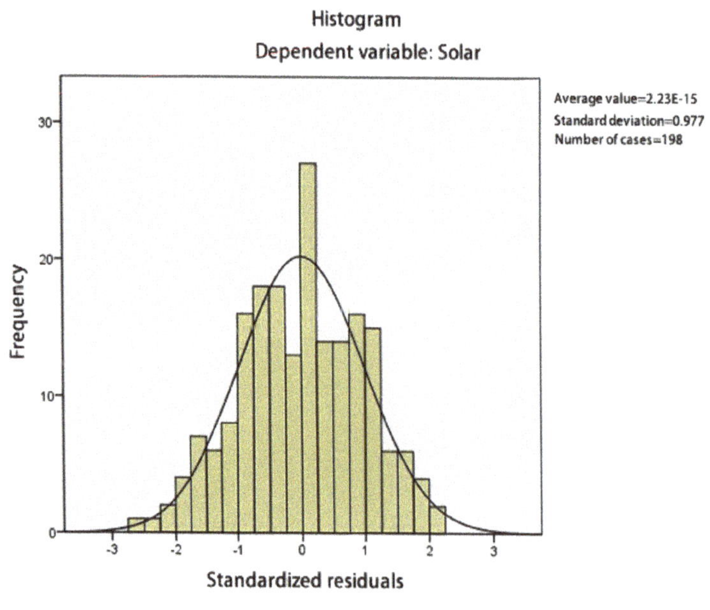

Figure 10. Residual histogram.

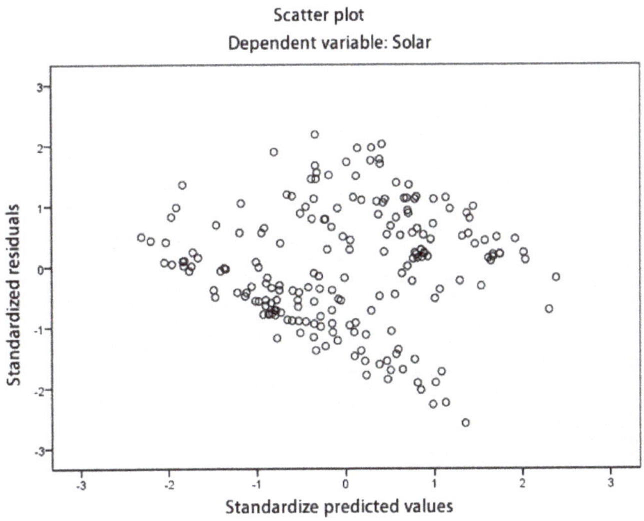

Figure 11. Residual histogram.

4.4.4. Summer Solar Radiation Regression Verification

Figure 12 displays the residual histogram of this linear regression model, revealing a close resemblance between the frequency distribution of residuals and the standard normal curve. Within the context of the multiple linear regression model, an assumption is made regarding the independence of residuals. Notably, no autocorrelation is present in this model, as affirmed by the scatter plot (Figure 13), which demonstrates a relatively uniform distribution of residuals, indicating the absence of significant heteroscedasticity. Based on a comprehensive diagnostic assessment of the linear regression model, we can confidently assert the validity of the results, underscoring the practical significance of the derived linear regression equation.

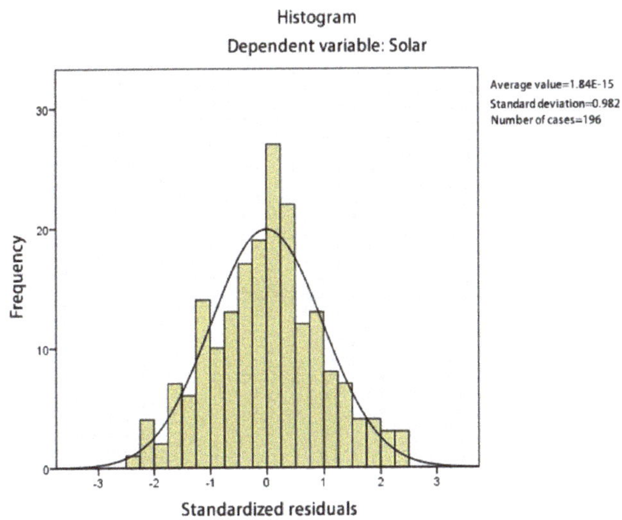

Figure 12. Residual histogram.

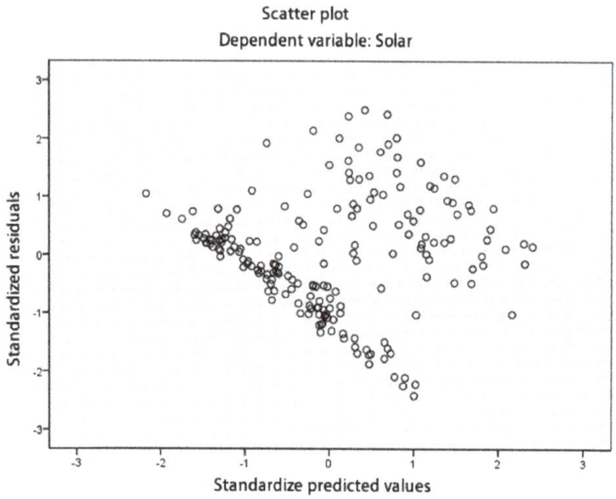

Figure 13. Residual histogram.

4.4.5. Spring Wind Speed Regression Verification

Figure 14 illustrates the residual histogram of this linear regression model. It is evident that the frequency distribution of residuals deviates from the strict adherence to the normal distribution curve. However, there is a noticeable increase in the frequency of residuals with a value of 0, suggesting proximity to a normal distribution. In the context of the multiple linear regression model, an assumption is made concerning the independence of residuals, and no autocorrelation is detected in this model. Additionally, the scatter plot of the regression equation (Figure 15) displays a relatively uniform distribution of residuals, indicating the absence of significant heteroscedasticity. Based on a comprehensive diagnostic evaluation of the linear regression model, we confidently conclude that the results obtained from this linear regression model are valid, underscoring the practical value of the derived linear regression equation.

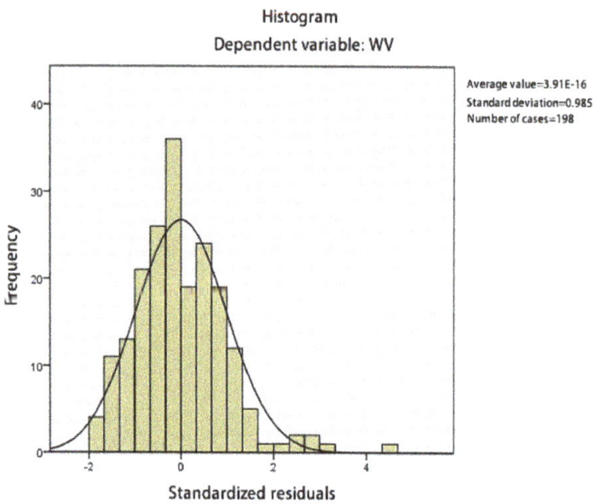

Figure 14. Residual histogram.

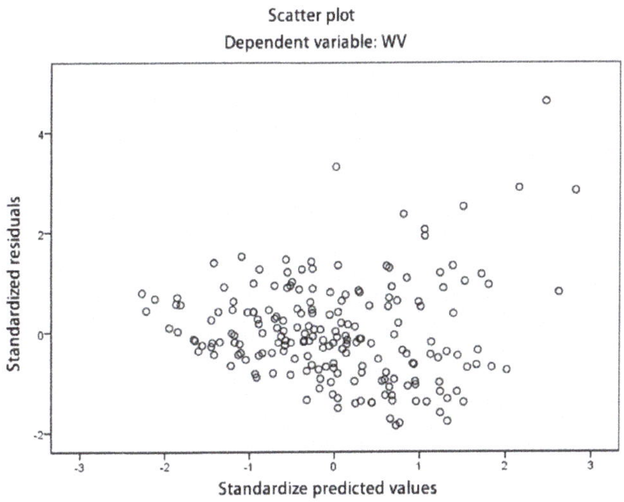

Figure 15. Residual histogram.

4.4.6. Summer Wind Speed Regression Verification

Figure 16 displays the residual histogram of this linear regression model. The frequency distribution of residuals is not strictly aligned with the normal distribution curve. Nevertheless, a noteworthy increase in the frequency of residuals with a value of 0 suggests proximity to a normal distribution. In the context of the multiple linear regression model, an assumption is made concerning the independence of residuals. In linear regression models, it is a standard assumption that the error terms are uncorrelated. Moreover, no autocorrelation is evident in this model. The scatter plot of this model (Figure 17) illustrates a relatively uniform distribution of residuals, indicating the absence of significant heteroscedasticity. A comprehensive diagnostic evaluation of the linear regression model leads us to confidently affirm the validity of the results obtained, emphasizing the practical significance of the derived linear regression equation.

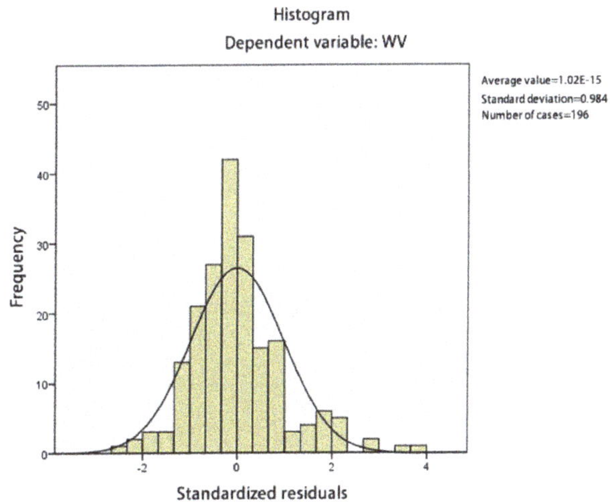

Figure 16. Residual histogram.

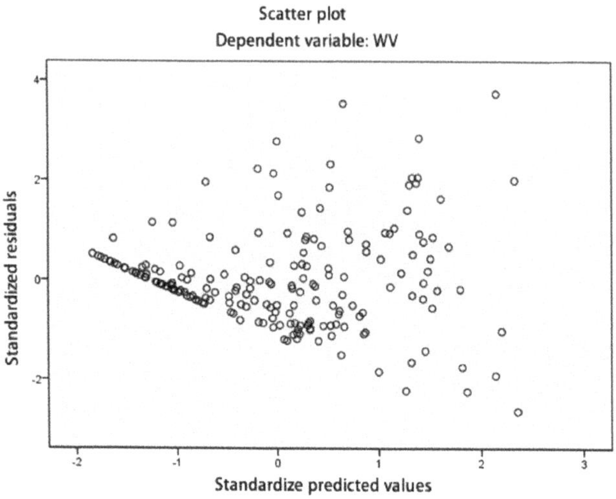

Figure 17. Residual histogram.

5. Discussion

Based on the comparison of the air temperature model during spring and summer, the study reveals the following key findings. Firstly, irrespective of the season, the sky view factor (SVF) exerts a significant influence on the temperature at the measurement points. Secondly, in spring, the floor area ratio (FAR) notably impacts the air temperature (Ta) at the measurement points, with higher FAR values associated with lower air temperatures. This effect is attributed to the increased FAR leading to reduced solar radiation received within the street canyon, resulting in lower temperatures. Conversely, during summer, the green planting ratio (GnPR) significantly affects the air temperature (Ta) at the measurement points, with higher GnPR values being correlated with lower air temperatures. Consequently, it can be deduced that reducing the floor area ratio proves beneficial in elevating environmental air temperatures and enhancing comfort levels in early spring. Furthermore, during summer, enhancing green areas, particularly with higher leaf area index (LAI) values for trees, serves as an effective approach to lowering environmental air temperatures and improving comfort levels.

Based on the comparison of the solar radiation models during spring and summer, it is evident that the sky view factor (SVF) significantly influences solar radiation in both seasons, with weights of 0.444 and 0.483, respectively. A larger SVF corresponds to higher solar radiation received at the measurement points. In summer, the average building height (L) also exhibits statistical significance, with a weight of −0.12, indicating that taller buildings result in reduced solar radiation at the measurement points. Additionally, the greenery parameters "Tree crown area" and "Green Planting Ratio (GnPR)" demonstrate significant correlations with solar radiation during both spring and summer. However, they fail to pass the significance test. This discrepancy arises from the fact that, at a specific measurement point, the presence of tree crown shading directly impacts solar radiation more than the tree crown area within a 50 m radius around the point. Consequently, the SVF inherently accounts for information about both buildings and trees obstructing the measurement points, rendering the inclusion of greenery parameters unnecessary in the solar radiation models for both spring and summer.

Based on the comparison of wind speed (WV) models during spring and summer, it is evident that, under calm wind conditions, the environmental meteorological factor, specifically the reference humidity (RefRH), significantly impacts the wind speed at the measurement points. The weights for RefRH are 0.251 and 0.384 in spring and summer, respectively. Among the urban morphological factors, the "average building height (L)" exerts the most substantial influence on wind speed at the measurement points in both seasons, with impact weights of 0.333 and 0.313, indicating that higher L values are associated with greater wind speeds. The "average building height-to-area ratio (HBDG)" also significantly affects wind speed at the measurement points in both seasons, with weights of −0.214 and −0.150, respectively, suggesting that higher HBDG values correspond to lower wind speeds. In spring, the "tree crown area (Tree)" notably impacts wind speed at the measurement points, with a weight of −0.198, indicating that larger tree crown areas are associated with lower wind speeds. In summer, both the "Sky View Factor (SVF)" and the "building shape coefficient (BSC)" significantly influence wind speed at the measurement points, with weights of 0.244 and −135, respectively. Hence, higher SVF values and lower BSC values are correlated with higher wind speeds at the measurement points.

While this paper has made preliminary research progress, further comprehensive investigations are necessary to address several aspects. The key points for further discussion are as follows:

(1) The study has successfully established a predictive model for urban microclimate and comfort based on real measurements. However, the model's training demands a substantial amount of long-term data. Due to constraints related to time, weather conditions, personnel, budget, and safety, the measurements were limited to clear and slightly cloudy weather with calm winds during spring and summer, utilizing mobile measurement methods. In future research, it is imperative to augment meteorological data for autumn and winter, as well as daytime, by utilizing an adequate number of fixed meteorological stations.

(2) As the study's investigations into microclimate and subjective comfort are rooted in empirical studies conducted in the Shanghai region, the drawn conclusions are applicable solely to Shanghai and areas with similar climates. For regions with differing climates, further comparative studies are essential.

(3) Considering the growing popularity of artificial neural network models, it is essential to explore the utilization of smart city data to acquire urban meteorological big data and to compare the fitting robustness of multiple linear regression analysis models using alternative modeling approaches.

6. Conclusions

This study utilizes meteorological data obtained through field measurements and urban morphology data acquired from a geographic information system (GIS) to analyze the impact of urban morphology factors on microclimate parameters. Partial correlation analysis and multiple linear regression analysis were performed using the SPSS 27 software, with measured temperature, solar radiation, and wind speed as dependent variables and meteorological parameters at reference points and urban morphology factors at measurement points as independent variables. Predictive models were developed accordingly. Ultimately, a set of seven urban morphology factor indicators that influence the microclimate was established, including sky view factor (SVF), floor area ratio (FAR), average building height (L), greenery ratio (GnPR), height-to-built area ratio (HBDG), building site coverage (BSC), and tree canopy area (Tree).

The findings are as follows: (1) Open spaces with high spatial openness, such as squares, playgrounds, and lawns, exhibit better comfort levels during early spring and other transitional seasons, while the opposite is true during summer. Therefore, planting deciduous trees in these areas is advisable to mitigate high summer temperatures. (2) To maximize solar radiation intake during transitional seasons such as spring, permanent structures such as buildings should be designed to increase the sky view factor. Simultaneously, the dense canopies of deciduous trees provide effective shading during summer. (3) The correlation between greenery and wind speed is more significant in spring, whereas the relationship between building morphology factors and wind speed is more significant in summer. Outdoor public spaces should be strategically positioned around buildings in sheltered locations to minimize wind exposure.

In conclusion, this empirical study confirms the substantial influence of urban morphology factors on urban microclimate. In the future, it is important to consider specific factors based on regional environmental characteristics, combining building morphology factors such as floor area ratio and exterior wall surface area with greening factors. This integration will improve the natural climatic conditions of urban spaces, enhance the thermal comfort for users, and provide a scientifically guided approach to urban design.

Author Contributions: Conceptualization, R.W. and C.X.; methodology, R.W.; software, R.W.; validation, R.W. and Z.C.; formal analysis, R.W. and C.X.; investigation, R.W.; resources, R.W.; data curation, R.W. and C.X.; writing—original draft preparation, C.X. and R.W.; writing—review and editing, D.S. and H.T.; project administration, R.W. and H.T.; funding acquisition, H.T. All authors have read and agreed to the published version of the manuscript.

Funding: Shandong Jianzhu University Doctoral Fund Project, Research on the Correlation between Urban Morphological Factors and Microclimate (X20006Z). China Ministry of Housing and Urban-Rural Development 2022 Scientific and Technological Project Plan (2022-K-184, 2022-K-148). China National Natural Science Foundation General Project (52078341, 51878392). China Ministry of Education Humanities and Social Sciences Research Youth Fund Project (21YJC840038).

Data Availability Statement: No new data were created.

Conflicts of Interest: The authors declare no conflict of interest.

References

1. Oke, T.R. *Boundary Layer Climates*; Routledge: London, UK, 1987.
2. Masson, V.; Heldens, W.; Bocher, E.; Bonhomme, M.; Bucher, B.; Burmeister, C.; de Munck, C.; Esch, T.; Hidalgo, J.; Kanani-Sühring, F.; et al. City-descriptive input data for urban climate models: Model requirements, data sources and challenges. *Urban Clim.* **2020**, *31*, 100536. [CrossRef]
3. Ding, W.; Hu, Y.; Dou, P. Research on the correlation between urban morphology and urban microclimate. *J. Archit.* **2012**, *7*, 16–21.
4. Yang, F.; Chen, L. Developing a thermal atlas for climate-responsive urban design based on empirical modeling and urban morphological analysis. *Energy Build.* **2016**, *111*, 120–130. [CrossRef]
5. Chen, S.; Wong, N.H.; Zhang, W.; Ignatius, M. The impact of urban morphology on the spatiotemporal dimension of estate-level air temperature A case study in the tropics. *Build. Environ.* **2023**, *228*, 109843. [CrossRef]
6. Yu, Z.; Chen, S.; Wong, N.H.; Ignatius, M.; Deng, J.; He, Y.; Hii, D.J.C. Dependence between urban morphology and outdoor air temperature: A tropical campus study using random forests algorithm. *Sustain. Cities Soc.* **2020**, *61*, 102200. [CrossRef]
7. Xu, M.; Hong, B.; Jiang, R. Study on the impact of campus roadside trees on outdoor pedestrian thermal comfort in summer. *Chin. Landsc. Archit.* **2020**, *36*, 139–144.
8. Li, R.; Zeng, F.; Zhao, Y.; Wu, Y.; Niu, J.; Wang, L.L.; Gao, N.; Shi, X. CFD simulations of the tree effect on the outdoor microclimate by coupling the canopy energy balance model. *Build. Environ.* **2023**, *230*, 109995. [CrossRef]
9. Yang, Y.; Gatto, E.; Gao, Z.; Buccolieri, R.; Morakinyo, T.E.; Lan, H. The "plant evaluation model" for the assessment of the impact of vegetation on outdoor microclimate in the urban environment. *Build. Environ.* **2019**, *159*, 106151. [CrossRef]
10. Liu, J.; Tang, H.; Zheng, B. Simulation study of summer microclimate in street space of historic conservation areas in China—A case study in Changsha. *Front. Environ. Sci.* **2023**, *11*, 1146801. [CrossRef]
11. Hu, X.; Wei, D.; Li, B. Study on the correlation between urban spatial morphological indicators and street wind environment. *New Archit.* **2020**, 139–143. [CrossRef]
12. Cai, Z.; Tang, Y.; Liu, C.; Matthias, D. Planning responses to the three-dimensional urban spatial morphological evolution and surface heat island effect: A case study of Beijing. *Int. Urban Plan.* **2021**, *5*, 61–68.
13. Krüger, E.; Givoni, B. Outdoor measurements and temperature comparisons of seven monitoring stations: Preliminary studies in Curitiba, Brazil. *Build. Environ.* **2007**, *42*, 1685–1698. [CrossRef]
14. Giridharan, R.; Lau, S.S.Y.; Ganesan, S. Nocturnal heat island effect in urban residential developments of Hong Kong. *Energy Build.* **2005**, *37*, 964–971. [CrossRef]
15. Shashua-Bar, L.; Tzamir, Y.; Hoffman, M.E. Thermal effects of building geometry and spacing on the urban canopy layer microclimate in a hot-humid climate in summer. *Int. J. Climatol.* **2004**, *24*, 1729–1742. [CrossRef]
16. Giridharan, R.; Ganesan, S.; Lau, S.S.Y. Daytime urban heat island effect in high-rise and high-density residential developments in Hong Kong. *Energy Build.* **2004**, *36*, 525–534. [CrossRef]
17. Eliasson, I.; Svensson, M.K. Spatial air temperature variations and urban land use—A statistical approach. *Meteorol. Appl.* **2003**, *10*, 135–149. [CrossRef]
18. Alcoforado, M.J.; Andrade, H. Nocturnal urban heat island in Lisbon (Portugal): Main features and modelling attempts. *Theor. Appl. Climatol.* **2006**, *84*, 151–159. [CrossRef]

19. Tong, S.; Wong, N.H.; Jusuf, S.K.; Tan, C.L.; Wong, H.F.; Ignatius, M.; Tan, E. Study on correlation between air temperature and urban morphology parameters in built environment in northern China. *Build. Environ.* **2018**, *127*, 239–249. [CrossRef]
20. Lu, J.; Li, C.; Yu, C.; Jin, M.; Dong, S. Regression analysis of the relationship between urban heat island effect and urban canopy characteristics in a mountainous city, Chongqing. *Indoor Built Environ.* **2012**, *21*, 821–836. [CrossRef]
21. Vuckovic, M.; Kiesel, K.; Mahdavi, A. Toward advanced representations of the urban microclimate in building performance simulation. *Sustain. Cities Soc.* **2016**, *27*, 356–366. [CrossRef]
22. Yan, T.; Jin, H. Study on urban heat island effect in severe cold regions based on WRF/UCM numerical simulation. *Build. Sci.* **2020**, *36*, 107–113.
23. Zhang, H.; Tang, J.; Guo, F. WRF/UCM assessment method for the impact of urbanization on thermal environment. *J. Dalian Univ. Technol.* **2019**, *59*, 372–378.
24. Zhang, H.; Tang, J.; Guo, F.; Lu, X. Heat island mitigation strategies for high-density historic blocks based on ventilation corridors. *J. Archit.* **2020**, 17–21.
25. Tang, Y.; Chen, K. Research on urban design strategies for green open spaces based on climate adaptation zoning. *Urban Archit.* **2017**, 33–37. [CrossRef]
26. Liu, S.; Yu, B.; Song, D.; Wang, S.; Wu, B. A preliminary exploration of the construction method for urban ventilation systems based on climate analysis: A case study of Xiamen City. *Urban Archit.* **2019**, *16*, 61–66.
27. You, X.; Li, Q.; Meng, Q. Research on spatial pattern and differentiated regulation strategies of urban heat island: A case study of downtown Guangzhou. *Landsc. Archit.* **2021**, *28*, 74–79.
28. Huang, Y.; Liu, J.; Chen, F.; Shen, L.; Stewart, I. Study on urban morphology and its heat island intensity based on local climate zoning theory. *New Archit.* **2019**, 126–131. [CrossRef]
29. Yuan, X.; Meng, Q. Analysis and evaluation of urban underlying surface temperature based on thermal infrared remote sensing. *Build. Energy Effic.* **2020**, *48*, 112–117+146.
30. Yuan, X.; Meng, Q. Study on regional surface temperature distribution based on low-altitude thermal infrared remote sensing. *Build. Energy Effic.* **2019**, *47*, 145–151+182.
31. Li, L.; Meng, Q.; Zhang, L. Research on large-scale solar radiation simulation system based on climatic wind tunnel. *Build. Energy Effic.* **2020**, *48*, 75–79+138.
32. Jiang, Y.; Zhao, L.; Meng, Q. Research on thermal comfort of urban pedestrian spaces in hot and humid regions during summer. *J. Civ. Environ. Eng.* **2020**, *42*, 174–182.
33. Jin, H.; Qiao, L.; Cui, P. Research on winter thermal comfort of campus in severely cold regions based on outdoor activities. *Build. Sci.* **2021**, *37*, 87–92+103.
34. Sun, C.; Luo, X.; Liu, L. Study on the correlation between microclimate thermal comfort and crowd behavior in cold region commercial pedestrian streets: A case study of Harbin Central Street. *Build. Sci.* **2022**, *38*, 240–249.
35. Taleghani, M.; Crank, P.J.; Mohegh, A.; Sailor, D.J.; Ban-Weiss, G.A. The impact of heat mitigation strategies on the energy balance of a neighborhood in Los Angeles. *Sol. Energy* **2019**, *177*, 604–611. [CrossRef]
36. Morakinyo, T.E.; Lau, K.L.; Ren, C.; Ng, E. Performance of Hong Kong's Common Trees Species for Outdoor Temperature Regulation, Thermal Comfort and Energy Saving. *Build. Environ.* **2018**, *137*, 157–170. [CrossRef]
37. Santamouris, M.; Haddad, S.; Saliari, M.; Vasilakopoulou, K.; Synnefa, A.; Paolini, R.; Ulpiani, G.; Garshasbi, S.; Fiorito, F. On the Energy Impact of Urban Heat Island in Sydney. Climate and Energy Potential of Mitigation Technologies. *Energy Build.* **2018**, *166*, 154–164. [CrossRef]
38. Sasaki, Y.; Matsuo, K.; Yokoyama, M.; Sasaki, M.; Tanaka, T.; Sadohara, S. Sea breeze effect mapping for mitigating summer urban warming: For making urban environmental climate map of Yokohama and its surrounding area. *Urban Clim.* **2017**, *24*, 529–550. [CrossRef]
39. Chao, R.; Chao, Y.; He, Z.; Wu, E. Research on Urban Ventilation Corridors and Their Planning Application. *Urban Plan. J.* **2014**, *3*, 52–60.
40. Ng, E. Towards planning and practical understanding of the need for meteorological and climatic information in the design of high-density cities: A case-based study of Hong Kong. *Int. J. Climatol.* **2012**, *32*, 582–598. [CrossRef]
41. Zang, X. *Research on Green Block Urban Design Strategies and Methods*; Tianjin University: Tianjin, China, 2014.
42. Song, D.; Wei, R. Green Building Design Strategies for High-Density Cities in Tropical Climates: The Case of WOHA Architecture Firm's Hostel Practice in Singapore. *Huazhong Archit.* **2017**, 78–87.
43. Ministry of Housing and Urban-Rural Development of the People's Republic of China, *Thermal Design Code for Civil Buildings*; China Planning Press: Beijing, China, 1993.
44. Li, Z. *Research on Urban Street Layout Index System: Taking 1000 m × 1000 m Grids as an Example*; Nanjing University: Nanjing, China, 2016.
45. Dong, C. *High-Density Architecture*; China Architecture & Building Press: Beijing, China, 2012.
46. Yuan, C. Exploring Planning Methods to Alleviate the Urban Heat Island Effect in High-Density Cities: A Case Study of Hong Kong. *J. Archit.* **2010**, 120–123.
47. Chen, J.M.; Black, T.A. Measuring leaf area index of plant canopies with branch architecture. *Agric. For. Meteorol.* **1991**, *57*, 1–12. [CrossRef]

48. Ong, B.L. Green plot ratio: An ecological measure for architecture and urban planning. *Landsc. Urban Plan.* **2003**, *63*, 197–211. [CrossRef]
49. Jusuf, S.K.; Wong, N.H. Development of empirical models for an estate-level air temperature prediction in Singapore. In Proceedings of the Second International Conference on Countermeasures to Urban Heat Islands, Berkeley, CA, USA, 21–23 September 2009; pp. 21–23.
50. Jin, H.; Cui, P.; Wong, N.H.; Ignatius, M. Assessing the Effects of Urban Morphology Parameters on Microclimate in Singapore to Control the Urban Heat Island Effect. *Sustainability* **2018**, *10*, 206. [CrossRef]

buildings

MDPI

Article

Unveiling the Urban Morphology of Small Towns in the Eastern Qinba Mountains: Integrating Earth Observation and Morphometric Analysis

Xin Zhao and Zuobin Wu *

Department of Urban and Rural Planning, School of Architecture, Xi'an University of Architecture and Technology, 13 Yanta Road, Xi'an 710055, China; kuker@alumni.tongji.edu.cn
* Correspondence: ben-wzb@xauat.edu.cn

Abstract: In the context of the current information age, leveraging Earth observation (EO) technology and spatial analysis methods enables a more accurate understanding of the characteristics of small towns. This study conducted an in-depth analysis of the urban morphology of small towns in the Qinba Mountain Area of Southern Shaanxi by employing large-scale data analysis and innovative urban form measurement methods. The U-Net3+ model, based on deep learning technology, combined with the concave hull algorithm, was used to extract and precisely define the boundaries of 31,799 buildings and small towns. The morphological characteristics of the town core were measured, and the core areas of the small towns were defined using calculated tessellation cells. Hierarchical clustering methods were applied to analyze 12 characteristic indicators of 89 towns, and various metrics were calculated to determine the optimal number of clusters. The analysis identified eight distinct clusters based on the towns' morphological differences. Significant morphological differences between the small towns in the Qinba Mountain Area were observed. The clustering results revealed that the towns exhibited diverse shapes and distributions, ranging from irregular and sparse to compact and dense forms, reflecting distinct layout patterns influenced by the unique context of each town. The use of the morphometric method, based on cellular and biological morphometry, provided a new perspective on the urban form and deepened the understanding of the spatial structure of the small towns from a micro perspective. These findings not only contribute to the development of quantitative morphological indicators for town development and planning but also demonstrate a novel, data-driven approach to conventional urban morphology studies.

Keywords: urban morphology; spatial analysis; small town; deep learning; urban morphometrics

1. Introduction

Small towns throughout the world exhibit vastly diverse characteristics, and in contrast to large cities, their scope has not been the subject of extensive research. Small towns serve as a transition between urban and rural areas, representing the lowest tier of the urban system [1]. Nevertheless, they cannot be classified as "micro-cities" [2] but are closer to a hybrid form with both urban and rural characteristics [3]. Defining small towns is by no means an easy task. Internationally, the definition and perception of small towns vary due to differences in population size, economic activities, and geographical location. The criteria for the definition of small towns differ widely between countries. For instance, Germany considers towns with a population between 5000 and 20,000 as small towns [4], whereas the definition adopted in the United States also refers to "urban clusters", with a total range of 25,000 to 50,000 [5]. Chinese small towns are significantly different from those in other countries. In China, small towns are typically considered administrative regions, encompassing both the town center and the surrounding rural areas [6,7]. These small towns represent the lowest level in China's urban system, situated between cities and villages, and exhibit the characteristics of both urban and rural areas. The towns extracted and studied in this work are administrative units known as "established towns". For a long time, small towns have been neglected, whether as research subjects or in public policy [8]. Research measuring the spatial characteristics of small towns is even scarcer.

In *A Theory of Good City Form* , Lynch argues that the quality of a place results from the interaction between the urban form and the society that inhabits the city [9]. Urban morphology refers to the physical characteristics of the urban form, such as the spatial arrangement of buildings, urban landscapes, and their transformations [10]. Scholars have found that planning strategies designed to optimize the physical forms of cities play a crucial role in promoting sustainable urban development [11]. It is thus essential to investigate the traits of urban morphology and uncover the underlying drivers of these transformative strategies. From a comparative perspective, Oliveira et al. [12] detail different approaches to urban morphology, emphasizing the importance of combining methods to enhance our understanding of the urban form and its evolution. In morphological studies, one of the most important methods is morphometrics—measuring the geometric shapes or relations of urban elements [13,14]. The measurement and classification of urban forms have been some of the most important themes in the field of urban geography and urban planning in recent years. For example, Fleischmann et al. [15] categorized 450 research articles identifying urban morphology indicators into six types: dimensions, shapes, spatial distribution, intensity, connectivity, and diversity. The identified indicators can be measured at diverse scales and scopes. Statistical clustering techniques were utilized by Berghauser Pont et al. [16] to describe contemporary urban forms in five European cities and allowed them to perform a quantitative city comparison.

By numerically describing urban elements such as buildings, streets, and enclosed spaces, it is possible to reveal the patterns of urban organization and the potential relationships among different forms. The measurement of the urban form is linked with the measurement of the cellular and organismal morphology. From this point of view, cities can be considered analogously to organisms: streets and buildings play the roles of cells and tissue, and the city itself is a "tissue structure". In this case, most of the concepts used in urban morphology, such as the "urban fabric" and "urban cells", are essentially morphological concepts in biology [17]. Fleischmann et al. [18] drew an analogy between this approach and the methods of early biologists, who sought to classify biological species based on their morphological similarities. This analogy is not limited to the morphological structure but also includes a response to environmental factors, such as the cities' ability to adapt to climate change and the mechanisms' evolution [19]. As emphasized by Lobo and others in their research [20], progress in cell morphology analysis highlights the need for standardized morphological indicators, including the cell area, aspect ratio, convex hull perimeter, convex hull area, roundness, and ellipticity, to assess the significance of the cell morphology in the context of regenerative medicine. Cell shapes can be seen as the result of the interaction between internal and external constraints, a "structural" outcome [21]. The concept of fractal dimensions applied to the demarcation of small-town boundaries is inspired by insights from the measurement of organisms and cell morphometry. The central regions of small towns can be seen as analogous to the "nuclei" of biological cells.

Research on small towns is relatively limited, despite significant strides having been made through the quantitative study of the urban morphology. The reasons for this mainly lie in the fact that collecting relevant data pertinent to those small towns is very complicated, which constrains research both in scope and depth. In addition, the morphological characteristics and developmental tendencies are relatively high, which makes this area of study challenging. With the continuous enhancement of Earth observation technology and its data acquisition capabilities, these obstacles are expected to be overcome, leading to a better understanding and description of the urban morphology of small towns [22]. The increasing availability of spatial data and the emergence of new geospatial tools offer tremendous potential for the analysis of the urban form [23]. Earth observation (EO) has recently been proven to be an essential tool in capturing the urban morphology. Thus far, different methods including pixel-based or object-based analysis, machine learning, and visual image interpretation have been used, and the spatial-scale variations across urban areas, cities, settlements, or individual buildings have been applied in EO research [24]. EO offers a less expensive and rapid means of data acquisition that provides planners with records of the number, size, footprint, density, and layout of buildings [25]. This study focuses on building layers, as they are the primary human-made elements. Their quantitative morphological indicators, such as their density and shape, largely reflect their socioeconomic characteristics [26].

Considering these factors, studying the urban morphology of the small towns in the eastern part of the Qinba Mountain Area in Southern Shaanxi could be interesting from multiple perspectives. Currently, there is a notable lack of quantitative analyses of the spatial morphology of small towns, and data collection for small towns in mountainous areas with complicated terrain is challenging. Moreover, when urban theorists discuss the "urban form", they predominantly refer to large cities, overlooking small towns and their unique characteristics. Specifically, at the small-townlevel, the methods of morphological analysis are distinct from those used for large cities and rural areas. With few studies available nowadays, particularly in the quantitative analysis of small towns' spatial forms, discussions relating to the urban morphology in small towns may thus fill this gap and enrich the literature on urban morphology. Discussions of such topics would help to identify the characteristics and features of the urban morphology in small towns and formulate corresponding planning strategies to optimize the physical forms of small cities toward sustainable development. The study of the urban morphology at the small-townlevel has become possible through various improvements, including the current levels of technology in Earth observation and data acquisition. Spatial data based on geospatial tools provide a data-driven approach to decision support for planning and development activities in these small towns. The main objective of this study is to conduct an in-depth analysis of the urban morphology of the small towns in the Qinba Mountain Area of Southern Shaanxi using large-scale data analysis and innovative urban form measurement methods. This will provide a comprehensive understanding of the spatial structure and characteristics unique to this region, contributing valuable insights for town development and planning.

2. Materials and Methods

2.1. Study Area

The main study area covers the eastern region of the Qinba Mountains in Southern Shaanxi, Shangluo City (Figure 1). The Qinba Mountain Area boasts a diverse and rich array of natural vegetation types and biological resources, serving as a critical ecological security barrier in Central China. The Qinba Key Ecological Function Zone was among the first areas in China designated as a critical environmental function zone. However, the region's ecological environment is fragile, with deficiencies in land resources as well as serious soil erosion and frequent natural disasters, which all restrict the sustainable development of the Qinba Mountain Area. Shangluo is one of several cities in the Qinba Mountain Area, characterized by abundant natural resources and a complex geography [27–29]. Shangluo City is located on the southern edge of the Qinling Mountains and is characterized by extremely high bio-geomorphic diversity and complexity, including mountains, hills, and plains that form an intricate net. Such a complicated natural background provides numerous examples and a varied town background that is well suited to the research of small towns in this region. Shangluo City includes towns and villages that have developed across different terrains from adjacent mountains to hills. Research has demonstrated that the spatial growth models of rural settlements in the Qinba Mountain Area remain distinct, exhibiting a variety of terrains and differences in economic development intensity, transportation accessibility, etc. [30]. The intertwined mountains and rivers in the Qinba Mountain Area have had a profound impact on the development of urban and rural spaces [31]. Due to their unique topographical conditions, mountain towns face specific restrictions and challenges in their development. For example, to some extent, the complexity of the terrain and the heterogeneity of the space limit the development of the transportation network and the effective use of land resources, affecting the spatial layout and economic development of these towns. At the same time, the natural resources and environmental conditions of mountainous areas have a profound impact on the lifestyles and economic activities of the local residents. These factors together influence the formation and evolution of the urban form [28].

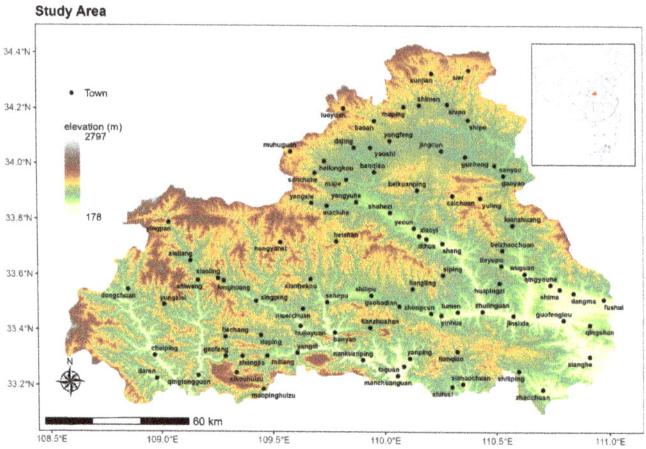

Figure 1. Study area: Eastern Qinba Mountains in Southern Shaanxi, Shangluo City.

2.2. Research Methodology

In this study, as illustrated in Figure 2, a novel research methodology was utilized for the large-scale measurement and classification of urban forms in small towns within the Qinba Mountain Area of Southern Shaanxi. This approach enabled a comprehensive analysis of the urban morphology, facilitating a deeper understanding of the spatial structures and characteristics unique to this region.

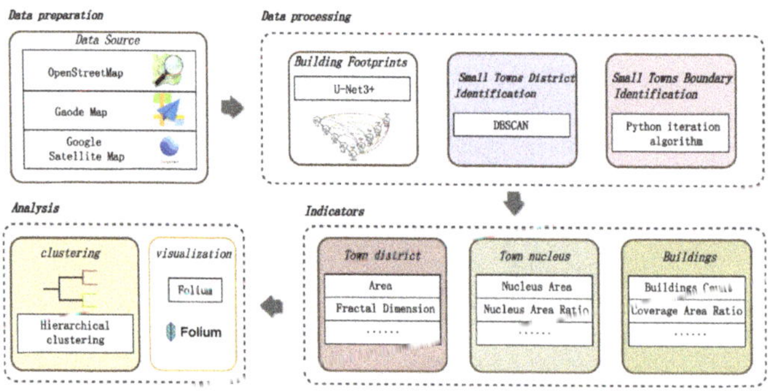

Figure 2. Methodological framework for the large-scale measurement and classification of urban forms in small towns within the Qinba Mountain Area.

For the accurate targeting of the analysis within the small towns, town government coordinates linked with Gaode Maps and OpenStreetMap were obtained, serving as the central points and boundaries for the study area. OpenStreetMap (OSM) is a global mapping community and online geographic information system that plays an increasingly important role in various academic studies. It provides free, high-quality data on global street networks and other urban infrastructure. Users can directly access the OSM database to download, edit, and use these map data [32,33]. This study processes and analyzes the data using Python and Jupyter Notebook.

There are a number of publicly available Earth observation image datasets throughout the world. Google Earth provides high-resolution satellite images in the form of tiles, with each tile representing a specific area. These tiles are integrated from high-resolution composite images sourced from various satellites, such as Maxar, Airbus, Terrametrics, QuickBird, and WorldView. Various approaches can be used to download these data. Google Earth Pro can directly download the images at maximum resolution, while third-party open tools such as SAS Planet facilitate data extraction from the Google Earth portal. SAS Planet is a free program, widely used in research, that allows the download of georeferenced images (or maps) from systems such as Google Earth, Bing Maps, and ESRI. Cheng et al. [34] used SAS Planet to obtain high-resolution Google Earth image tiles for peach orchard segmentation. Murillo-García and Alcántara-Ayala [35] used SAS Planet to download georeferenced images from systems such as Google Earth and Bing Maps, aiding in their creation of a multi-temporal landslide inventory map. Vasavi et al. [36] used SAS Planet to acquire high-resolution satellite images to train a U-Net- and ResNet-based model to classify buildings in Nashik. To obtain data relevant to this study, the SAS Planet software(v.230909) was used to download Google Earth image tiles at zoom level 19 with a spatial resolution of 0.6 m. Since our study area focused on the small towns of Shangluo City, the volume of the downloaded data did not qualify as a "large-scale data download" and thus did not violate the data usage guidelines.

Due to the lack of suitable datasets for small towns, a dataset was manually created. This dataset comprised 16,128 manually annotated training images, sourced from high-resolution Google Earth imagery. The specifically selected U-Net3+, known for its successful implementation in image segmentation [37], was used to precisely extract building footprint data, allowing for the identification of the outlines of structures within the small towns. The combination of geographic information from Gaode Maps and OpenStreetMap with U-Net3+ analysis enabled the capturing of numerous detailed urban forms from different sources.

The next step involved using OpenCV for data optimization and processing. This step was aimed at clarifying the boundaries of the buildings to increase the accuracy of the subsequent analysis. As an open-source library, OpenCV has powerful image processing capabilities and can effectively support the optimization and processing of images [38]. The optimized data were processed using the DBSCAN algorithm to determine the town's district boundaries. As a density-based spatial clustering algorithm, DBSCAN can identify areas with similar densities, thereby effectively delineating the areas of the small town.

At the same time, this study also adopted methods from biology and the method of cell morphology measurement, and through iteration calculation and the concave hull algorithm, it accurately obtained the small town's boundaries. With a series of indicators, the boundaries of the small-town and their characteristics were analyzed. We conducted a comprehensive analysis of the small town's spatial characteristics and the composition of its structure. The measurement method was inspired by the measurement and classification of cell and organism forms in biological morphology. In urban form analysis, the use of this method also contributes to interdisciplinary innovation [37].

This study conducted detailed measurements and calculations of the urban morphology of the small towns through the aforementioned data processing and analysis methods, and it performed a clustering analysis based on the morphological characteristics.

2.3. Morphometric Characteristics

In assessing urban morphology, researchers have identified and applied various metrics to capture the complexity and evolution of urban forms. Basaraner and Cetinkaya [39] refined the analysis of building footprints in GIS by employing shape indices such as circularity, convexity, and rectangularity and introducing new indices such as the Equivalent Rectangular Index and Roughness Index to enhance the classification schemes. Song et al. [40] applied indicators such as the fractal dimension, compact ratio, and shape factor to examine the urban morphology at different scales, providing insights into both the overall and localized urban characteristics. Additionally, Triantakonstantis and Stathakis [41] used spatial metrics to track urban sprawl across Europe, noting a substantial increase in urban areas and identifying trends toward simpler, more circular urban forms. These studies underscore the utility of various metrics in detailing the structural complexities of urban environments. The field of urban morphology research is continually advancing with the development of innovative technological tools, which significantly enhance the methods used to analyze urban forms. Serrano et al. [42] utilized GIS techniques to classify European urban settlements, demonstrating the detailed delineation of the urban form from high-density clusters to small towns. Zhang et al. [43] utilized remote-sensing and segmentation techniques for the mapping of urban informal settlements. The integration of open-source tools such as Python and OpenStreetMap has significantly advanced the methodologies used for the analysis of urban forms. Ma et al. [44] leveraged OpenStreetMap data to develop the OSMsc framework, creating semantic 3D city models that enhance urban analysis in fields such as morphology and microclimates.

In this study, a comprehensive set of morphological indicators was adopted. This system not only covered the dimensions and shape characteristics of the town district but also included a detailed analysis at the level of tessellation cells, defined by Martin Fleischmann et al. [45] as distinct spatial units formed via Voronoi-based spatial partitioning techniques and buildings. Notably, our selection of indicators was greatly inspired by the field of biology and cell morphology studies. Progress in cell morphology analysis highlights the need for standardized morphological indicators, including the cell area, aspect ratio, convex hull perimeter, convex hull area, roundness, and ellipticity, to assess the significance of the cell morphology in the context of regenerative medicine. These indicators, while evaluating the morphology of biological cells, also provide us with new perspectives and tools for the assessment of the spatial forms of small towns. A range of morphological feature indicators was utilized to conduct the detailed measurement and analysis of the urban morphology of the small towns. These indicators ranged from the whole town district to specific buildings, aiming to fully capture and illustrate the spatial structure and morphological characteristics of the small towns.

Fleischmann et al. [15] considered morphological aspects in urban research that were quantifiable in nature, such as the shape of the building and its basic dimensions. Dibble et al. [46] analyzed morphological urban evolution by measuring the building and spatial shape and distribution, where the best urban morphological indicators seemed to be the features of the shape and dimension. In this context, Wentz et al. [47] illustrated six basic features conceptualizing a multidimensional urban form, eventually including the characteristics of the spatial distribution and shape of buildings, thus addressing the need for such categorizations. This shows the importance of measures that describe the properties and dimensions, such as the area and perimeter; those that describe the scale, size, and shape, such as the circular compactness and number of corners; those that describe the geometric features of buildings and their intensity, such as the coverage area ratio and density; and those that describe the distribution density and intensity of buildings in space. Through a holistic analysis of such features, it is possible for the spatial structure and the morphological evolution of urban areas to be more effectively characterized and understood.

A range of morphological feature indicators were utilized to conduct the detailed measurement and analysis of the urban morphology of the small towns. These indicators ranged from the whole town district to specific buildings, aiming to fully capture and illustrate the spatial structure and morphological characteristics of the small towns. As detailed in Table 1, these indicators were categorized into three levels: the town district, the tessellation cell, and the building. Each level was further divided into three categories, dimension, shape, and intensity, enabling a more systematic and comprehensive analysis of the spatial features of the small towns.

At the town district level, indicators such as the area, perimeter, fractal dimension, length of the longest axis, roundness, average distance from the centroid to corner points, shell perimeter, and aspect ratio are included to describe the overall size and shape of the town. The tessellation cell level focuses on the town's core district, using indicators such as the core district size, number of core buildings, and core district ratio to depict the characteristics of the town's internal spatial units. The building level considers individual buildings and their clusters, with indicators such as the number of buildings, the building area, and the ratio of covered areas, reflecting the town's building distribution and density.

Table 1. Morphological feature indicators for small towns' urban forms in the Qinba Mountain Area of Southern Shaanxi.

Index	Element	Level	Category
Area	Town district	L	Dimension
Perimeter	Town district	L	Dimension
Fractal Dimension	Town district	L	Shape
Longest Axis Length	Town district	L	Shape
Circularity	Town district	L	Shape
Centroid–Corner Mean Distance	Town district	L	Shape
Hull Perimeter	Town district	L	Dimension
Aspect Ratio	Town district	L	Shape
Nucleus Area	Tessellation cell	M	Dimension
Nucleus Buildings Count	Tessellation cell	M	Dimension
Nucleus Area Ratio	Tessellation cell	M	Intensity
Building Count	Building	S	Dimension
Area	Building	S	Dimension
Coverage Area Ratio	Building	S	Intensity

2.4. Morphological Measurement of Small Towns in Shangluo City, Qinba Mountains, Southern Shaanxi

2.4.1. Small-Town District Identification

In the process of identifying small-town districts, machine learning algorithms such as DBSCAN (Density-Based Spatial Clustering of Applications with Noise) offer an innovative approach. This method not only effectively identifies the boundaries of towns but also explores the spatial structures and functional areas within them. By analyzing geospatial data, this approach can reveal the morphological characteristics of small towns, thus providing a new perspective for a deeper understanding of towns' features [48]. Research by Cetinkaya et al. [49] further demonstrates the utility of DBSCAN in urban blocks, highlighting its ability to effectively distinguish building groups based on proximity.

By applying the DBSCAN algorithm to define the vector edges of structures, the recognition of the ranges of small-town areas can be automated. By excluding the neighboring villages, it is possible to determine the study subjects more precisely. This enables us to obtain clustering results reflecting the small-town regions. After calculating them on a map, it is also possible to identify the main spatial zones of these study subjects, providing useful information for further research.

Figure 3 depicts the results of the clustering with the DBSCAN algorithm and the visualization of these results on a map. It clearly marks the boundaries of the small towns, showing the automatic clustering process after converting the building vector boundaries into points, effectively distinguishing the town areas from the surrounding village areas.

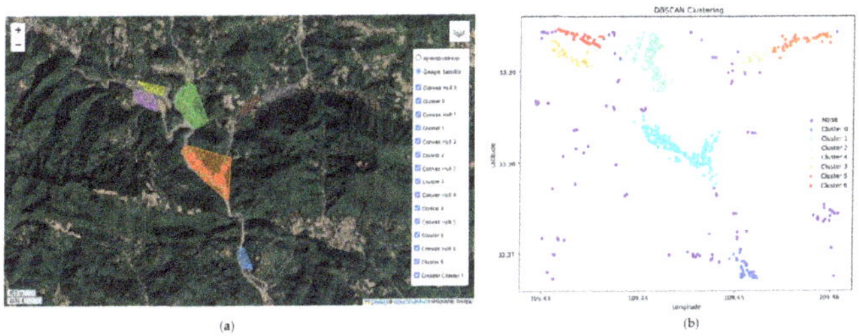

(a) (b)

Figure 3. (**a**) Visualization of clustering for town areas and rural areas. (**b**) DBSCAN clustering process.

2.4.2. Small-Town Boundary Identification

In traditional methods of determining the boundaries of small towns in research, the administrative boundaries formulated by the government are always emphasized. This is an essential part of national spatial planning. However, data collection for small towns in mountainous areas with complicated terrain is challenging, and it is essential for the town boundaries to be calculated using geospatial data and computational geometry.

The convex hull and concave hull represent the two central concepts associated with describing the shape of a set of points. The convex hull is the smallest convex polygon encapsulating the set of points, whereas the concave hull is a non-convex polygon that closely fits around the shape of the set of points. The definition and algorithms of the convex hull have been extensively studied. Edelsbrunner et al. [50] proposed an algorithm based on the shape of planar point sets, which has significant applications in information theory. Convex hulls are widely used in image processing, geographic information systems (GIS), and computer graphics due to their simplicity and efficiency. However, it has been noticed that convex hulls fail to accurately describe point sets with complex boundaries. As a result, researchers proposed the concept of the concave hull. Duckham et al. [51] proposed that a simple polygon generated by an efficient algorithm can capture the shape of the planar point set more accurately. Therefore, this method provides an effective solution for the generation of a concave hull due to the balance between computational efficiency and the accuracy of shaping.

By iteratively determining the locations of small towns' boundaries using the convex hull and concave hull algorithms, this study is able to increase the accuracy of boundary recognition and better represent a real geographical small town. By exploring the possibility of identifying the boundaries of the small towns in the Shangluo City region within the Qinba Mountains of Southern Shaanxi, this research introduces a novel approach to boundary recognition based on concave hull algorithm calculations. Not only does it consider the continuity and integrity of the geographical space of a small town, but it also overcomes issues related to data acquisition in disadvantaged mountainous areas, contributing a novel solution for the accurate recognition of small towns' boundaries.

With the advent of powerful computing tools, the calculation of convex and concave hulls has become more accessible and efficient. Python, in particular, offers a range of libraries that simplify the computation of these geometric shapes. This study uses the shapely [52] library in Python to compute the convex hull and the alphashape [53] library to compute the concave hull, providing a robust framework for boundary recognition in geographical data analysis. Additionally, the relevant dataset and codes used in this study are publicly available. The convex hull algorithm generates the smallest convex polygon that can enclose all points. This provides a simple yet effective way to outline the general shape of a set of points, forming the foundation for boundary recognition. The algorithm first extracts all of the points from the provided GeoDataFrame and then uses the MultiPoint class from the shapely library to combine these points into a MultiPoint object, computing the convex hull by adopting the method that generates the smallest convex polygon that can enclose all of the points. The concave hull algorithm extends this concept by allowing the boundary to follow the actual contours of the points more closely, capturing the nuances of the geographical shape. The iterative approach ensures that the algorithm fine-tunes the boundary to avoid splitting it into multiple parts, thus maintaining the integrity of the geographical space. The algorithm extracts all points from the provided GeoDataFrame, initializes the parameters with an initial alpha value (controlling the shape of the concave hull) and other iteration parameters, and then iteratively calculates the alpha shape (an approximation of the concave hull) for the current alpha value using the alphashape function. If the resulting shape is a valid polygon, it is saved as the last valid concave hull, and the alpha value is incremented by a certain factor to gradually adjust the shape. This process continues until the shape is split into multiple parts or the maximum number of iterations is reached.

By applying these algorithms to the Shangluo City region within the Qinba Mountains of Southern Shaanxi, this research demonstrates a novel approach to boundary recognition. The combination of convex and concave hull calculations enhances the accuracy of delineating small towns' boundaries, addressing the challenges related to data acquisition in mountainous areas.

Figure 4 displays the process of iteratively identifying the boundaries of small towns using the concave hull algorithm, where image (a) shows the convex hull boundary and the generated compact concave boundary, while image (b) details the method of identifying the concave boundary through the iterative process. The key aspect of this method is that it begins with an initially small alpha value and gradually increases this value, iterating the calculation until the concave hull begins to split into multiple parts, thereby identifying the final effective concave boundary. This process effectively captures the true geographical boundaries of small towns.

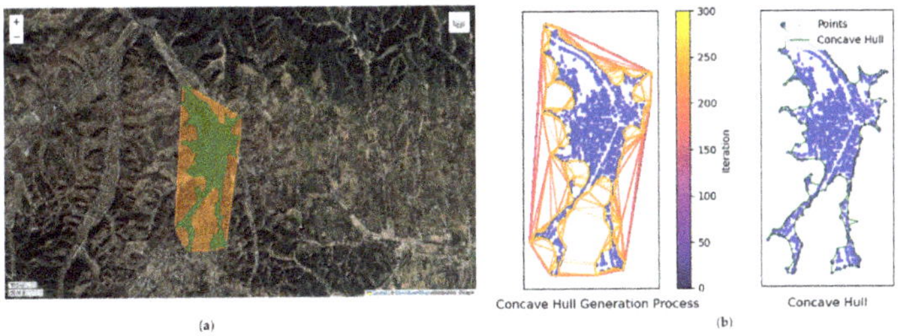

Figure 4. Application of the iterative concave hull algorithm in identifying the boundaries of small towns: (**a**) visualization of convex and concave hull boundaries; (**b**) detailed explanation of the concave hull iterative identification process.

2.4.3. Morphological Measurement

In this section, the unique morphological features of the small towns in mountainous areas were demarcated and assessed through quantitative measurement. Given that the morphology of these small towns tends to be more organic, morphometric methods used in cell biology and ecology were applied.

The metrics measured included the area and perimeter of the town region, the fractal dimension, the length of the longest axis, the roundness, the average distance from the centroid to the vertices, the perimeter of the convex hull, the aspect ratio, the core area of the tessellation units, the number of core buildings, and the core area ratio, among others. To estimate the longest axis, the Minimum Bounding Rectangle (MBR) is used, which encompasses the shape and aligns with the shape's main direction. The length of the longest axis is then equal to the length of the longest side of this rectangle. These indicators cover not only dimensions and shapes but also aspects such as density and intensity, providing a comprehensive description of the small town's morphology. One of the metrics included in the calculation of various indicators is circularity, which is calculated as follows.

$$\text{Circularity} = \frac{4\pi \times \text{Area}}{\text{Perimeter}^2} \quad (1)$$

where π is the constant Pi; Area represents the area of the town region; and Perimeter is the perimeter of the town region.

Following this, the calculation methods for the fractal dimension and indicators related to the tessellation cell part (i.e., the nucleus part) are also detailed. The application of the concept of the fractal dimension to the demarcation of small-town boundaries is inspired by insights gained in the measurement of organisms and cell morphometry. By viewing geospatial data as analyzable morphological structures, a similar approach can be adopted to explore the spatial complexity and structural characteristics of small towns. The fractal dimension quantifies the complexity of spaces or structures. Although there are several techniques for its calculation, the box-counting algorithm emerges as the predominant method [54]. The pivotal formula in this calculation is the computation of the fractal dimension, which is ascertained through the slope of the linear regression, as calculated according to the formula described below.

$$D = \lim_{\varepsilon \to 0} \frac{\log(1/\varepsilon)}{\log N(\varepsilon)} \tag{2}$$

where D is the fractal dimension; ϵ is the size of the box used in the grid; and $N(\epsilon)$ is the number of boxes of size ϵ that contain part of the shape.

The fractal dimensions of 89 towns were calculated, and, for a comprehensive case study, a particular focus was placed on Rencun Town, Tiechang Town, and Shimen Town. Their fractal dimensions are 1.223, 1.406, and 1.57, which means that, in these towns, there is a difference in the complexity and self-similarity of the spatial forms. The complexity increases with higher values of the fractal dimension and a highly fragmented urban form, while a simple urban form is characterized by a low value for the fractal dimension.

Figure 5 demonstrates the three distinct phases of fractal dimension calculation by superimposing grids of varying sizes onto a map. The dimensions and arrangement of these grids intuitively display the shifts in spatial scale associated with the application of the box-counting algorithm, as well as the interactions between the town boundaries and these grids.

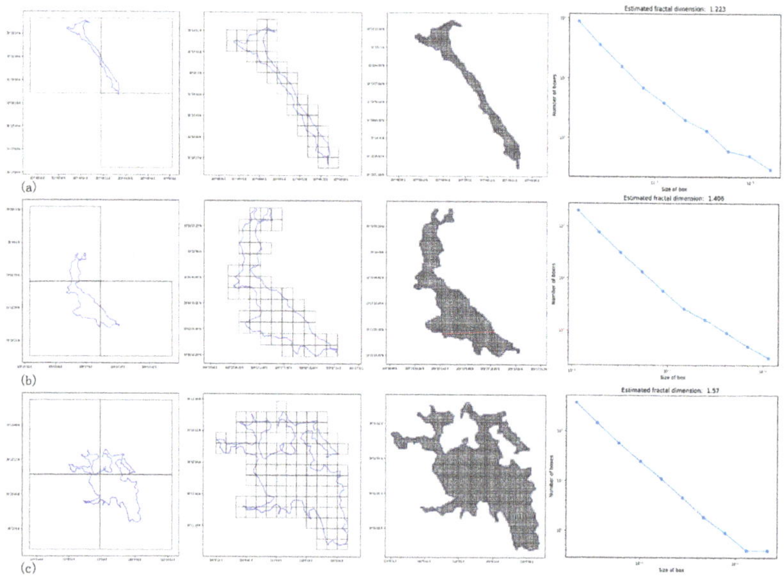

Figure 5. Calculation of fractal dimension for small towns. (**a**) Rencun Town, (**b**) Tiechang Town, and (**c**) Shimen Town.

Nuclear morphology analysis and classification methods were adopted to discern the features of the central regions in the small towns, analogous in concept to the "nuclei" of biological cells. The principal methodology was inspired by the concept of enclosed tessellation, as proposed by Martin Fleischmann et al. [45], defined as specific spatial units created through Voronoi-based spatial partitioning techniques. The morphological and functional attributes of these enclosed tessellation units were assessed within their respective spatial contexts. To generate tessellation cells, the momepy library [55], an open-source toolkit tailored to urban form measurements, was utilized. These tessellation cells were then categorized via cluster analysis.

Having calculated the coverage area and radius of proximity for the enclosed tessellation cells, and using the natural breaks method of spatial classification, the above approach allowed us to identify core regions in the small towns with distinctive spatial-related peculiarities for the further study of their structure and form. Figure 6 presents the results of our methodology's application, where, after the detailed investigation of the enclosed tessellation units, a comparison not only of the number of core regions in the small towns but also of their abundance and shape across different towns is possible. This enables a more in-depth consideration of the spatial structures and unique urban morphologies of each small town.

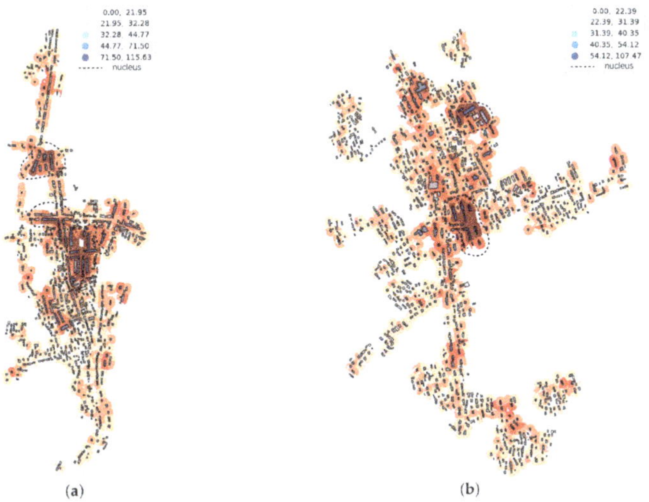

Figure 6. Spatial distribution and classification of enclosed tessellation cells in two small towns. (**a**) The tessellation category counts for Sanyao Town. (**b**) The tessellation category counts for Longtou Town.

3. Results

3.1. Building Footprint Extraction utilizing Open EO Data in the Qinba Mountain Area, Shangluo City

This study, conducted in the Qinba Mountain Area of Shangluo City, has led to significant advancements in the extraction of building footprints and town boundaries using open EO data. The U-Net3+ network was utilized to train deep learning models on high-resolution image data acquired through remote sensing. With the trained model, 31,799 buildings were successfully recognized. The redesigned full-scale skip connection mechanism in the U-Net3+ architecture notably enhanced the multi-scale feature aggregation and deep learning supervision. The novel mechanism could better promote the learning of hierarchical representations [56,57]. OpenStreetMap and Gaode Maps were used to obtain the town center points and the extent of the towns in the study area. We applied for a free API from Gaode Maps, which can be accessed at (https://lbs.amap.com/api/webservice/summary). Then, the optimization of the extracted building footprints was carried out with OpenCV, aiming to improve the clarity and accuracy of the outlines of the buildings. The purpose of this step was to delineate the building outlines obtained from the remote sensing imagery so that they corresponded more precisely and unambiguously to their actual forms. The quality of this information was vital to the accuracy of the subsequent analysis. Subsequently, the concave hull algorithm was iteratively applied to accurately describe the boundaries of the town forms.

In Figure 7, the results of building footprint and boundary extraction are illustrated. The superimposition of various small-town forms onto the map of the Shangluo area reveals the rich morphological diversity present within the district. The detailed examination of six representative towns, each situated in different landscapes ranging from mountains to plains, underscores the importance of recognizing and understanding the unique characteristics and diversity of small towns' forms.

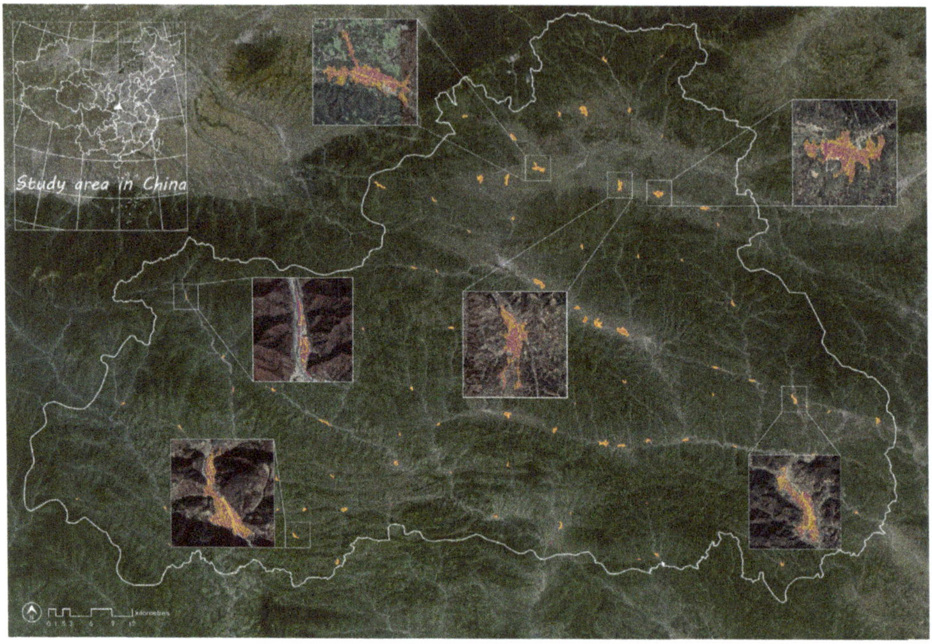

Figure 7. Building footprint and boundary extraction results for small towns in Shangluo City, Qinba Mountains.

3.2. *Morphological Measurement of Small Towns in Shangluo City, Qinba Mountain Area*

Morphological Clustering

When performing hierarchical cluster analysis on the 12 characteristic indicators of the 89 towns, three of the most important clustering evaluation metrics—the Davies–Bouldin Index, the Calinski–Harabasz Index, and the Silhouette Score—were calculated to find the optimal number of clusters. These three indices adopt different calculation methods and offer different perspectives in measuring the effectiveness of clustering, so they provide a scientific and rational basis for determining the optimal number of clusters. Additionally, limitations in computational resources and software capabilities also led us to adopt these three easily computable and widely used indices. Python has several convenient libraries, such as scikit-learn, that can be used to calculate these indices, enabling us to efficiently perform a clustering analysis and evaluation.

The Davies–Bouldin Index calculates the cluster cohesion and separation, where lower scores indicate tightly grouped clusters that are clearly differentiated from others, signaling effective clustering [58]. As depicted in Figure 8, our analysis found that a cluster count of eight yielded a relatively low Davies–Bouldin Index, marking it as a judicious choice. The Calinski–Harabasz Index assesses a cluster's compactness relative to the spread between clusters at any given count, which is why a higher score means a more defined cluster and independent delineation between two distinct canopies. At the same time, the Silhouette evaluates the difference between the similarity of a sample to the samples in its own cluster and the similarity of a sample to the samples in the nearest other cluster. Considering the classification results obtained, eight was selected as the best cluster number because it enabled a positive trade-off between the intra-cluster similarity and inter-cluster dissimilarity.

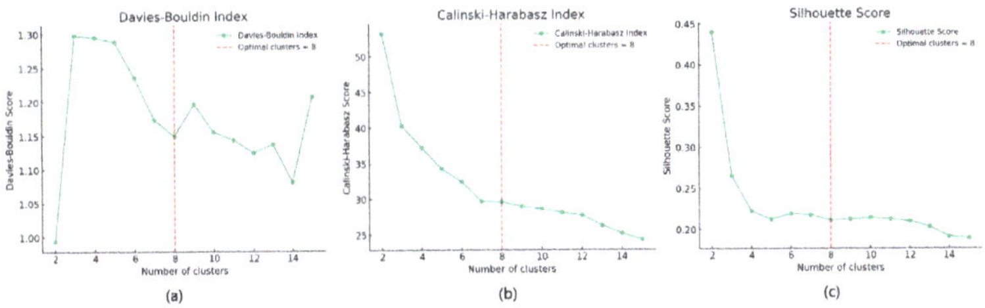

Figure 8. Evaluation of clustering metrics: (**a**) Davies–Bouldin Index, (**b**) Calinski–Harabasz Index, and (**c**) Silhouette Score for optimal number of clusters in morphological clustering of 89 towns using 12 feature indices.

The hierarchical clustering method was used to study the morphological features of 89 towns. These algorithms progressively combine smaller clusters into larger ones based on the similarity in the length or correlation between samples, continuing this process until all samples are paired, thereby creating a clustering structure that resembles a tree [59].

As shown in Figure 9, the data were first standardized to eliminate the differences between the scales of various features. The "Ward" method of hierarchical clustering was used, which is optimal in the sense that it minimizes the increment in within-cluster variance as the criterion for merging, and it can thus effectively identify clusters with significant structural differences [60]. A dendrogram is a tree-like diagram that shows the merging of clusters. At each fusion event in hierarchical clustering, a vertical line represents the distance between the two clusters being merged. The length of this vertical line indicates the magnitude of the difference between the two clusters. By examining the dendrogram and setting a threshold line, shown as the red dashed line in Figure 9, a reasonable number of clusters can be determined based on the distance threshold.

Based on the hierarchical classification of the small towns in the Qinba Mountain Area of Southern Shaanxi, the towns themselves were segregated into eight distinguishable categories with different morphological characteristics. A general summary of each cluster is provided in Table 2. This table differentiates the clusters into multiple groups and describes their most prominent features, including the form of the geographical entities, the concentration of constructions receiving the same measurement, and the total area.

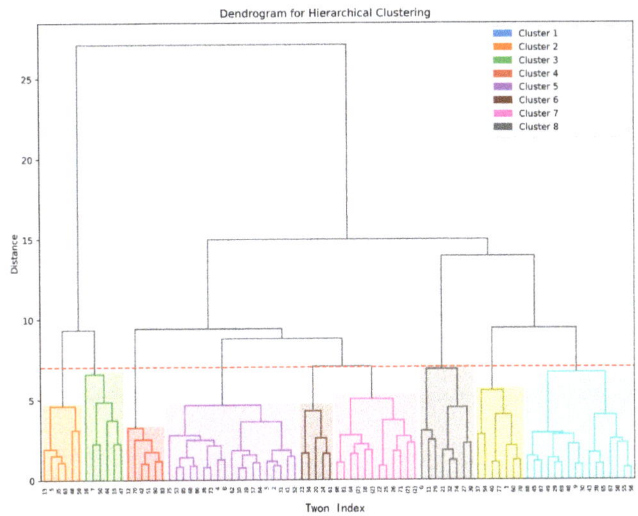

Figure 9. Dendrogram illustrating the hierarchical clustering of 89 towns based on 12 morphological features.

Table 2. Summary of each cluster's characteristics based on its geographical and architectural features. Clusters are analyzed to provide insights into the underlying patterns of the dataset.

Cluster Number	Shape	Density	Size
1	Irregular	Low	Large
2	Irregular	Low	Large
3	Circular	Moderate	Small
4	Long–Narrow	High	Small
5	Regular	Moderate	Medium
6	Compact	Very High	Very Small
7	Long–Narrow	Low	Large
8	Irregular	High	Medium

Category 1 towns, notably Zhongcun and Jingcun, featured larger perimeters and areas, irregular shapes, and sparse buildings. They were distributed in the northern and central regions, usually located in relatively flat terrain areas. Category 2 comprised towns such as Bao'an and Shangzhen, which differed from Category 1, with even larger perimeters and areas, a low building density, and higher fractal dimensions. This suggests that the town boundaries are more complex and take on more unique morphologies determined by geography. These towns are primarily concentrated in the central region, with a few located in the northern flat terrain areas. As for Category 3 towns, characterized by Banqiao and Xianghe, the mean perimeters and areas were smaller and the circularity was higher, with little difference between the perimeters and areas. This could point to their closeness to a circle-like shape, and it may indicate more centralized and planned development. These towns are mostly scattered in the northern and southeastern regions. For Category 4 towns, including Xiaoliang, Dongchuan, and Liangling, the geographical entities were narrow and elongated, which could have resulted from the mean perimeters and areas being lower and also the combination of the high aspect ratio and building area to concave area ratio. They are dispersed throughout the entire region. Category 5 towns, such as Yungaisi and Nankuanping, had a medium size and regular shape, which corresponded to a medium geographical scale, regular terrain, and medium building density. These towns are mostly distributed in the southern region. Category 6 towns, such as Sanchahe and Beizhaochuan, were extremely small but also had the highest aspect ratios, which indicates the development of either small or narrow strips with densified buildings. They are primarily located in the southeastern region. Category 7 towns, such as Sanyao and Xingping, show similar mean perimeters and areas but the highest aspect ratio, meaning that the constitutive geographical entities are very long and narrow and likely formed due to a river or valley, exhibiting valley areas with sparse building layouts. These towns have a very distinctive shape and are distributed in the central region. Finally, Category 8 towns, including Fenghuang and Luanzhuang, show a moderate area and building density but an irregular shape, and they are also expected to be located in complex terrain. They are dispersed throughout the entire region. In Figure 10, the overall distribution of the clusters is shown, along with the representative Clusters 1, 2, 7, and 8. Table A1 in the Appendix A details the classification of each small town within these eight morphological clusters.

The boxplots in Figure 11 provide a comprehensive overview of the morphological characteristics of the towns across the eight identified clusters in the Qinba Mountain Area of Southern Shaanxi.

The perimeter feature boxplots suggested that, for towns in Clusters 1 and 2, there is the largest variability in the size of the perimeters. Moreover, towns in Cluster 1 exhibited particularly large perimeters, which could mean that these towns are more irregularly shaped and more sprawling. In contrast, Clusters 3 and 6 exhibited much tighter distributions and smaller values for the perimeters, which implies that they are more compact. Regarding the area, Clusters 1 and 2 were once again the most prominent clusters with larger areas. It can be inferred that their towns occupy more extensive terrain. Meanwhile, Clusters 4 and 6 covered smaller areas but had denser settlements. The fractal dimension is an indicator of boundary complexity, and Clusters 2 and 3 showed the highest median value for this feature. This could mean that their boundaries are more complex and they may exhibit the natural features of the terrain or be structured in a complex pattern due to historical constraints. Clusters 7 and 8 have much smaller fractal dimensions, suggesting towns with more straightforward boundaries. In the case of the aspect ratio, the towns in Cluster 7 had a noticeably higher spread and median. It could be hypothesized that towns stretching along geographical features such as a river or a valley have a large amount of space occupied in a single direction. Clusters 1 and 3 had lower values, meaning that their towns were possibly very expansive in two dimensions, which could imply more regular shapes. The circularity feature also reflects the towns' shapes, and Clusters 3 and 5 tightly enclosed towns that closely approximated a circular shape, which implies that the towns' geometries were more centralized. Clusters 1 and 8, especially the latter cluster, with the most extensive spread, are towns of different shapes that vary from a standardized geometry, from round to square or between these two, and they represent towns that have been shaped by very different influencing factors. The building area/concave area ratio boxplot reveals that Cluster 6 has the town with the highest median ratio, which implies that it is very densely built.

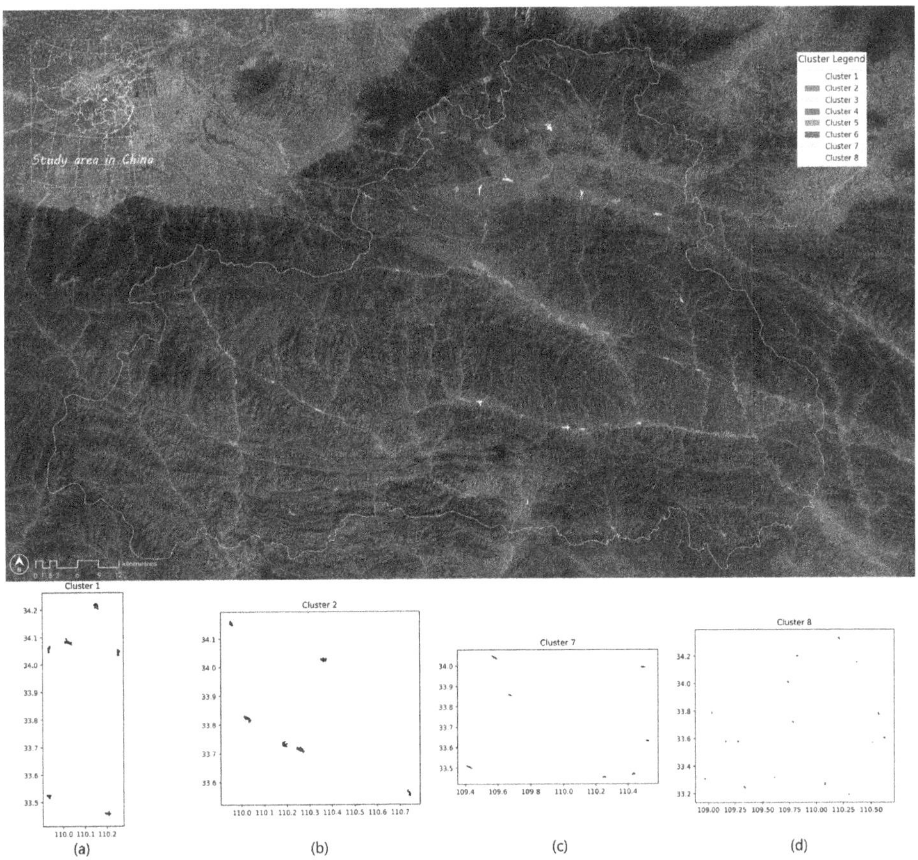

Figure 10. Hierarchical clustering of towns: (**a**) Cluster 1—expansive, irregular shapes; (**b**) Cluster 2—large, complex boundaries; (**c**) Cluster 7—elongated, high aspect ratios; (**d**) Cluster 8—medium size, irregular shapes.

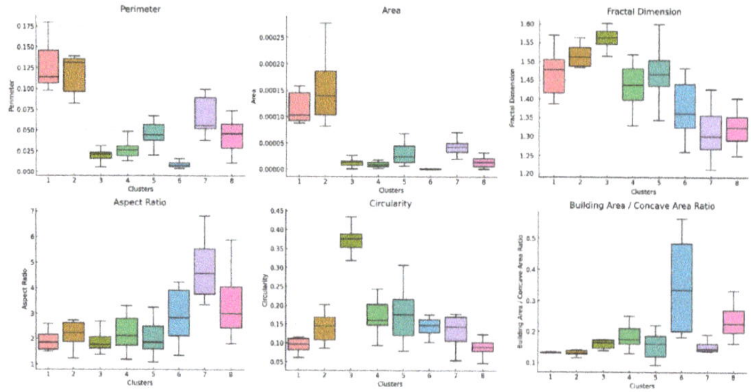

Figure 11. Boxplots providing a comparative visualization of the variance and central tendency for each morphological feature across the clusters.

4. Discussion

This study utilizes Earth observation data, addressing the data scarcity challenge typically encountered in small-town research. In most cases, urban morphology research data are acquired from government-provided cartographic information. This information is relatively easily accessible for large- and medium-sized cities. For instance, Jiménez-Espada et al. [61] utilized geographic information systems and aerial images provided by official sources like the National Statistics Institute, the City Council, and the National Geographic Information Center while studying the medium-sized World Heritage City of Cáceres in Spain. However, data scarcity is a common challenge for small towns. Gu et al. [62] demonstrated that while small towns are essential to the Chinese urban system, data collection and analysis are challenging due to the complicated and diverse functions and changing roles of these towns in urban–rural relations. Our approach uses Earth observation data and the deep learning model of U-Net3+ for extracting the building footprints in small towns lying in mountainous areas with complex terrain. Informed by Earth observation data, our research straddles the limitations of traditional data sources and provides new data references and approaches that contribute to the study of small towns.

Defining the boundaries and scope of small towns has been a persistent challenge due to the lack of precise boundary data. This study addresses this issue through several innovative approaches. Firstly, the DBSCAN algorithm, which has been previously applied in urban analysis research, such as Cetinkaya et al.'s work on building proximity grouping [49], is innovatively employed to determine the scope of small towns. By combining geospatial data with the DBSCAN algorithm, a more precise differentiation between towns and surrounding rural districts is achieved, effectively separating towns from their rural surroundings and enhancing the accuracy of the analysis. Secondly, an iterative computation that embeds both convex and concave hull algorithms is introduced to increase the accuracy in delineating complex boundaries of small towns. Ogle et al. [63] employed the convex hull algorithm to assess urban compactness in Pocatello. Despite having low computational complexity, the convex hull algorithm cannot process complex shapes with many concave boundaries. This paper incorporates the concave hull algorithm to resolve this limitation. The concave hull algorithm has the advantage that it captures irregular boundaries and hence makes up for the deficiencies of the convex hull algorithm, though it is computationally more complex. By integrating these two algorithms, a more precise definition of the complex boundaries of small towns can be realized, hence providing a reliable data foundation for further morphological analysis.

Drawing from biological cell morphometry, this study presents approaches and indices to offer a new perspective toward an urban morphology analysis of small towns from multiple dimensions. Fleischmann et al. [18] drew an analogy between this approach and the methods used by early biologists to classify species based on morphological similarities, discussing the effectiveness of morphological methods in understanding complex structures and highlighting the potential applications of these methods in urban structure analysis. The application of specific indicators remains limited in existing research. Most studies have only used fundamental indicators like size and area for in-depth morphological features, and while Basaraner and Cetinkaya [39] calculated several indicators, such as circularity, convexity, and rectangularity in their work about building shapes, very little attention was given to indicators like fractal dimensions, which better reflect the complexity of shape structures. This paper proposes some indicators of the morphology of small towns, including fractal dimension, circularity, and convex hull, from different viewpoints concerning complexity and compactness. In particular, introducing fractal dimensions makes it possible to measure boundary complexity. Based on systematic measurement with 12 indicators of 89 small towns in the Qinling-Daba Mountains, this paper displays the richness of morphological differences among them, including irregular sprawling morphologies, compact high-density morphologies, and elongated strip-like morphologies. These differences reflect the role of unique geographical environment conditions and development paths in shaping mountain towns' morphology. It is through morphological clustering that eight significantly different small-town types have been identified, thus providing scientific proof for the formulation of town planning strategies under local conditions. This paper has provided a more detailed and in-depth analysis of the morphology of small towns with methodological innovation and large-scale empirical data and enriched the perspectives and methods of research in urban morphology.

This paper focuses on small towns in Qinba mountainous areas, which have complex terrain conditions and unique morphological features. Compared with larger cities, which have consistently been focal points of study, this paper seeks to fill a significant gap in urban morphology studies. Through precise morphometric measurements and classification analysis, the research aims to provide a scientific basis for the development and planning of these areas. This study reveals significant morphological diversity among the small towns in the Qinba Mountain Area. This aligns with the views of Wang et al. [64], who posited that the spatial distribution and morphology of mountain towns are influenced by multiple factors, such as the terrain, level of economic development, and transport accessibility. Through the large-scale extraction of building outlines and the precise delineation of town boundaries, the classification of small-town morphologies was obtained that aligned with the geographical characteristics of the Qinba Mountain Area. This finding supports the assertion made by Shang et al. [30] that the terrain significantly impacts the land use and town layout among mountain areas.

There are still many challenges and unresolved issues in the research on small-town morphology measurement and classification. The morphological characteristics, such as the fractal dimension, building density, and spatial configuration of small towns are interlinked with the geographical environment [65]. For example, understanding and interpreting the relationships between the morphological characteristics of small towns and their ecological environments, as well as how to apply the research findings in small-town planning and management, require further exploration and innovation in future studies. Additionally, the choice of the value of k in clustering can sometimes be subjective, adding another layer of complexity to the analysis.

5. Conclusions

This study performed an in-depth analysis of the urban morphology of small towns in the Qinba Mountain Area of Southern Shaanxi by innovatively integrating Earth observation data and advanced morphometric methods. The use of the U-Net3+ deep learning model to extract the building footprints was quite beneficial in defining the boundaries of the small towns with great accuracy. Such precise delineation allowed a detailed morphological analysis, which was found to be crucial given the complex topographical conditions of the area under study. The hierarchical cluster analysis of 12 morphological indicators in 89 towns made it possible to divide them into eight clusters, each of which differed from the others because of the peculiarities of their spatial features. The clustering results highlighted that the towns exhibited a variety of shapes and distributions, such as large irregular perimeters, compact high-density clusters, and narrow elongated forms. These variations reflect distinct layout patterns unique to each town. The revealed diversity, on a large scale, is caused by various factors, including relief, economic development, and transportation accessibility. This diversity calls for more tailored urban planning strategies rooted in the peculiar geographical and socioeconomic contexts of each city. This study also introduces an innovative interdisciplinary methodology inspired by biological cell morphometry. The possibility of viewing the towns through the lens of morphometric analysis allowed us to obtain new knowledge on the spatial organization, as well as the structural complexity, of the urban forms. In other words, through the analysis of specific data regarding the urban structures of various towns, we were able to point out the core area of the city and capture its morphological characterization at a micro-scale, similar to the nucleus of a biological cell. This extensive analysis will allow a detailed understanding of urban dynamics, contributing to better-informed decision-making regarding urban planning. Additionally, due to the use of data from EO and advanced geometric algorithms in this research, the integration of big data and artificial intelligence technologies should be considered in future studies of the urban morphology. Apart from an increase in the accuracy of this type of morphological measurement, such a data-driven approach also makes the analysis of large-scale spatial patterns possible, enabling a greater contribution to the field of urban studies. The identified morphological clusters provide a scientific basis for the study of similar problems in the design of tailor-made planning strategies that promote sustainable urban development.

Author Contributions: Conceptualization, X.Z.; methodology, Z.W. and X.Z.; software, X.Z.; original draft preparation, X.Z.; review and editing, Z.W. and X.Z.; supervision, Z.W.; funding acquisition, Z.W.; investigation, X.Z.; project administration, Z.W. All authors have read and agreed to the published version of the manuscript.

Funding: This study was funded by the National Natural Science Foundation of China, grant number 52178055, and the National Key Research and Development Program under grant 2022YFC3802803.

Data Availability Statement: The datasets in this study are publicly available and can be accessed at Figshare via the following link: https://figshare.com/articles/dataset/QinbaTownMorphologyDataset/26022310/1 (accessed on 7 June 2024).

Acknowledgments: We are grateful to the anonymous reviewers for their valuable comments.

Conflicts of Interest: The authors declare no conflicts of interest.

Appendix A

Table A1. List of small towns in Shangluo City within the Qinba Mountain Area of Southern Shaanxi by cluster.

Cluster	Towns
1	Zhongcun Town, Shilipu Town, Jingcun Town, Yongfeng Town, Chimen Town, Yaoshi Town
2	Bao'an Town, Gucheng Town, Shang Town, Dihua Town, Shahezi Town, Qingyouhe Town
3	Shiliping Town, Banqiao Town, Xianghe Town, Xizhaochuan Town, Qingtongguan Town, Gaoyao Town
4	Xialiang Town, Dongchuan Town, Liangling Town, Dangma Town, Beikuanping Town, Daping Town, Yuling Town, Muwang Town, Banyan Town, Manchuanguan Town, Niu'echuan Town, Shipo Town, Hongyansi Town, Caichuan Town, Daren Town, Jinsixia Town, Maping Town, Machihe Town, Majie Town
5	Yungaisi Town, Nankuanping Town, Tumen Town, Yecun Town, Dajing Town, Tianzhushan Town, Xiaoyi Town, Fushui Town, Siping Town, Sier Town, Xiaohekou Town, Zhangjia Town, Hujia Yuan Town, Milang Town, Sehepu Town, Maoping Hui Town, Shim Town, Zhaochuan Town, Tiechang Town, Gaobadian Town, Gaofeng Town

Table A1. *Cont.*

Cluster	Towns
6	Sanchahe Town, Beizhaochuan Town, Tianqiao Town, Xiaoling Town, Yanping Town, Yangyuhe Town, Guofenglou Town, Qingshan Town
7	Sanyao Town, Xingping Town, Yangxia Town, Mu Hu Guan Town, Zhulin Guan Town, Tieyupu Town, Yinhua Town
8	Fenghuang Town, Luanzhuang Town, Xunjian Town, Yangdi Town, Chaiping Town, Wuguan Town, Faguan Town, Luoyuan Town, Shifosi Town, Shipu Town, Shiwen Town, Huapingzi Town, Yingpan Town, Xikou Hui Town, Heishan Town, Heilongkou Town

References

1. Stoica, I.V.; Tulla, A.F.; Zamfir, D.; Petrişor, A.I. Exploring the Urban Strength of Small Towns in Romania. *Soc. Indic. Res.* **2020**, *152*, 843–875. [CrossRef]
2. Vaishar, A.; Zapletalová, J.; Nováková, E. Between Urban and Rural: Sustainability of Small Towns in the Czech Republic. *Eur. Countryside* **2016**, *8*, 351–372. [CrossRef]
3. Zamfir, D.; Tălângă, C.; Stoica, I.V. Romanian Small Towns Searching for Their Identity. *J. Urban Reg. Anal.* **2009**, *1*, 41–53. [CrossRef]
4. Wolff, M.; Haase, A.; Leibert, T. Contextualizing Small Towns—Trends of Demographic Spatial Development in Germany 1961–2018. *Geogr. Ann. Ser. B Hum. Geogr.* **2021**, *103*, 196–217. [CrossRef]
5. Urbanska, W.; Levering, F. *Moving to a Small Town: A Guidebook to Moving from Urban to Rural America*; Simon and Schuster: New York, NY, USA, 1996.
6. Wu, Y.; Chen, Y.; Deng, X.; Hui, E.C.M. Development of Characteristic Towns in China. *Habitat Int.* **2018**, *77*, 21–31. [CrossRef]
7. Liu, Y. Introduction to Land Use and Rural Sustainability in China. *Land Use Policy* **2018**, *74*, 1–4. [CrossRef]
8. Servillo, L.; Atkinson, R.; Smith, I.; Russo, A.; Sýkora, L.; Demazière, C. *TOWN Small and Medium Sized Towns in their Functional Territorial Context-Inception Report*; European Union: Brussels, Belgium, 2012; p. 47.
9. Lynch, K. *A Theory of Good City Form*; MIT Press: Boston, MA, USA, 1981.
10. Chen, F. Urban Morphology and Citizens' Life. In *Encyclopedia of Quality of Life and Well-Being Research*; Springer: Berlin/Heidelberg, Germany, 2021; pp. 1–6.
11. Hatvani-Kovacs, G.; Bush, J.; Sharifi, E.; Boland, J. Policy Recommendations to Increase Urban Heat Stress Resilience. *Urban Climate* **2018**, *25*, 51–63. [CrossRef]
12. Oliveira, V.; Monteiro, C.; Partanen, J. A Comparative Study of Urban Form. *Urban Morphology* **2014**, *19*, 73-92. [CrossRef]
13. Berghauser Pont, M.; Haupt, P. *Spacemate: The Spatial Logic of Urban Density*; Delft University Press Science: Delft, The Netherlands, 2004.
14. Bocher, E.; Petit, G.; Bernard, J.; Palominos, S. A Geoprocessing Framework to Compute Urban Indicators: The MApUCE Tools Chain. *Urban Climate* **2018**, *24*, 153–174. [CrossRef]
15. Fleischmann, M.; Romice, O.; Porta, S. Measuring Urban Form: Overcoming Terminological Inconsistencies for a Quantitative and Comprehensive Morphologic Analysis of Cities. *Environ. Plan. B Urban Anal. City Sci.* **2021**, *48*, 2133–2150. [CrossRef]
16. Berghauser Pont, M.; Stavroulaki, G.; Bobkova, E.; Gil, J.; Marcus, L.; Olsson, J.; Sun, K.; Serra, M.; Hausleitner, B.; Dhanani, A.; Legeby, A. The Spatial Distribution and Frequency of Street, Plot and Building Types Across Five European Cities. *Environ. Plan. B Urban Anal. City Sci.* **2019**, *46*, 1226–1242. [CrossRef]
17. Kristjánsdóttir, S. Roots of Urban Morphology. *ICONARP Int. J. Archit. Plan.* **2019**, *7*, 15–36. [CrossRef]
18. Fleischmann, M.; Feliciotti, A.; Romice, O.; Porta, S. Methodological Foundation of a Numerical Taxonomy of Urban Form. *Environ. Plan. B Urban Anal. City Sci.* **2022**, *49*, 1283–1299. [CrossRef]
19. Rojas-Cortorreal, G.; Navés, F.; Peña, J.; Roset, J.; López-Ordóñez, C. Climate and Urban Morphology in the City of Barcelona: The Role of Vegetation. In *Mediterranean Identities—Environment, Society, Culture*, 1st ed.; ; Fuerst-Bjelis, B., Ed.; InTech: Rijeka, Croatia, 2017; p. 426. [CrossRef]
20. Lobo, J.; See, E.Y.S.; Biggs, M.; Pandit, A. An Insight into Morphometric Descriptors of Cell Shape That Pertain to Regenerative Medicine. *J. Tissue Eng. Regen. Med.* **2016**, *10*, 539–553. [CrossRef]
21. D'Anselmi, F.; Valerio, M.; Cucina, A.; Galli, L.; Proietti, S.; Dinicola, S.; Pasqualato, A.; Manetti, C.; Ricci, G.; Giuliani, A.; Bizzarri, M. Metabolism and Cell Shape in Cancer: A Fractal Analysis. *Int. J. Biochem. Cell Biol.* **2011**, *43*, 1052-1058. [CrossRef]
22. Porta, S.; Venerandi, A.; Feliciotti, A.; Raman, S.; Romice, O.; Wang, J.; Kuffer, M. Urban MorphoMetrics + Earth Observation: An Integrated Approach to Rich/Extra-Large-Scale Taxonomies of Urban Form. *Projections* **2022**, *16*. Available online: https://projections.pubpub.org/pub/vfwe0fdj (accessed on 7 June 2024).
23. Yap, W.; Janssen, P.; Biljecki, F. Free and Open Source Urbanism: Software for Urban Planning Practice. *Comput. Environ. Urban Syst.* **2022**, *96*, 101825. [CrossRef]
24. Taubenböck, H.; Esch, T.; Wurm, M.; Roth, A.; Dech, S. Object-based Feature Extraction Using High Spatial Resolution Satellite Data of Urban Areas. *J. Spatial Sci.* **2010**, *55*, 117–132. [CrossRef]

25. Geiß, C.; Taubenböck, H.; Wurm, M.; Esch, T.; Nast, M.; Schillings, C.; Blaschke, T. Remote Sensing-Based Characterization of Settlement Structures for Assessing Local Potential of District Heat. *Remote Sens.* **2011**, *3*, 1447–1471. [CrossRef]
26. Salazar Miranda, A. The Shape of Segregation: The Role of Urban Form in Immigrant Assimilation. *Cities* **2020**, *106*, 102852. [CrossRef]
27. Wang, Z.; Su, Y. Assessment of Soil Erosion in the Qinba Mountains of the Southern Shaanxi Province in China Using the RUSLE Model. *Sustainability* **2020**, *12*, 1733. [CrossRef]
28. Shang, H.; Hu, Y.; Fan, J.; Song, N.; Su, F. Analysis of Farm Household Livelihood Sustainability Based on Improved IPAT Equation: A Case Study of 24 Counties in 3 Cities in the Qin-Ba Mountain Region of Southern Shaanxi. *Land* **2023**, *12*, 980. [CrossRef]
29. Ma, Q.; Zhou, M.; Liu, J.; Zhao, J.; Xi, M. Coupling Relationship between Ecosystem Service Value and Socioeconomic Development in the Qinba Mountains, China. *Diversity* **2022**, *14*, 1105. [CrossRef]
30. Chen, S.; Mehmood, M.S.; Liu, S.; Gao, Y. Spatial Pattern and Influencing Factors of Rural Settlements in Qinba Mountains, Shaanxi Province, China. *Sustainability* **2022**, *14*, 10095. [CrossRef]
31. Wu, Z.; Cheng, G.; Wang, D.; Guo, Q. Urban and Rural Space Development in Qinba Mountain Area Based on Basins. *Strateg. Study Chin. Acad. Eng.* **2020**, *22*, 56–63. [CrossRef]
32. Jokar Arsanjani, J.; Zipf, A.; Mooney, P.; Helbich, M. An Introduction to OpenStreetMap in Geographic Information Science: Experiences, Research, and Applications. In *OpenStreetMap in GIScience*; Jokar Arsanjani, J., Zipf, A., Mooney, P., Eds.; Springer: Cham, Switzerland, 2015; pp. 1–15. [CrossRef]
33. Barrington-Leigh, C.; Millard-Ball, A. Correction: The World's User-Generated Road Map is More Than 80% Complete. *PLoS ONE* **2019**, *14*, e0224742. [CrossRef]
34. Cheng, J.; Zhu, Y.; Zhao, Y.; Li, T.; Chen, M.; Sun, Q.; Gu, Q.; Zhang, X. Application of an Improved U-Net with Image-to-Image Translation and Transfer Learning in Peach Orchard Segmentation. *Int. J. Appl. Earth Obs. Geoinf.* **2024**, *130*, 103871. [CrossRef]
35. Murillo-García, F.G.; Alcántara-Ayala, I. Landslide Inventory, Teziutlán Municipality, Puebla, México (1942–2015). *J. Maps* **2017**, *13*, 767–776. [CrossRef]
36. Vasavi, S.; Somagani, H.S.; Sai, Y. Classification of Buildings from VHR Satellite Images Using Ensemble of U-Net and ResNet. *Egypt. J. Remote Sens. Space Sci.* **2023**, *26*, 937–953. [CrossRef]
37. He, L.; Li, M.; Wang, X.; Wu, X.; Yue, G.; Wang, T.; Zhou, Y.; Lei, B.; Zhou, G. Morphology-Based Deep Learning Enables Accurate Detection of Senescence in Mesenchymal Stem Cell Cultures. *BMC Biol.* **2024**, *22*, 1. [CrossRef]
38. Chen, S.; Biljecki, F. Automatic Assessment of Public Open Spaces Using Street View Imagery. *Cities* **2023**, *137*, 104329. [CrossRef]
39. Basaraner, M.; Cetinkaya, S. Performance of Shape Indices and Classification Schemes for Characterising Perceptual Shape Complexity of Building Footprints in GIS. *Int. J. Geogr. Inf. Sci.* **2017**, *31*, 1952–1977. [CrossRef]
40. Song, Y.; Long, Y.; Wu, P.; Wang, X. Are All Cities with Similar Urban Form or Not? Redefining Cities with Ubiquitous Points of Interest and Evaluating Them with Indicators at City and Block Levels in China. *Int. J. Geogr. Inf. Sci.* **2018**, *32*, 2447–2476. [CrossRef]
41. Triantakonstantis, D.; Stathakis, D. Examining Urban Sprawl in Europe Using Spatial Metrics. *Geocarto Int.* **2015**, *30*, 1092–1112. [CrossRef]
42. Serrano Giné, D.; Russo, A.; Brandajs, F.; Pérez Albert, M.Y. Characterizing European Urban Settlements from Population Data: A Cartographic Approach. *Cartogr. Geogr. Inf. Sci.* **2015**, *43*, 442–453. [CrossRef]
43. Zhang, Y.; Lu, C.; Wang, J.; Du, F. A Large-Scale Extraction Framework for Mapping Urban Informal Settlements Using Remote Sensing and Semantic Segmentation. *Geocarto Int.* **2024**, *39*, 1. [CrossRef]
44. Ma, R.; Chen, J.; Yang, C.; Li, X. OSMsc: A Framework for Semantic 3D City Modeling Using OpenStreetMap. *Int. J. Geogr. Inf. Sci.* **2023**, *38*, 1–26. [CrossRef]
45. Fleischmann, M.; Arribas-Bel, D. Geographical Characterisation of British Urban Form and Function Using the Spatial Signatures Framework. *Sci. Data* **2022**, *9*, 546. [CrossRef] [PubMed]
46. Dibble, J.; Prelorendjos, A.; Romice, O.; Zanella, M.; Strano, E.; Pagel, M.; Porta, S. On the Origin of Spaces: Morphometric Foundations of Urban Form Evolution. *Environ. Plan. B Urban Anal. City Sci.* **2019**, *46*, 707–730. [CrossRef]
47. Wentz, E.A.; York, A.M.; Alberti, M.; Conrow, L.; Fischer, H.; Inostroza, L.; Jantz, C.; Pickett, S.T.A.; Seto, K.C.; Taubenböck, H. Six Fundamental Aspects for Conceptualizing Multidimensional Urban Form: A Spatial Mapping Perspective. *Landsc. Urban Plan.* **2018**, *179*, 55–62. [CrossRef]
48. Xiao, T.; Wan, Y.; Jin, R.; Qin, J.; Wu, T. Integrating Gaussian Mixture Dual-Clustering and DBSCAN for Exploring Heterogeneous Characteristics of Urban Spatial Agglomeration Areas. *Remote Sens.* **2022**, *14*, 5689. [CrossRef]
49. Cetinkaya, S.; Basaraner, M.; Burghardt, D. Proximity-Based Grouping of Buildings in Urban Blocks: A Comparison of Four Algorithms. *Geocarto Int.* **2014**, *30*, 618–632. [CrossRef]
50. Edelsbrunner, H.; Kirkpatrick, D.; Seidel, R. On the Shape of a Set of Points in the Plane. *IEEE Trans. Inf. Theory* **1983**, *29*, 551–559. [CrossRef]
51. Duckham, M.; Kulik, L.; Worboys, M.; Galton, A. Efficient Generation of Simple Polygons for Characterizing the Shape of a Set of Points in the Plane. *Pattern Recognit.* **2008**, *41*, 3224–3236. [CrossRef]
52. Gillies, S. *The Shapely User Manual*, 2013. Available online: https://sethc23.github.io/wiki/Python/The_Shapely_User_Manual_%E2%80%94_Shapely_1.2_and_1.3_documentation.pdf (accessed on 26 August 2022).

53. Bellock, K.E. *Alpha Shape Toolbox*; Version v1.3.1, 2021. Available online: https://alphashape.readthedocs.io/en/latest/alphashape.html (accessed on 16 April 2021).
54. Ai, T.; Zhang, R.; Zhou, H.W.; Pei, J.L. Box-Counting Methods to Directly Estimate the Fractal Dimension of a Rock Surface. *Appl. Surf. Sci.* **2014**, *314*, 610–621. [CrossRef]
55. Fleischmann, M. momepy: Urban Morphology Measuring Toolkit. *J. Open Sour. Softw.* **2019**, *4*, 1807. [CrossRef]
56. Philippi, D.; Rothaus, K.; Castelli, M. A Vision Transformer Architecture for the Automated Segmentation of Retinal Lesions in Spectral Domain Optical Coherence Tomography Images. *Sci. Rep.* **2023**, *13*, 517. [CrossRef]
57. Qin, C.; Wu, Y.; Liao, W.; et al. Improved U-Net3+ with Stage Residual for Brain Tumor Segmentation. *BMC Med. Imaging* **2022**, *22*, 14. [CrossRef]
58. Ashari, I.F.; Nugroho, E.D.; Baraku, R.; Yanda, I.N.; Liwardana, R. Analysis of Elbow, Silhouette, Davies-Bouldin, Calinski-Harabasz, and Rand-Index Evaluation on K-Means Algorithm for Classifying Flood-Affected Areas in Jakarta. *J. Appl. Inform. Comput.* **2023**, *7*, 89–97. [CrossRef]
59. Reddy, C.K.; Vinzamuri, B. A Survey of Partitional and Hierarchical Clustering Algorithms. In *Data Clustering*; Chapman and Hall/CRC: Boca Raton, FL, USA, 2018; pp. 87–110.
60. Ramos, M.C. Divisive and Hierarchical Clustering Techniques to Analyse Variability of Rainfall Distribution Patterns in a Mediterranean Region. *Atmos. Res.* **2001**, *57*, 123–138. [CrossRef]
61. Jiménez-Espada, M.; Cuartero, A.; Breton, M.L. Sustainability Assessment through Urban Accessibility Indicators and GIS in a Middle-Sized World Heritage City: The Case of Cáceres, Spain. *Buildings* **2022**, *12*, 813. [CrossRef]
62. Gu, C.; Li, Y.; Han, S.S. Development and Transition of Small Towns in Rural China. *Habitat Int.* **2015**, *50*, 110–119. [CrossRef]
63. Ogle, J.; Delparte, D.; Sanger, H. Quantifying the Sustainability of Urban Growth and Form Through Time: An Algorithmic Analysis of a City's Development. *Appl. Geogr.* **2017**, *88*, 1–14. [CrossRef]
64. Wang, X.; Liu, S.; Sykes, O.; Wang, C. Characteristic Development Model: A Transformation for the Sustainable Development of Small Towns in China. *Sustainability* **2019**, *11*, 3753. [CrossRef]
65. Wei, W.; Bao, Y.; Wang, Z.; Chen, X.; Luo, Q.; Mo, Y. Response of Habitat Quality to Urban Spatial Morphological Structure in Multi-Mountainous City. *Ecol. Indic.* **2023**, *146*, 109877. [CrossRef]

buildings

MDPI

Review

Development and Research Regarding Stormwater Runoff Management: Bibliometric Analysis from 2001 to 2021

Weihao Hao [1], Dong-Wook Sohn [1,*] and Da Wan [2,*]

1 Department of Architecture and Architectural Engineering, The Lab of Architectural & Urban Space Design, Yonsei University Seoul Campus, Seoul 03722, Republic of Korea; 2021313131@yonsei.ac.kr
2 School of Architecture, Tianjin Chengjian University, Tianjin 300380, China
* Correspondence: sohndw@yonsei.ac.kr (D.-W.S.); wanda@tcu.edu.cn (D.W.)

Abstract: As a result of climate change and urbanization, human activities are placing increasing pressure on nature, including with regard to urban stormwater runoff; consequently, various concepts related to urban stormwater runoff management have been proposed to tackle this problem in multiple countries. In this study, the latest research and techniques related to stormwater runoff management are reviewed in detail. A bibliometric analysis of proposed stormwater runoff management concepts developed from 2001 to 2021 was conducted based on a screening of 1771 studies obtained from the Web of Science (WoS). Bibliometric analysis is a research method that can be used to quantitatively analyze academic literature. Visualization of the data obtained from the literature using CiteSpace software and subsequent analysis of patent data through S-curve prediction were performed. The United States, China, and Australia were the top three countries from which publications on this issue were sourced. Each country tends to study its own most relevant issues and has a particularly clear understanding of its own research landscape. The development of stormwater runoff management concepts was analyzed using reference emergence analysis. This was followed by keyword clustering and keyword emergence analysis to identify current research hotspots, trends, technological developments, and limitations. The limitations and emerging trends related to current stormwater runoff management concepts are discussed thoroughly, and suggestions for future studies are provided.

Keywords: urban stormwater management; low-impact development (LID); sustainable urban drainage system (SUDS); water-sensitive urban design (WSUD); sponge city

1. Introduction

Global urbanization and population growth are placing increasing pressure on nature [1]. In conjunction with urban expansion and damage to the natural environment, the varieties of pollutants in runoff are increasing, along with their concentrations, resulting in a significant overall increase in urban water pollution [2]. As global climate change has intensified, frequent urban flooding and droughts have brought increased attention to the management, use, and harvesting of rainwater around the globe [3,4]. The increase in impermeable surfaces due to high rates of urbanization, especially in developed countries with rapid urban growth, has affected ecological environments surrounding major cities [5]. Therefore, methods and concepts for managing urban stormwater runoff have been proposed and developed over the last few decades to address the environmental problems that arise from urbanization and climate change. Approaches to urban stormwater runoff management include low-impact development (LID) in the USA and Canada, sustainable urban drainage systems (SUDSs) in the UK, and water-sensitive urban design in Australia (WSUD) [6,7]. LID focuses on managing stormwater at the source through the use of tailored site-design techniques and green infrastructure practices, such as rain gardens, green buildings, and permeable pavement, to promote infiltration, evaporation, and filtration of runoff. WSUD is a similar approach that emphasizes the use of landscape-based solutions to manage urban water while also considering the social and ecological context of the urban environment in which it is implemented. SUDSs are used in the United Kingdom and are similar to LID, focusing on managing surface water runoff in a sustainable manner while also considering the needs of the community [8,9]. China has grown rapidly over the past decade, which has led to an increase in environmental problems due to urbanization [10]. In 2013, a new type of urban stormwater management was proposed in China known as the "sponge city" approach. Sponge cities are cities that are well-adapted to the environmental and rainwater changes brought about by natural disasters, employing absorption, storage, infiltration, and purification of rainwater followed by release of water stores, with waiting periods between each period of use. This approach is an effective way to improve urban stormwater runoff and manage urban stormwater runoff problems in China [11]. Each of these approaches has unique benefits; however, the original intent and design philosophy of sponge cities are similar to those of other proposed methods [12].

Directions of research on stormwater runoff management vary between countries due to differences in policy and their natural environments. A comparative analysis of stormwater runoff management in the UK and China was previously conducted by Lashford et al. and built upon by Wang et al. They applied these concepts to the development of techniques with the capability of managing pollution caused by water runoff [11,13]. A study published in the *Journal of Environmental Management* in 2021 compared the effectiveness of LID and WSUD practices for the management of stormwater runoff within an urban catchment in Australia [14]. Another study published in the *Journal* by Zeng et al. in 2020 evaluated the sustainability of SUDS practices in the United Kingdom [15]. However, there are few articles summarizing the developmental trends and research hotspots for rainfall runoff management concepts across different periods of time. The aims of this study were to analyze trends and research hotspots in the field through a review of the existing literature, to provide guidance on current research trends and hotspots, and to recommend future studies.

This study provides an overview of the development of stormwater runoff management concepts and associated technologies based on an analysis of literature patents in the field within the WoS and presents trends and research hotspots with regard to these concepts in a clear and simple manner through CiteSpace mapping visualization, which will help researchers quickly access relevant research trends and information. The discussion of the literature using the above methodologies led to the following observations: Countries prefer to study localized stormwater runoff problems and have a clear understanding of their own research directions and problems. Countries and regions with longer histories of research and more publications have closer relationships with other regions in terms of the amount of cooperation. Geographical constraints, technical limitations, and data uncertainty limit the development and dissemination of stormwater runoff management. In the future, multifunctional stormwater runoff design can be used to better address urban issues, and increased interdisciplinary research collaboration can keep the project sustainable

2. Materials and Methods

The data used in this study were collected from the WoS due to the fact that the work in the WoS represents the international research frontier for the relevant scientific fields and much of it has high international impact. Additionally, the WoS has been cited as the primary data source in many recent bibliometric reviews on building structures [16]. The literature search was conducted on 14 July 2022. In the first phase (Figure 1), 1771 articles related to stormwater runoff management concepts were identified. These articles were broken down into the following document types: articles (number = 1651, percentage = 93%), reviews (number = 120, percentage = 6%), conference papers (number = 350, percentage = 16%), other (<100 documents total). Bibliometric analysis and visualization of the information was performed in stage two using the comprehensive data exported from the WoS (fully recorded author information, citations, references, and other textual information).

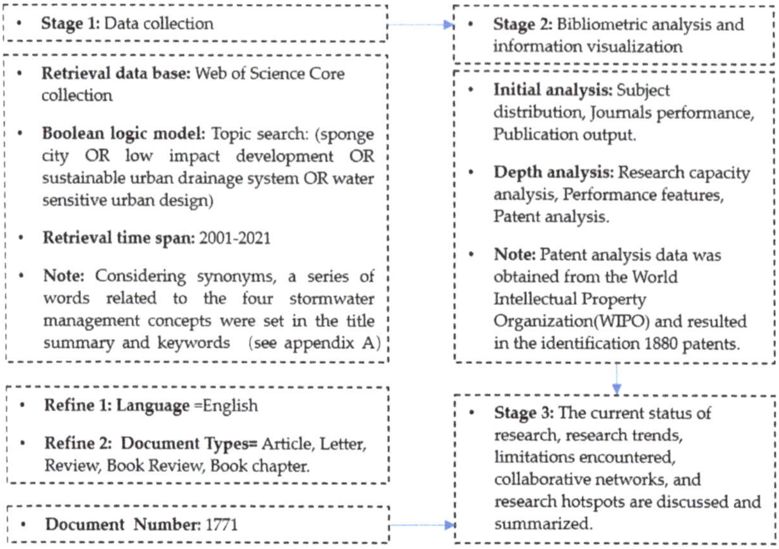

Figure 1. Stages of the bibliometric analysis of urban runoff management (Appendix A).

CiteSpace, a visual analysis tool, was adopted in this study to map multiple knowledge maps and spectrums. Therefore, both research trends and hot topics in the field at large were represented through elements such as node size, network connectivity, and keyword co-occurrence. One of the core functions of CiteSpace is to detect and analyze frontiers in research and knowledge relationships [17]. Therefore, this software allows scientists to determine the extent to which changes are occurring in a field through the global publication distribution, research institution networks, and analysis of co-cited studies.

Furthermore, a specialist database, that of the World Intellectual Property Organization (WIPO), was applied as a data source in this study to analyze and predict the future direction of stormwater runoff technology. The patent analyses provided an understanding of current and future trends, as well as strategic targets for this field [18]. A total of 1880 relevant patents were identified.

A graph known as the s-curve was mainly used for the analysis of patent data. The s-curve is often adopted to describe future prospects and trends in both economies and technologies [19]. In general, technological growth occurs in a cyclical pattern, and the relationship between slow and quick stages of technological development is generally captured by an s-curve. Furthermore, the life cycles of technologies and products can be divided into four phases: the exploration phase (introduction), the growth phase, the maturity phase, and the stagnation phase (saturation) [20]. In this study, a patent trend analysis was performed in Loglet Lab 4 as a means of analyzing and predicting the range of technological developments related to stormwater runoff.

3. Results

3.1. Subject Categories and Their Distribution

The SCI-E and SSCI databases revealed 94 subjects from 1771 reviewed articles, which included agriculture, biology, and energy. These data suggest that the field of stormwater runoff management is interdisciplinary, with the predominant research areas being related to environmental sciences, ecology, and water resources, comprising more than three quarters of all combined categories.

3.2. Journal Performance

Of the 1771 articles that were analyzed, 40% were published in the top ten most recognized journals (Figure 2), with a sudden spike in the number of articles pertaining to "water" after 2015. The majority of journals showed a decline in publications with regard to this topic between 2019 and 2020, which may have been due to the influence of external factors. The journal impact factor (JIF) quartile of a journal indicates the impact of that journal on its field and whether it is considered cutting edge. Of the top 30 journals containing the highest numbers of published articles, 36.7% of the journals were in quartile one. These data suggest that this field is highly influential, and further studies in stormwater runoff management would be valuable.

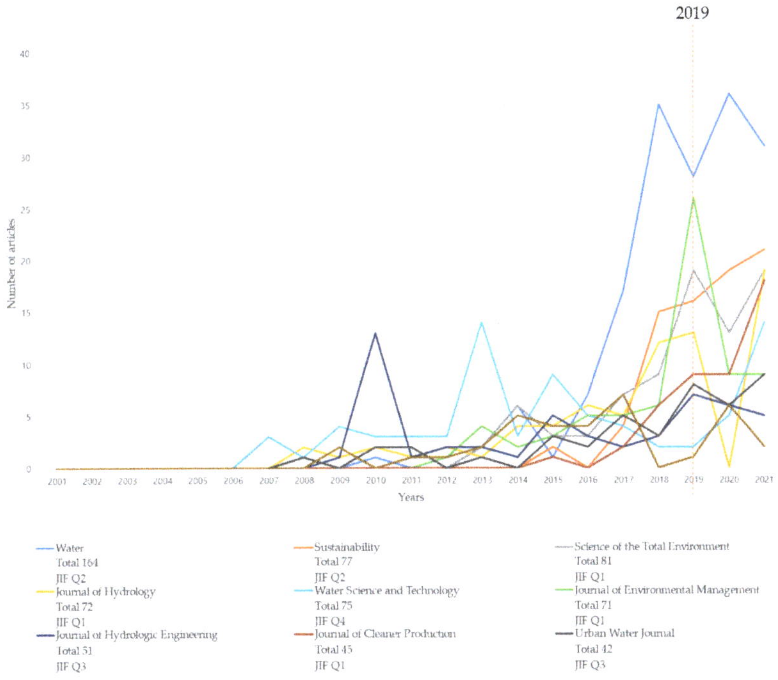

Figure 2. The performance of the top ten most impactful journals.

3.3. Publication Output

Figure 3 shows the trend for annual publications between 2001 and 2021 in this field. Excluding the initial years from 2001 to 2003, the growth rates for publications in this field in 2008, 2013, and 2017 were high at 80%, 103%, and 74%, respectively. These data suggest that potential external interventions, such as government policies or the emergence of new technologies, affected the number of publications. Furthermore, the data show that the number of publications continued to increase after 2016. The growth rate for publications in the field began to slow after 2017; however, the overall trend remained positive. Furthermore, the growth in publications in this field drastically increased between 2020 and 2021, indicating that research efforts in this field have not diminished and can be expected to intensify in the future, partly due to growing environmental concerns.

Figure 3. The yearly numbers of articles published from 2001 to 2021.

3.4. Research Capacity Analysis

3.4.1. Country Distribution

Figure 4 shows the global distribution of publications in the field at large, with China taking first place with 35.42%, followed by the USA and Australia with 28.28% and 11.69%, respectively. Countries and territories such as South Korea, Iran, and Europe have also contributed to the development of urban runoff concepts but to a lesser extent than the big three.

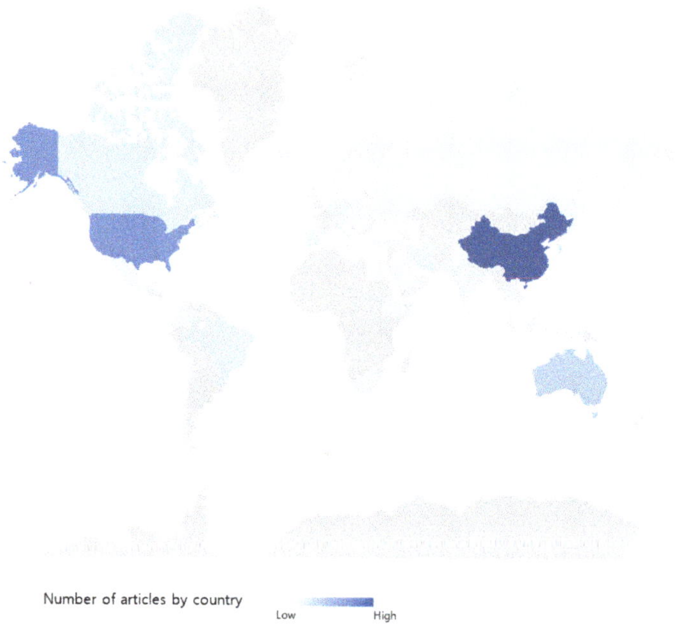

Figure 4. Global geographical distribution of authors' locations.

3.4.2. Global Research Network Trends

Figure 5 demonstrates the collaborative links between countries through a chronogram generated in CiteSpace software. The overall size of the wheel reflects the number of citations, while the wheel color corresponds to the year and the wheel thickness indicates the number of articles published within the corresponding timeframe. A total of 67 countries were involved in research in this field, with approximately 297 lines showing the cooperation between countries and a density score of 0.1343, which refers to the average level of cooperation for urban runoff management. The data show that a longer study period corresponded to a higher number of publications and connections with other countries, indicating a higher degree of cooperation. Conversely, countries with fewer publications tended to have lower levels of cooperation with other countries.

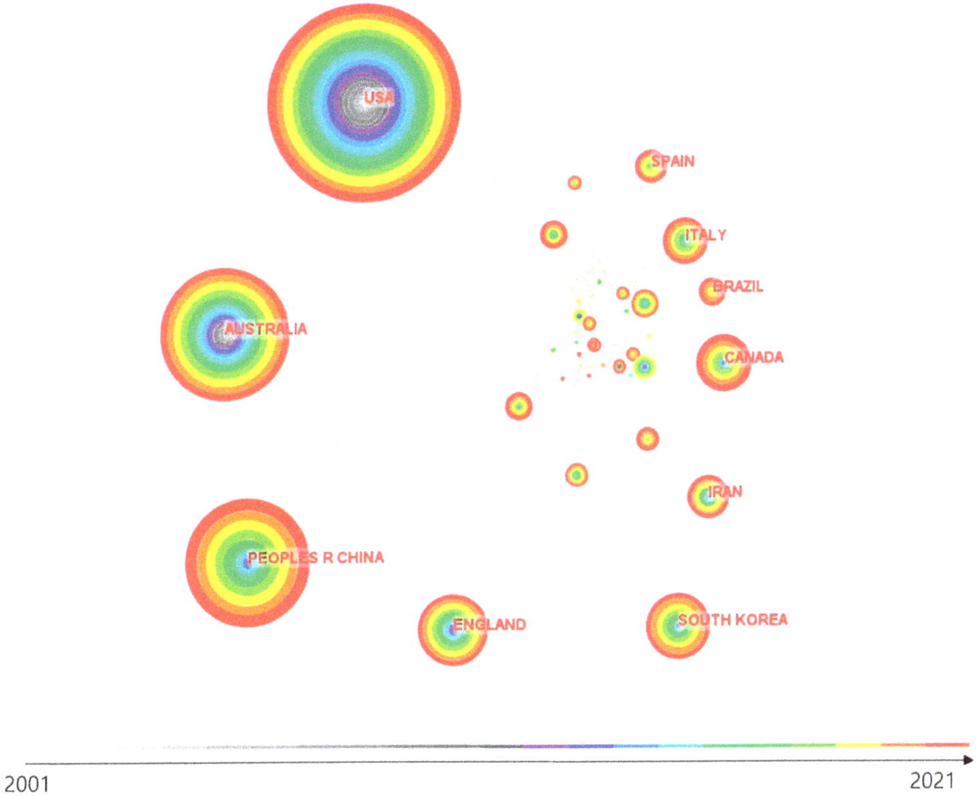

Figure 5. Chronogram cooperation chart for countries employing stormwater runoff management practices.

Figure 6 represents the academic publishing trends (number of articles) for the top five countries employing stormwater runoff management practices. The United States had the highest share of publications until 2017, with publications declining in 2020 and rising in 2021. All other countries showed a steady increase, with China more than doubling the number of publications from 43 in 2017 to 91 in 2018. A similar upward trend was observed in the UK and Australia, with only South Korea showing a decline in the past year. These data suggest that China has invested significant amounts of money in stormwater runoff management research. These data also show that many other countries and regions have maintained their interest in the field.

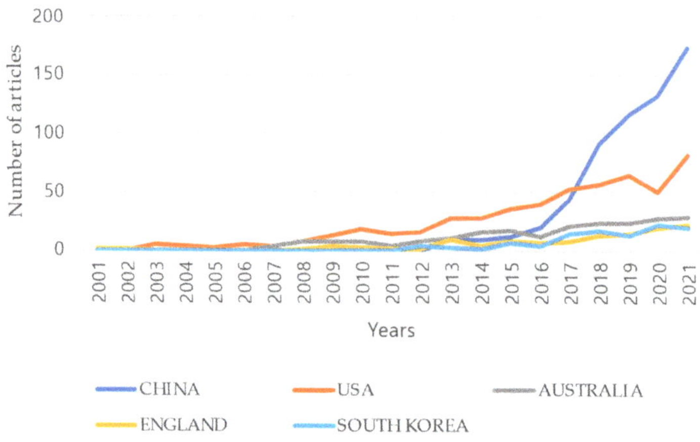

Figure 6. Publishing trends for the top five countries undertaking stormwater runoff management between the years 2001 and 2021.

3.4.3. Analysis of Research Institutions

A network mapping analysis of institutional cooperation identified research institutions at the forefront of research in urban runoff management. There were 236 institutions involved in this area of research (Figure 7) with 372 links and an overall cooperation density of 0.0134, making the density at the institutional level lower than that of international cooperation. Universities accounted for 83.3% of the institutions in the top ten. For example, Monash University in Australia ranked first in terms of the number of articles published and was the first institution to invest in this scientific field in 2007. The institution's top two most cited articles are by Roy, A. (2008) and Coutts, A. (2012), which have 373 and 215 citations, respectively. The study by Roy, A. (2008) summarized seven key barriers to stormwater management, drawing on lessons learned from traditional green infrastructure failures in Australia and the USA, with the aim of slowing the ecological degradation of rivers and proposed solutions to improve river ecology by promoting urban stormwater management [21]. Furthermore, Coutts, A. (2012) proposed to combine water-sensitive and climate-sensitive urban designs to harvest and retain rainwater in the context of an urban landscape in order to reduce urban temperatures and improve human comfort [22]. Table 1 shows the top 20 research institutions with the most publications in the field. China accounted for 55.5% of these institutions; however, it recently entered the field in 2014. The top two most cited articles from Chinese research institutions are by Xia, J. (2017) and Sang, Y. (2016), with 185 and 42 citations, respectively. Xia, J. (2017) studied the urban water issues plaguing China and concluded that the construction of sponge cities in China is confined to low-impact development, demonstrating relatively weak integration with other disciplines, such as urban hydrology and ecological landscaping, and thus requiring improvement [23]. Furthermore, Sang, Y. (2016) also focused on urban flooding and argued that current runoff management approaches have not met expectations. Therefore, urban flooding and its associated problems have not been resolved and require both structural and non-structural measures to improve sponge city construction [24]. It is clear that institutions prefer to focus on problems within their own countries, and Chinese institutions began researching these technologies later than other countries. Articles from China are mainly based around major questions and constructive comments. Australian institutions were the first to begin research in this field and are in a better position to suggest concrete solutions and direct future studies. On the whole, these data suggest that countries have a clear understanding of their research landscapes, as well as the areas they need to develop in the future.

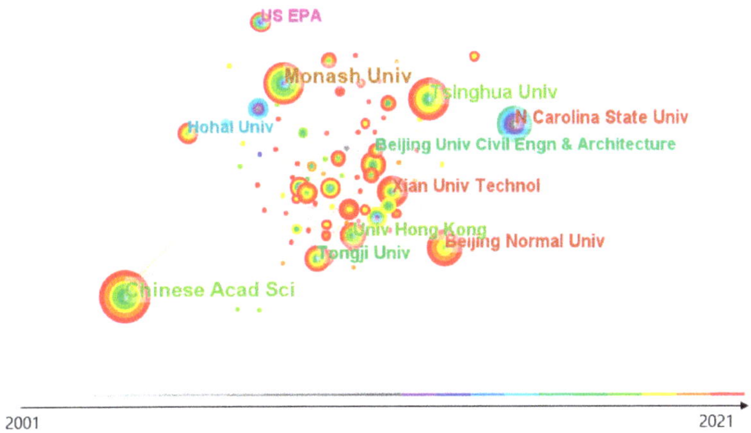

Figure 7. Chronological chart of international research institutions.

Table 1. Top 20 research institutions.

Serial No.	Name of Institution	Year of Debut	Number of Articles
1	Monash University	2007	45
2	Chinese Academy of Sciences	2016	43
3	Tsinghua University	2014	39
4	Beijing Normal University	2018	29
5	Xi'an University of Technology	2016	28
6	North Carolina State University	2009	27
7	Tongji University	2015	26
8	Kongju National University	2012	24
9	University of Hong Kong	2015	24
10	Beijing University of Civil Engineering and Architecture	2013	21
11	Hohai University	2017	21
12	US EPA	2009	21
13	Peking University	2017	20
14	University of Melbourne	2013	20
15	Wuhan University	2017	20
16	Purdue University	2015	17
17	University of South Australia	2009	17
18	China Institute of Water Resources and Hydropower Research	2017	16
19	Texas A&M University	2017	16
20	University of Copenhagen	2012	16

3.4.4. Analysis of Co-Cited Studies

The clustering of co-cited studies reflects the extent of knowledge in a field, which is fundamental to a field's development and overall level of theoretical support, as shown in Figure 8. The "green roof" categories (#0 "green roof" and #2 "extensive green roof") are technical and are included in every runoff management approach. Green roofs include different national and climate characteristics and can have positive effects with regard to runoff, rainwater harvesting, and rainwater quality filtration [25–28]. Category #4, "international practice", refers to the comparison and evaluation of different international standards and technologies for stormwater management. In this work, researchers showed a positive tendency towards proposing new stormwater management practices and suggested more effective solutions and improvements by studying and comparing relevant technologies and management frameworks [29–31]. In the two categories "urban stormwater reclamation" (#6) and "compact growth" (#9), the role of drainage facilities with regard to urban stormwater management is mentioned extensively because sustainable drainage has the potential to improve urban water management and can reduce problems involving road surface runoff and pollutant loads through bioretention facilities and permeable pavement materials [32–34].

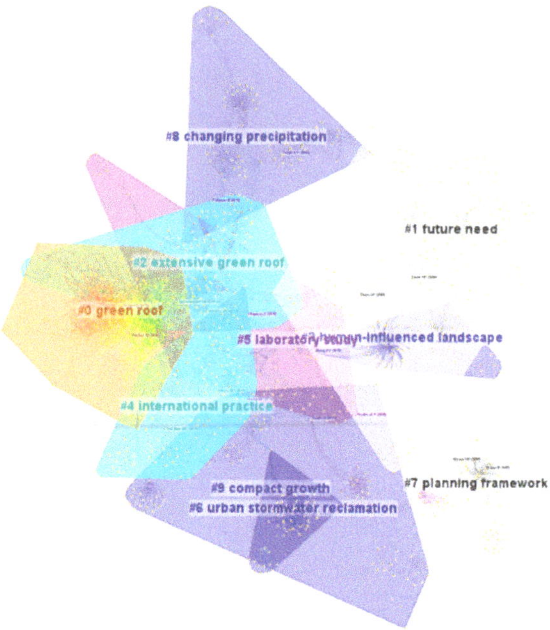

Figure 8. Cluster map for urban rainfall runoff management research.

3.4.5. Reference Emergence Analysis

Citation bursts are sudden increases in citations that occur over a short period of time; they are analyzed to check the range of and trends in the research impacted by the cited studies within a given cycle, as well as to identify influential topics in terms of burst intensity and duration. For this study, burst intensity was analyzed for the top ten publications, with a total of 60 burst items ranked according to intensity, as shown in Figure 9. The color red indicates the burst time period and the color blue indicates the unaffected period. For each stormwater management research area, both emergence intensity and duration were analyzed.

References	Year	Strength	Begin	End	2001 - 2021
Ahiablame LM, 2012, WATER AIR SOIL POLL, V223, P4253, DOI 10.1007/s11270-012-1189-2, DOI	2012	23.49	**2014**	2017	
Qin HP, 2013, J ENVIRON MANAGE, V129, P577, DOI 10.1016/j.jenvman.2013.08.026, DOI	2013	31.4	**2015**	2018	
Ahiablame LM, 2013, J ENVIRON MANAGE, V119, P151, DOI 10.1016/j.jenvman.2013.01.019, DOI	2013	17.43	**2015**	2018	
Fletcher TD, 2013, ADV WATER RESOUR, V51, P261, DOI 10.1016/j.advwatres.2012.09.001, DOI	2013	15.28	**2015**	2018	
Fletcher TD, 2015, URBAN WATER J, V12, P525, DOI 10.1080/1573062X.2014.916314, DOI	2015	54.16	**2017**	2021	
Palla A, 2015, J HYDROL, V528, P361, DOI 10.1016/j.jhydrol.2015.06.050, DOI	2015	28.54	**2017**	2021	
Baek SS, 2015, WATER RES, V86, P122, DOI 10.1016/j.watres.2015.08.038, DOI	2015	21.47	**2017**	2021	
Rosa DJ, 2015, J AM WATER RESOUR AS, V51, P746, DOI 10.1111/jawr.12272, DOI	2015	18.17	**2017**	2021	
Ahiablame L, 2016, J ENVIRON MANAGE, V171, P81, DOI 10.1016/j.jenvman.2016.01.036, DOI	2016	17.96	**2017**	2021	
Chui TFM, 2016, J HYDROL, V533, P353, DOI 10.1016/j.jhydrol.2015.12.011, DOI	2016	16.32	**2017**	2021	

Figure 9. Top ten references with the strongest citation bursts.

In terms of emergence intensity, the most intense article was determined to be the article by Fletcher, T. (2015), which was a study on the evolution and application of urban drainage terminology with an impact period between 2017 and 2021. In summary, the article provides an overview of the development of existing urban stormwater management concepts and describes the evolution of the different terms, as well as their scope of coverage. Furthermore, it also analyzes shortcomings in the field of "uniform terminology" and states that the local development of terminology has an important role to play in advancing the profession and facilitating future communication between different disciplines on the approach to stormwater management [7].

From a time-of-appearance perspective, the earliest cited article was published in 2012, indicating that research on stormwater management began to move from exploration to development. The development period can be divided into two phases: 2014–2017 and 2017–2021. Furthermore, the papers can be divided into two categories. The first category provide favorable evidence regarding the effectiveness of low-impact development practices, demonstrating the effectiveness of the technical facilities that have been completed for flood control, as well as rainwater harvesting and purification, while offering recommendations for future studies [35–37].

The second category of papers attempt to improve the way urban runoff is managed by using simulation models to predict rainfall and runoff and to develop new technologies [35]. This phase of stormwater runoff management research has been proven to be effective, with initial attempts to integrate new technologies and enhance original concepts being largely successful. Five previous studies refer to the development of simulation models to demonstrate the usefulness of low-impact development in extreme climates, such as those with large storm floods, in addition to the evolution of the terminology and applications of urban drainage mentioned above, illustrating the ongoing integration in the field with regard to new technologies and other disciplines [2,38–41].

The development of stormwater runoff management theory can be described as follows. First, a concept is proposed, and then it is clarified and standardized. Further studies demonstrate its feasibility in practice while simultaneously solving problems, optimizing and innovating the technology used for the proposed concept. The final step is to promote the innovations that have been demonstrated through interdisciplinary research cooperation.

3.5. *Performance Features*

In this section, by combining the Environmental Performance Index (EPI) framework and the concept of stormwater runoff management, trends in research on stormwater runoff management in various environmental performance areas, as well as research hotspots, are examined through keyword timeline analysis and keyword emergence analysis.

As a benchmark index, the EPI framework can be easily adopted by policymakers, environmental scientists, environmental advocates, and the general public [42]. In the context of reducing the impact of environmental stress on human health, as well as enhancing ecosystem vitality and promoting better management of natural resources, the introduction of this performance framework can be an effective way to demonstrate the effect of urban stormwater runoff management on the environment, as well as its effect on the direction of future research.

The EPI framework contains 3 policy objectives and 11 issue categories (Table 2) [43]. Based on this framework, 944 relevant papers were selected from an initial set of 1756 papers, using the concept of stormwater runoff management as a background.

Table 2. Literature screening with the EPI framework.

Policy Objectives	Issue Categories	Article Numbers
Climate	Climate change mitigation	311
Environmental health	Air Quality Waste management Water and sanitation Heavy metals	145
Ecosystem vitality	Biodiversity and habitat Ecosystem services Fisheries Agriculture Acid rain	527

3.5.1. Climate

1. Keyword clustering timeline analysis

The keywords are the primary themes of a written work and can provide clear indications about the topic while serving as summaries of the main subject. They can effectively reflect the areas covered in the literature and, thus, facilitate investigation into the research hotspots and developments in the field, as well as the knowledge background.

The keyword clustering time shows the time span, the evolution of each clustering hotspot, and the relationship between each cluster and the others. The longer the horizontal line is, the greater the time span is, with the cluster coming into existence earlier and lasting longer. The six clusters are plotted in Figure 10. For the climate policy objective, most clusters were closely linked, which indicates a high level of cross-sectional research. The strongest relationships in the study period from 2010 to 2021 were between clusters #0 (source control), #1 (water sensitive urban design), and #2 (bioretention cell), indicating that stormwater runoff management is primarily a response and adaptation to runoff and flooding caused by environmental changes. As can be seen, stormwater runoff management is primarily concerned with responding and adapting to runoff caused by environmental changes, such as flooding.

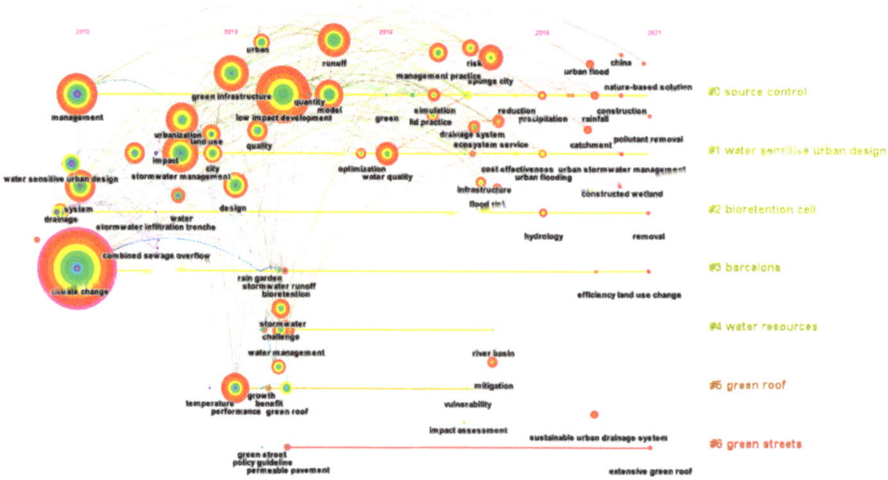

Figure 10. Climate keyword clustering timeline analysis.

Cluster #0, "source control", included keywords such as "low impact development", "green infrastructure", "management run off", and "model". This cluster was mainly concerned with the prediction of the coping capacity of urban rainwater runoff management facilities when facing possible climate changes in the future. By building a prediction model that can simulate climate change, future urban waterlines, floods, and other problems can be predicted and evaluated. Using these evaluative methods, it has been shown that urban runoff management facilities can have a positive impact on the future of climate change management. Some positive effects of these facilities (including biological retention pool facilities, porous pavement facilities, and others) are that they can effectively reduce surface runoff volume and velocity and the value of the volume capture ratio for annual rainfall can be significantly increased, thus reducing the rate of occurrence of urban floods [44–46]. However, in this cluster, no measures to solve existing climate problems have been proposed.

Cluster #1, "water sensitive urban design", included "rainwater management", "urbanization", "ecosystem service", "urban flooding", and other keywords. It was mentioned in the studies from this cluster that the effective implementation of rainwater management can improve the sustainability of cities and urban water circulation. In addition, rainwater runoff management facilities can help alleviate the effects of climate change on urban environments. The measures mentioned above can be implemented through facilities such as green roofs and rainwater wetlands, which can not only control rainwater runoff but also reduce air pollution and alleviate the effects of climate change [47]. Furthermore, some countries have effectively reduced greenhouse gas emissions resulting from urban water circulation through the optimization of urban water circulation systems, as well as the filtering, storage, and reuse of rainwater made possible by rainwater management facilities [48]. However, these facilities have strict requirements for location, making the implementation of these methods impossible on a large scale.

Cluster #2, "bioretention cell", included keywords such as "system", "water sensitive urban design", "drainage", "hydrology", "strategy", etc. This cluster evaluated and predicted the performance of rainwater runoff management facilities in dealing with rainwater runoff caused by climate change. Runoff management facilities play a significant role in dealing with urban floods caused by smaller storms, but their performance may worsen when faced with the challenge of larger urban floods in the future [49]. The combination of a permeable pavement facility and a detention basin facility is the best arrangement of facilities, providing higher flexibility in the face of major urban floods. However, the flexibility of this combination may be limited depending on the geographical location and the extent of rainwater management facilities [50].

2. Keyword emergence analysis

The emergence of keywords refers to the appearance of terms that are frequently used or closely followed by scholars in a field. It indicates the hotspots and research frontiers in the field and areas that have good research potential, and it can serve as a basis for analyzing development trends.

In total, there were 12 emergent items. The top five were selected, and the top two to three words in terms of intensity were analyzed. As shown in Figure 11, "water sensitive urban design" (WSUD) was not only the most frequent term but also the longest lasting emergent item. This is attributable to the number of measures, many of which are multifunctional, that WSUD offers. These facilities are not only better able to cope with the negative effects of climate change but can also help to address the urban heat island effect and improve the microclimate [51–53].

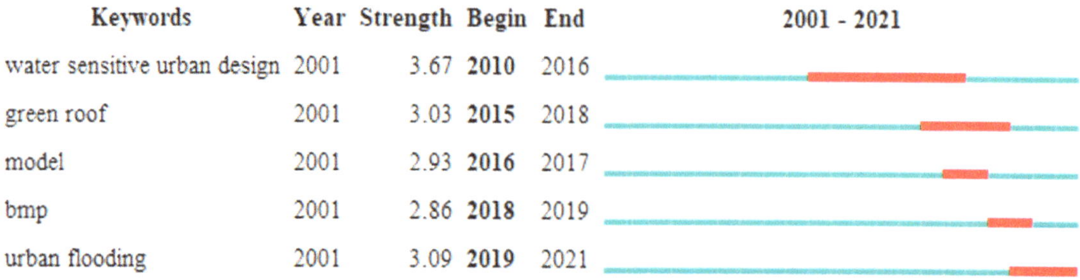

Figure 11. Top five emergent climate keywords.

Green roofs are a kind of popular green infrastructure in highly urbanized areas [54] that have a very good rainwater runoff reduction effect on both inclined and flat impervious roofs [55]. However, the construction of green roof facilities is widely restricted by local environment and climate conditions. Very few green roofs have been implemented in Australia, and their application is still in the developmental stage [56]. However, in other countries during the same period, such as the United States, Germany, and Japan, green roofs have been installed on a large scale with significant results [57]. In addition to environmental factors, there may be difficulties in implementing these facilities on a large scale due to the unavoidable high cost of green roofs, the lack of appropriate regulatory systems among regional agencies, and the lack of local environmental monitoring data [58]. In addition, green roofs can effectively reduce most of the pollutants in rainwater runoff, but material selection and material aging issues should be considered. For example, aging roofs can become a source of metal pollution [59,60].

The most recent keyword was "urban flooding", indicating that it is a frontier research topic in the field of climate research. The problem of urban flooding is becoming increasingly disturbing with the changing climate. It is of particular concern for low-lying cities and has garnered recent attention in several countries [61,62]. In addition, due to the development of rainwater management modeling and monitoring technology, future urban flood problems can be predicted and simulated, and corresponding planning regulations and guidelines can be formulated in order to deal with possible urban flooding problems in the future [63,64]. Rainwater management facilities have performed well in controlling small and medium-sized urban floods, while in the face of serious urban floods, their effect has been limited.

3.5.2. Environmental Health

1. Keyword clustering timeline analysis

Eight clusters were extracted (Figure 12) spanning the period from 2007 to 2021. The field of environmental health has existed longer than the field of climate research. Cluster #1, "polycyclic aromatic hydrocarbons", was most closely associated with clusters #2, "LID", and #4, "sensitivity analysis", and it showed a longer research history.

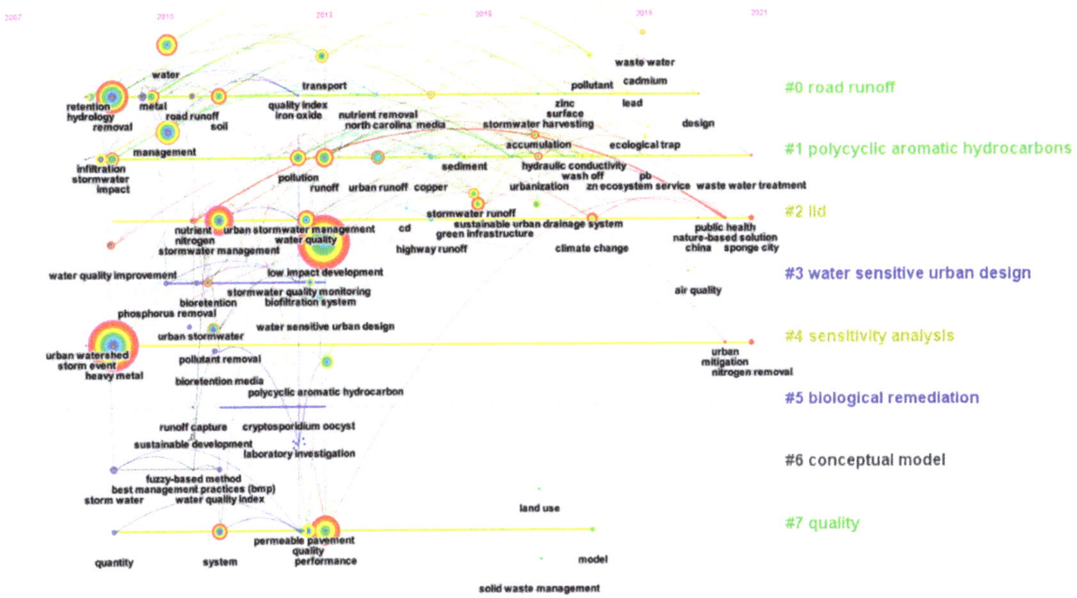

Figure 12. Environmental health keyword clustering timeline analysis.

Cluster #1, "polycyclic aromatic hydrocarbons", included keywords such as "plant", "phytoextraction", "phytoremediation", "land", etc. In this cluster, researchers studied the removal of heavy metals, organic substances, and other pollutants dissolved in rainwater mainly by means of biological interception [65], which improves environmental health. Plants play an important role in filtration, and different plants have different capabilities for removing different pollutants [66]. However, the majority of successful cases come from a small number of specific countries, and there are differences in biological interception systems in different geographical locations, which has caused failures in the installation of rainwater management facilities in other areas [67].

Cluster #2, "LID", included keywords such as "adsorption", "zinc", "copper", "sediment removal", etc. The adsorption of pollutants in water by physical means was the central topic of this cluster, and this is a very important remediation measure in rainwater runoff management facilities [68]. The selection of filter media leads to different effects for different pollutants, and factors such as temperature, humidity, and the flow rate of rainwater runoff can affect the adsorption effect [69]. Therefore, more accurate data sources are needed for the monitoring of local rainwater pollutants, as well as on-site experimental simulation results, prior to the installation of rainwater runoff management facilities.

The keywords included in cluster #4, "sensitivity analysis", were "uncertainty", "rainwater quality", "urban flooding", "hierarchy of hazard control", etc. The research in this cluster mainly discussed how sensitivity analysis and uncertainty analysis can be used to determine the great influence climate change can have on the removal of rainwater pollutants by rainwater runoff management facilities. Due to the limited interception capacities of these facilities, their performance in extreme weather conditions is highly uncertain [70] and can vary depending on geographical location and land-cover distribution [71]. In addition, in the construction of rainwater runoff facilities, the construction of buildings and the transportation of materials can have extremely adverse impacts on the environment [72]. Therefore, prior to the construction of rainwater runoff facilities, there should be extensive early planning, and the benefits should be weighed against the uncertainty regarding the facilities' function and any possible adverse impact on the environment.

2. Keyword emergence analysis

There were five emergent terms in total, all of which were included in the analysis (Figure 13). In general, stormwater management concepts concern the removal of pollutants, such as heavy metals, from urban runoff and have a concentrated time cycle with a short timeframe. The absence of new, emerging terms in the field in recent years suggests that no major new research trends have emerged and that past research directions have continued up to the present. "WSUD" is still the prevalent term. This is due to the multifunctional facilities it offers and the fact that it makes an important contribution to enhancing the inhabitability of cities and improving urban environments through its many amenities [73].

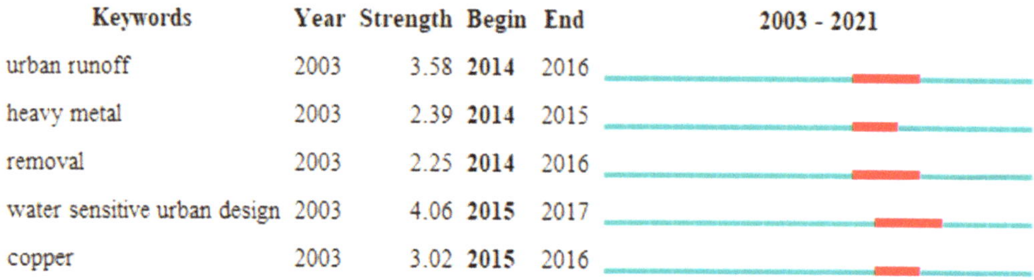

Figure 13. Top five emergent environmental health keywords.

Urban runoff ranked second in the keywords. Any method of rainwater runoff management can be considered an effective measure to reduce urban runoff and improve water quality. With regard to environmental health, many pollutants carried by rainwater runoff are discharged into streams, rivers, and other natural water sources without treatment, causing environmental pollution. It has been proved that runoff management facilities are effective in removing runoff pollutants [74]. However, problems such as the overflow of facilities in cases of heavy rain, the blockage of pipelines after long-term use, and the difficulty in maintaining these facilities are being continuously studied [75,76].

3.5.3. Ecosystem Vitality

1. Keyword clustering timeline analysis

A total of eight clusters were generated and the first five were selected (Figure 14). In these clusters, which span the period from 2004 to 2021, stormwater runoff management concepts have the longest history of study in terms of ecosystem vitality. All studies included have continued to the present day and are closely interlinked. Group #2, the "SWMM model", was the cluster with the longest history of studies being published. The Storm Water Management Model (SWMM) is a software application developed by the University of Florida, USA, which has been widely applied in the analysis of the impact of rainfall on urban runoff. It is helpful in planning how urban drainage systems can be optimized in order to reduce economic losses, such as those caused by urban flooding [77]. This software allows for the effective analysis and assessment of the impact of stormwater runoff management on urban runoff control [78]. More simulation models are currently being applied to rainwater runoff management, including the simulation of pollutants and permeability. However, due to the limitations of pollutants and the lack of concentration data regarding the process of pollutant diffusion, it is impossible to verify the simulation data results. Thus, more systematic data collection and the unification of relevant numerical models could be helpful in improving diffusion and anti-diffusion process related to pollutants [79].

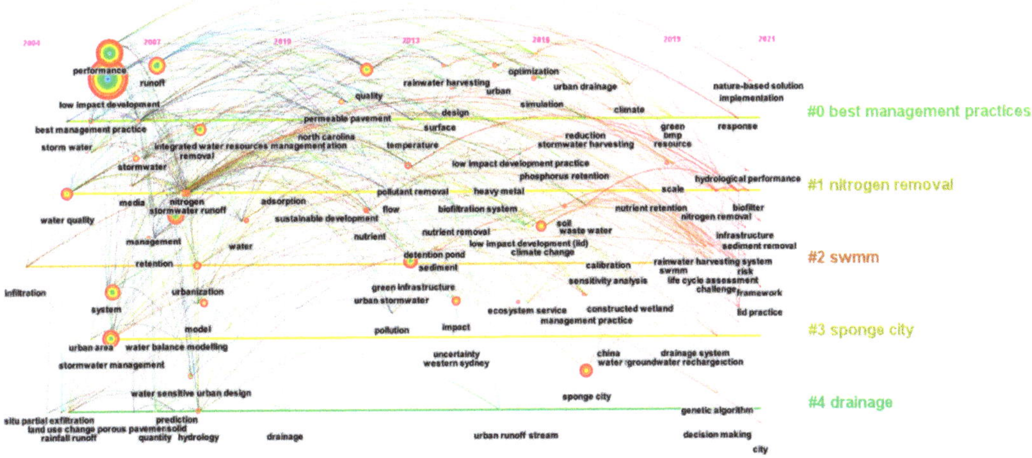

Figure 14. Ecosystem vitality keyword clustering timeline analysis.

Cluster #0, "best management practices", consisted of "low impact development", "performance", "rainwater management", "run off", and other keywords. In this cluster, discussion was mainly focused on urban rainwater runoff management facilities that can provide ecosystem services [80], as well as other benefits in terms of culture, aesthetics, and biodiversity [81]. Although ecosystem services are not the primary function of stormwater management facilities, providing a variety of ecosystem services can increase social acceptability and be conducive to replication.

2. Keyword emergence analysis

There were nine emergent terms in total, and the top five were selected (Figure 15). No new terms have emerged recently, and, therefore, there were no new research trends in the field. The term "best management practice" (BMP) had the longest time span and the greatest strength, and it covered two aspects: industrial wastewater control and municipal wastewater control. Low-impact-development stormwater management measures are referred to as subsets of BMPs and BMP facilities [82]. The combination of LIP and BMP measures can effectively address the complexities of the urban hydrological environment and social factors [83] and provide a more diverse range of ecosystem services.

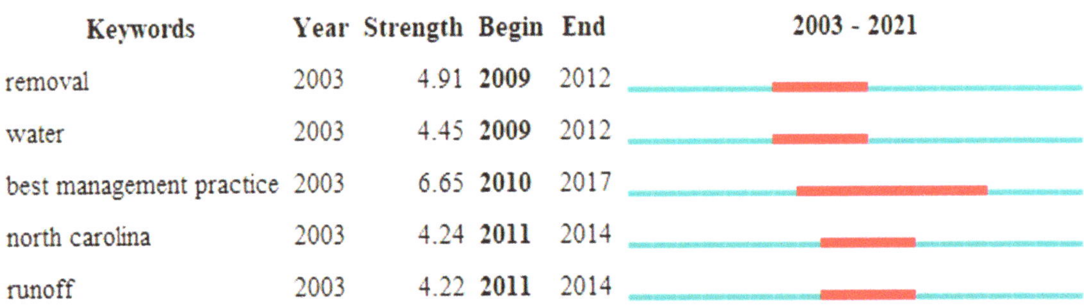

Figure 15. Top five emergent ecosystem vitality keywords.

"Removal" was the second ranking keyword in terms of frequency. Rainwater runoff management facilities can significantly reduce concentrations of urban runoff pollutants, with the treatment process being carried out using chemical, biological, and physical means, including sedimentation, filtration, adsorption, reduction, plant absorption, and microbial biomass assimilation [82]. A single removal method cannot solve all of the problems faced, and a combination of multiple means can be more effective in achieving the removal of pollutants than any single method. In addition, the combination of treatment methods should be adjusted according to the requirements of the particular location, making the promotion of standardized methods less applicable.

3.6. Patent Analysis

Compared to article analysis, patents can often better reflect the results of actual technical applications. For this paper, patent records for stormwater runoff technologies were extracted for the period from 2001 to 2021 using the World Intellectual Property Organization's database. Technological development follows a pattern—namely, precursors–invention–development–maturity, which is known as the S-shaped growth pattern [69,84]. A total of 1880 patents were found covering 520 items in the International Patent Classification (IPC) database. Employing the IPC codes for the top 20, this article classified the top patents into four fields for further S-curve analysis. These were "Drainage", "Collection", "Treatment", and "Testing and Simulation" (Table 3).

Table 3. Top 20 keywords in classified patents.

Class	IPC Index	Keywords	Number of Patents
Drainage	E03F1/00	Draining off of sewage or storm water	209
	E03F5/04	Drainage wells	126
	E03F3/04	Drainage accessories	65
	E03F3/02	Piping system arrangement	44
	E04D13/08	Roof drainage fitting	40
	E03F5/00	Sewerage structures	36
	E04D13/00	Roof drainage facilities	24
	E01C11/22	Gutters and curbs	87
Collection	E03F5/10	Sewage collection	176
	E03B3/02	Rainwater collection	103
	A01G9/02	Planter, flower box, or tree bed water collection	41
Treatment	C02F3/32	Plant filtration	91
	C02F9/14	Biofiltration	102
	C02F1/00	Wastewater and water treatment	57
	C02F1/28	Adsorption	41
	C02F3/34	Microbial filtration	38
	E03F5/06	Filtration structure	37
	C02F101/16	Filtered nitrogen	34
	E03F5/14	Solid–liquid separation	171
Runoff prediction and modelling	G06Q50/26	Runoff simulation and detection	53

3.6.1. Drainage

The patents for drainage were divided into three categories: road drainage fittings and facilities; drainage structures and plans; and roof drainage fittings and facilities. As is shown in Figure 16, generally speaking, this field is at an intermediate to late stage in its development, with both road drainage and other drainage structures having been in precursor (emerging) periods for a long time and roof drainage entering the invention stage early on. Urban road drainage is very complex and is not just a single discipline but rather involves many aspects [85]. Its complexity has resulted in a long period of exploration.

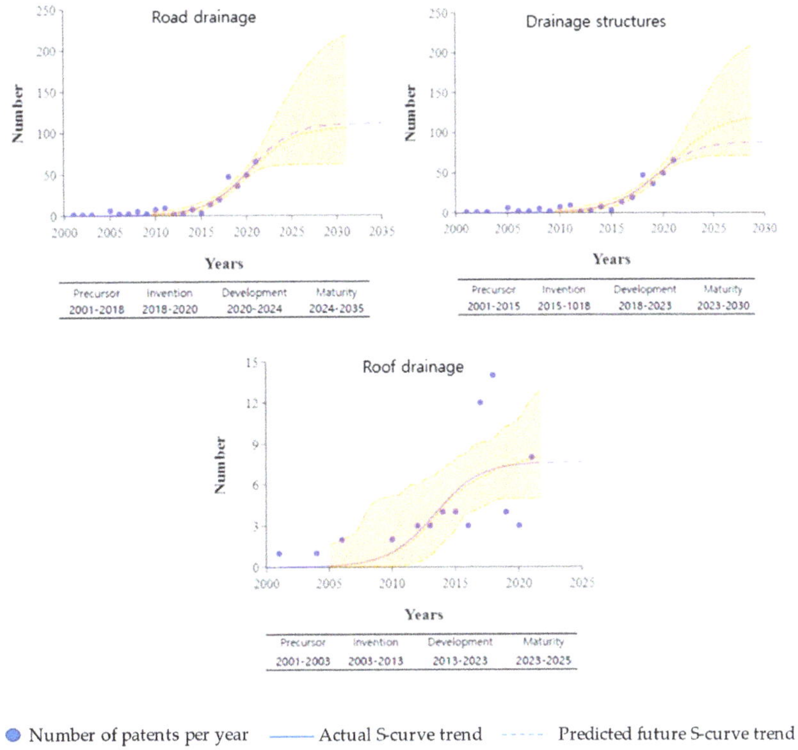

Figure 16. S-curves for drainage in patents.

3.6.2. Collection

Collection patents fell into two categories, wastewater collection technology and rainwater collection technology (Figure 17), with wastewater collection technology being at an advanced stage of development and rainwater collection at the beginning stage of development, demonstrating a long maturity cycle. Rainwater harvesting is considered to be a part of a rainwater harvesting system and includes collection, storage, and use of rainwater [46]. At present, rainwater harvesting techniques include not only collection but also consideration of later use in order to ensure safety [86].

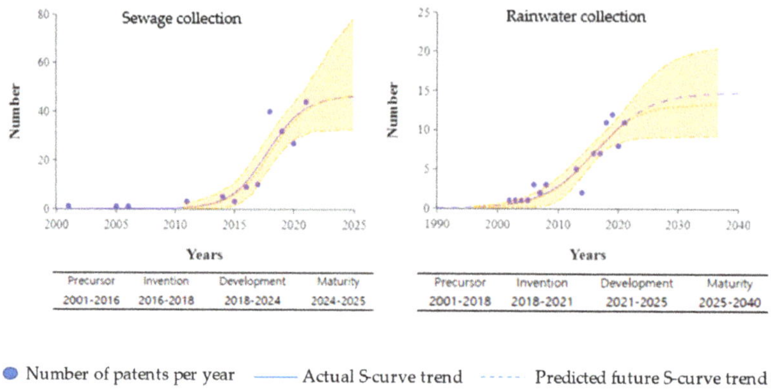

Figure 17. S-curves for collection in patents.

3.6.3. Treatment

Treatment patents were divided into three categories (Figure 18): sedimentation, adsorption, and biofiltration. Sedimentation is in the early stages of development, while adsorption and biofiltration technologies are in the early stages of maturity. The periods of development for these two technologies were also periods of growth in the literature on these technologies, and there was a positive correlation between the two technologies. In the context of stormwater runoff management, there has been a drive to use natural means to remove harmful substances and suspended solids from stormwater. The development of biofiltration systems, which combine technologies from a variety of fields, is constantly advancing [87].

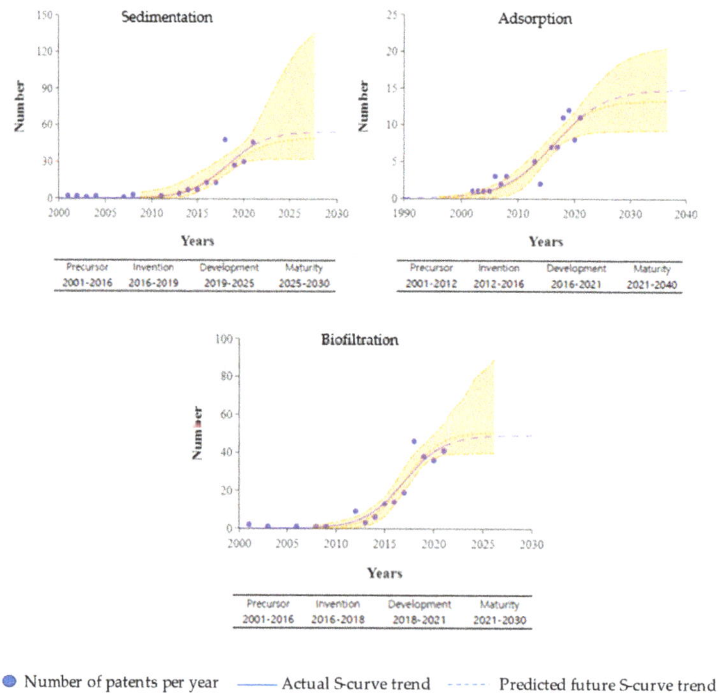

Figure 18. S-curves for treatment in patents.

3.6.4. Runoff Prediction and Modeling

Runoff prediction and modeling technology is the latest of these patented technologies and is still in the invention phase (Figure 19), with significant potential for future development. Modeling techniques can make cities more resilient and better able to predict and cope with extreme weather and urban flooding in the future [40].

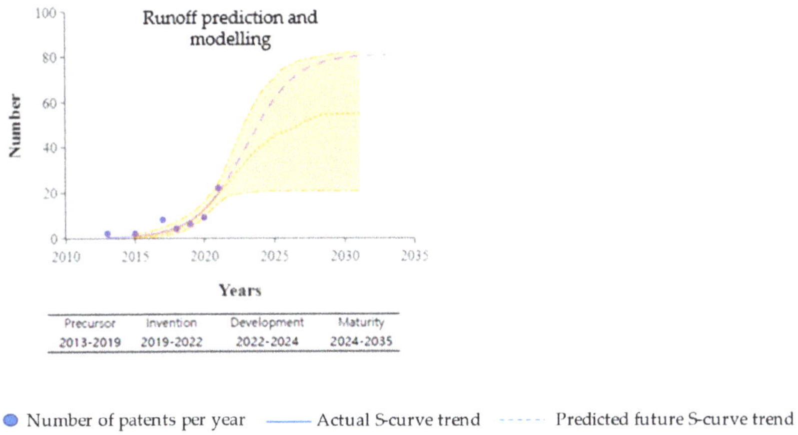

Figure 19. S-curve for runoff prediction and modeling in patents.

4. Discussion

4.1. Constraints on Development

Geographic constraints. There is wide variability between one geographic area and another due to local climate factors, other environmental factors, and localized urban runoff problems [50,69], and these factors can have a great impact on the effectiveness of stormwater runoff facilities [67,69,82]. In addition, the construction of stormwater runoff facilities itself has many constraints regarding site selection, including financial investment, local policies, and land area [58]. In summary, fixed models and concepts are difficult to replicate on a large scale in different regions.

Technological limitations arise mainly from the fact that technological developments related to stormwater runoff management facilities are at an advanced stage of development or even fully mature, but there are still uncertainties as to the performance of the facilities in the face of extreme weather. In Figures 13 and 15, the word "removal" is highlighted, indicating that the removal of pollutants from rainwater has figured prominently in the development of stormwater runoff management, with chemical and biological pollutants being of particular importance [88,89], the main filtration methods for them being adsorption and biofiltration. From a technological perspective (Figure 18), the growth period in the development of adsorption and biofiltration has come to an end. However, when the runoff velocity exceeds the threshold, problems such as the pollution caused by effluent overflow and limited interception capacity remain [70,75]. In addition, there is still much room for progress in the sustainable reuse of materials, cost control, and research and development regarding new materials and systems [87,90].

Uncertainty of data. The scarcity of front-end data can affect the proper planning of stormwater runoff facilities, which can produce secondary pollution, as well as affect the utility of the facilities [69,72]. Since the post-simulation data cannot be verified [79], there are questions about the authenticity of the final results, as well as the sustainability at later stages.

4.2. Future Development Trends

The trend for stormwater runoff management will tend toward multifunctional designs. Versatile stormwater runoff management facilities can more effectively address multi-disciplinary urban issues compared to single-method facilities. In order to cope with the changing climate and urban development, approaches to urban stormwater runoff management should face and solve more problems, not only limited to the stormwater runoff aspect but also involving more fields. The most prominent of these multifunctional approaches is the WSUD approach (Figures 11 and 13), which is implemented through both structural and non-structural measures [91]. WSUD has established a complete evaluation platform incorporating multiple disciplines in order to evaluate space feasibility through various approaches, such as modeling, reviews, and planning simulations, making it possible to build WSUD facilities in the most suitable place [92]. In addition, multiple targets that are originally processed separately are integrated for unified processing in order to achieve several side functions while simultaneously realizing the main functions. Strict evaluation methods and multifunctionality not only reduce expenses but also greatly improve practicality [93]. The above approaches have put forward new methods for the improvement of runoff rainwater management facilities that have previously failed to achieve the desired effect [24]. Functions have been improved through the combination of stormwater runoff management facilities and other facilities. Stormwater runoff management itself offers an improvement over previously existing urban planning methods [35] and can effectively solve multiple problems when used in combination with traditional grey infrastructure, drainage systems, and green infrastructure [94–97]. For example, the combination of LID and BMP (Figure 16) facilities can more effectively solve multiple complex problems [84]. This combination is bidirectional and can not only improve the functionality of stormwater runoff management but can also contribute to the repair of the original facilities [98]. The mutual combination of functions also merits further evaluation, as the situation is more complex than a simple combining of two methods but rather requires the consideration of various factors in order to maximize the beneficial effects [99]. Efficient synergistic effects can be produced through the correct combination of facilities [100]. Increased interdisciplinary research collaboration can maintain the sustainability of the project. In recent years, integrated models for addressing environmental issues have been developed and applied extensively to urban water management practices [101]. As research in each direction continues, the uncertainty in model predictions can be reduced [102,103]. For example, one of the hot research topics in recent years, "urban flooding" (Figure 12), not only relies on the integrated simulation of models but also requires a large amount of rainfall data and extensive data acquisition [104]. The application and acquisition of these data are closely related to the development of smart cities, collection and analysis using big data networks, the development of new sensors, and various other new technologies. In the digital age, with the maturity of the concept of and technology for stormwater runoff management, there is great value in the maintenance and updating of facilities in the later stages of the development of these technologies [58–60,76]. With cutting-edge technologies, more effective early planning for and prediction of future economic and social benefits can be achieved [81].

5. Conclusions

A review and analysis of 1756 English-language articles published from 2001 to 2021 in the WoS databases were conducted. The results of this study show that the literature on urban stormwater runoff management concepts has grown rapidly over the last 20 years. This indicates that research in this field is expanding and developing rapidly, generating a great deal of interest from academia and industry. This suggests that issues involved in the field of stormwater runoff management are becoming increasingly important and require more research in order to understand and address them in depth. Furthermore, the fast growth of the literature may reflect the innovative and forward-looking nature of the research within this field, as well as the wide recognition from and interest of researchers.

According to the distribution of publications and articles by country and institute, China, the United States, and Australia have produced the most publications in this field. However, as the research period has extended and more articles have been published, greater cooperation with other countries has occurred, which may indicate that the research in this field has reached a certain level of maturity and scale and that international cooperation has become more important in order to promote development and innovation. International cooperation can provide researchers with more research resources and expertise, increasing the depth and comprehensiveness of the research.

Reference emergence analysis showed that the development of stormwater runoff management concepts has evolved from theoretical formulation toward the current need to promote innovation and application through interdisciplinary research collaboration, which indicates that the problems in this field have become increasingly complex, thus requiring cooperation and communication between various disciplines. This provides more opportunities for collaboration and innovation among researchers from various disciplines and promotes the interdisciplinary development of academic research.

As can be seen from Sections 3.5 and 3.6 on performance features and patent analysis, the implementation of stormwater runoff management has undoubtedly improved the urban environment. The development of technologies such as green roofs, permeable pavements, adsorption and sedimentation techniques, and simulation and prediction has brought benefits to the urban climate environment, improving urban flood control and urban water quality, but deficiencies remain. "Urban flooding" is a current research hotspot, and the "sponge city" concept has significant advantages in the field of urban flood control. The advantages of the WSUD concept include its favoring of the establishment of multifunctional stormwater runoff facilities, which is in line with the future development direction for this research.

In the future of stormwater runoff management, it will be necessary to solve the problem of limited promotion due to geographical problems and the bottleneck in the development of practical technology. The collection and detection of urban data are critical to future development, and advanced technological tools, such as artificial intelligence, big data, cloud computing, and the Internet of Things, should be used in order to achieve digitalized smart cities. Multifunctionality and sustainability are the primary development directions for stormwater runoff management. The popularity of multifunctional stormwater runoff facilities, as well as greater consideration of the future sustainability of and concept planning for these facilities, can facilitate further development in this field.

In this study, we mainly searched for papers in English and, therefore, papers in other languages were not included, but this does not negate the fact that research in this area has been carried out in non-English-speaking regions; in recent years, there has been an increase in research in the Middle East and the Arab region in the field of stormwater runoff. Therefore, this paper has some limitations in terms of research orientation, and in subsequent research, the scope of the multilingual paper search can be expanded in order to better summarize and examine current hot topics and research trends.

Author Contributions: Conceptualization, W.H.; methodology, W.H.; software, W.H.; validation, W.H.; formal analysis, W.H.; investigation, W.H.; resources, W.H.; data curation, W.H.; writing—original draft preparation, W.H.; writing—review and editing, D.W.; visualization, W.H.; supervision, D.-W.S.; project administration, W.H.; funding acquisition, D.-W.S. All authors have read and agreed to the published version of the manuscript.

Funding: This research received no external funding.

Data Availability Statement: Data available on request due to restrictions; e.g., privacy or ethics.

Acknowledgments: We are very grateful to Xinyu Shi (Qingdao University of Technology, China) for his generous advice and help in the writing process.

Conflicts of Interest: The authors declare no conflict of interest.

Appendix A

"Sponge city" or "sponge cities" or "low impact development" or "low impact development(lid)" or "low-impact development" or "low-impact development(lid)" or "water sensitive urban design" or "water-sensitive urban design" or "water sensitive urban design(wsud)" or "water-sensitive urban design(wsud)" or "sustainable urban drainage systems" or "sustainable urban drainage".

References

1. Donaghy, K.P. Urban environmental imprints after globalization. *Reg. Environ. Chang.* **2011**, *12*, 395–405. [CrossRef]
2. Fletcher, T.D.; Andrieu, H.; Hamel, P.; Reid, M. Understanding, management and modelling of urban hydrology and its consequences for receiving waters: A state of the art. *Adv. Water Resour.* **2013**, *51*, 261–279. [CrossRef]
3. Kim, R.; Lee, S.; Kim, Y.; Lee, J.; Kim, S.; Kim, S. Pollutants in Rainwater Runoff in Korea: Their Impacts on Rainwater Utilization. *Environ. Technol.* **2005**, *26*, 411–420. [CrossRef]
4. Schets, F.; Italiaander, R.; van den Berg, H.; de Roda Husman, A. Rainwater harvesting: Quality assessment and utilization in The Netherlands. *J. Water Health* **2009**, *8*, 224–235. [CrossRef] [PubMed]
5. 5Spromberg, J.; Baldwin, D.; Damm, S.; McIntyre, J.; Huff, M.; Sloan, C.; Anulacion, B.; Davis, J.; Scholz, N. Coho salmon spawner mortality in western US urban watersheds: Bioinfiltration prevents lethal storm water impacts. *J. Appl. Ecol.* **2015**, *53*, 398–407. [CrossRef] [PubMed]
6. Hoyer, J. *Water Sensitive Urban Design*; Jovis Verlag: Berlin, Germany, 2011; pp. 15–16.
7. Fletcher, T.; Shuster, W.; Hunt, W.; Ashley, R.; Butler, D.; Arthur, S.; Trowsdale, S.; Barraud, S.; Semadeni Davies, A.; BertrandKrajewski, J.; et al. SUDS, LID, BMPs, WSUD and more the evolution and application of terminology surrounding urban drainage. *Urban Water J.* **2009**, *12*, 525–542. [CrossRef]
8. Liu, J.; Du, H.; Yuan, S.; He, W.; Yan, P.; Liu, Z. Alkaline deoxygenated graphene oxide as adsorbent for cadmium ions removal from aqueous solutions. *Water Sci. Technol.* **2015**, *71*, 1611–1619. [CrossRef]
9. Low Impact Development (LID) Guidance. Low Impact Development (LID) Guidance—Washington State Department of Ecology. Available online: https://ecology.wa.gov/Regulations-Permits/Guidance-technical-assistance/Stormwater-permittee-guidance-resources/Low-Impact-Development-guidance (accessed on 19 March 2023).
10. Li, Y.; Zhao, C.; Liu, Y.; Huang, Q.; Zhang, J. Urbanization and water pollution: A case study of China. *Sci. Total Environ.* **2020**, *729*, 138891. [CrossRef]
11. Wang, Z.; Qi, F.; Liu, L.; Chen, M.; Sun, D.; Nan, J. How do urban rainfall-runoff pollution control technologies develop in China? A systematic review based on bibliometric analysis and literature summary. *Sci. Total Environ.* **2021**, *789*, 148045. [CrossRef] [PubMed]
12. European Commission. *Green Infrastructure and Flood Management: Promoting Cost-Efficient and Resilient Flood Risk Reduction*; Publications Office of the European Union: Luxembourg, 2019.
13. Lashford, C.; Rubinato, M.; Cai, Y.; Hou, J.; Abolfathi, S.; Coupe, S.; Charlesworth, S.; Tait, S. SuDS & Sponge Cities: A Comparative Analysis of the Implementation of Pluvial Flood Management in the UK and China. *Sustainability* **2019**, *11*, 213. [CrossRef]
14. Feng, D.; Song, C.; Mo, W. Environmental, human health, and economic implications of landfill leachate treatment for per- and Polyfluoroalkyl Substance Removal. *J. Environ. Manag.* **2021**, *289*, 112558. [CrossRef] [PubMed]
15. Yazdani, M.; Kabirifar, K.; Frimpong, B.E.; Shariati, M.; Mirmozaffari, M.; Boskabadi, A. Improving construction and demolition waste collection service in an urban area using a simheuristic approach: A case study in Sydney, Australia. *J. Clean. Prod.* **2021**, *280*, 124138. [CrossRef]
16. Liu, G.; Xiao, M.; Zhang, X.; Gal, C.; Chen, X.; Liu, L.; Clements-Croome, D. A review of Air Filtration Technologies for sustainable and Healthy Building Ventilation. *Sustain. Cities Soc.* **2017**, *32*, 375–396. [CrossRef]
17. Jia, G.; Ma, R.; Hu, Z. Review of Urban Transportation Network Design Problems Based on CiteSpace. *Math. Probl. Eng.* **2019**, *2019*, 5735702. [CrossRef]
18. Zeng, J.; Zhang, H.G.; Dong, Y.Q.; Tan, R.H.; Yu, F. Application Research of SWOT Analysis in Enterprise Patent Portfolio Strategy. *Adv. Mater. Res.* **2011**, *418–420*, 2219–2225. [CrossRef]
19. Liu, C.; Wang, J. Forecasting the development of the biped robot walking technique in Japan through S-curve model analysis. *Scientometrics* **2009**, *82*, 21–36. [CrossRef]
20. Priestley, M.; Sluckin, T.; Tiropanis, T. Innovation on the web: The end of the S-curve? *Internet Hist.* **2020**, *4*, 390–412. [CrossRef]
21. Roy, A.; Wenger, S.; Fletcher, T.; Walsh, C.; Ladson, A.; Shuster, W.; Thurston, H.; Brown, R. Impediments and Solutions to Sustainable, Watershed-Scale Urban Stormwater Management: Lessons from Australia and the United States. *Environ. Manag.* **2008**, *42*, 344–359. [CrossRef]
22. Coutts, A.; Tapper, N.; Beringer, J.; Loughnan, M.; Demuzere, M. Watering our cities. *Prog. Phys. Geogr. Earth Environ.* **2012**, *37*, 2–28. [CrossRef]
23. Xia, J.; Zhang, Y.; Xiong, L.; He, S.; Wang, L.; Yu, Z. Opportunities and challenges of the Sponge City construction related to urban water issues in China. *Sci. China Earth Sci.* **2017**, *60*, 652–658. [CrossRef]

24. Sang, Y.; Yang, M. Urban waterlogs control in China: More effective strategies and actions are needed. *Nat. Hazards* **2016**, *85*, 1291–1294. [CrossRef]
25. Razzaghmanesh, M.; Beecham, S.; Kazemi, F. Impact of green roofs on stormwater quality in a South Australian urban environment. *Sci. Total Environ.* **2014**, *470–471*, 651–659. [CrossRef]
26. Peng, Z.; Jinyan, K.; Wenbin, P.; Xin, Z.; Yuanbin, C. Effects of Low-Impact Development on Urban Rainfall Runoff under Different Rainfall Characteristics. *Pol. J. Environ. Stud.* **2018**, *28*, 771–783. [CrossRef] [PubMed]
27. Gnecco, I.; Palla, A.; Lanza, L.; La Barbera, P. The Role of Green Roofs as a Source/sink of Pollutants in Storm Water Outflows. *Water Resour. Manag.* **2013**, *27*, 4715–4730. [CrossRef]
28. Zubelzu, S.; Rodríguez-Sinobas, L.; Andrés-Domenech, I.; Castillo-Rodríguez, J.; Perales-Momparler, S. Design of water reuse storage facilities in Sustainable Urban Drainage Systems from a volumetric water balance perspective. *Sci. Total Environ.* **2019**, *663*, 133–143. [CrossRef]
29. Sage, J.; Berthier, E.; Gromaire, M. Stormwater Management Criteria for On-Site Pollution Control: A Comparative Assessment of International Practices. *Environ. Manag.* **2015**, *56*, 66–80. [CrossRef]
30. Winston, R.J.; Al-Rubaei, A.M.; Blecken, G.T.; Viklander, M.; Hunt, W.F. Maintenance measures for preservation and recovery of permeable pavement surface infiltration rate—The effects of street sweeping, vacuum cleaning, high pressure washing, and milling. *J. Environ. Manag.* **2016**, *169*, 132–144. [CrossRef]
31. Lucke, T.; Nichols, P. The pollution removal and stormwater reduction performance of street-side bioretention basins after ten years in operation. *Sci. Total Environ.* **2015**, *536*, 784–792. [CrossRef]
32. Semadeni-Davies, A.; Hernebring, C.; Svensson, G.; Gustafsson, L. The impacts of climate change and urbanisation on drainage in Helsingborg, Sweden: Suburban stormwater. *J. Hydrol.* **2008**, *350*, 114–125. [CrossRef]
33. Grebel, J.; Mohanty, S.; Torkelson, A.; Boehm, A.; Higgins, C.; Maxwell, R.; Nelson, K.; Sedlak, D. Engineered Infiltration Systems for Urban Stormwater Reclamation. *Environ. Eng. Sci.* **2013**, *30*, 437–454. [CrossRef]
34. Li, Y.; Deletic, A.; McCarthy, D. Removal of E. coli from urban stormwater using antimicrobial-modified filter media. *J. Hazard. Mater.* **2014**, *271*, 73–81. [CrossRef]
35. Ahiablame, L.; Engel, B.; Chaubey, I. Effectiveness of Low Impact Development Practices: Literature Review and Suggestions for Future Research. *Water Air Soil Pollut.* **2012**, *223*, 4253–4273. [CrossRef]
36. Qin, H.; Li, Z.; Fu, G. The effects of low impact development on urban flooding under different rainfall characteristics. *J. Environ. Manag.* **2013**, *129*, 577–585. [CrossRef]
37. Ahiablame, L.; Engel, B.; Chaubey, I. Effectiveness of low impact development practices in two urbanized watersheds: Retrofitting with rain barrel/cistern and porous pavement. *J. Environ. Manag.* **2013**, *119*, 151–161. [CrossRef] [PubMed]
38. Rosa, D.; Clausen, J.; Dietz, M. Calibration and Verification of SWMM for Low Impact Development. *JAWRA J. Am. Water Resour. Assoc.* **2015**, *51*, 746–757. [CrossRef]
39. Ahiablame, L.; Shakya, R. Modeling flood reduction effects of low impact development at a watershed scale. *J. Environ. Manag.* **2016**, *171*, 81–91. [CrossRef] [PubMed]
40. Palla, A.; Gnecco, I. Hydrologic modeling of Low Impact Development systems at the urban catchment scale. *J. Hydrol.* **2015**, *528*, 361–368. [CrossRef]
41. Baek, S.; Choi, D.; Jung, J.; Lee, H.; Lee, H.; Yoon, K.; Cho, K. Optimizing low impact development (LID) for stormwater runoff treatment in urban area, Korea: Experimental and modeling approach. *Water Res.* **2015**, *86*, 122–131. [CrossRef] [PubMed]
42. Zanella, A.; Camanho, A.; Dias, T. Benchmarking countries' environmental performance. *J. Oper. Res. Soc.* **2013**, *64*, 426–438. [CrossRef]
43. Yale Center for Environmental Law and Policy. 2020 Environmental Performance Index. 2020. Available online: https://epi.yale.edu/epi-results/2020/component/epi (accessed on 7 October 2021).
44. Wang, M.; Zhang, D.; Lou, S.; Hou, Q.; Liu, Y.; Cheng, Y.; Qi, J.; Tan, S.K. Assessing Hydrological Effects of Bioretention Cells for Urban Stormwater Runoff in Response to Climatic Changes. *Water* **2019**, *11*, 997. [CrossRef]
45. Guo, X.; Guo, Q.; Zhou, Z.; Du, P.; Zhao, D. Degrees of hydrologic restoration by low impact development practices under different runoff volume capture goals. *J. Hydrol.* **2019**, *578*, 124069. [CrossRef]
46. Ahammed, F. A review of water-sensitive urban design technologies and practices for sustainable stormwater management. *Sustain. Water Resour. Manag.* **2017**, *3*, 269–282. [CrossRef]
47. Li, C.; Peng, C.; Chiang, P.-C.; Cai, Y.; Wang, X.; Yang, Z. Mechanisms and applications of green infrastructure practices for stormwater control: A review. *J. Hydrol.* **2018**, *568*, 626–637. [CrossRef]
48. Rodríguez-Sinobas, L.; Zubelzu, S.; Perales-Momparler, S.; Canogar, S. Techniques and criteria for sustainable urban stormwater management. The case study of Valdebebas (Madrid, Spain). *J. Clean. Prod.* **2018**, *172*, 402–416. [CrossRef]
49. Jiang, A.Z.; McBean, E.A. Performance of lot-level low impact development technologies under historical and climate change scenarios. *J. Hydro-Environ. Res.* **2021**, *38*, 4–13. [CrossRef]
50. Mei, C.; Liu, J.; Wang, H.; Yang, Z.; Ding, X.; Shao, W. Integrated assessments of green infrastructure for flood mitigation to support robust decision-making for sponge city construction in an urbanized watershed. *Sci. Total Environ.* **2018**, *639*, 1394–1407. [CrossRef] [PubMed]
51. Razzaghmanesh, M.; Beecham, S.; Salemi, T. The role of green roofs in mitigating Urban Heat Island effects in the metropolitan area of Adelaide, South Australia. *Urban For. Urban Green.* **2016**, *15*, 89–102. [CrossRef]

52. Lerer, S.; Arnbjerg Nielsen, K.; Mikkelsen, P. A Mapping of Tools for Informing Water Sensitive Urban Design Planning Decisions—Questions, Aspects and Context Sensitivity. *Water* **2015**, *7*, 993–1012. [CrossRef]
53. Richards, D.; Edwards, P. Using water management infrastructure to address both flood risk and the urban heat island. *Int. J. Water Resour. Dev.* **2017**, *34*, 490–498. [CrossRef]
54. Getter, K.L.; Rowe, D.B. The Role of Green Roofs in Sustainable Development. *Hortscience* **2006**, *41*, 1276–1285. [CrossRef]
55. Zhou, K. A comparative study on rainfall runoff control indicators of green roof. *Water Supply* **2020**, *20*, 2036–2042. [CrossRef]
56. Wilkinson, S.; Feitosa, R.C.; Kaga, I.T.; De Franceschi, I.H. Evaluating the thermal performance of retrofitted lightweight green roofs and walls in Sydney and Rio de Janeiro. *Procedia Eng.* **2017**, *180*, 231–240. [CrossRef]
57. Besir, A.B.; Cuce, E. Green roofs and facades: A comprehensive review. *Renew. Sustain. Energy Rev.* **2018**, *82*, 915–939. [CrossRef]
58. Xie, Y.; He, Z.; Li, Y.; Wang, D.; Li, C. Research and development status of sponge city planning in China. *J. Clean. Prod.* **2021**, *312*, 127881. [CrossRef]
59. Gregoire, B.G.; Clausen, J.C. Effect of a modular extensive green roof on stormwater runoff and water quality. *Ecol. Eng.* **2011**, *37*, 963–969. [CrossRef]
60. Speak, A.; Rothwell, J.; Lindley, S.; Smith, C. Metal and nutrient dynamics on an aged intensive green roof. *Environ. Pollut.* **2014**, *184*, 33–43. [CrossRef]
61. Caprario, J.; Rech, A.S.; Tasca, F.A.; Finotti, A.R. Influence of Drainage Network and Compensatory Techniques on Urban Flooding Susceptibility. *New Trends Urban Drain. Model.* **2018**, *79*, 717–722. [CrossRef]
62. Song, J.; Chang, Z.; Li, W.; Feng, Z.; Wu, J.; Cao, Q.; Liu, J. Resilience-vulnerability balance to urban flooding: A case study in a densely populated coastal city in China. *Cities* **2019**, *95*, 102381. [CrossRef]
63. Piyumi, M.; Abenayake, C.; Jayasinghe, A.; Wijegunarathna, E. Urban Flood Modeling Application: Assess the Effectiveness of Building Regulation in Coping with Urban Flooding Under Precipitation Uncertainty. *Sustain. Cities Soc.* **2021**, *75*, 103294. [CrossRef]
64. Bao, Z.; Yang, H.; Li, W.; Xu, Y.; Huang, G. A Low-Impact Development–Based Multi-Objective Optimization Model for Regional Water Resource Management under Impacts of Climate Change. *Front. Earth Sci.* **2020**, *8*, 527388. [CrossRef]
65. Davis, A.P.; Shokouhian, M.; Sharma, H.; Minami, C.; Winogradoff, D. Water quality improvement through bioretention: Lead, copper, and zinc removal. *Water Environ. Res.* **2003**, *75*, 73–82. [CrossRef] [PubMed]
66. Ma, W.; Zhao, B.; Ma, J. Comparison of heavy metal accumulation ability in rainwater by 10 sponge city plant species. *Environ. Sci. Pollut. Res.* **2019**, *26*, 26733–26747. [CrossRef] [PubMed]
67. Vijayaraghavan, K.; Biswal, B.K.; Adam, M.G.; Soh, S.H.; Tsen-Tieng, D.L.; Davis, A.P.; Chew, S.H.; Tan, P.Y.; Babovic, V.; Balasubramanian, R. Bioretention systems for stormwater management: Recent advances and future prospects. *J. Environ. Manag.* **2021**, *292*, 112766. [CrossRef] [PubMed]
68. Allen, D.; Haynes, H.; Arthur, S. Contamination of Detained Sediment in Sustainable Urban Drainage Systems. *Water* **2017**, *9*, 355. [CrossRef]
69. Laurenson, G.; Laurenson, S.; Bolan, N.; Beecham, S.; Clark, I. The Role of Bioretention Systems in the Treatment of Stormwater. *Adv. Agron.* **2013**, *120*, 223–274. [CrossRef]
70. Kim, H.; Yoon, J.; Koo, J.; Shin, H.S. Performance and Uncertainty Analysis of Rainwater Management Facilities Con-sidering Climate Change Impacts. *Water* **2018**, *10*, 1557. [CrossRef]
71. Yang, W.; Zhang, J.; Mei, S.; Krebs, P. Impact of antecedent dry-weather period and rainfall magnitude on the performance of low impact development practices in urban flooding and non-point pollution mitigation. *J. Clean. Prod.* **2021**, *320*, 128946. [CrossRef]
72. Zhang, Q.; Zhang, H.; Wang, Y.; Sun, J. Environmental impact assessment of urban rainwater management facilities based on life cycle assessment: A case study in China. *J. Clean. Prod.* **2019**, *222*, 222–232. [CrossRef]
73. Urrutiaguer, M.; Lloyd, S.; Lamshed, S. Determining water sensitive urban design project benefits using a multi-criteria assessment tool. *Water Sci. Technol.* **2010**, *61*, 2333–2341. [CrossRef]
74. Helmreich, B. Rainwater Management in Urban Areas. *Water* **2021**, *13*, 1096. [CrossRef]
75. Liu, J., Sample, D.J.; Bell, C.; Guan, Y. Review and Research Needs of Bioretention Used for the Treatment of Urban Stormwater. *Water* **2014**, *6*, 1069–1099. [CrossRef]
76. Qin, Y. Urban Flooding Mitigation Techniques: A Systematic Review and Future Studies. *Water* **2020**, *12*, 3579. [CrossRef]
77. Gironás, J.; Roesner, L.; Rossman, L.; Davis, J. A new applications manual for the Storm Water Management Model (SWMM). *Environ. Model. Softw.* **2010**, *25*, 813–814. [CrossRef]
78. Wang, Y. Study on the Simulation and Evaluation of LID Adaptation Measures Based on SWMM. *J. Water Resour. Res.* **2020**, *9*, 22–32. [CrossRef]
79. You, X.; Liu, S.; Dai, C.; Guo, Y.; Zhong, G.; Duan, Y. Contaminant occurrence and migration between high- and low-permeability zones in groundwater systems: A review. *Sci. Total Environ.* **2020**, *743*, 140703. [CrossRef]
80. Wong, T.H.; Brown, R.R. The water sensitive city: Principles for practice. *Water Sci. Technol.* **2009**, *60*, 673–682. [CrossRef]
81. Jose, R.; Wade, R.; Jefferies, C. Smart SUDS: Recognising the multiple-benefit potential of sustainable surface water management systems. *Water Sci. Technol.* **2014**, *71*, 245–251. [CrossRef]
82. Hager, J.; Hu, G.; Hewage, K.; Sadiq, R. Performance of low-impact development best management practices: A critical review. *Environ. Rev.* **2019**, *27*, 17–42. [CrossRef]

83. Zhang, K.; Chui, T. A comprehensive review of spatial allocation of LID-BMP-GI practices: Strategies and optimization tools. *Sci. Total. Environ.* **2018**, *621*, 915–929. [CrossRef]
84. Bejan, A.; Lorente, S. The physics of spreading ideas. *Int. J. Heat Mass Transf.* **2012**, *55*, 802–807. [CrossRef]
85. Kuller, M.; Bach, P.; Roberts, S.; Browne, D.; Deletic, A. A planning-support tool for spatial suitability assessment of green urban stormwater infrastructure. *Sci. Total Environ.* **2019**, *686*, 856–868. [CrossRef]
86. Zhang, L.; Seagren, E.; Davis, A.; Karns, J. The Capture and Destruction of *Escherichia coli* from Simulated Urban Runoff Using Conventional Bioretention Media and Iron Oxide-coated Sand. *Water Environ. Res.* **2010**, *82*, 701–714. [CrossRef] [PubMed]
87. Liu, A.; Li, Y.; Li, H.; Li, C.; Xie, Y.; Chen, X. Advances in biofiltration systems for urban stormwater runoff treatment: A review. *J. Clean. Prod.* **2020**, *266*, 121914. [CrossRef]
88. Liu, T.; Lawluvy, Y.; Shi, Y.; Yap, P. Low Impact Development (LID) Practices: A Review on Recent Developments, Challenges and Prospects. *Water Air Soil Pollut.* **2021**, *232*, 344. [CrossRef]
89. Roy-Poirier, A.; Champagne, P.; Filion, Y. Review of Bioretention System Research and Design: Past, Present, and Future. *J. Environ. Eng.* **2010**, *136*, 878–889. [CrossRef]
90. Wang, Q.; Zhao, Y.; Chen, X.; Sun, Y.; Shen, Z. Sustainable stormwater management: A review of challenges and opportunities.
91. Butler, D.; Memon, F.A. (Eds.) *Water Demand Management*; Iwa Publishing: London, UK, 2005.
92. Fryd, O.; Dam, T.; Jensen, M. A planning framework for sustainable urban drainage systems. *Water Policy* **2012**, *14*, 865–886. [CrossRef]
93. Kuller, M.; Bach, P.; Ramirez-Lovering, D.; Deletic, A. Framing water sensitive urban design as part of the urban form: A critical review of tools for best planning practice. *Environ. Model. Softw.* **2017**, *96*, 265–282. [CrossRef]
94. Liu, Z.; Xu, C.; Xu, T.; Jia, H.; Zhang, X.; Chen, Z.; Yin, D. Integrating socioecological indexes in multiobjective intelligent optimization of green-grey coupled infrastructures. *Resour. Conserv. Recycl.* **2021**, *174*, 105801. [CrossRef]
95. Dong, X.; Yuan, P.; Song, Y.; Yi, W. Optimizing Green-Gray Infrastructure for Non-Point Source Pollution Control under Future Uncertainties. *Int. J. Environ. Res. Public Health* **2021**, *18*, 7586. [CrossRef]
96. Alves, A.; Sanchez, A.; Vojinovic, Z.; Seyoum, S.; Babel, M.; Brdjanovic, D. Evolutionary and Holistic Assessment of Green-Grey Infrastructure for CSO Reduction. *Water* **2016**, *8*, 402. [CrossRef]
97. Wang, M.; Zhang, Y.; Zhang, D.; Zheng, Y.; Li, S.; Tan, S. Life-cycle cost analysis and resilience consideration for coupled grey infrastructure and low-impact development practices. *Sustain. Cities Soc.* **2021**, *75*, 103358. [CrossRef]
98. Kvitsjøen, J.; Paus, K.; Bjerkholt, J.; Fergus, T.; Lindholm, O. Intensifying rehabilitation of combined sewer systems using trenchless technology in combination with low impact development and green infrastructure. *Water Sci. Technol.* **2021**, *83*, 2947–2962. [CrossRef] [PubMed]
99. Lähde, E.; Khadka, A.; Tahvonen, O.; Kokkonen, T. Can We Really Have It All?—Designing Multifunctionality with Sustainable Urban Drainage System Elements. *Sustainability* **2019**, *11*, 1854. [CrossRef]
100. Zeng, S.; Guo, H.; Dong, X. Understanding the synergistic effect between LID facility and drainage network: With a comprehensive perspective. *J. Environ. Manag.* **2019**, *246*, 849–859. [CrossRef] [PubMed]
101. Bach, P.M.; Rauch, W.; Mikkelsen, P.S.; McCarthy, D.T.; Deletic, A. A critical review of integrated urban water modelling—Urban drainage and beyond. *Environ. Model. Softw.* **2014**, *54*, 88–107. [CrossRef]
102. Nguyen, T.; Ngo, H.; Guo, W.; Wang, X. A new model framework for sponge city implementation: Emerging challenges and future developments. *J. Environ. Manag.* **2019**, *253*, 109689. [CrossRef]
103. Gnecco, I.; Tscheikner-Gratl, F.; Murla, D.; Achleitner, S.; Nachtnebel, H.P. Uncertainty analysis of urban drainage mod-eling: Review of techniques and application to a large scale urban catchment. *Water Sci. Technol.* **2019**, *80*, 240–252. [CrossRef]
104. Hu, C.; Liu, C.; Yao, Y.; Wu, Q.; Ma, B.; Jian, S. Evaluation of the Impact of Rainfall Inputs on Urban Rainfall Models: A Systematic Review. *Water* **2020**, *12*, 2484. [CrossRef]

MDPI AG
Grosspeteranlage 5
4052 Basel
Switzerland
Tel.: +41 61 683 77 34

Buildings Editorial Office
E-mail: buildings@mdpi.com
www.mdpi.com/journal/buildings